Assessment of Risk from Low-Level Exposure to Radiation and Chemicals

A Critical Overview

BASIC LIFE SCIENCES

Alexander Hollaender, General Editor

Council for Research Planning in Biological Sciences, Inc., Washington, D.C.

Recent volumes in the series:

Volume 23 INDUCED MUTAGENESIS: Molecular Mechanisms and Their Implications for Environmental Protection
Edited by Christopher W. Lawrence

Volume 24 ORGAN AND SPECIES SPECIFICITY IN CHEMICAL CARCINOGENESIS
Edited by Robert Langenbach, Stephen Nesnow, and Jerry M. Rice

Volume 25 BASIC BIOLOGY OF NEW DEVELOPMENTS IN BIOTECHNOLOGY
Edited by Alexander Hollaender, Allen I. Laskin, and Palmer Rogers

Volume 26 GENETIC ENGINEERING OF PLANTS: An Agricultural Perspective
Edited by Tsune Kosuge, Carole P. Meredith, and Alexander Hollaender

Volume 27 NEUTRONS IN BIOLOGY
Edited by Benno P. Schoenborn

Volume 28 GENETIC CONTROL OF ENVIRONMENTAL POLLUTANTS
Edited by Gilbert S. Omenn and Alexander Hollaender

Volume 29 SISTER CHROMATID EXCHANGES: 25 Years of Experimental Research
Part A: The Nature of SCEs
Part B: Genetic Toxicology and Human Studies
Edited by Raymond R. Tice and Alexander Hollaender

Volume 30 PLASMIDS IN BACTERIA
Edited by Donald R. Helinski, Stanley N. Cohen, Don B. Clewell, David A. Jackson, and Alexander Hollaender

Volume 31 GENETIC CONSEQUENCES OF NUCLEOTIDE POOL IMBALANCE
Edited by Frederick J. de Serres

Volume 32 TISSUE CULTURE IN FORESTRY AND AGRICULTURE
Edited by Randolph R. Henke, Karen W. Hughes, Milton J. Constantin, and Alexander Hollaender

Volume 33 ASSESSMENT OF RISK FROM LOW-LEVEL EXPOSURE TO RADIATION AND CHEMICALS: A Critical Overview
Edited by Avril D. Woodhead, Claire J. Shellabarger, Virginia Pond, and Alexander Hollaender

---

A Continuation Order Plan is available for this series. A continuation order will bring delivery of each new volume immediately upon publication. Volumes are billed only upon actual shipment. For further information please contact the publisher.

Assessment of Risk from Low-Level Exposure to Radiation and Chemicals

A Critical Overview

Edited by

AVRIL D. WOODHEAD
CLAIRE J. SHELLABARGER
VIRGINIA POND
Brookhaven National Laboratory
Upton, New York

and

ALEXANDER HOLLAENDER
Council for Research Planning in Biological Sciences, Inc.
Washington, D.C.

PLENUM PRESS • NEW YORK AND LONDON

Library of Congress Cataloging in Publication Data

Main entry under title:

Assessment of risk from low-level exposure to radiation and chemicals.

(Basic life sciences; v. 33)

Papers presented at a meeting held at Brookhaven National Laboratory, Upton, N.Y., May 20–23, 1984 jointly sponsored by the Biology and Medical Departments.

Includes bibliographies and indexes.

1. Ionizing radiation—Toxicology—Congresses. 2. Toxicology—Congresses. 3. Health risk assessment—Congresses. I. Woodhead, Avril D. II. Brookhaven National Laboratory. Biology Dept. III. Brookhaven National Laboratory. Medical Dept. IV. Series.

RA1231.R2A78 1985 616.9′89707 85-6312

ISBN-13: 978-1-4684-4972-3 e-ISBN-13: 978-1-4684-4970-9

DOI: 10.1007/978-1-4684-4970-9

Proceedings of a meeting on Assessment of Risk from Low-Level Exposure to Radiation and Chemicals: A Critical Overview, held May 20–23, 1984, at Brookhaven National Laboratory, Upton, New York

A Division of Plenum Publishing Corporation
233 Spring Street, New York, N.Y. 10013

PREFACE

The present workshop had its origins in discussions among Alexander Hollaender, Oddvar Nygaard, Donald Borg, Richard Setlow and Victor Bond on the need for a symposium that would deal with a broad spectrum of pressing subjects related to the physics, chemistry, and biological actions of ionizing radiations, and the theoretical and practical problems of risk assessment. It soon became apparent that the spectrum of subjects considered was too broad for the desired depth of coverage; in fact, it seemed unlikely that the conference participants would have the background knowledge to span the gamut. Therefore, two separate meetings were decided upon, the first of which, Comparison of Mechanisms of Carcinogenesis by Radiation and Chemical Agents, was held at the National Bureau of Standards, Gaithersburg on December 6-7, 1983. The meeting was sponsored by the NCI. The second meeting has emerged as the present workshop at Brookhaven National Laboratory. An interface between the two conferences has been provided by Michael Fry and Donald Borg, who have summarized the salient points emanating from the NCI Symposium.

We intended that the first conference should focus on the basic mechanisms of radiation and chemical carcinogenesis, while the second, the present meeting, would emphasize exposure-response relationships, particularly the theoretical and practical similarities and differences between exposure to chemical carcinogens compared to exposure to ionizing radiation. We hoped that ultimately the perspectives of the two conferences would provide us with information from fundamental physics and chemistry to the practical aspects of risk assessment vital in the setting of judicious exposure standards. We also hoped that the inclusion of our discussions would aid substantially in facilitating a convergence towards that goal.

Symposium Committee:

A. D. Woodhead,
Chairperson
C. J. Shellabarger,
Chairperson
V. P. Bond
D. C. Borg
R. B. Setlow
R. R. Tice

ACKNOWLEDGEMENTS

This volume is the outcome of a meeting held at Brookhaven National Laboratory, Upton, New York, May 20-23, 1984, jointly sponsored by the Biology and Medical Departments. It is a pleasure to acknowledge financial support provided by the U.S. Department of Energy and the National Cancer Institute, and the valued assistance of agency representatives, Drs. Charles W. Edington and Bruce W. Wacholz. The Meeting Coordinator, Ms. Helen Kondratuk, did a superb job in running the meeting efficiently and with zest. Our thanks are due to Ms. Kathy Kissel for her untiring help in putting this book together. We also wish to record our gratitude to Dr. Alexander Hollaender, General Editor of this series, for his unceasing support throughout all phases of the planning and holding of the meeting. Finally, we especially wish to express our warmest thanks to Dr. Eric J. Hall for a most delightful and entertaining symposium banquet speech.

CONTENTS

Charles Little Dunham
1906 — 1975

DEDICATION: TO OUR COLLEAGUE AND FRIEND DR. CHARLES LITTLE DUNHAM

This volume is dedicated to the memory of Dr. Charles Little Dunham, Director of the Division of Biology and Medicine, U. S. Atomic Energy Commission from 1955 until 1967 and Chairman of the Division of Medical Sciences, National Academy of Sciences from 1967 to 1974.

Dr. Dunham was a scientist, scholar, teacher and administrator, a man of unusual foresight and an almost uncanny ability to predict scientific trends. He advocated large-scale programs in ecology long before the importance of the subject was generally realized. He understood the importance of basic research, and vigorously defended the tenet that progress in science or technology is most rapid when good people are given free rein to exploit their insights to the fullest extent. Although Dr. Dunham recognized the importance of work on animals and lower organisms, he also appreciated the significance of having data on humans; accordingly, he gave unqualified support to the study of long-term effects of radiation exposure in Hiroshima and Nagasaki.

Under Dr. Dunham's overall direction, the first "BEIR Committee" was set up, and much of the work he promoted and sponsored provided crucial evidence for the development of defensible national and international standards for radiation protection. Dr. Dunham was more influential than any other individual in advancing the Division of Biology and Medicine of the Atomic Energy Commission into one of the largest, and most broadly based research organizations in the country. In many ways the present Workshop has sprung from his many initiatives, and he would have been pleased to have been present at our meeting to see the degree to which his efforts have borne fruit.

Dr. Dunham was a good friend of mine for many years, and was well known to most on the Workshop Organizing Committee as well as to many of the participants at the Workshop. We are pleased to have this opportunity to acknowledge our indebtedness to an outstanding man.

Victor P. Bond

KEYNOTE ADDRESS

EPIDEMIOLOGICAL RESERVATIONS ABOUT RISK ASSESSMENT

Richard Peto
Radcliffe Infirmary
Oxford OX2 6HE
England

INTRODUCTION

The term risk assessment is used to describe three quite different types of activity, two of which are reasonably well-founded, and one of which is not. The fundamental problem is, of course, that reliable evidence of human hazard exists for only a few dozen of the thousands of chemicals that are of potential concern. Where it does exist, there should, in principle, be no great difficulty in using it, although the long-running arguments about the risks associated with various levels of radiation illustrate the gap that can exist between principle and practice.

USE OF HUMAN EVIDENCE DERIVED FROM COMPARABLE INTENSITIES OF EXPOSURE

This is the first type of risk assessment. If, for example, patients with one cancer are being referred for curative radiotherapy, it is possible to assess directly, and fairly reliably, the risk that this amount of radiation will induce a second cancer ten or twenty years later simply by turning to the accumulated experience of tens of thousands of other patients previously exposed to comparable doses (IARC, 1984). The risk is not negligible, nor is it overwhelmingly large, and in circumstances where radiotherapy has a reasonable prospect of eradicating the first cancer, its use may be thoroughly justified.

The situation with cytotoxic chemotherapy for the first type is not yet as clear-cut. Chemotherapy can eradicate some of the less common types of cancer, but it may not be until early in the next century that a reliable assessment of its long-term risks will emerge. At present, there are still some orders of magnitude of uncertainty: the long-term risk of a second tumor from some particular chemotherapy could be order unity, 10^{-1}, 10^{-2}, or 10^{-3}, and no currently available types of animal experiment could narrow this range of uncertainty very much.

USE OF HUMAN EVIDENCE FROM MUCH HIGHER DOSE LEVELS

A second situation where reasonable risk assessment may be possible is where there is direct, reliable evidence from much higher dose levels. For example, suppose it has been established that exposure throughout a working life to an average of one asbestos fibre per ml of air produces a lifelong cancer risk of about 1 percent, and suppose that one is trying to decide whether to spend vast sums of money reducing the levels in government offices down from 10^{-5} fibres/ml. It would then probably be reasonable to proceed as if the dose-response relationship were one of direct proportionality, even though no direct check on this assumption is likely to become available. Likewise, in estimating the lung cancer risks conferred by passive exposure to other people's cigarette smoke, it would be reasonable to assume that since passive exposure from a smoking spouse results in the absorption of only about 1 percent as much nicotine as active exposure does, the non-smoker will have about 1 percent of the smoker's lung cancer risk. If, however, all that were available to us was data from animal experiments, the problems of making any kind of practically useful assessment of the risks of passive smoking would be overwhelming; one would not even know whether the risks were of order unity, or of order 10^{-6}!

Indeed, it is salutary to reflect on what would be known about the two largest reliably known causes of neoplastic death in the U.S. (that is, tobacco and alcohol) if animal experiments were the only source of data available. Smoking causes about 30 percent of all U.S. cancer deaths, and alcohol causes about three percent. Yet, for many years, the tobacco industry tried to use the failure of tobacco smoke inhalation to cause malignant neoplasms in animals as evidence that it was not an important cause of human cancer. Alcohol alone still does not cause any cancers in standard animal feeding experiments!

USE OF ANIMAL EXPERIMENTS TO PREDICT HUMAN EXPERIENCE: THE CONTRASTING PERSPECTIVES OF THE HYGIENIST AND THE EPIDEMIOLOGIST

Some fairly reliable human evidence was used as the basis of the foregoing two types of risk assessment, and as a result, both seem to be quite reasonable procedures that should be exploited as widely as possible. The reservations expressed in the remainder of this chapter will relate only to the third type of risk assessment, where, in the absence of any reliable human data, attempts are made to predict human hazards from observations on short-lived animal species, or from some other laboratory system. Indeed, henceforth in this chapter risk assessment will be used exclusively to refer to this third type of activity.

Two quite different perspectives on risk assessment exist. One is that of the industrial hygienist, who in general will be worrying about a very large number of different chemicals, for most of which no direct evidence of human risk exists; hence, only indirect risk assessments will be possible. The other perspective is that of the epidemiologist, who in general will be worrying about the few factors that have been directly shown to cause substantial hazards in certain human populations; his chief concern is to see that reasonable steps are taken to limit the impact of these known causes, and to progress towards discovery of a few other major causes.

From the viewpoint of the hygienist, the epidemiologist may often seem to be more hindrance than help. For nearly all of the chemicals of concern, no useful epidemiological information is available, yet epidemiologists may arrogantly assert that no assessments of risk are at all reliable unless based reasonably directly on epidemiology.

Conversely, from the epidemiologist's viewpoint, the hygienist, with a long list of worrisome chemicals, may seem at best an irrelevancy and, at worst, a thoroughly diversionary influence. On our present knowledge, the chief priority for cancer prevention obviously is control of the effects of smoking (Tables 1, 2), yet incessant coverage in the mass media of one real or suspect carcinogen after another makes it unnecessarily difficult to get the public to appreciate the unique importance of smoking. (Indeed, for many years the health warning on saccharin in the United States was at least as strong as that on cigarettes!) Moreover, in the search for knowledge about other major causes of human cancer in developed countries, the hypotheses that appear most promising to many epidemiologists are perhaps those involving infective, nutritional or hormonal factors, rather than those involving exposure to traces of man-made chemicals. So, even within the world of research, the industrial hygienists' interests may be irrelevant or diversionary to the epidemiologist.

The fundamental source of divergence may be that the real purposes of many epidemiologists are not the same as those of many hygienists. The essential purpose of epidemiology is to understand (and, if possible, control) some major avoidable cause(s) of human disease. The fundamental purposes of many hygienists are rather

Table 1. Future Perfect: Estimate of the Proportions of Cancer Deaths that will be Found to be Attributable to Various Factors[a]

	Percent of All US Cancer Deaths: Best Estimate	Percent of All US Cancer Deaths: Range of Acceptable Estimates
Tobacco	30	25-40
Alcohol	3	2-4
Diet	35	10-70
Food additives	<1	-5[b]-2
Sexual behavior	1	1
Yet-to-be-discovered hormonal analogues of reproductive factors	-6	-12-0
Occupation	4	2-8
Pollution	2	1-5
Industrial products	<1	<1-2
Medicines and medical procedures	1	0.5-3
Geophysical factors (mostly natural background radiation and sunlight)	3	2-4
Infective processes	10?	1-?
Unknown	?[a]	?
Total	200% or more[a]	

[a]Since one cancer may have two or more causes, the grand total in such a table will probably, when more knowledge is available, greatly exceed 200%. (It is merely a coincidence that the suggested figures in the present table happen to add up to nearly 100%.)

[b]The net effects of food additives may be protective, e.g. against stomach cancer.

Table 2. Present Imperfect: Reliably Established (as of 1981), Practicable[a] Ways of Avoiding the Onset of Life-Threatening Cancer

	Percent of All US Cancer Deaths Known to be Thus Avoidable
Avoidance of tobacco smoke	30%
Avoidance of alcoholic drinks or mouthwashes	3%
Avoidance of obesity	2%
Regular cervical screening and genital hygiene	1%
Avoidance of inessential medical use of hormones or radiology	<1%
Avoidance of unusual exposure to sunlight	<1%
Avoidance of known effects of current levels of exposure to carcinogens (for which there is epidemiological evidence of human hazard) in	
(i) occupational context	<1%
(ii) food, water or urban air	<1%

[a]Excluding ways such as prophylatic prostatectomy, mastectomy, hysterectomy, oophorectomy, artificial menopause or pregnancy.

Both Tables from: Doll and Peto, 1981b.

different. To be sure, many are indeed chiefly concerned with the alleviation of human disease (though, if so, why don't they put at least as much political effort into the discouragement of smoking or of saturated fat intake as into industrial hygiene?), but others are not; whatever their original motives may have been, some appear to be chiefly concerned either with defending or with disliking modern industrial society. Out of their repeated confrontations is emerging the current approach to risk assessment. Because of their roots in adversarial politics, risk assessment procedures sometimes appear to be designed chiefly to be legalistically defensible rather than epidemiologically sensible. My present purpose is, therefore, first to introduce a fairly standard epidemiological perspective, and then to consider how this might affect the ways in which we view data from laboratory studies of individual chemicals.

AN EPIDEMIOLOGICAL PERSPECTIVE ON THE AVOIDANCE OF CANCER

My perspective will be summarized fairly briefly as it has been presented in detail, with full references, in our earlier report on the causes of cancer (Doll and Peto, 1981a,b). As a prerequisite, it must be appreciated that cancers at different anatomic sites are very different diseases, and factors that importantly affect the incidence of one type may have little effect on the incidence of another.

Wide International Variation in Site-Specific Cancer Rates

Although there has been no general increase in nonrespiratory cancer over the past few decades (Fig. 1), there is no type of cancer for which a high incidence is inevitable; every type that is currently common in America could be made substantially less so by practicable changes in the way people live. The fundamental evidence for this statement comes from comparing different populations with each other (Fig. 2): if these differences are not due to genetic factors, but chiefly due to differences in the ways people live, then comparison of the U.S. with Japan would suggest that at least 80-odd percent of all U.S. breast cancer and intestinal cancer might be avoidable, at least in principle. Likewise, comparison of the U.S. with countries where prolonged smoking is uncommon would suggest that most lung cancer should be avoidable (a conclusion strongly reinforced by comparison of the overall U.S. lung cancer rates with those of U.S. lifelong nonsmokers: see Fig. 1).

Confirmation that these large differences in site-specific cancer rates are not chiefly genetic comes from the study of people of Japanese (or West African) descent who have lived their entire lives in the U.S. By middle age, the incidence rates of colon

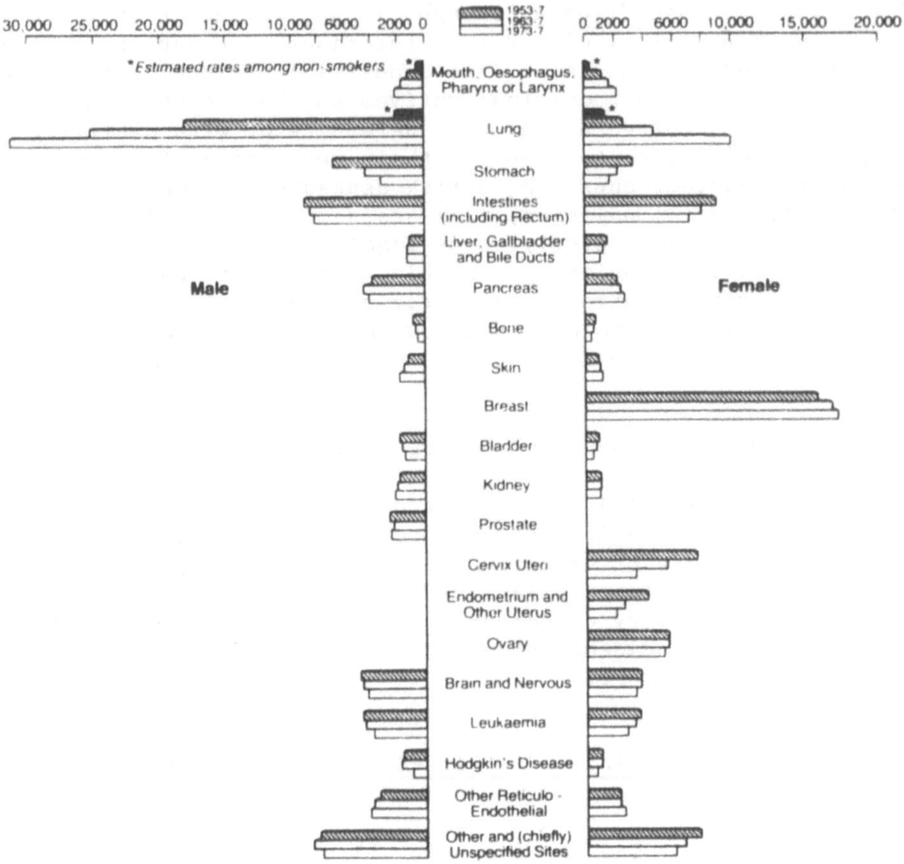

Fig. 1. Trends during the past quarter century in mortality from various types of cancer in the United States. Note the large increases in lung cancer mortality. Most of these trends (except for leukemia and Hodgkin's disease) are not much affected by changes in treatment, for treatments for the common cancers were as likely to effect cure in the 1950s as in the 1970s. All these death certification rates are "age standardized," that is, they are adjusted for any effects of changes in the age structure of the U.S. population. The cited rates are per 100 million people under 65; since there were about 100 million Americans of each sex aged under 65 in 1981, the cited rates are similar to the actual numbers of middle-aged Americans that died of these diseases that year. From Doll and Peto, 1981b.

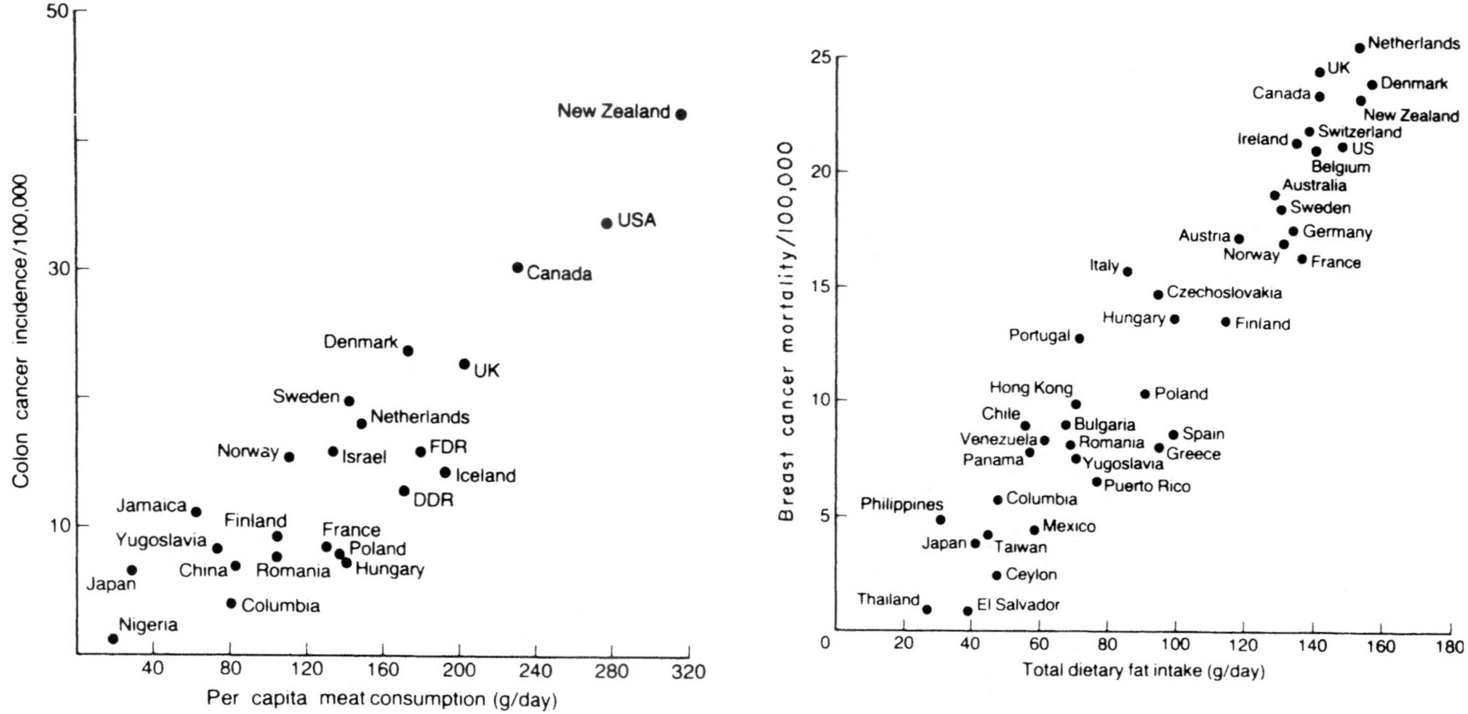

Fig. 2. In these graphs, the cancer rates per 100,000 women relate to women of similar age, and so are not materially affected by the greater risks of premature death from other causes in poor countries. The countries (generally prosperous) where meat or fat consumption is highest are those where women of a given age are at greatest risk of developing cancer of the colon or breast. But, although the explanation of this remains obscure, this is not strong evidence that either fat or meat are important causes of these cancers, merely strong evidence that these cancers have causes. From Doll and Peto, 1981b.

cancer and breast cancer among them are much more like those among U.S. whites than those among Japanese who are still living in Japan (or among blacks who are still living in West Africa). This is true for most types of internal cancer, although this is not true for skin cancer since skin color affects skin cancer risks so strongly.

Other evidence that various types of cancer have potentially avoidable causes comes from the fact that (1) there are strong correlates of the incidence rates of some types of cancer in different countries (Fig. 2); (2) there are rapid changes in the incidence of some types of cancer (the large increase in lung cancer is chiefly due to the delayed effects of past increases in cigarette smoking, but the large decreases in stomach and uterine cancer are not properly understood); and (3) a few dozen causes of human cancer have already been established beyond reasonable doubt, so presumably others still await discovery.

Arguments based on such international comparisons suggest that in total about three-quarters of all U.S. cases of internal cancer could be avoided by means that some populations have already adopted (Doll and Peto, 1981a). Since there may also exist practicable preventive strategies that none of the populations studied happens to have adopted, perhaps over 75 percent of cancer deaths in countries such as the U.S. might <u>in principle</u> be avoidable. Apart from the avoidance of tobacco (which is currently responsible for about one-third of all U.S. cancer deaths), however, the chief means of achieving this are not yet known. The proportion avoidable by <u>practicable</u> means may be smaller, for there are many aspects of the lifestyle in poor countries that most Americans would not willingly adopt. It is proving difficult enough to control the effects of tobacco although a little progress has been made. Over the past 30 years, U.S. cigarette tar deliveries have decreased by more than half, and over the past ten years, cigarette sales per U.S. adult have decreased by about one-sixth. It may, therefore, prove even more difficult to influence any dietary factors that turn out to be important determinants of cancer, unless the required changes involve prescription of some beneficial foodstuff(s) rather than restriction of some harmful one(s).

No Generalized Increase in Nonrespiratory Cancer

The foregoing arguments suggest that much cancer is avoidable, but they do not suggest how. Examination of U.S. national mortality trends for various types of cancer (Doll and Peto, 1981a,b; Peto, 1981) may indicate whether those types that are currently common are arising chiefly because of some newly introduced factor(s), or chiefly because of factors that have characterized the U.S. way of life throughout much of this century (or even earlier). There are two main difficulties in interpreting such trends. One problem is getting reasonably reliable data comparing cancer onset rates in different decades, and the other is the extraordinarily long delay that may be involved between the widespread introduction (or control) of a cause and emergence of its full effect on national cancer death rates. For example, the vast increase in lung cancer that is currently occuring is due principally to the delayed effect of the large increase in U.S. cigarette consumption that took place between 1915 and 1945 (Doll and Peto, 1981a,b). Indeed, the increase in lung cancer caused by the pre-1945 increase in cigarette consumption will probably continue throughout the remainder of the century, even though cigarette tar deliveries have been falling in recent years, and there has been no further large rise in cigarette consumption since 1945.*

The possibility of such long delays means, of course, that any causes that have been introduced widely only in recent decades may as yet have no clear effect on national trends, even if they eventually will have a large effect in the next century. So national trends can never be used to guarantee that recently introduced exposures are harmless, though it can be useful to bear them in mind when seeking causes for recent cases of cancer.

For most of the types of cancer that are currently common in adult life, there has been disappointingly little improvement in curative treatment over the past few decades. For these types, the national trends may be estimated most reliably by trends in age-specific death certification rates, rather than by the trends in the age-specific numbers of new cases recorded (for the recording of non-fatal cases may have become increasingly thorough). When examining trends in U.S. cancer death rates, two approaches are possible. On the one hand, bias may be minimized by restricting attention to middle-age, and to the period since 1950 (for mortality trends in old age since 1950 and at all ages before 1950 may be less reliable). This approach yields the trends already illustrated in Fig. 1, where very large increases in lung cancer may be seen, along with moderately large decreases in cancer of the uterus and stomach.

Alternatively, if we try to go further back and examine data from the 1930s, substantial uncertainties arise that preclude detailed analysis. But it is clear that among males, cancer of the stomach was then by far and away the commonest neoplastic cause of

*Between 1915 and 1939, cigarette consumption per U.S. adult per day increased from less than one up to five, and then, during the 1939-45 war, it suddenly increased to about 10, after which it has remained roughly constant at 10-12.

death (and cancer of the lung was quite rare), while among females the commonest causes of neoplastic death were cancers of the uterus and of the stomach.

It is particularly interesting to note that cancers of the breast and large intestine were about as common then as they are now. For these two diseases, the chief causes of the large differences between the U.S. and Japan (Fig. 2) presumably involve differences between the lifestyles that have existed throughout most of the present century. It is difficult to imagine what such factors might involve--nutritional and hormonal differences are perhaps the most promising hypotheses--but risk assessments for new synthetic chemicals do not appear to be an immediately attractive way of starting to investigate the problem. A few years ago, we prepared an extensive review of the standard epidemiological evidence about the avoidability of cancer (Doll and Peto, 1981a). Our main conclusions are summarized in Table 1; briefly, the only factor that is both large and reliably known is smoking, particularly of cigarettes. All other factors are either much smaller, or much less certainly known. Shortly afterwards we prepared a related review, the conclusions of which are summarized in Table 2 (Doll and Peto, 1983). There, we listed only those current exposures that could actually be modified and whose modification was known to be effective. Some aspects of these tables involve matters of judgement, where opinions might legitimately differ, and so certain details might justifiably be disputed, but not, I think, their overall pattern. In retrospect, the only change of substance I would now make would be to increase the probable importance of hormonal factors.

Consider, for example, breast cancer. In the U.S., the disease has been about as common throughout the past half century as it is now, but it is much less common in certain other countries. Reproductive factors have a vast effect on incidence rates in late middle and old age: for example, women who have their first child at 16 have only half the later risk of those who have their firstborn at 36; women who started to menstruate before 12 have more than double the risk of those who started to menstruate a few years later; early menopause is protective; the amounts of free estrogen in the blood appear to correlate directly (and strongly) with the subsequent incidence of the disease; and even after the disease has developed, its rate of growth can be modified by removing the ovaries or by anti-estrogen drug therapy. In experimental animals hormonal and nutritional manipulations can greatly alter the incidence of the disease, and there is every reason to believe the same will be true of humans, too, even though the relevant hormonal and nutritional factors may differ. In these circumstances, the most promising lines of enquiry about the etiology of breast cancer do not appear to be risk assessments for particular chemicals, but instead a more precise characterization of the exact hormonal factors that affect breast cancer incidence, coupled with a search for the chief extrinsic determinants of these hormonal factors (or for artificial ways of modifying them). The same is probably true for cancers of the endometrium, ovary and prostate. In any case, the pursuit of risk assessment for particular chemicals looks much less promising than does a continued investigation of the causes and effects of hormonal factors.

Hormonal factors may also be relevant for cancer of the cervix, but the most exciting possibilities for research at present appear to be those involving some newly discovered type(s) of human papilloma virus (HPV). For many years it has been clear that some agent--presumably a virus--passed between partners during sexual intercourse was a cause of most causes of carcinoma of the cervix, uteri (and, perhaps of the vagina, vulva and penis, though these are numerically less important diseases). But it has proved surprisingly difficult to pin down any one etiologic agent. Suspicion has centered first on one virus (Herpes simplex), then on another (type 6 or type 11 HPV), but results have remained elusive. Suddenly, however, with the discovery of various new types of HPV (particularly, type 16 HPV), much clearer correlations are emerging, and again, the most promising lines of research are not those involving risk assessments for particular chemicals, but those involving infective factors.

For cancer of the large intestine, the most promising lines of research are the nutritional ones, involving various specific types of fiber (particularly the pentosan-rich fibers found in many cereals), and the next most promising perhaps are those involving various specific types of fats, bile acids and hormones. Again, subjecting hundreds of chemicals to risk assessments seems a rather unpromising enterprise.

And so on: if "cancer" is to be prevented, then cancers of some specific sites must be prevented, and for none of these does an approach based on risk assessments for mass of chemical contaminants look particularly plausible. Even for lung cancer (where dusts, mists and fumes of chemicals gain direct access to the target cells, and where some extrinsic chemical exposures have already been shown to be carcinogenic), the chief problem in cancer control still is not in controlling synthetic chemicals, but in reducing the impact of tobacco. One way of doing this is to decrease the hazard per cigarette; indeed U.S. male lung cancer rates in early middle age (that is, under 50) are already beginning to fall, perhaps because of the large decreases in tar delivery that have taken place over the past 25 years. The most important way of reducing the impact of tobacco, however, is to discourage people from smoking, by helping smokers really to understand how uniquely large is the contribution of smoking to cancer risks (Table 2). In addition, smoking probably kills more people by chronic lung and heart disease than by cancer. Here, some positive disadvantages of a series of widely publicized risk assessments may begin to make themselves felt, for few members of the general public would guess, from a typical year's media coverage of one cancer scare after another, that boring old nonnewsworthy tobacco accounted for far more cancer deaths than all other reliably known effects put together.

Thus, the reason why many epidemiologists are unhappy with the wide use of risk assessments is because the few important things that are really established about the causes of human cancer may be swamped by a mass of pseudo-knowledge, based on unsubstantiated extrapolations from animals to humans--and there is no good evidence why risks in long-lived humans should remotely resemble those in short-lived animals (Doll and Peto, 1981a).

This brings us back, however, to the hygienists' central complaint against epidemiology--for most chemicals of immediate concern, no useful human data on risk has been produced, and there is little likelihood of the epidemiologists producing any within the next few years. For each substance, there is a fairly urgent need to make a decision (whether it be no restriction, partial restriction or complete banning), and if epidemiological information does not exist, there will be an obvious temptation to use risk assessments to invent it. Inventing it may, however, be going unnecessarily far. In our discussion of the types of evidence that laboratory studies can generate, we concluded that quantitative human risk assessment, as currently practiced, is so unreliable, suffering not only from random but also from large systematic errors of unknown direction and magnitude, that it should definitely be given another name: "priority setting" might be a more honest, although less saleable, name (Doll and Peto, 1981a). So many thousands of chemicals are active to some extent in one laboratory test or another that it is difficult to know what, if any, practicable regulations to enact on this basis. It has been recommended for some tests that the regulations which are promulgated should be based only on whether or not the chemical is active, irrespective of its quantitative degree of activity. But if no explicit use is to be made of the degree of activity of each chemical, then instead of effective reduction of the total of all human cancer the chief result may be complete paralysis (either of the regulators or of the "regulatees").

A more proper use of each particular laboratory test might be to multiply the potency of each chemical studied by whatever crude estimate is available of the degree of human exposure to that chemical to yield some sort of index of human hazard. When this has been done for long-term tests and each separate short-term test, it is likely that one or a few chemicals will stand out head and shoulders above the rest with respect to these indices of human hazard. The best use of the various laboratory tests might be to identify, study, and, if possible, reduce these few with the most extreme degrees of activity (together with any more moderate exposures that can be cheaply controlled), without necessarily requiring direct human evidence of harm. Although in many cases the benefits might be illusory, a few prudent restrictions against the apparent extremes with respect to each type of test might be nearly as effective as broad action against all apparently active chemicals. This is true, however, only if a really serious effort has been made to seek as many sources of exposure as possible to agents active in that test including endogenous formation of mutagens and other active agents in the gastrointestinal tract and exogenous absorption of such agents from our "normal" diet (Ames, 1983) and from involuntary or deliberate inhalation of smoke.

In deciding how rigorously to regulate one particular source of environmental contamination by dioxins, for example, it may be helpful to bear in mind the background extent to which such agents may be formed wherever organic matter is burned. In deciding how rigorously to regulate one particular source of dietary nitrosamines, it may be helpful to bear in mind the background extent to which nitrosamines are formed anyway in the human digestive tract. And, in deciding how rigorously to regulate some other source of inhaled or ingested mutagens, it may be helpful to bear in mind the background extent to which the blood and urine of cigarette smokers are contaminated by such agents anyway; massive regulatory efforts against quantitatively less important targets may, in their total effects, be diversionary. (Indeed, when the chairman of America's largest cigarette manufacturer was asked a few years ago why the cancer scare no longer seemed to be hitting cigarette sales the way it used to, he reportedly said that because so many things were getting linked to cancer, people were beginning to take a "more objective" view of the health issues!)

In the context of background exposures, a particularly interesting recent paper is that of Ames and the resulting correspondence (Ames, 1983), in which he points out that many natural plant and other foodstuffs contain large quantities of natural chemicals that would be active in various laboratory test systems. His thesis is not that these are known to be harmful or harmless, but simply that it makes little sense to pursue traces of man-made substances vigorously because of some type of activity that already is found in far greater quantities in natural products. He also pointed out that one reason why so many man-made chemicals are found to be carcinogenic is simply because we tend to test far more man-made chemicals than natural products. In the absence of direct evidence of quantitative relevance to humans of the findings in long-term animal tests or in any of the short-term tests, blanket restrictions on very large numbers of minor chemical pollutants may be unacceptably expensive, and the approach suggested above might turn out to be a socially acceptable alternative way of setting a few sensible priorities.

REFERENCES

Ames, B.N., 1983, Dietary carcinogens and anti-carcinogens, Science, 221:1256; also subsequent correspondence (Science, 1984, 224:668, 757).

Doll, R., and Peto, R., 1981a, The causes of cancer: quantitative estimates of avoidable risks of cancer in the United States today, J. Natl. Cancer Inst., 66:1191.

Doll, R., and Peto, R., 1981b, Why cancer? The causes of cancer in developed countries, in: "The Times Health Supplement," November 6, 1981, pp. 12-14, The Times, London.

Doll, R. and Peto, R. (1983) The epidemiology of cancer, in: "The Oxford Textbook of Medicine," pp. 4.51-4.79, D.J. Weatherall, J.D. Ledingham, and D.A. Warrell, eds., Oxford University Press, Oxford.

International Agency for Research on Cancer, 1984, "Second Cancers in Relation to Radiation Treatment: Results of a Cancer Registry Collaboration," Scientific publication No. 52, N.E. Day and J.C. Boice, Jr., eds., IARC Publications, Lyon.

Peto, R. 1981, Trends in U.S. cancer onset rates, in: "Quantification of Occupational Cancer," pp. 269-284, R. Peto and M.A. Schneiderman, eds., Cold Spring Harbor Publications, Cold Spring Harbor, New York.

REPORT OF NATIONAL CANCER INSTITUTE SYMPOSIUM: COMPARISON OF MECHANISMS OF CARCINOGENESIS BY RADIATION AND CHEMICAL AGENTS
I. COMMON MOLECULAR MECHANISMS

Donald C. Borg

Medical Department
Brookhaven National Laboratory
Upton, New York 11973

INTRODUCTION

The present workshop is the second of two coordinated conferences on mechanisms of carcinogenesis. The first meeting highlighted the molecular mechanisms common to radiation and chemical agents. Here we will place greater emphasis on epidemiological considerations and on dose-response models used in risk assessment to extrapolate from experimental data obtained at high doses to the effects from long-term, low-level exposures. Nevertheless, we feel so strongly that the sorting out of biologically realistic dose-response models must follow from mechanistic insight that we will open Session I with reports from the NCI symposium/workshop. It is my charge to deal with common molecular mechanisms; Michael Fry will be the rapporteur on cellular and animal models.

I shall present the material from the previous conference in a thematic way rather than summarizing each presentation as such. Accordingly, I shall not attempt to give a comprehensive overview of the fully packed program of the NCI symposium, but the names of the speakers on molecular mechanisms and the titles of their presentations are given in the list of references at the end of this chapter. I have picked one theme, DNA damage, that struck me as especially relevant to the needs of this meeting, because it is important to an understanding of differences between radiation and chemical carcinogenesis. I shall touch on some other topics that seem fundamental in clarifying the picture of carcinogenic mechanisms, while others I have chosen to discuss remain controversial or are otherwise appealing. For good measure, I will remark on one or two features of carcinogenic mechanisms that were not developed at the NCI conference but which I consider worth bringing to the attention of this audience.

The opening address at the NCI symposium, given by Mortimer Mendelsohn, presented the scope of the problem (Mendelsohn, 1983). He set the scene with a few pithy remarks that are equally applicable to this workshop and which express succintly the sentiments of the BNL organizing committee:

"Carcinogenesis is an immense subject, long on history and detail, and short on mechanistic understanding. Radiation carcinogenesis and chemical carcinogenesis are likewise immense, are studied by different people, are interpreted by different mechanisms and even funded by different agencies. An approach to mechanism through the comparison of these two modalities of carcinogenesis is timely and appropriate. Ideas and people from the two sub-fields should be merged, conceptual similarities and differences identified, and new insights into--or points of attack on--the overall process of carcinogenesis sought."

Different Aims of the NCI and BNL Conferences

At present, carcinogenesis preoccupies the body politic in its assessment of risk from low-level exposures to radiation and chemicals; therefore carcinogenesis is the major health concern of this conference as well. However, whereas the NCI meeting provided guidance for the direction of research programs and did not deal directly with risk assessment, this workshop actually focuses on the latter, and we will lay heavy emphasis on the dose-response functions that underlie extrapolation models. Such models are needed to project findings from the region of relatively high doses where experiments are feasible and epidemiology is most clear cut to the near-background exposure regimes that are of greatest concern to environmental health scientists and regulators.

We consider it imperative that trustworthy extrapolation models be biologically realistic as well as statistically sound. With that orientation in mind, I shall comment on some of the aspects of molecular mechanisms common to radiation and chemical carcinogenesis that were developed at the NCI meeting, particularly upon the DNA damage done by these agents.

COMMON AND DIFFERENT ASPECTS OF MOLECULAR EFFECTS

Routes of Exposure and Pharmacodynamic Considerations

Before beginning my discussion, it is important to point out that neither the NCI nor this conference deals very explicitly with routes of exposure, assimilation, and physiological distribution which apply to chemicals and to internal radiation sources (i.e., to radioisotopes taken up by the body) but not to external irradiation. Nor is there direct treatment in either program of metabolic activation and detoxification, which, of course, apply primarily to chemicals. However, there is now much confidence (although not absolute conviction) that DNA of somatic cells, especially stem cells, is the primary target for at least the so-called initiation step of both chemical and radiation carcinogenesis, a point to which I will return later. Therefore, although chemical carcinogenesis is clearly different from radiation carcinogenesis in its strong dependence on chemical uptake and metabolism prior to any effects on DNA targets, this realization is so widely accepted that no comprehensive overview of the topic was necessary for the NCI audience.

Characterization of Exposure

In outlining the scope of the problems to be addressed by the NCI conferees, Mendelsohn sought to compare and contrast aspects of the exposure and response of organisms to external ionizing radiation and to carcinogenic chemicals. In Fig. 1 he grouped a number of items related to characterizing and quantifying exposure. At the outset major differences are clear: radiation exposure is conceived and measured in extensive units, while chemical exposure, although complex and highly variable, is more usually viewed as an intensive burden or challenge. Hence there are huge, irreconcilable discrepancies between the ways one deals with the two modalities.

Interaction of Agents with the Organism

The interaction of the agent with the exposed organism is also extremely different in the two cases (Fig. 2). Pharmacodynamic factors play an important role in determining organotropic

EXTERNAL RADIATION	EXPOSURE	CHEMICAL
R, rads, rem, ? volume	Units	Grams, moles, per kg, per M^2
LET; type, energy	Quality	Chemical; chemical class
Single dose, dose-rate	Acute	Single dose, C x t
Fractionated, continuous	Chronic	Fractionated, continuous
Partial or whole-body	Route	Multiple

Fig. 1. Contrasting aspects of exposures to radiation and chemicals. (From Mendelsohn, 1983).

EXTERNAL RADIATION	CHEMICAL
Penetration of primary particles	Absorption
Formation of secondary electrons	Transport Diffusion
Ionization Excitation	Metabolic modification
Chemical change in target molecule	Chemical change in target molecule

Fig. 2. Contrasts in the interactions of radiation and chemicals with the organism. (From Mendelsohn, 1983).

carcinogenesis by chemicals and in explaining aspects of the differing sensitivities of species or of individuals, and even of different tissues or organs. The response depends strongly upon the metabolism of specific cells, tissues and organs, and in each case there are marked species and strain differences. The levels and inducibilities of enzymes capable of repairing premutagenic lesions may also show organ, species and strain specificity.

Metabolic modification should be taken into account when evaluating dose-response functions and developing extrapolation models. Metabolic processing of chemicals is also important in considering modifiers of mutagenesis and carcinogenesis and in evaluating the suitability of cellular and animal models as surrogates for effects on humans. External radiation really has no comparable counterpart to the metabolic conversion of precarcinogens to detoxified catabolites or to electrophiles capable of conjugating with DNA.

Mendelsohn also called attention to differences in the rates of interactions of radiation and chemicals with host tissues. Radiation events occur within microseconds, whereas the reactions of carcinogenic chemicals are measured by the much longer time scale of physiology and metabolism.

Distribution of Dose Effect

There are major differences in the distribution of dose at organismal, cellular and molecular levels following exposure to radiation or chemicals (Fig. 3). Pharmacodynamic considerations affect organ and tissue distribution in the case of chemicals, but local chemical reactions also largely determine dose distribution at cellular and molecular dimensions. However, as the chains of events are followed down to the level of target biomolecules,

EXTERNAL RADIATION		CHEMICAL
	DOSE DISTRIBUTION	
Square law, absorption	Macro	Enterohepatic, vascular, pulmonary Storage and excretion
Microdosimetry, track interaction Local chemistry	Micro	Size, charge, lipid solubility Half-life Receptors Local chemistry
Primary ionization or excitation of target	Direct	Primary adduction of target
Secondary effect on target, free radicals	Indirect	Adduction of target after activation Secondary adduction, free radicals
Remote effect	Abscopal	Remote effect
O_2, SH	Dose Modifiers	O_2, SH

Fig. 3. Contrasts in dose distribution and modification of radiation and chemical effects. (From Mendelsohn, 1983).

there is increasing evidence of commonality between radiation and chemical exposure. The profiles of direct effects on DNA, the primary target molecule for carcinogenic initiation, are usually highly distinctive for the two classes of agents, being largely strand breaks and chromosome-level changes for radiation and mostly adduct or conjugate formation for the reactive, electrophilic forms of chemicals, the so-called ultimate carcinogens. In terms of ultimate carcinogenic effect, these are very important distinctions. Despite these contrasts in the direct effects of radiation and chemicals on DNA, the so-called indirect effects may involve free radicals in both cases, albeit in very different ways.

I want to digress now for a moment to discuss biochemistry involving free radicals, a subject dear to my heart. The usual connotation of the phrase, indirect effects of ionizing irradiation, refers to the secondary reactions of free radicals resulting from the ionization of tissue water. With low-LET ionizing irradiation, the oxyradicals produced in this way are responsible for the predominant share of molecular damage, both to DNA and to most other functionally or structurally important macromolecules. But the metabolism of some chemical carcinogens can also give rise to reactive free radical intermediates, so that sometimes the ultimate carcinogens may be free radical metabolites of the carcinogenic chemicals themselves. In other cases autoxidation reactions that are secondary to the metabolism of carcinogens may afford significant fluxes of oxyradicals, including the extraordinarily electrophilic hydroxyl radicals that are the dominant toxic species mediating the indirect effects of radiation damage. On mechanistic grounds, therefore, one might expect this latter class of chemicals to be more radiomimetic in terms of DNA damage and thus, ultimately, in terms of carcinogenic impact as well.

Abscopal effects are those changes produced in organs or tissues distant from the site of primary cancer whose impact on carcinogenesis is mediated indirectly through endocrine, immunological or other physiological systems. Abscopal mechanisms are qualitatively quite similar for radiation and chemicals, and the effects of dose modifiers are also roughly comparable for both classes of carcinogens. These correspondences stand in sharp contradistinction to the discrepancies between radiation and chemicals which characterize most of the other features of exposure and response that I have discussed.

Emphasis on DNA as the Critical Target

The extraordinarily complicated network of intertwined physiological and metabolic pathways that intervenes between the exposure of an individual or an organism to chemical carcinogens and the ultimate dosing of DNA in target organs or tissues by reactive metabolites acts as a highly variable and often unpredictable filter strongly modulating the carcinogenic impact of any exposure regimen. Therefore, when chemical dosage is quantified in terms of environmental exposure, correlations with carcinogenesis are notoriously poor and usually unrevealing of underlying mechanisms. However, if DNA is considered the molecular target for carcinogenic initiation, then one may relieve the confounding influence of highly variable pharmacodynamics and obtain more predictable and understandable dose-response relationships. There is a growing consensus (and near unanimity at the NCI workshop) that it would be desirable to express chemical dose-response relationships in terms of the molecular dosimetry of DNA.

Both the NCI conference and this workshop emphasize the fundamental importance of DNA damage and its consequences in comparing and contrasting the mechanisms of carcinogenesis by ionizing radiation and chemical agents. Radiation and chemicals both damage DNA, but they give rise to very different kinds of lesions (Fig. 4). There was almost unanimous agreement at the NCI symposium that despite major gaps and uncertainties in the

EXTERNAL RADIATION	LESION	CHEMICAL
Strand breaks	Molecular dosimetry	Adducts
	Repair - type - error proneness - completeness	
	Biological dosimetry - mutation - aberration - transformation	

Fig. 4. Differences in the dominant DNA lesions caused by radiation and chemicals and certain considerations common to both. (From Mendelsohn, 1983).

understanding of carcinogenic mechanisms, it is clear that the different profiles of DNA lesions produced by radiation and by chemicals have profound significance, probably giving rise to characteristically different dose-response curves and possibly causing interfering or synergistic carcinogenic effects from some combinations of radiation and chemical exposures.

EFFECTS ON DNA

In offering some selective conclusions about the complicated matter of how the lesions produced in DNA by radiation and by chemicals compare, I have drawn mostly upon the contributions of Peter Brookes, Mortimer Elkind, Mortimer Mendelsohn, Julian Preston, and John Ward; and I have also taken account of some remarks by Lawrence Marnett and Malcolm Paterson.

Primary Interactions

At the outset, the initial damaging reactions in DNA caused by radiation and by chemicals are very differently distributed in both space and time. Radiation can alter all components of DNA: by hydrogen abstraction or by hydroxyl or hydrogen addition to the bases, by hydrogen abstraction from the carbohydrate, and by oxidation of the phosphate. Furthermore, even with low-LET irradiation, the clustered pattern of primary ionizations and the submicrosecond time scale of the primary events mean that at the molecular level secondary free radicals are formed virtually simultaneously in a lumpy or bunched distribution. Such local concentrations of rather indiscriminantly reactive intermediates increase the likelihood of there being multiply damaged regions of DNA where cooperative phenomena like strand breaks can occur, especially double-strand breaks. At such sites, with loss of nucleotides at each locus, there may also be loss of genetic information, because no effective repair is possible from complementary strands, as occurs in the repair of singly damaged sites. Furthermore, although there is initially much more base adduction than strand breakage, almost all of the hydroxyl or hydrogen adducts produced by free radicals from ionizations in water are repaired within ten minutes or so, yet most of the hydrogen abstractions from deoxysugars and many of the apurinic or apyrimidinic sites resulting from base repair give rise to strand breaks within minutes to about an hour.

A significant consequence of these events is that ionizing radiation produces high ratios of strand breaks to base changes. The breaks are mostly single-strand breaks, but some double-strand breaks are formed either directly (especially with high-LET radiations) or by coincidental overlap of single-strand breaks on complementary DNA strands. There is quick and nearly error-free repair of single-strand breaks, but the rapid repair of double-strand breaks seems always to leave a residuum of chromatid or chromosome aberrations, perhaps because there is a high likelihood of DNA exchange occurring during the coincidental repair of two nearby breaks. These events overlap in time but are brief in comparison to the time for cell cycling and DNA synthesis, so radiation can induce chromosomal or chromatid aberrations in all stages of the cell cycle.

By way of contrast to the nearly instantaneous effects of radiation, the reactive forms of chemical carcinogens are presented to DNA over the relatively long times required for necessary physiological transport and/or metabolic activation. Unlike the chemically indiscriminate but spatially heterogeneous attacks on DNA by both primary ionizations and the products of radiolysis, the electrophilic ultimate carcinogenic forms of most chemicals interact more homogeneously with DNA, but they are highly selective with regard to the molecular sites with which they react. The initial DNA lesions from chemicals are usually base adducts, and there are preferred configurations for many of them.

Repair of DNA Damage

The repair of chemical damage to DNA is, perhaps, even more complicated than its initial causation. The source of my selected remarks about it are the contributions to the NCI meeting of Peter Brookes, Mortimer Elkind, Robert Haynes, Mortimer Mendelsohn, Malcolm Paterson, Julian Preston, and Katherine Sanford.

Normal cells have several error-free modes of restoring integrity to DNA altered by chemical adducts. There may be direct reversal in situ, such as the alkyl transfer from O^6-alkylguanine residues (Setlow, this volume). And there are at least two classes of excision repair in higher organisms: base excision of alkyl and other small adducts, with retained chain continuity, and nucleotide excision with chain scission and ligation for bulky adducts, such as polyaromatic hydrocarbons which favor the more nucleophilic sites, especially ring nitrogens N-7 of guanine and, secondarily, N-3 of adenine. Some participants at the NCI symposium felt that repair of certain kinds of chemical adducts may be complete at low doses, giving rise to thresholds for damage. This idea is consistent with some dose-response data for mutagenesis, which also suggest the presence of effective thresholds.

Sanford presented evidence that normally there is a competent postreplication excision repair in the G-2 or prophase stage of the cell cycle, but the ligation step of this excision system is faulty in cancer cells. Deficient excision repair causes accumulation of single-strand breaks, which then convert to double-strand breaks by endonuclease action and give rise to the well-known karyotypic instability of cancer. This repair defect may be a precondition for carcinogenic initiation.

It is uncertain whether the error-prone postreplication repair demonstrated in prokaryotes occurs in eukaryotes. Some error-prone or SOS repair occurs in higher organisms, but it is not clear whether this is similar to the postreplication repair of prokaryotes or to the repair described by Sanford. Paterson contends that so-called error-prone repair is really what he would term a tolerance mechanism for by-passing DNA lesions and not a true repair in any sense.

Preston noted that unlike the rapid repair of radiation damage at all stages of the cell cycle most chemical damage to DNA is repaired very slowly except in the S phase of DNA synthesis. Only during S phase is there much likelihood that coincidental repair of adjacent but independent lesions will interact to produce aberrations. Hence most chemicals are much weaker clastogens than is radiation, and their effects are manifest primarily in S and will give rise mostly to chromatid aberrations, as opposed to chromosome aberrations.

Brookes took into account the highly selective nature of chemical adduct repair and correlated it with mutagenicity. Mutagenicity is more selective in its dependency on specific DNA adducts than is toxicity, and only particular adducts are premutagenic, such as the 6-alkylguanines and the exocyclic N-2 adducts of some large hydrocarbons, both of which may be effectively repaired in some circumstances but not in others (for example, not in certain organs of some species). Since equal levels of overall damage correlate with equal toxicity, while mutation rather than toxicity correlates with carcinogenicity, Brookes proposed that earlier hypotheses relating tumor initiation with the extent of DNA adduction should be modified. Rather, the correlation is with those DNA reactions leaving residual lesions that induce mutations in mammalian cells.

Cell Transformation and Carcinogenesis

The last item from the NCI symposium that I will discuss is transformation (Fig. 4). It remains unclear what, if any, aspects of carcinogenesis *in vivo* are modeled by transformation *in vitro*. For years there has been a feeling that most transformable cell lines are already initiated or otherwise precommitted, as, for example, by the activation of an oncogene for immortalilty. Mendelsohn expressed doubt that transformation, as a generality, deals with the same kinds of primary events as are involved in spontaneous or low-level induced carcinogenesis. Brookes made a supporting observation that although mutagenicity correlates well with carcinogenicity *in vivo*, it correlates poorly with transformation.

Conclusions

I have attempted to cover a complex subject in a rather brief and simplified way. However, I have provided sufficient mechanistic insight into DNA damage and repair so that I can formulate several important phenomenological generalizations:

1. Radiation is a strong clastogen and a weak mutagen, and it is common for radiation to cause changes in chromatin that are gross on the molecular scale, although often submicroscopic. Conversely, with chemical carcinogens, base changes and frame shifts are the most frequent residual lesions and are correlated with mutagenicity. The fact that many mutagens are weak clastogens can also be understood.

2. The two-strand lesions more efficiently formed by radiation, such as crosslinks and double-strand breaks, tend to be lethal. Conversely, one-strand lesions, such as base alterations and, possibly, single-strand breaks, tend to be mutagenic. As a result, the ratio of cell killing to mutagenicity is far higher for radiation than for chemicals. Since cell death precludes carcinogenic expression, the intrinsic carcinogenicity of radiation may be modulated, especially at higher doses.

3. A very important observation with regard to extrapolation models is that there are plausible grounds to expect real or practical thresholds for the mutagenicity and carcinogenicity of some chemicals, at least at sufficiently low rates or levels of

exposure. On the other hand, a no-threshold, linear dose-effect response at very low doses or dose rates of radiation is predicted. Elkind presented the stochastic, target-theory argument for this latter prediction based on the absolute, all-or-none quality of the so-called hit or critical radiation-induced event (presumably some intranuclear change). However, the physical chemical nature of the target undergoing such a discrete and absolute change of state remains elusive. I prefer the explanation based on the competitive kinetics of the repair of double-strand breaks as one providing more mechanistic insight.

4. Finally, a query regarding the biological reality of extrapolation models. Preston noted that with high-LET radiation and with low-LET radiation below about 50 rad, chromosomal aberrations are due mostly to spatially coincident double-strand breaks rather than the temporally coincident strand repair that dominates the quadratic region of low-LET dose-response. A change in mechanism with dose in this experimentally important range challenges the validity of models based on a single response function.

RELATIONSHIP OF MOLECULAR MECHANISMS TO DOSE-RESPONSE

In light of the mechanisms I have discussed, it is time to look briefly at some representative dose-response curves (Fig. 5). This afternoon's session and several other papers will deal explicitly with dose-response functions and extrapolation models, so I shall not describe the classes of models commonly used.

When all-or-none responses to ionizing radiation are measured, such as chromosome aberrations or survival of cells or of complex organisms, it appears indisputable that the dose-response functions are basically linear for high-LET radiations and linear-quadratic for low-LET radiations, with a significant no-threshold linear component in the low dose range. This generalized representation (Fig. 5) comes from the work of Victor Bond, one of the organizers and guiding spirits of this workshop, who, unfortunately, is unable to be present. The letter "H" denotes typical high-LET radiation behavior, and "L" denotes low-LET radiation. It is true that at sufficiently high doses, here denoted by "H'" and "L'", respectively, saturation sets in as the fraction of cells not yet responding quantally (that is, in an all-or-none fashion) becomes small. Nonetheless, these curves do represent the "classical" dose-response for ionizing irradiation. It is not certain that radiation carcinogenesis manifests similar linear and linear-quadratic responses, respectively, but there is much evidence that this is so. Linear regressions are often taken as the so-called conservative dose-response functions in representing chemical carcinogenesis, but this seems poorly justified on mechanistic grounds, as Fig. 6 shows.

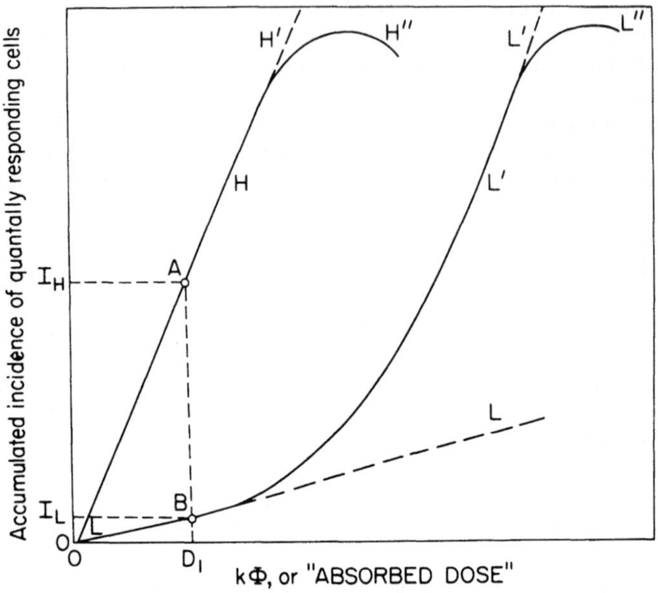

Fig. 5. Generalized representations of all-or-none responses in cells as a function of radiation dose at the target. I_H and I_L represent the incidence of hit cells with heavily and lightly ionizing irradiation at a given absorbed radiation dose corresponding to points A and B on curves H and L, respectively. (From Bond and Varma, 1983).

Some more recent evaluations of carcinogenic risk take into account nonlinear response functions that could be caused by substrate induction of enzymes participating in metabolic activation or detoxification. These models may also include competitive inhibition or the saturation kinetics of activation or repair systems. Strongly nonlinear and time-dependent damage functions are predicted for the low-dose regimes when saturation of DNA repair occurs at higher exposure levels. This gives rise to the so-called hockey stick damage function, as seen in panels c and d of Fig. 6. These dose-response curves utilize the concept of dose as the concentration of DNA adducts in the target organ rather than applied dose or exposure, much in the manner that I have discussed. Although the modification of restricting dose to those DNA lesions which induce mutations is not explicitly included, it may be thought of as part of the definition of effective dose in these generalized relationships. In any case, it is clear that these curve shapes are in serious conflict with the linear extrapolations often assumed to be conservative.

The organizing committee of this workshop contends that to choose proper models for extrapolating to low doses from experimental data obtained at high doses, it is of considerable importance to determine molecular mechanisms of damage to DNA by radiation and by chemicals and also to identify and characterize the repair enzymes that different DNA lesions call into play. In

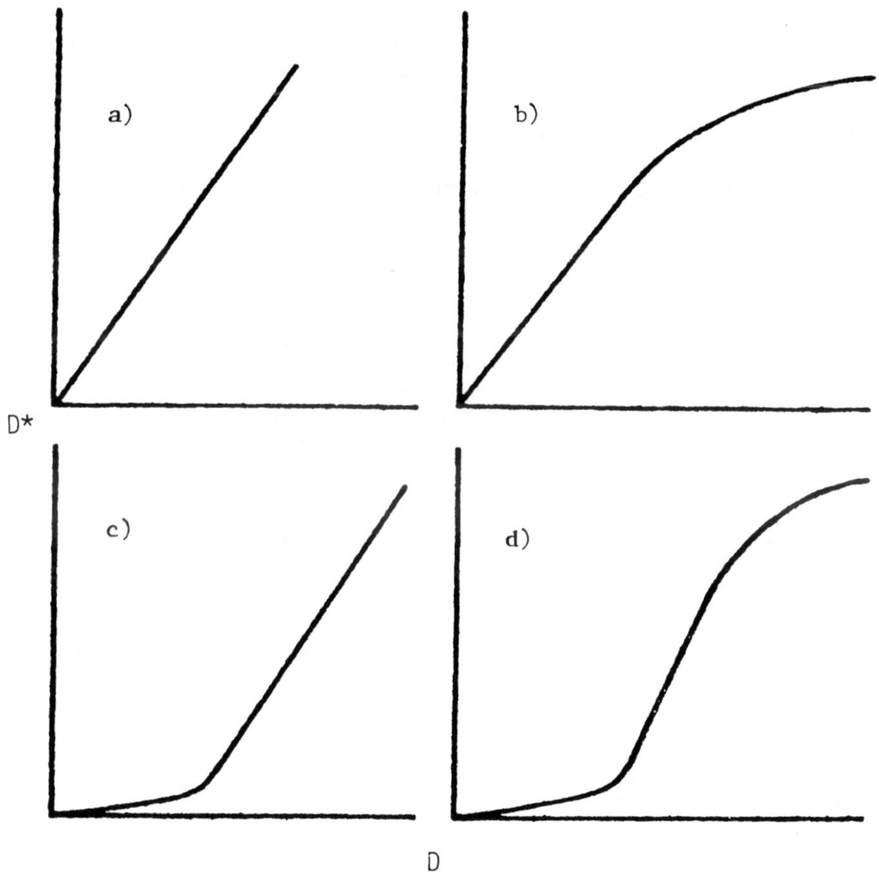

Fig. 6. Possible relations between administered dose, D, and effective dose at the target (e.g. DNA), D*, for several kinetic models. a), Simple first-order kinetics; b), saturation of the activation system; c), saturation of detoxification or repair systems; and d), combination of b) and c). (After Hoel et al., 1983).

this context, my report on the NCI conference has focussed on the molecular mechanisms associated with DNA damage in order to emphasize the similarities and differences in low-level exposures to radiation and to chemicals that seem most critical to risk assessment. At the NCI symposium Mendelsohn also concluded that sorting out the biologically realistic dose-response models must follow mechanistic insight.

There is much clarification to be gained in risk-assessment modeling by normalizing carcinogen dosimetry to DNA effects. Along the path from initial exposure to final manifestations of overt cancer there is increasing commonality between radiation and chemical carcinogenesis. There was a strong consensus at the NCI meeting that at the level of initiation radiation and chemical carcinogenesis have effectively converged. Since initiation, in turn, is thought to be intimately associated with a heritable change in DNA, I essentially limited my report to this workshop to that part of the NCI conference. In terms of distinguishing between radiation and chemical carcinogenesis, the remainder of the process affords little, if anything, "on which one can hang one's hat," as Mendelsohn put it.

OTHER ASPECTS OF COMMON MOLECULAR MECHANISMS

Much material on molecular mechanisms was presented at the NCI meetings that is of great importance to cancer research but which dealt with topics other than DNA damage. Indeed, much of the present excitement in experimental oncology lies in those other areas: even my own research interests have their greatest relevance there. Nonetheless, as rapporteur to this BNL workshop, I judged these other topics less pertinent in characterizing comparative risks from low-level exposures to radiation and chemicals.

Other main themes at the NCI conference that appertained to common molecular mechanisms included: the number and kind of stages or steps in the overall development of cancer; the nature of carcinogenic promotion and its apparent substages; the roles of oxygen free radicals, the arachidonic acid cascade, and membranes in promotion; the unlikelihood of there being a single theory of cancer that will prove to be correct; and the accelerated use of recombinant DNA and other new methods of contemporary molecular biology to obtain detailed information on DNA damage. Because of the importance of these topics, I shall make a few broad, summary remarks.

Once initiation has occurred, radiation and chemical carcinogenesis have largely converged (that is, promotion and progression are mechanistically the same for both). However, despite some measure of consensus amongst research scientists, there is still no broadly acceptable definition of what is meant by initiation, promotion, or hit (versus event). The NCI conferees could not agree as to whether progression is a genomic change separate from promotion or part of it.

It is known that oncogenes can be expressed by single base pair substitution, gene amplification, and by several kinds of genetic translocation involving regulatory sequences. These alterations to oncogenes appear to be likely steps for the convergence of radiation and chemical carcinogenesis, but there is no consensus of opinion on how to reconcile intermediation of oncogenes with the concepts of initiation or promotion.* Although expressed oncogenes have been identified with only a small fraction of human cancers, this situation may reflect mostly the inadequacy of present test systems. Hence it is premature to opine regarding the obligatory or the optional involvement of oncogenes in cancer.

Some recent evidence strongly suggests that promoters are clastogens, and it is tempting to relate chromosomal damage to oncogene activation. In the studies reported, however, clastogenesis may be an epiphenomenon, reflecting only the presence of hydrogen peroxide educed (from neutrophils) by some of the promoters used. It is important to resolve this issue, as well as to ascertain the roles of proliferation and of inflammatory cells in promotion other than in the skin, and to determine whether skin is a general model for promotion or a special case.

It will not suffice to explain aspects of promoter/antipromoter behavior as resulting from a pro-oxidant state or from active oxygen. The promoting roles of free radicals, oxidants and lipid peroxidation are presently unclear, and greater efforts must be made to identify the intermediates and to determine their reactivities under biologically significant conditions. The specific chemistry of oxidants/antioxidants should be defined to clarify how oxygen radicals work through a common, focussed, hormone-like pathway in promotion and to elucidate the reactions of superoxide anion and its dismutases with plasma clastogenic factors that may be active in promotion.

---

*Following the NCI Symposium, Hsiao and coworkers submitted a report suggesting "...that during multistage carcinogenesis, the initiating carcinogen might function by activating a cellular proto-oncogene, whereas tumor promoters might enhance outgrowth of the altered cells or the expression of other cellular genes that complement the function of the activated oncogene." (Hsiao et al., 1984)

Another area requiring biochemical clarification involves the arachidonic acid cascade in both carcinogenic initiation and promotion. The roles of several distinct but interrelated pathways involving oxygen-requiring free radical reactions in prostaglandin or leukotriene synthesis and action have yet to be distinguished. The importance of precarcinogen activation *in vivo* through co-oxidation by prostaglandin hydroperoxidase also warrants further investigation.

REFERENCES GIVEN IN THIS REPORT:

Bond, V. P., and Varma, M. N., 1983, A stochastic weighted hit size theory of cellular radiobiological action, *in*: "Radiation Protection, Eighth Symposium on Microdosimetry," p. 423, J. Booz and H. G. Ebert, eds., EUR 8395 en.

Hoel, D. G., Kaplan, N. L., and Anderson, M. W., 1983, Implications of nonlinear kinetics on risk estimation in carcinogenesis, *Science*, 219:1032

Hsiao, W.-L. W., Gattoni-Celli, S., and Weinstein, I. B., 1984, Oncogene-induced transformation of C3H 10T 1/2 cells is enhanced by tumor promoters, *Science*, 226:552.

Mendelsohn, M. L., 1983, Comparison of mechanisms of carcinogenesis by radiation and chemical agents: the scope of the problem, *at*: NCI symposium, National Bureau of Standards, Gaithersburg, Maryland, Dec. 6 (unpublished).

REFERENCES GIVEN BY SPEAKERS:

Oxy-Radical Chemistry: Radical Lifetimes; Autoxidation; Induced Decompositions of Peroxides - W. A. Pryor

Mill, T., and Hendry, D. G., 1980, Kinetics and mechanisms of free radical oxidation of alkanes and olefins in the liquid phase, *in*: "Chemical Kinetics," p. 1, Elsevier Publishing Company, New York.

Pryor, W. A., 1966, "Free Radicals," McGraw-Hill Book Company, New York.

Pryor, W. A., 1982, Free radical biology: xenobiotics; cancer; aging, *Ann. N.Y. Acad. Sci.*, 393:1.

Pryor, W. A., 1984, Free radicals in autoxidation and aging, *in*: "Free Radicals in Biology and Aging," D. Armstrong, et al., eds., Raven Press, New York.

Pryor, W. A., Hales, B. J., Premovic, P. I., and Church, D. F., 1983, The radicals in cigarette tar: their nature and suggested physiological implications, *Science*, 220:425.

Walling, C., 1957, "Free Radicals in Solution," John Wiley and Sons, New York.

Physiological Targets of Oxy Radicals and Carcinogenesis - L. J. Marnett

Basu, A., and Marnett, L. J., 1983, Unequivocal demonstration that malondialdehyde is a mutagen, Carcinogenesis, 4:331.

Coffey, R. G., and Hadden, J. W., 1983, Phorbol myristate acetate stimulation of lymphocyte guanylate cyclase and cyclic guanosine 3':5'-monophosphate phosphodiesterase and reduction of adenylate cyclase, Cancer Res., 43:150.

Esterbauer, H., 1982, Aldehydic products of lipid peroxidation, *in*: "Free Radicals, Lipid Peroxidation, and Cancer," p. 101, D. H. McBrien and T. F. Slater, eds., Academic Press, New York.

Jones, D.P., Thor, H., Smith, M. T., Jewell, S. A., and Orrenius, S., 1983, Inhibition of ATP-dependent microsomal Ca^{2+} sequestration during oxidative stress and its prevention by glutathione, J. Biol. Chem., 258:6390.

Marnett, L. J., and Eling, T. E., 1983, Cooxidation during prostaglandin biosynthesis: a pathway for the metabolic activation of xenobiotics, *in*: "Reviews in Biochemical Toxicology," p. 135, E. Hodgson, J. R. Bend, and R. M. Philpot, eds., Elsevier Science Publishing Co., New York.

Travis, J., and Salvesen, G. S., 1983, Human plasma proteinase inhibitors, Ann. Rev. Biochem., 52:655.

Ward, J. F., 1975, Molecular mechanisms of radiation-induced damage to nucleic acids, Adv. Radiat. Biol., 5:181.

The Role of DNA Alkylation in Chemical Carcinogenesis - P. Brookes

Balmain, A., and Pragnelli, I. B., 1983, Mouse skin carcinomas induced *in vivo* by chemical carcinogens have a transforming Harvey-ras oncogene, Nature (London), 303:72.

Brookes, P., Newbold, R. F., and Osborne, M. R., 1979, Mechanism of the carcinogenicity of polycyclic hydrocarbons, *in*: "Environmental Carcinogenesis," p.123, P. Emmelot and E. Kriek, eds., Elsevier/North Holland, Amsterdam.

Lawley, P. D., 1976, Carcinogenesis by alkylating agents, *in*: "Chemical Carcinogenesis," American Chemical Society Monograph 173, p.83, C. E. Searle, ed., Washington, D.C.

Magee, P. N., Montesano, R., and Preussman, R., 1976, N-nitroso compounds and related carcinogens, *in*: "Chemical Carcinogenesis," American Chemical Society Monograph 173, p.491, C. E. Searle, ed., Washington, D.C.

Newbold, R. F., and Overell, R. W., 1983, Fibroblast immortality is a prerequisite for transformation by EJ c Ha-*ras* oncogene, Nature (London), 304:648.

Ionizing Radiation Damage - J. F. Ward

Bernhard, W. A., 1981, Solid-state radiation chemistry of DNA: the bases, Adv. Radiat. Biol., 9:199.

Goodhead, D., 1979, "Radiation Biology in Cancer Research," Raven Press, New York.

Mattern, M., Hariharan, P., and Cerutti, P., 1975, Selective excision of gamma ray damaged thymine from the DNA of cultured mammalian cells, Biochim. Biophys. Acta, 395:48.

Ward, J. F., 1975, Molecular mechanisms of radiation induced damage to nucleic acids, Adv. Radiat. Biol., 5:182.

Ward, J. F., 1981, Some biochemical consequences of the spatial distribution of ionizing radiation-produced free radicals, Radiat. Res., 86:185.

The Role of Chromosomal Damage by Indirect Action in Carcinogenesis - P. Cerutti

Amstad, P., and Cerutti, P., 1983, DNA binding of aflatoxin B_1 by co-oxygenation in mouse embryo fibroblasts C3H/10T1/2, Biochem. Biophys. Res. Commun., 112:1034.

Cerutti, P., 1978, Repairable damage in DNA, in: "DNA Repair Mechanisms," p. 717, P. Hanawalt, E. Friedberg, and C. Fox, eds., Academic Press, New York.

Cerutti, P., Emerit, I., and Amstad, P., 1983, Membrane-mediated chromosomal damage, in: "Genes and Proteins in Oncognesis," p. 55, I. B. Weinstein, and H. Vogel, Eds., Academic Press, New York.

Emerit, I., Cerutti, P., 1981, Tumor promotor phorbol-12-myristate -13-acetate induces chromosomal damage via indirect action, Nature (London), 293:144.

Emerit, I., and Cerutti, P., 1982, Tumor promotor phorbol-12-myristate-13-acetate induces a clastogenic factor in human lymphocytes, Proc. Natl. Acad. Sci. USA, 79:7509.

Emerit, I., Levy, A., and Cerutti, P. A., 1983, Suppression of tumor promoter phorbol myristate acetate-induced chromosome breakage by antioxidants and inhibitors of arachidonic acid metabolism, Mutat. Res., 110:327.

Emerit, I., and Cerutti, P., 1983, Clastogenic action of tumor promotor phorbol-12-myristate-13-acetate in mixed human leukocyte cultures, Carcinogenesis, 4:1313.

Friedman, J., and Cerutti, P., 1983, The induction of ornithine decarboxylase by phorbol-12-myristate 13-acetate or by serum is inhibited by antioxidants, Carcinogenesis, 4:1425.

Zimmerman, R., and Cerutti, P., 1984, Active oxygen acts as a promotor of transformation in mouse embryo fibroblasts C3H/10T1/2/C18, Proc. Natl. Acad. Sci. USA, 81:(in press).

Role of DNA Repair in Radiation and Chemical Induced Carcinogenesis - M. C. Paterson

Hanawalt, P. C., Cooper, P. K., Ganesan, A. K., and Smith, C. A., 1979, DNA repair in bacteria and mammalian cells, Ann. Rev. Biochem., 48:783.
Lindahl, T., 1982, DNA repair enzymes, Ann. Rev. Biochem., 51:61.
Paterson, M.C., 1982, Heritable cancer-prone disorders featuring carcinogen hypersensitivity and DNA repair deficiency, in: "Host Factors in Human Carcinogenesis: Proceedings of a Symposium, Cape Sunion, Greece, June 8-11, 1981," IARC Sci. Publ. no. 39, p. 57, H. Bartsch and B. Armstrong, eds., IARC, Lyon.
Paterson, M. C., 1983, in: "Radiation Carcinogenesis: Epidemiology and Biological Significance," p.319, J. D. Boice, Jr., and J. F. Fraumeni, Jr., eds., Raven Press, New York.
Rajalakshmi, S., Rao, P. M., and Sarma, D. S. R., 1982, Chemical carcinogenesis: interactions of carcinogens with nucleic acids, in: "Cancer: A Comprehensive Treatise, Vol. 1," p. 335, F. F. Becker, ed., Plenum Press, New York.

Multiple Primary Targets for Induced Carcinogenesis - R. H. Haynes

Das, S. K., Benditt, E. P., and Loeb, L. A. 1983, Rapid changes in dNTP pools in mammalian cells treated with mutagens, Biochem. Biophys. Res. Comm., 114:458.
Haynes, R. H., Little, J. G., Kunz, B. A., and Barclay, B. J., 1982, Non-DNA primary targets for the induction of genetic change, in: "Environmental Mutagens and Carcinogens," p. 121, T. Sugimura, S. Kondo, and H. Takebe, eds., University of Tokyo Press, Tokyo.
Kunz, B. A., 1982, Genetic effects of dNTP pool imbalances, Environ. Mutagen., 4:695.
Land, H., Parada, L. F., and Weinberg, R. A., 1983, Cellular oncogenes and multistep carcinogenesis, Science, 222:771.
Loeb, L. A., and Kunkel, T. A., 1982, Fidelity of DNA synthesis, Ann. Rev. Biochem., 52:429.
Newman, C. N., and Miller, J. H., 1983, Mutagen-induced changes in cellular dCTP and dTTP in CHO cells., Biochem. Biophys. Res. Comm., 114:34.

Chromosome Rearrangements in Cancer - R. J. Preston

ar-Rushdi, A., Nishikura, K., Erikson, J., Watt, R., Rovera, G., and Croce, C. M., 1983, Differential expression of the translocated and the untranslocated c-myc oncogene in Burkitt lymphoma, Science, 222:390.

de Klein, A., van Kessel, A. G., Grosveld, G., Bartram, C. R., Hagemeijer, A., Bootsma, D., Spurr, N. K., Heisterkamp, N., Groffen, J., and Stephenson, J. R., 1982, A cellular oncogene is translocated to the Philadelphia chromosome in chronic myelocytic leukemia, Nature (London), 300:765.

Neel, B. G., Jhanwar, S. C., Chaganti, R. S. K., and Hayworth, W. S., 1982, Two human c-onc genes are located on the long arm of chromosome 8, Proc. Natl. Acad. Sci. USA, 79:7842.

Rowley, J. D., 1980, Chromosome abnormalities in human leukemia, Ann. Rev. Genet., 14:17.

Taub, R., Kirsch, J., Morton, C., Lenoir, G., Swan, D., Tronick, S., and Leder, P., 1982, Translocation of the c-myc gene into the immunoglobulin heavy chain locus in human Burkitt lymphoma and murine plasmacytoma cells, Proc. Natl. Acad. Sci. USA, 79:7837.

Promotion/Progression and Involvement of Non-DNA Targets - T. J. Slaga

(No references submitted)

Modifiers of Mutagenesis and Carcinogenesis - L. W. Wattenberg

Griffin, C., 1982, in: "Molecular Interrelations of Nutrition and Cancer," Raven Press, New York.

Medina, D., and Sheperd, F., 1981, Selenium-mediated inhibition of 7,12-dimethylbenz(a)anthracene-induced mouse mammary tumorigenesis, Carcinogenesis, 2:451.

Mirvish, S., 1981, "Inhibition of Tumor Induction and Development," Plenum Press, New York.

Narisawa, T., Sato, M., Tani, M., Kudo, T., Takahashi, T., and Goto, A., 1981, Inhibition of development of methylnitrosourea-induced rat colon tumors by indomethacin treatment, Cancer Res., 41:1954.

Sporn, M., and Roberts, A., 1983, Role of retinoids in differentiation and carcinogenesis, Cancer Res., 43:3034.

Wattenberg, L. W., 1983, Inhibition of neoplasia by minor dietary constituents, Cancer Res. (suppl.), 43:2448s.

Yavelow, J., Finlay, T., Kennedy, A., and Troll, W., 1983, Bowman-Birk soybean protease inhibitor as an anticarcinogen, Cancer Res. (suppl.), 43:2454s.

DISCUSSION

Holtzman: Would you care to comment on evidence by Michael Gould and co-workers (Gould, 1984; Mulcahy et al, 1984) and evidence from "so-called" synergistic interactions in our laboratory, that initiation events are much more common then can be determined from the results of carcinogenesis studies.

Borg: What is the comparison?

Holtzman: More lesions than you would see expressed as tumors.

Borg: Yes. That must be so. The overwhelming number of lesions are repaired for both radiation and chemicals. But even some of the lesions that are not chemically repaired do not seem to give rise to base-pair errors upon replication and therefore do not give rise to misinformation. As long as that particular strand of altered DNA is there, it has the lesion, but the lesion will not have much import. Brookes made the point that some lesions are selectively repaired in different organs and tissues, and it is those organs that cannot repair satisfactorily that show the impact, and other lesions are silent.

Clifton: I believe Dr. Holtzman is referring to Gould's suggestion that there are many initiation targets. He did not specify that the targets are DNA. Initiation on a cellular basis is far too frequent to be due to mutation in a few specific loci.

Borg: I am reminded of another line of evidence which was represented at the NCI conference by Anne Kennedy and Jack Little. Looking at their systems *in vitro* it would appear that essentially all cells are being initiated and, therefore, you can keep diluting them and diluting them in culture without changing the transformation frequency. What has occurred is that the bias toward transformation of all cells has been changed. Would you have some comments on this point Dr. Haynes?

Haynes: I do not want to speak directly about Dr. Kennedy's findings, but the problem nevertheless relates to what Dr. Clifton has just said and also to some of the comments that I made at that symposium. The first is, I think that we could be on dangerous ground if we put too much emphasis on the rather simple picture that these processes act by agents which interact with DNA either directly, or after metabolic activation, to produce changes in DNA, and then the resulting biologic effect. One also has to consider more generally errors or pathologies of DNA metabolism. Faulty DNA synthesis, in particular, can give rise to mutagenic changes. One particular category of changes are those relating to distrubances in DNA nucleotide precursor pools. It has been known for many years that if you perturb or bias precursor synthesis, it will alter pool balance, and may result in a range of genetic effects, such as mutation, recombination, and transformation. Furthermore, it has been shown recently that standard mutagens, including x-rays, UV, and nitrosoguanidine, may also produce pool disturbances. While there is no question that one can generate incorrect DNA or faulty DNA by direct damage, attack on the entire synthesizing system indirectly may cause DNA damage. The point that I was going to raise with respect to Dr. Clifton's remark concerning frequencies of the order of 10^{-3}, is that while these frequencies are far too high to be accounted for on the basis of mutations in individual loci, nonetheless, those orders of frequencies are quite typical of frequencies of mitotic recombination, such as mitotic crossing over. In view of the finding that genetic rearrangements are implicated in oncogenic activation, it is possible that these frequencies reflect that fact.

Borg: Let me ask a question of Dr. Haynes. The homeostasis of these nucleotide precursor pools in vivo looks remarkably robust. Also, it is quite surprising how much you can replace the bases with substitutes and distort those pools. You may know that there are a number of ideas for radiation therapy in which it is proposed to replace thymine, up to 5% or more, with one or another halogenated uracil. It is surprising how well animals tolerate this, and we have some similar experience in humans. There is a good deal of robustness, therefore, at the same time as there is sensitivity.

Thilly: This point may in fact not be relevant to the question of pools since substitutions of bromodeoxyuridine at 5% do not arise from concentrations which disturb pools. When the bromodeoxyuridine concentration is high enough to cause a TTP imbalance feedback on the reductase deoxycytidine diphosphate one sees this imbalance.

Bridges: Did I hear correctly that you stated that ionizing radiation was a strong clastogen and a weak mutagen? If so, I would question the latter statement. Calculated on a per DNA lesion basis, ionizing radiation in bacteria is a more potent inducer of base pair substitution mutations than is ultraviolet light, an agent that is normally considered to be a strong mutagen. The reason why ionizing radiation is so often (and erroneously) considered as a weak mutagen is because it is so cytotoxic that the majority of the cells are dead at doses below which large numbers of mutants can be identified.

Borg: I did not mean mutation per lesion; I meant as a fraction of the residual damage in the DNA.

Bridges: However, per lesion, it is not a weak mutagen.

Thilly: I'd like to echo Dr. Bridge's statement. A lot of confusion has arisen as a result of an article written by Cox in which he argued that ionizing radiation was not causing gene mutation in human skin or other mammalian cells; ouabain resistance was not induced. Further, in some of the 6-thioguanine mutants, chromosome breakage related to the X-chromosome was reported. However, we have been studying both oubain resistance and 6-thioguanine resistance in human cells for many years and cannot confirm these observations. An alternate and reasonable interpretation for existing data was that the particular set of mutations responsible for oubain resistance were not induced by the particular radiation that Cox used. Therefore a confusion has arisen within the literature about how good a gene mutagen is ionizing radiation. If you use the gene mutation-to-toxicity ratio (an interesting way of thinking about the severity of gene mutation) x-radiation is a pretty good mutagen in human cells.

Painter: I disagree with the views expressed by Dr. Thilly. X-ray-induced HPRT mutations are almost always accompanied by chromosomal changes in mammalian cells. At low doses this cannot be definitely determined because induced mutants cannot be distinguished from spontaneous ones. The early work of Chu showed that at high doses, essentially all mutants caused chromosomal changes. Nothing since then has changed that concept.

Peto: When you describe the effects of chemicals and radiation on DNA, you point out how many factors can multiply the extent to which damage is caused. You do not point out that there are many other things which multiply the extent to which cancer arises. The emphasis is placed just on the chemical carcinogen or radiation one is studying, but there is no mention of the modifying factors as being worthy of assessment in a risk assessment sense. They are regarded as interesting scientific ideas, yet they do not contribute to risk assessment. In terms of the prevention of human cancer, however, modifiers may be most critical. I think that there is an imbalance in our emphasis.

Borg: Let me say amen; I fully agree. This morning, however, I was focusing upon the difference between cancer initiation by radiation and by chemicals, and how this difference relates to model building and dose extrapolation models.

Abrahamson: I agree with Dr. Painter, but I am directing a question both to Dr. Bridges and to Dr. Thilly. First of all, what do you mean by mutation? With x-ray exposure, are you talking about base substitution events, deletions, and frame shifts? Is there good evidence that x-rays, at any dose, really produce base substitutions? In the case of ouabain resistance, does Dr. Thilly have data that show there is base substitution events induced by x-rays at low doses?

Bridges: My remarks were confined to bacteria where with UV we were producing base pair substitution mutations. In all our studies with human cells, we have used HGKRPT resistance, and we have never used ouabain resistance.

Thilly: No, we have not seen the induction by ionizing radiation of ouabain resistance. The point that I was trying to raise was that the absence of induced ouabain resistance represents a demonstration of the absence of a particular type of mutagenic exposure. Ouabain resistance would probably arise from some subset of nonsense mutations which would prevent the action of oubain on the ATPases, and thus, any particular mutagen might not produce the missense mutations which give rise to ouabain resistance. For instance, we have studied a series of loci: ouabain resistance, DRB resistance, and podomphylootoxin resistance in human lymphoblasts and found that for chemical mutagens there is a wide variation in their ability to cause particular subsets of putative nonsense mutations. We have not sequenced the genes of mutant ATPases nor the kinds of mutations which are induced in mammalian cells by the HGPRT gene. In the absence of this kind of data, I think we may have jumped the gun a little when interpreting the evidence that ouabain resistance is not influenced by x-rays, to mean that base-pair substitution mutations are not produced in mammalian cells by x-irradiation.

Borg: You raised a question which cannot be answered until the work is done. Some of the members of the workshop at the Washington conference felt very strongly that people working in these fields have not been very quick to take up the sequencing techniques that would give more definitive answers.

Lee: This question is in answer to the previous question by Drs. Thilly and Bridges in regard to point mutations induced by ionizing radiation. We have found in my laboratory that 7 out of 31 mutants induced by x-rays at the ADH locus in *Drosophila* sperm, have normal restriction maps at the ADH locus. Only 2 of these 7 mutants produce a protein of approximately normal molecular weight. I will discuss this data in a later paper.

Borg: You're talking about the *Drosophila* are you not?

Lee: Yes, this is induced in *Drosophila* sperm cells with x-rays at the alcohol dehydrogenase locus. I will go over that in the talk I am giving later.

Bridges: It is not widely realized that if you take the base pair substitution mutation rate per locus per rad with ionizing radiation in bacteria and calculate whether you would detect this rate in mammalian cells, you find you could not. You could not measure it, because the cells would all be dead. The rate would have to be at least 10 times greater in order to be detected. The reason we do not have any convincing evidence of base pair substitution mutations in mammalian cells with ionizing radiation is simply because even if they were there, the systems that we have to measure them are insufficiently sensitive. As long as we restrict ourselves to cells which have to form viable foci, we will probably never see such point mutations induced by ionizing radiation in mammalian cells.

Thilly: Just a point of clarification. When speaking of gene mutations, are we referring to all changes which can be thought to occupy a space within the size of a gene but not those changes involving clastogenesis, where there is a transfer of much larger pieces of information, either from or among chromosomes. My previous remarks on gene mutation included frame shifts, small deletions, small additions, and not only base pair substitutions, which presumably are the only events detected by assays involving ouabain resistance. On this basis I interpret Dr. Lee's remarks to mean that seven out of 31 mutants, did, in fact, have normal restriction maps after x-radiation. These data are direct evidence that a fair portion of the mutations induced were in fact the normal gene structure and, therefore, would be called gene mutations.

Borg: One point about this: I do not know whether a missense that is caused by radiation is more or less potent because of its particular structure. But there are the reasons I reviewed in my report for expecting that the initial yields of chemical lesions on DNA following radiation are quite different than those from most chemicals. I say "most chemicals", because there are some, and bleomyocin is one example, that get into DNA at preferred locations and stay there, being what I would call hydroxyl radical-manufacturing factories. They act very much like the strongly interacting ionization centers from high-LET radiation in that they keep popping off hydroxyl radicals (until they commit suicide). Therefore in a small locus there are the intensely interactive free-radical effects which are so characteristic of high-LET radiation. However, when you look at the initial reactions from low-LET radiation, there are the reasons I cited in my report to expect profiles of lesions on DNA different from those induced by most chemicals, so the ratios of persistent and unrepaired lesions ought also to be different. But this does not mean that once you have a mutational change, say from a submicroscopic deletion, its genetic consequences are necessarily different from similar changes effected by different mechanisms.

REFERENCES

Gould, M.N., 1984, Radiation initiation of carcinogenesis *in vivo*: a rare or common cellular event, *in*: Radiation Carcinogenesis, Epidemiology and Biological Significance, 347-358, J.D. Boyce, Jr., and Fraumeni, J.F. Jr., eds., Raven Press, New York.

Mulcahy, R.T., Gould, M.N., and Clifton, K.H., 1984, Radiogenic initiation of thyroid cancer: a common cellular event, *Intern. J. Radiat. Biol.*, 45:419.

REPORT OF NATIONAL CANCER INSTITUTE SYMPOSIUM: COMPARISON OF MECHANISMS OF CARCINOGENESIS BY RADIATION AND CHEMICAL AGENTS II. CELLULAR AND ANIMAL MODELS

R. J. M. Fry

Biology Division
Oak Ridge National Laboratory
Oak Ridge, Tennessee

INTRODUCTION

This segment of the report of the proceedings of the National Cancer Institute symposium is devoted to the presentations about studies with *in vitro* cell systems, *in vitro-in vivo* systems, and whole animals including humans. The NCI symposium was designed to cover many aspects of carcinogenesis so that the similarities and differences of the manner in which ionizing radiation and chemical carcinogens initiate cancer and complete its expression could be examined. The hope was that the identification of both the common and the clearly distinct features would help elucidate mechanisms and indicate areas for new research.

The epidemiological differences in cancer induced by chemicals and ionizing radiation were considered by Upton and Miller, and both held that ionizing radiation induced a broader spectrum of cancers than any chemical carcinogen. Miller pointed out that very few chemical agents caused cancer in childhood whereas radiation caused leukemia and thyroid tumors. There are tissues that are significantly more susceptible to radiation induction of cancer than others, and it was noted that the breast and thyroid appeared to be susceptible to radiation but cancers of these organs are not usually associated with the direct action of chemical carcinogens. The question is not settled of whether individuals with inherited diseases, such as ataxia telangectasia (AT) in which the patients' cells show hyperradiosensitivity, face an increased risk of radiation-induced cancer. We could make an important distinction if such susceptibility were restricted to a particular agent, but it is not clear if the type of inherited defect may be important in determining whether the increased susceptibility is general or specific. In the case of patients with xeroderma pigmentosum (XP) the question of excess risks for tumors other than skin cancer may be settled soon. If the incidence of nondermal tumors is increased, it will presumably indicate a lack of specificity. It is of interest that cells from AT and XP patients that are hypersensitive to ionizing radiation and ultraviolet radiation, respectively, are not hypersensitive to a number of chemical mutagens and carcinogens.

Miller pointed out that specific cancers were produced in persons who had been exposed *in utero* to chemicals that were teratogens. For example, treatment of the mother with diethylstilbestrol during pregnancy increased considerably the probability of the normally rare cancer of the vagina in the daughter. Similarly, diphenylhydantoin treatment of pregnant women may induce neuroblastomas in the offspring. Radiation, which is also a teratogen, has not been shown to produce any pathognomonic cancer, although the induction of an excess risk of leukemia and thyroid tumors has been reported.

There are a number of tumor types that appear to occur much more frequently with radiation than with chemical carcinogens. For example, carcinomas of the mastoid epithelial lining are pathognomic for radium. The tumors probably occur because of high doses from radon gas trapped in the air spaces for protracted periods. Osteogenic sarcoma is an interesting case of a cancer that can be induced by radiation, whereas, as yet, bone tumors have not been causally associated with chemical carcinogens. Once again it may be a matter of dose level to the target cells.

Tumors in some tissues are more readily induced experimentally with certain chemicals than with ionizing radiation. For example, papillomas of the mouse skin are induced in large numbers by exposure to chemical carcinogens, especially polycyclic hydrocarbons, either alone or followed by 12-*O*-tetradecanoyl-phorbol-13-acetate (TPA); in contrast, ionizing radiation is far less effective. Of equal interest is the marked difference in susceptibility of different tissues in highly inbred strains of mice to whole body or local irradiation.

Little pointed out that chemical carcinogens and radiation could cause different types of tumors in the lung because of differences in their localization after intratracheal instillations. Radiation from ^{210}Po induced mainly epidermal carcinomas and adenocarcinomas in the periphery of the lung and benzo(*a*)pyrene induced tracheal and bronchial epidermal carcinomas. This is a good example of the fact that the cells at risk in an organ may differ depending on the nature and distribution of the inducing agent.

The great problem of comparing the carcinogenic potency of different agents, or the tissue susceptibility to the individual carcinogens is that the dosimetry for the agents is usually neither adequate nor appropriate. Until we know what are the salient molecular events in the induction process, there can be no unequivocal quantitative comparisons. The use of external ionizing radiation in investigations of the comparative susceptibility of different cell types has the advantage that equivalent doses can be delivered to the target cells.

Multistage Carcinogenesis

A recurring theme of the NCI symposium was the multistage nature of carcinogenesis. Features of the multistage characteristics were described in *in vitro* systems by M. M. Elkind and A. R. Kennedy, in the *in vivo-in* vitro systems by R. L. Ullrich and M. Terzaghi, in whole animals by F. J. Burns, J. B. Little and H. C. Pitot.

Evidence from varied sources, in particular the findings from initiation promotion studies of skin carcinogenesis, has led to a general acceptance of the idea that cancer is a multistage process. Epidemiologists consider that a multistage model of carcinogenesis provides a useful framework for understanding the relationship between cancer incidence and time. They can distinguish differences resulting from exposures affecting early stages of carcinogenesis from those affecting the late stages from two types of evidence. First, whether risk of excess cancer decreases following cessation of exposure, and second whether the age at exposure influences the risk. As yet, it is not clear how radiation and chemicals compare in their action on the early and late stages. The stage dependency may in fact be more a feature of the specific tissue than the specific agent. In the case of breast cancer, radiation is considered to act at the early stages, since exposure at ages greater than about 40-45 years does not cause excess risk. The carcinogenic agents in tobacco on the other hand, may act on the lung at any stage.

This concept of early and late stages has not yet been translated into precise mechanistic terms or models that might be tested. The shapes of the dose-response curves are influenced by the nature of the stages, their number, and in the case of at least some carcinogens further exposures to the same or other agents.

The action of radiation or chemical carcinogens cannot be tested on a pristine population, be it cellular or whole animal, because there is always some probability that a cell will transform or an animal will develop a tumor without exposure to the carcinogenic agent. Since the probability of cancer increases with age in humans and animals, we may ask what effect does exposure have in a population since some must have tumor cells in various stages of development. Despite the example of the breast given above and the claim that the older patients with ankylosing spondylitis that were irradiated were at greater risk than the younger patients, we know very little about the effects of carcinogens at the different ages. The effect, on what the epidemiologists refer to as the late stage, is assumed to be on a cell population that have already undergone some pathological change.

I have started my report at the epidemiological level, as did the speakers concerned with humans, because eventually what we learn at the molecular, cellular and tissue level will be melded together with the epidemiological evidence to elucidate the nature of induction of cancer in humans. At the epidemiological level it is only possible to distinguish and divide the sequential process of carcinogenesis into two broad categories--early and late. Experimental work on whole animal, tissue, and cellular models must supply the finer details of the stages of carcinogenesis.

The schematic in Fig. 1 indicates a possible form of the sequential process of carcinogenesis and the points at which the comparative effects of ionizing radiation and chemical carcinogens might be examined. The simplicity of the outline conceals the fact that facets of the mechanisms involved in carcinogenesis are different in different tissues and organs. One problem is that we are attempting to deduce a complex process from very little pertinent information. A further problem is that many of the changes that can be detected may not be relevant in carcinogenesis. There is a reasonable understanding of what a cancer cell is and is not and a considerable inventory of its characteristics, but many of the important individual changes in the development of the cancer cell are difficult to detect. Altered gene expression plays an important role in the development of cancer, and although understanding of oncogene control is becoming clearer, we have a very fuzzy picture about gene control or how to distinguish the gene products that reflect a change that is central to induction of cancer.

The delineation of normality is itself not easy. In certain experimental animals the heritable aspects of susceptibility seem clear. Thus, the response to an agent will be determined by the genetic makeup of the specific cells. Such heritible factors might range from fragile sites of chromosomes that, in turn, correlate with breakpoints involved in chromosomal rearrangements and oncogene expression to the type of mutational change that reduces the number of further mutations required for transformation (for example, retinoblastoma). Although the information is scanty,

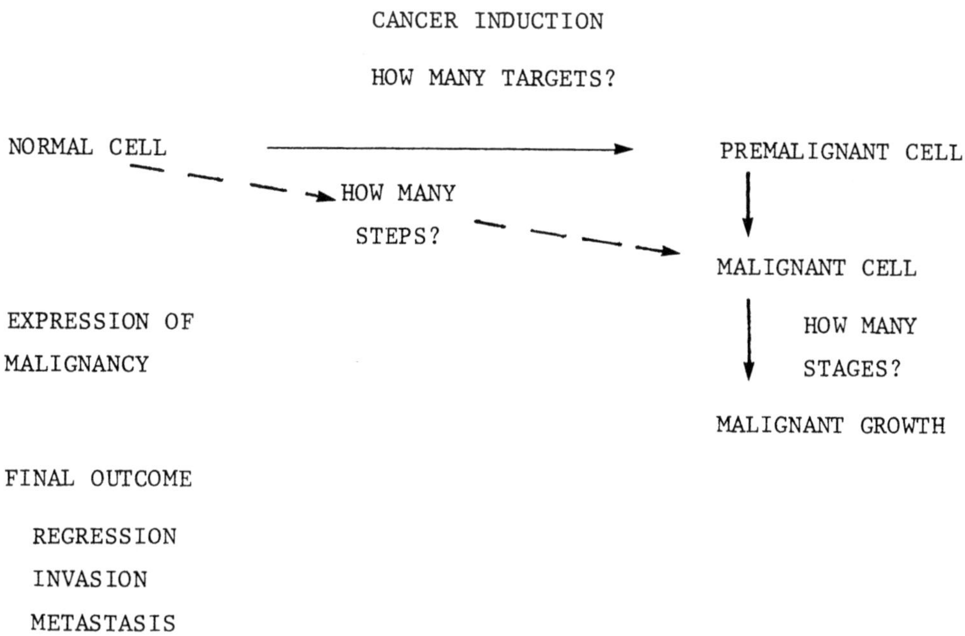

Fig. 1. Schematic diagram of the events involved in tumor induction and the development of malignancy.

in some strains of mice there appears to be a positive correlation between the natural incidence of specific tumors and the susceptibility to the induction by either radiation or chemical carcinogens.

Pitot defined the stages that could be identified in the development of most tumors as initiation, promotion and progression. Initiation appears to have no threshold and to obey single hit kinetics. The resulting change(s) are heritable and irreversible. Promotion tends to be defined as much by what it is not as by what it is--a clear indication of the lack of understanding about the process. Slaga considered the major effect of promoters in the skin to be the specific clonal expansion of the initiated cells that may involve both direct and indirect mechanisms. The direct action of promoters on the target cell alters the differentiation capability of the cell. Not surprisingly, experimental findings are not consistent with the idea of a single stage of promotion. Gene amplification and epidermal cell proliferation are thought to be important in the second stage of promotion. Pitot considered that progression resulted from genomic changes that could range from gene amplification to chromosomal translocations.

The carcinogenic process can be discontinuous. A good example described by Miller is radiation-induced breast cancer in the survivors of the atomic bomb. In humans under 10 years of age a single exposure to radiation induces the changes that eventually result in breast cancer, but after a dormancy of about 30 years. Apparently the young breast is not overtly altered by the presumed oncogene changes. Radiation-induced cancers do not appear until the age is reached at which the incidence of breast cancer starts to rise in the unexposed members of the population. It is possible that an age-dependent hormonal imbalance is involved. The period of dormancy (latent period) decreases with age at exposure. Ullrich and Terzaghi presented experimental evidence showing that expression of initiated mammary cells can be brought about by altering the cell-cell interrelationship, which suggests that expression of initiated cells is controlled at the tissue level as well as the systemic level.

A single exposure to radiation at a high dose rate can induce the changes in all the targets required to convert a normal cell to a malignantly transformed cell. Recent findings about oncogene activation and the requirement for alteration at two loci is consistent with two or more targets. Any further changes in the target cell may come as a consequence of these initial changes in the target cell. The development of variant cells would be an example.

The identification of the steps in the development of a transformed cell depends on the rate of transition. If the transition takes place very rapidly it may appear that the number of steps is less than if the process is spread out in time. With chemical carcinogens, especially at relatively low dose levels, the transitions between stages can be relatively slow and so-called premalignant states have been identified and studied by serial sampling, particularly in liver as noted by Pitot. In Fig. 2 the possible changes in the pathway of normal cells to premalignant cells to malignant cells are shown in more detail than in Fig. 1. It is neither clear whether the type of carcinogenic agent influences the nature of the stage-to-stage transition nor whether different agents act in a qualitatively different manner on the various stages. Agents classified as promoters do act very effectively on the stages between the initiated or altered cell and the appearance of frank malignancy.

Slaga described the evidence showing that promotion consists of at least two stages, and that there was some specificity related to the stage affected by various promoters. The demonstration that free radicals play a role in promotion by chemicals has raised the question of whether free radical production could

CARCINOGENESIS

NORMAL CELL

↓

ALTERED CELL

↓

CLONAL EXPANSION

↓

DEVELOPMENT OF VARIANTS

↓

MALIGNANT CELL

↓

FURTHER DEVELOPMENT OF VARIANTS

Fig. 2. Flow diagram of the changes at a cellular level in carcinogenesis.

be a mechanism in carcinogenesis that was common to ionizing and ultraviolet radiation and chemical carcinogens. However, certain chemical agents appear to be much more effective than ionizing radiations as promoters. The role of ionizing radiation on any stage other than initiation has not been studied systematically. Perhaps the reason for this emphasis is that radiation is most effective in single doses and, when given in multiple fractions or is protracted over a long time, its effectiveness is reduced. In the case of ultraviolet radiation (UVR), it is believed that many of the later exposures in a fractionation regime are "promoting" the lesions induced by earlier exposures. Also, UVR enhances or promotes the expression of cells initiated by other agents including radiation. As Little pointed out, promotion or enhancement of tumorigenesis can be due to nonspecific agents. He gave the example of the increase in lung tumors in hamsters exposed to ^{210}Po followed by saline instillations into the trachea.

Much of what is referred to as progression appears to be the development of cell variants, and proliferation must play a role in this process. Selection due to chromosomal changes and proliferative advantage can change the characteristics of the tumor cell population markedly. These alterations appear to increase the probability that the features we denote as malignancy will occur, namely, local invasion and distant metastases. The rapidity of the development of highly malignant variants is probably dose-dependent. Since one assumes the changes in the gene loci that result in initiation are all or none events, the difference in rate of development of variants, and therefore the malignancy, may reflect damage to other DNA sites and particularly chromosome aberrations with subsequent instability of the genome. Some investigators would agree that the degree of malignancy (in itself a rather vague parameter) is dose-dependent but this effect is difficult to quantify satisfactorily.

Weinstein summarized the current state of information about the effects of different carcinogenic agents on the various stages of carcinogenesis (Table 1). It can be seen that the believes the gaps in the information lie with ionizing and ultraviolet radiation.

The temporal patterns of cancer incidence have been used to support the thesis that carcinogenesis involves multiple stages. The number of stages has been determined from the exponent of the power function that relates time and cancer yield. Burns outlined the use of both this model and one based on the exponent of the dose-response function that provides an estimate of the number of dose-dependent stages to compare radiation and chemical carcinogenesis. In both models cells are assumed to progress from stage to stage as a result of spontaneous alterations or those changes caused by carcinogens. The altered cells are considered to be viable and capable of clonal growth. Experiments on the induction of skin tumors in the rat by low-LET radiation supported the model that only two events are involved in transition between stages and that one of the events was repairable with a halftime of about 3.5

Table 1. Multistage Carcinogenesis

	Carcinogenic Agent		
Stage	Chemicals	UVR	Ionizing Radiation
Initiation	+	+	+
Promotion	+	+	?
Progression	+	?	?
Complete carcinogenesis	+	+	+

hours. When the effects of multiple doses of radiation were compared with the effects of single doses, the time exponent increased from about 2 to 6 which was similar to the value obtained with multiple doses of chemical carcinogens in the same experimental system. Burns suggested that the increase in the time exponent reflects clonal growth of cells at an early stage of transformation and not an increase in the number of stages.

An important difference between multiple exposures to radiation and chemical carcinogens that is suggested by these studies is that repair occurs after radiation exposure but does not take place in the case of chemical carcinogens. There appears to be additivity of the carcinogenic effects of multiple exposures to certain chemical carcinogens but less than additivity for exposures to gamma or x-rays.

Papillomas induced on mouse skin by single applications of the chemical carcinogen benzo(*a*)pyrene (B[*a*]P) followed by repeated applications of TPA were used to study the early stage clones. It is thought that some of the papillomas are clonal expansions of cells in the early stages of the carcinogenic process, perhaps the first carcinogen-dependent stage. When TPA was added to the weekly B(*a*)P applications, the dose exponent dropped from 2 to 1. In terms of the multistage model this reduction can be explained if the inherent amplification in clonal growth has caused the spontaneous transitions to exceed the number of carcinogen-induced transitions. There has not been a sufficient number of appropriate experiments on different tissues to allow analyses of how the stages vary and also whether the nature of the process is dependent on the type of carcinogen.

The investigation of dose-response relationships has been a cornerstone of studies of radiation-induced carcinogenesis. Also, variations of the conditions of exposure, such as fractionation and dose rate, have been more extensively studied using radiation than with chemical carcinogens both in experimental animals and with *in vitro* cell systems. Elkind showed that both a reduction in the dose rate and dividing a dose of gamma radiation (low LET) into five daily fractions resulted in a significantly lower transformation frequency. These results are consistent with repair of sub-transformation lesions. In contrast reducing the dose rate of fission neutron radiation (high LET) increased the frequency of transformation. As Elkind indicated, it is of interest that the initial slopes of the dose-response curves for both low- and high-LET radiation are linear but the effect of reducing the dose rate of the two types of radiation is different. Linear or single-track responses are commonly considered to be dose-rate independent but Elkind and his colleagues' findings suggest that this may not be so. In the case of low-LET radiation error-free repair reduces the effect but with high-LET radiation the lower dose rate increases it. Elkind speculated on whether the neutrons induced an error-prone repair or whether the protracted exposure acted as a promoter and increased the expression of the initiated cells.

Carcinogenic agents are cytotoxic and cell killing will reduce the number of cells that can express a transformation. In the case of radiation, Elkind pointed out that tumorigenesis is the net effect of a low probability induction process and a high probability of cell killing. Cell killing may play a role in carcinogenesis in a number of different ways. First, a reduction in carcinogenic effect may be caused by the loss of potential cancer cells. Second, an increase in effect may result from (a) disruption of a tissue with loss of cell-cell communication; (b) loss of cells with regenerative cell proliferation which may fix an induced lesion, or add an error, or assist the expansion of a transformed cell clone; and (c) uptake of DNA from killed cells by untransformed and viable cells that theoretically could lead to incorporation and activation of proto-oncogenes. The probability of the latter occurring especially with low doses must be very small. In a nonrenewal or very slowly renewing system such as the liver cell killing and repair of damage appears to be important. The relative effectiveness of cell killing and malignant transformation in vitro and in vivo is different for different types of agents, such as chemicals, ultraviolet radiation, and ionizing radiation, and examples were given by Kennedy, Elkind and Ullrich.

The development of in vitro cell systems suitable for quantitative studies of malignant transformation has made it possible to dissect the carcinogenic process at a cellular level; Kennedy and Elkind discussed the results of such studies. In vitro cell systems consist of cell lines, such as C3H 10T1/2 and 3T3 cells, as well as primary cultures or cell strains such as Syrian hamster embryo (SHE) cells. If the establishment of a cell line involves one of the major changes in the development of a malignant cell, comparative studies on cell lines and cell strains should be extremely informative since in the cell lines, the change to immortality has occurred. Furthermore, some cell lines such as C3H 10T1/2 are aneuploid, a change that appears to predispose them to further changes. In the diploid SHE cell system the susceptibility to transformation by X-rays decreases dramatically with the first few passages. The reasons for this intriguing change in susceptibility is not known.

If the immortal state of cells indicates that one of the targets for transformation has been altered, comparison of dose responses between cell lines and cell strains should be useful. As Kennedy pointed out, the dose responses to radiation appear to be qualitatively similar in both systems. Experiments at the molecular level should be designed with the target theory in mind in order to confirm or establish the number of targets or steps involved in transformation; in discussion, Borek referred to such experiments.

The ability to manipulate cells in culture and to expose them to agents at various stages of a sequential process, such as in vitro transformation, has proven very useful in the dissection of the transformation process and in showing how it may be modified. The approaches and the interpretations of the results of experiments that illustrated the range of agents that had been used to initiate and to modify the expression of the early events were given by Kennedy who presented an impressive catalogue.

Kennedy used Fig. 3 to illustrate a working model of the process of in vitro transformation. In this model at least two stages are required for transformation and both enhancement or inhibition of transformation can be carried out with agents applied between the two stages. Kennedy indicated the similarities between induction of malignant transformation by radiation and chemical carcinogens and the apparent similarities between the in vitro and in vivo systems. Her presentation made it clear that even in cell strains malignant transformation is a complex multistage process but that this very fact presented opportunities for intervention, and therefore, prevention of the completion of the process. Kennedy described how protease inhibitors had been used to carry out such prevention in cells exposed to either radiation or chemical carcinogens.

Kennedy also discussed the evidence showing that the initial event in the transformation of C3H 10T1/2 cells is a rather common occurrence. There is a great deal of evidence, some of it quite old, that in humans and animals the presence of initiated cells can be demonstrated or inferred, and that they occur much more frequently than do cancers in the same tissues.

The fact that a large number of in vitro experiments had all been carried out by a relatively small number of research groups illustrated a practical, if not scientific, difference between in vitro and in vivo methods, namely, the number of in vitro experiments that can be carried out in months rather than years exceeds greatly the number of animal experiments. However, cancer is not just a cellular disease and the role of the tissue organization and the influential systemic factors can only be investigated using both in vivo-in vitro systems and whole animals. Ullrich presented some information that has been obtained using two of a number in vitro-in vivo systems. Experiments using tracheal or mammary cells and an epithelial focus assay, plus assays of the emergence of the malignant phenotype have identified stages in the development of malignancy and the effects of different agents on them. An advantage of these epithelial systems is that treatments can be carried out in vivo and the effects assayed after manipulation in vitro; if required, the cells can be returned to appropriate sites in animals in order to study host factors.

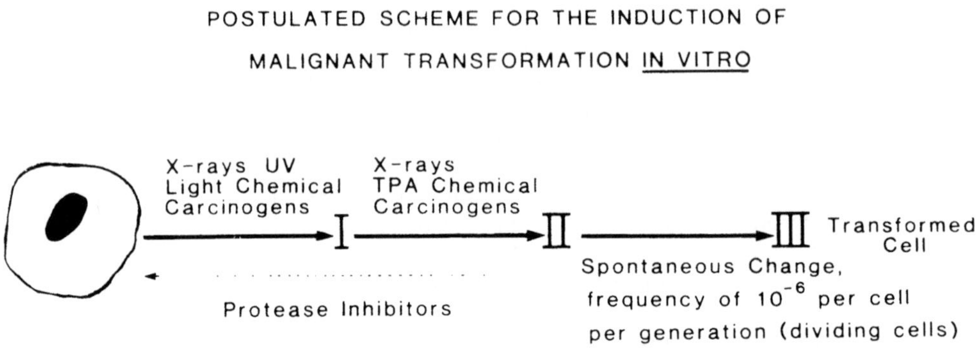

Fig. 3. Schematic diagram of the stages in the induction of *in vitro* malignant transformation. Supplied kindly by A. R. Kennedy.

Assays have been used to identify and quantitate three phenotypic changes in growth of clonogenic cells after exposure to chemical carcinogens and ionizing radiation: 1) the clonogenic unit gives rise to an epithelial focus, 2) an epithelial focus that escapes senescence and can be subcultured, and 3) gives rise to foci that can be subcultured and are tumorigenic when injected into the mammary pad in the case of mammary cells or into the trachea stripped of its epithelium in the case of tracheal cells. Thus, the stages of development of malignant cells from the time of treatment can be studied sequentially in epithelial cells.

Ullrich showed that mammary tumors in mice can be induced by both radiation and 7,12-dimethylbenz(a)anthracene (DMBA). Comparable incidences of cancer were produced by doses of DMBA that killed a few cells but in the case of radiation required doses that resulted in marked cell killing. In the tracheal cell system DMBA appears to be much more effective than radiation in the induction of the initial events. Exposure to x-rays after DMBA treatment reduced the latent period or period required for expression.

CONCLUSION

We have not identified the point at which the common final pathway for induction of cancer by chemical carcinogens and ionizing radiation occurs. Although common molecular targets are suggested by recent findings about the role of oncogenes, the mechanisms may be quite different by which the deposition of radiation energy and the formation of adducts or other DNA lesions induced by chemicals cause the changes in the relevant targets. Damage to DNA that plays no part in the transformation events influence the stability of the genome, and therefore, the probability of subsequent changes that influence tumorigenesis may be more readily induced by some agents than others. Similarly, the extent of cytotoxic effects that disrupt tissue integrity and increase the probability of expression of initiated cells may be dependent on the type of carcinogen. Also, evidence was presented that repair of the initial lesions could be demonstrated after exposure to low-LET radiation but not after exposure to chemical carcinogens. In short, there are a number of ways in which radiation and chemical carcinogens may differ qualitatively that influence their carcinogenic effectiveness.

There are specific questions about the mechanisms of carcinogenesis that can be answered either more easily or more quantitatively with specific carcinogens. Some of these opportunities emerged from the interchange between workers devoted to one particular class of carcinogen and hopefully the cross fertilization will provide the catalyst for new experimental approaches.

ACKNOWLEDGMENT

Research sponsored by the Office of Health and Environmental Research, U. S. Department of Energy under contract DE-AC05-84OR21400 with the Martin Marietta Energy Systems, Inc.

REFERENCES GIVEN BY SPEAKERS:

<u>Comparative Mechanisms of Carcinogenesis by Radiation and Chemicals: Implications of the Human Experience</u> - A. C. Upton

International Agency for Research on Cancer: Evaluation of the Carcinogenic Risk of Chemicals to Humans, 1982, IARC Monographs Supplement 4, IARC, Lyon.

National Research Council, Committee on the Biological Effects of Ionizing Radiations, 1980, "The Effects on populations of Exposure to Low Levels of Ionizing Radiation," National Academy of Sciences, Washington, DC.

United Nations Scientific Committee on the Effects of Atomic Radiation: Sources and Effects of Ionizing Radiation, 1977,

Report to the General Assembly, with Annexes, 32 Session, Suppl. 40 (A/32/40), United Nations, New York.
Upton, A. C., 1981, Principles of Cancer Biology: Etiology and Prevention, in: "Principles and Practices of Oncology," pp. 33-58, V. T. DeVita, S. Hellman, and S. A. Rosenberg, eds., J. B. Lippincott Company, Philadelphia.
Weinstein, I. B., 1980, Molecular and cellular mechanisms of chemical carcinogenesis, in: "Cancer and Chemotherapy," Vol. I, pp. 169-196, S. T. Crooke, and A. W. Prestako, eds., Academic Press, New York.

Chemicals vs. Ionizing Radiation in Carcinogenesis: Human Experience - R. W. Miller

Miller, R. W., 1978, Environmental causes of cancer in childhood, Adv. Pediatr., 25:97.
Miller, R. W., 1979, Transplacental chemical carcinogenesis in man, Natl. Cancer Inst. Monogr., 52:13.
Miller, R. W., 1982, Radiation effects: Highlights of a meeting, J. Pediatr., 101:887.
Miller, R. W., and Beebe, G. W., Radiation leukemia and lymphoma in man, in: "Radiation Carcinogenesis," A. C. Upton, ed., Elsevier North-Holland, New York, in press.
Miller, R. W., and Boice, J. D., Jr., Radiogenic cancer after prenatal or childhood exposure, in: "Radiation Carcinogenesis," A. C. Upton, ed., Elsevier North-Holland, New York, in press.
Tomatis, L., Agthe, C., Bartsch, H., Huff, J., Montesano, R., Saracci, R., Walker, E., and Wilbourn, J., 1978, Evaluation of the carcinogenicity of chemicals, Cancer Res., 38:877.

Mechanisms of Carcinogenesis in vivo - F. J. Burns

Burns, F. J., Albert, R. E., Altshuler, B., and Morris, E., 1983, Approach to risk assessment for genotoxic carcinogens based on data from the mouse skin initiation-promotion model, Environ. Health Perspect., 500:309.
Burns, F. J., Strickland, P., Vanderlaan, M., and Albert, R. E., 1978, Rat skin tumors following single and fractionated exposures to proton radiation, Radiat. Res., 74:152.
Burns, F. J., and Vanderlaan, M., 1975, Split-dose recovery for radiation-induced tumors in rat skin, Int. J. Radiat. Biol., 32:135.
Druckery, H., 1967, Quantitative aspects of chemical carcinogenesis, in: "Potential Carcinogenic Hazards from Drugs Evaluation of Risks," VICC Monograph Series, Vol. 7, pp. 60-78, R. Truhart, ed., Springer-Verlag, New York.
Whittemore, A. S., 1978, Quantitative theories of oncogenesis, in: "Advances in Cancer Research," Vol. 27, pp. 55-88, G. Klein and S. Weinhouse, eds., Academic Press, New York.

In Vitro Studies with Radiation and Chemicals - M. M. Elkind

Gray, L. H., 1965, Radiation biology and cancer, in: "Cellular Radiation Biology," pp. 7-25, The Williams & Wilkins Company, Baltimore.

Han, A., and Elkind, M. M., 1982, Enhanced transformation of mouse 10T1/2 cells by 12-O-tetradecanoylphorbol-13-acetate following exposure to X-rays or to fission spectrum neutrons, Cancer Res., 42:477.

Han, A., Hill, C. K., and Elkind, M. M., 1980, Repair of cell killing and neoplastic transformation at reduced dose rates of ^{60}Co gamma-rays, Cancer Res., 40:3328.

Maher, V. M., and McCormick, J. J., 1976, Effect of DNA repair on the cytotoxicity and mutagenicity of UV irradiation and chemical carcinogens in normal and xeroderma pigmentosum cells, in: "Biology of Radiation Carcinogenesis," pp. 129-145, J. M. Yuhas, R. W. Tennant, J. D. Regan, eds., Raven Press, New York.

Susuki, F., Han, A., Lankas, G. R., Utsumi, H., and Elkind, M. M., 1981, Spectral dependencies of killing, mutation, and transformation in mammalian cells and their relevance to hazards caused by solar ultraviolet radiation, Cancer Res., 41:4916.

Stages in Radiation and Chemical Carcinogenesis - H. C. Pitot

Bohrman, J. S., 1983, Identification and assessment of tumor-promoting and cocarcinogenic agents: State-of-the-art in vitro methods, CRC Crit. Rev. Toxicol., 11:121.

Boutwell, R. K., 1974, Function and mechanism of promoters of carcinogensis, CRC Crit. Rev. Toxicol., 2:419.

Emerst, I., and Cerutti, P. A., 1982, Tumor promoter phorbol 12-myristate 13-acetate induces a clastogenic factor in human lymphocytes, Proc. Natl. Acad. Sci. USA, 79:7509.

Pitot, H. C., 1984, Neoplastic development and human cancer, Cancer Surveys, 2(4):519.

Experimental Lung Cancer Induced in Hamsters by Ionizing Radiation and Chemical Carcinogens - J. B. Little

Kennedy, A. R., and Little, J. B., 1974, Transport and localization of benzo(a)pyrene-hematite and ^{210}Po-hematite in the hamster lung following intratracheal instillation, Cancer Res., 34:1344.

Kennedy, A. R., and Little, J. B., 1975, Localization of polycyclic hydrocarbon carcinogens in the lung following intratracheal instillation in gelatin solution, Cancer Res., 35:1563.

Kennedy, A. R., Worcester, J., and Little, J. B., 1977, Deposition and localization of polonium-210 intratrachealy instilled in the hamster lung as determined by autoradiography of freeze-dried sections, Radiat. Res., 69:553.

Little, J. B., and Kennedy, A. R., 1979, Evaluation of alpha radiation-induced respiratory carcinogenesis in Syrian hamsters: total dose and dose rate, Prog. Exp. Tumor Res., 24:356.
Little, J. B., Kennedy, A. R., and McGandy, R. B., 1975, Lung cancer induced in hamsters by low doses of alpha radiation from polonium-210, Science, 188:737.
Little, J. B., Kennedy, A. R., and McGandy, R. B., 1978, Effect of dose distribution on the induction of experimental lung cancer by alpha radiation, Health Phys., 35:595.
Little, J. B., McGandy, R. B., and Kennedy, A. R., 1978, Interactions between polonium-210 alpha radiation, benzo(a)pyrene and 0.9% NaCl solution instillations in the induction of experimental lung cancer, Cancer Res., 38:1929.
Little, J. B., and O'Toole, W. F., 1974, Respiratory tract tumors in hamsters induced by benzo(a)pyrene and polonium-210 alpha radiation, Cancer Res., 34:3026.
Shami, S. G., Thibideau, L. A., Kennedy, A. R., and Little, J. B., 1982, Proliferative and morphological changes in the pulmonary epithelium of the Syrian golden hamster during carcinogenesis initiated by ^{210}Po alpha-radiation, Cancer Res., 42:1405.

Induction of In Vitro Transformation by Chemical and Radiation - A. R. Kennedy

Barrett, J. C., Hesterberg, T. W., and Thomassen, D., Use of cell transformation systems for carcinogenicity testing and mechanistic studies of carcinogenesis, Pharmacol. Rev., in press.
Bertram, J. S., Mordan, L. J., Domanska-Janik, K., and Bernacki, R. J., 1982, Inhibition of in vitro neoplastic transformation by retinoids, in: "Molecular Interrelationships of Nutrition and Cancer," pp. 315-335, M. S. Arnott, J. Van Eys, and Y.-M. Wang, eds., Raven Press, New York.
Borek, C., 1982, Radiation oncogenesis in cell culture, Adv. Cancer Res., 37:159.
DiPaolo, J. A., 1983, Relative difficulties in transforming human and animal cells in vitro, J. Natl. Cancer Inst., 70:3.
Elkind, M. M., Han, A., Hill, C. K., and Buonaguro, F., 1983, Repair mechanisms in radiation-induced cell transformation, in: "Proceedings of 7th International Congress of Radiation Research," pp. 33-42, J. J. Broerse, G. W. Barendson, H. B. Kal, A. J. Van der Kogel, eds., Martinus Nijhoff Publishers, Amsterdam.
Hall, E. J., and Miller, R. C., 1981, The how and why of in vitro oncogenic transformation, Radiat. Res., 87:208.
Heidelberg, C., 1980, Mammalian cell transformation and mammalian cell mutagenesis in vitro, J. Exp. Pathol. Toxicol., 3:69.
Huberman, E., 1978, Mutagenesis and cell transformation of mammalian

cells in culture by chemical carcinogens, J. Environ. Pathol. Toxicol., 2(1):29.
Kakunaga, T., 1981, Cell transformation as a system for studying mechanisms of carcinogenesis, in: "Gann Monograph on Cancer Research," Vol. 27, pp. 231-242, "Mutation, Promotion and Transformation In Vitro," Japan Scientific Society, Tokyo.
Kennedy, A. R., 1982, Antipain, but not cycloheximide, suppresses radiation transformation when present for only one day at five days postirradiation, Carcinogenesis, 3:1093.
Kennedy, A. R., 1984, Promotion and other interactions between agents in the induction of transformation in vitro in fibroblast, in: "Mechanisms of Tumor Promotion," Vol. III, "Tumor Promotion and Cocarcinogenesis In Vitro," pp. 13-55, T. J. Slaga, ed., CRC Press, Boca Raton.
Kennedy, A. R., Prevention of radiation-induced transformation in vitro, in: "Vitamins, Nutrition and Cancer," K. N. Prasad, and J. V. Sutherland, eds., S. Karger AG, Basel, in press.
Kennedy, A. R., Cairns, J., and Little, J. B., 1984, The timing of the steps in transformation of C3H/10T1/2 cells by X-irradiation, Nature(London), 307:85.
Kennedy, A. R., Fox, M., Murphy, G., and Little, J. B., 1980, Relationship between X-ray exposure and malignant transformation in C3H 10T1/2 cells, Proc. Natl. Acad. Sci. USA, 77:7262.
Kennedy, A. R., and Little, J. B., 1980, An investigation of the mechanism for the enhancement of radiation transformation in vitro by TPA, Carcinogenesis, 1:1039.
Kennedy, A. R., Murphy, G., and Little, J. B., 1980, The effect of time and duration of exposure to 12-O-tetradecanoyl-phorbol-13-acetate (TPA) on X-ray transformation of C3H/10T1/2 cells, Cancer Res., 40:1915.
Little, J. B., 1981, Radiation transformation in vitro: Implications for mechanisms of carcinogenesis, in: "Advances in Modern Environmental Toxicology," Vol. I, "Mammalian Cell Transformation by Chemical Carcinogens," pp. 383-426, N. Mishra, V. Dunkel, and M. Mehlman, eds., Senate Press, Inc., New Jersey.
Sivak, A., Charest, M. C., Rudenko, L., Silveira, D. M., Simons, I., and Wood, A. M., 1981, BALB/c-3T3 cells as target cells for chemically induced neoplastic transformation, in: "Advances in Modern Environmental Toxicology," Vol. I, "Mammalian Cell Transformation by Chemical Carcinogens," pp. 133-180, N. Mishra, V. Dunkel, and M. Mehlman, eds., Senate Press, Inc., New Jersey.
Ts'o, P. O. P., 1980, Neoplastic transformation, somatic mutation and differentiation, in: "Carcinogenesis: Fundamental Mechanisms and Environmental Effects," pp. 297-310, B. Pullman, P. O. P. Ts'o, and H. Gelboin, eds., D. Reidel Publishing Co., Hingham, MA.
Yang, T. C. H., and Tobias, C. A., 1982, Radiation and cell

transformation *in vitro*, *Adv. Biol. Med. Phys.*, 17:417.
Yavelow, J., Finlay, T. H., Kennedy, A. R., and Troll, W., 1983, Bowman-Birk soybean protease inhibitor as an anticarcinogen, *Cancer Res.*, 43:2454.

Neoplastic Development After Exposure to Radiation and Chemical Carcinogens - R. L. Ullrich

Ethier, S. P., and Ullrich, R. L., 1981, Detection of ductal dysplasia in mammary outgrowths derived from carcinogen-treated virgin female BALB/c mice, *Cancer Res.*, 41:1808.
Terzaghi, M., and Nettesheim, P., 1979, Dynamics of neoplastic development in carcinogen-exposed tracheal mucosa, *Cancer Res.*, 39:4003.
Terzaghi, M., Klein-Szanto, A., and Nettesheim, P., 1983, Effect of the promoter TPA on the evolution of carcinogen-altered cell populations in tracheas initiated with DMBA, *Cancer Res.*, 43:1461.
Ullrich, R. L., 1980, Interaction of radiation and chemical carcinogens, *in*: "Carcinogenesis - A Comprehensive Survey," Vol. 5, "Modifiers of Chemical Carcinogenesis: An Approach to the Biochemical Mechanisms and Prevention," pp. 169-184, T. J. Slaga, ed., Raven Press, New York.

DISCUSSION

Peto: I am very happy to see the science pursued along the lines you have discussed, but I am not terribly happy to see it pursued in the context of the formality of risk assessment. It would be beneficial if one could only divorce risk assessment from the scientific issues and have a separate conference about each: risk assessment is nowhere on the horizon of these scientific discussions. We can take your final discussion as an example, about what actually makes mice pregnant; this may tell you very little about how to prevent pregnancy in your daughter. The causes are different and, while these findings may provide useful insights, you cannot obtain direct quantitative estimates of the prevalence of pregnancy in different categories of human females from a knowledge of what determines pregnancy in mice. Cannot one just separate them, talk seriously about science and not keep relating it back to risk estimates every time?

Clifton: I suggest that the way one prevents pregnancy in one's daughter is based on an enormous amount of work on hormonal control of reproduction in mice and rats.

Fry: If you are talking about quantitative risk estimates, at the moment, there is not much argument that extrapolation of risk estimates across species is not yet accepted in general. Animal experiments are certainly important for establishing certain general principles. If we take radiation, there is practically no information about the effects of low dose-rate exposures, or fractionations in humans, but it can be obtained from experiments with animals. Further, there is good reason to believe that the results can be applied qualitatively to humans. There are all sorts of examples that everyone in this room can think of where you can at least find whether there are certain consistent general principles that apply to humans as well as experimental animals. I would point out that certainly in radiation risk, estimation is largely a matter of judgment, and judgment is experienced judgment, and that experience is going to be influenced by how much you understand, even if you don't understand it completely. So I think you can learn a lot that stops you making mistakes, perhaps more than making you correct.

Borg: This is more in the nature of a comment than a question. In response to Dr. Peto, the last slide that Dr. Fry chose makes the point that under some experimental conditions, an agent whose dose-response curve under other experimental conditions runs through the origin may manifest a response curve that does not. Now, to the extent that we want to look at the big picture, it does seem that society is spending a lot of time and effort looking at the effects of few molecules of known animal carcinogens in the environment, while it ignores most of the other environmental factors that really make a big difference in the incidence of cancer. A lot of social resources and dollars are going into the business of trying to determine what small amounts of a putative carcinogen or mutagen in the environment are going to do. The modelers have a lot to do before they can tell us how much we should worry. I say again the reason that the people in modeling and in research on mechanisms of damage have to get together is to determine how much that is known about mechanisms of higher levels of exposure is relevant to the effects occurring at low doses. If the response at low levels differs significantly from that at high doses, that is important, and we have to get a handle on the differences to formulate realistic dose-response curves. I would like people who are familiar with some of these models to cross the bridge and help those of us who study damage mechanisms so that we may give good advice to the modelers.

Thilly: After some thought, I am moved to wonder if this Y4 statement which Dr. Peto presents as a result of curve-fitting is not being generally mistaken as a mechanistic model implying the necessity of "n" independent events. Often times I find the models are all put in positive terms such that independent functions of dose are all forcing functions towards cancerous events. However, there is the opportunity for both positive and negative effects of forces. For example, if exposure to a chemical were linearly increasing the probability of a single necessary mutation with time, and aging was linearly decreasing the probability that a precancerous mutant cell would be suppressed, the result would be a t-squared function. I think these models are too restrictive when used generally in scientific conversation, and they lead students to look for the easiest conception, rather than to point out the need for consideration and separation of variables.

Carsten: This is a general comment relating to your slide and discussion of high- and low-LET radiation effects, particularly at low doses. I believe it should be emphasized, as it was in NCRP Report No. 63, that at very low doses from external exposure or internal emitters, average tissue dose has little meaning in terms of individual cell nuclei dose. As the concentration of a beta emitter such as tritium decreases, the average tissue dose decreases, while the average dose to the hit cell nucleus does not decrease below a constant. This constant will be between 0.27 and 0.34 rads, depending on whether the tritium disintegration is intranuclear (giving the 0.27 rad dose) or extranuclear. Indeed, a maximum permissible dose of 5 rem per year to the reference nucleus (as defined in NCRP No. 63) is delivered by 18 tritium decays a year, or only one decay approximately every twenty years. When considering cells at risk and the possible effects at very low doses, it must be noted that while average tissue dose may be reduced, the dose to the nucleus of affected cells may not be, and the shape of the dose-response curve at these low doses may or may not reflect this heterogeneity.

WHAT ARE WE DOING WHEN WE THINK WE ARE DOING RISK ANALYSIS?

John C. Bailar III[1] and Stephen R. Thomas[2]

1. Office of Disease Prevention and Health Promotion
Department of Health and Human Services
468 N Street, S.W.
Washington, DC 20024

and

Department of Biostatistics
Harvard School of Public Health
677 Huntington Avenue
Boston, MA 02115

2. Department of Health Policy and Management
Harvard School of Public Health
677 Huntington Avenue
Boston, MA 02115

INTRODUCTION

Risk analyses matter. Sometimes they matter a lot, so it is important for persons who contribute their expertise to the debate to know how both their efforts and their individual specialties fit into the whole.

We set the stage for a discussion of risk analysis by stating four related points that shape our later discussion: that different definitions of risk analysis have different important consequences; that certain primary questions have a profound effect on risk analysis; that a single risk analysis may have several competing or complementary conceptual loci; and that risk analysis tends to center on the management of data.

Definition of Risk Analysis

While definitions of fields concerned with risks are not standardized, risk assessment in a narrow sense may be taken to be the domain of experts of various disciplines working in close collaboration. In statistical terms, risk assessment is the development of conditional (subjective, or Bayesian) probabilities: if exposure A occurs, then outcomes B will occur with some estimated probability distribution. The familiar dose-response curve is simply a convenient way to summarize a lot of these conditional probabilities. Risk analysis understood more narrowly than we do here follows the model of mathematical decision theory and its distinctive (Bayesian and game theoretic) norms, by no means universally accepted. In turn, risk-benefit analysis is dominated by economists seeking to organize information about risks so that decision makers can act in ways consistent with economic efficiency. To these terms we would add risk appraisal, which includes evaluation of the possible effects of risk but is not directly linked to the making of decisions.

Our own broader definition of risk analysis is the study and interpretation of activities, processes, and situations that pose a hazard to human health and welfare. While some definitions of risk analysis would omit the interpretation phase, we believe that it should be included for two reciprocal reasons: interpretation is very heavily influenced by the structure, process and outcome of the risk analysis, and, conversely, the risk analysis depends (or should depend) on the way it is to be integrated into an interpretation and decision. Risk analysis includes the study of such questions as whether any hazard exists, assessments of the likelihood that various events will occur under one or another set of exposure conditions, and the translation of these assessments into terms relevant to private and public decision-making. As a practical activity, risk analysis thus includes elements of political theory, law, economics, and human behavior as well as the more familiar sciences of statistical analysis and biology. It is not in itself, however, equivalent to the making of decisions; as we shall show, it is one input to a larger process. Risk analysis has a near-term purpose--to help make judgments and come to decisions--so its component parts are "applied" disciplines.

Our definition calls attention to the norms characteristic of conceptions of risk that are in dispute among various students. These norms include such things as the ways that uncertainty is handled in quantitative assessments of risk, how various kinds of evidence are to be interpreted (are benign neoplasms evidence of carcinogenic potential?), the relative weights given to various adverse outcomes, how far to go in the study of more and more remote effects, and the role to be accorded economic and other effects not related to health. Our point is that the ways that institutions produce decisions bring these and other norms to the fore. To say that a decision made by a legislature or an official is wrong is to argue for norms different from those exhibited in the decision, and to argue for the authority of neglected expertise. It is to join the debate by trying to alter the terms of the debate (Graham, 1982). Such alterations can, of course, be good or bad in the eyes of various actors and bystanders, but we will not consider that aspect of the matter.

Primary and Secondary Questions

It is characteristic of risk analysis, as we broadly define it, that its practitioners disagree among themselves about its nature. A definition may imply that a certain set of activities is to be expanded or contracted to meet definitional boundaries, that certain kinds of data are to be used or ignored, or that certain methods are to be used. Definitions are thus likely to be one means used by astute managers, scientists, and other disputants to advance their own substantive views: that is, definitions are likely to be used as a means to settle questions that other practitioners may want to keep open, or to close in different ways. (We recognize this problem with definitions and we have tried to avoid it, but we do not guarantee our own success.) Examples of questions that may be prematurely or improperly closed by definition are: Whose business is it to see that one or another specific issue is settled? Who is qualified, and by what criteria, to be heard as an expert in discussions about the size or the meaning of certain risks? Who is to arbitrate, and how, the disagreements over policy that will flow from different risk assessments? And which debates should include the participation of individuals unversed in matters of risk assessment (affected persons, or "ordinary" citizens) and in what manner and degree? The lack of a single widely accepted definition of the subject matter and methods of risk analysis thus has practical consequences. For this reason it may reflect conscious or subliminal political strategies as well as simple confusion and private jockeying for position.

One or another set of answers to these broad definitional questions defines a logically secondary set of questions. These questions are generally familiar to practitioners and users of risk analysis and figure in the ethical, methodological, and organizational debates that surround the field(s). Many of these secondary questions hardly make sense except in a context of general agreement (perhaps tacit), about the primary issues. Examples follow.

Decision theorists will want to examine carefully the probability estimates of various adverse events and the decision rules that reflect one's attitudes about being wrong in one or another direction and degree. Statisticians and other scientists will favor one or another mathematical and/or biological model for moving with some specifiable degree of confidence from what is known to what is not known. Scientists will dispute among themselves the boundaries, the structure, and the content of a specific risk analysis, depending upon whether their primary orientation is towards toxicology, biochemistry, chemical engineering, clinical medicine, or any of the other relevant technical disciplines. And, since risks, costs, and benefits have to be not only estimated, but also distributed, distributional concerns will be important. Physical and biomedical scientists will thus be joined by economists and others in advocating one or another way of organizing the several activities related to data-gathering, estimation and inference, so as to promote the formulation of good analysis suitable for public decision-makers.

Risk analysis is likely to be a prominent part of any organized method for dealing with risk. One part of such an organizing structure is risk-benefit analysis; some of the other parts are cost-effectiveness analysis, establishment of rules for zero-risk decisions (exemplified by the Delaney amendment), and laisser-faire management of risk according to the legal doctrines governing private litigation.

Conceptual Loci

The distinction between primary and secondary questions calls attention to the importance of the political issues that seep into the work of scientists and technical specialists, and which may have unexpected effects on the course of "scientific" analysis of risks. Political considerations seep in because the practical consequences of research often favor some people at the expense of others, and the nature and size of the benefits and losses, as well as the identity of the winners and losers, can be profoundly affected by the terms in which debates about risk are shaped (Fischhoff et al., 1981).

After a risk analysis has been completed under some set of answers to the primary questions, the losers, so to speak, can carry their case not only to a different institutional forum--for example, from a losing legislative effort to ban a use of some chemical, to attempts to mobilize public opinion, or to litigation under laws already in place (e.g., to sue for negligence, or to challenge an environmental impact statement)--but also to a different conceptual locus of discussion.

By conceptual locus we mean something more subtle than misrepresentation, invective, or grandstanding. Rather, we mean that various facts, opinions, decisions, and conclusions can be linked in different ways to different underlying concepts, and those critical underlying concepts can vary from forum to forum, person to person, or time to time. Shifts in conceptual locus, sometimes unperceived by participants, are facilitated because risk analysis tends to combine several kinds of discourse that may occur simultaneously, even in the same documents, paragraphs, and words.

Multilevel discourse is characteristic of diplomacy, rhetoric, and poetry, from which "pure science" has quite insistently sought to free itself, after the manner of Bacon and Descartes. Risk analysis, however, even in its more scientific parts, is neither all pure nor all science, and multilevel discourse (using contestable concepts (Gallie, 1955/56)) should be expected, perhaps even sometimes welcomed. Thus, some of the debates over nuclear power plants are simultaneously debates over the role of high technology in modern society; the vigor of argument about the health effects of Agent Orange is enhanced by more general perceptions about our attitudes and practices regarding veterans of military service in Viet Nam; and decisions about designing and siting a new toxic waste dump may be affected more by who controls the risk analysis than by the risks themselves. The structure and process of risk analysis, as well as its outcome, are central to debate over each of these issues.

Management of Data

It seems to be characteristic of the big substantive targets of risk analysis (such as formaldehyde carcinogenesis, mammography for breast cancer screening, ambient air pollution, or the health consequences of various dietary practices) that they require and center on the management and integration of information. That information tends to have four features: it is vast, complex, too often of poor quality, and drawn from a wide variety of disciplines--for example, science, economics, law, politics, and cost accounting, to name some. Thus, it is understandable that much formal risk analysis focuses on techniques for collecting and processing information. The matter is much broader than questions of files and computer programs, which may not enter the picture at all. We refer rather to the concepts and procedures by which information will be collected, processed, and made available to the decision maker. Examples are WASH-1400, EPA's policy on the assessment of risks from carcinogens in drinking water, and the Army Corps of Engineers' formal procedures for benefit-cost analyses. We will not dwell here on the information-management aspects of risk analysis, but we note that attempts to rationalize risk assessment by rationalizing the management of information can sometimes come apart under the pressure of debate. This is not because of any lack of sophistication on the part of the analysts but because political controversy is inherent in risk analysis, and political controversy cannot always be contained by methods of organizing and processing information, no matter how clever they are.

These matters set our terms of reference. We turn now to a discussion of how approaches to risk have changed, so that we may understand better the disagreements that may grow up around an issue deemed to require a risk analysis.

BROAD, POLITICAL CHANGES

Socialization of Risk

Our first point here is that in recent years there has been an accelerated trend toward socializing risk: that is, toward treating risk as a social (public, or collective) matter rather than a private matter. This trend has been expressed in part in the recent establishment or revision of regulatory programs aimed at reducing or preventing harms (EPA, OSHA, CPSC (Consumer Product Safety Commission), NHTSA (National Highway Traffic Safety Administration), etc.) supplanting older common-law approaches. Like the socialization of financial security in old age (social security) and the socialization of employee interests (collective bargaining), the socialization of risk is a major development in public policy. This change mirrors a shift in our basic ideas about the proper functions of government as well as the increased complexity of our ideas about how the benefits and burdens of social existence may be managed, distributed, and compensated.

Socialization of a major issue rearely follows a smooth course; controversies abound. Some disputes are fundamentally ethical: how should we accommodate the interests--perhaps the lives--of generations not yet born, or how should we balance great but diffuse public benefits against smaller but very specific damage to identified persons? Some issues are organizational, like the distinction between risk assessment and risk management that have led the OSTP (Office of Science and Technology Policy) in 1983 to propose that these functions be assigned to different agencies of the federal government. (These problems with the assumptions behind such proposals are spelled out by Whittemore, 1983. In our view, neither side of that debate is yet supported by compelling general arguments.) Some problems are methodologic, such as the push and the counter-push over whether exponential (one-hit) models of carcinogenesis should be regarded as "conservative."

These and other ethical, methodological, and organizational controversies that surround risk analysis in general, and specific analyses in particular, reflect continuing differences in presuppositions about the way a political community is to be organized and the ends that are to be served by a community's basic institutional arrangements. These presuppositions may be designated "constitutional," in a broad and old-fashioned sense of the term. Here we sketch the ways that such ideas may figure in how analysts think about risk.

As Beer and others have pointed out, the growth of a technically proficient governmental service, based on a scientific professionalism with ties to universities and other research bases outside of government, has given rise to an increasingly autonomous role of government itself in problem definition (Beer, 1977; Brown, 1983). Government has come to define the problems to which it then applies its more traditional functions of problem resolution. This trend is especially clear with the discipline(s) of risk analysis, which reflect the growing institutionalization and growing strength and efficacy of means to alert citizens to hazards, not merely to respond to anxieties. (Of course, sounding public alarms can in general advance the public welfare only when something can be done about them, but we cannot examine here the role of analysis of (presently) unavoidable risks.)

Like technology assessment more generally, risk assessment and analysis, as it is self-consciously and explicitly pursued by government, has encouraged the development of new combinations of applied science and applied mathematics. Thus, our first professional society (the Society for Risk Analysis) and its organ (Risk Analysis), our first professional journal devoted to this field, are each barely four years old. These new fields fill a practical need. However, they are both creatures of and contributors to the activism of the state. They are not merely responsive to public demands for protection, but they put permanently on the public agenda the question of what hazards to protect against, and even the degree of protection to be afforded. We continue to disagree about what is acceptable risk; about whether certain injuries are compensable and by whom; or about whether policies are consistent, fair, or economically efficient, but we no longer argue very much about whether such issues are a proper concern of government.

These disagreements themselves are far from resolved, not only because they invoke different technical views about such things as probability of causality (a technical matter of science), adequate margins of safety (of politics), or of where the burden of proof is to be placed (of law). Broader than these technical disputes is the lack of consensus about the principles that should guide policy, or indeed about the social relations that appropriate principles should express and nurture (Ronge, 1982). Formal analysis of risks can contribute to the way we manage collective decisions, but it can hardly be expected that a methodology, however systematic, can create consensus.

Realignments of Power

Our second point in this account of the "constitutional" context of risk analysis is that these fundamental changes in our society's institutional arrangements have both caused and resulted from new alignments among the Congress, administrative agencies, and the courts. In identifying, assessing, and managing risks, the Congress has delegated powers--broad, often vague, and frequently internally inconsistent--to regulatory agencies, with the expectation (indeed, with the certain knowledge) that battles over actual consequences will be fought out first in the executive branch, then in the courts. The operational meaning of statutes reveals itself only on implementation (Melnick, 1983).

One result of this procedure has been that the risk analyst has an identifiable client in need--the "Agency Administrator," who must exercise powers of administrative discretion in doing his job. Secondary clients are the courts, which may or may not be willing to defer to Agency discretion. Congress delegates to experts: that is the underlying theory, modified by the widespread suspicion of expertise in general as well as by the more specific suspicion that agencies cannot be trusted because they may be "captured," sometimes by the interests they are created to protect, and sometimes by the interests they are created to regulate. The need for Congress to delegate is accompanied by the imperative to be suspicious, and to oversee. This combination has underlain "the new administrative law" as a rationale for an activist judiciary eager to scrutinize agencies so as to assure that they meet often unrealistic deadlines, do their duty by "significant risks," and act "rationally" in light of the evidence (Melnick, 1983; Stewart, 1975).

With the action shifted by Congress to the agencies and the courts, and with diverse kinds of uncertain evidence brought to bear on discretionary decisions, the framework of risk analysis has laid claim to being the required multidisciplinary and multi-objective language. But it is important to see that the necessity that is felt for such a framework is the product of political change.

It is also important to notice how debates about the adequacy of the framework are sometimes ways of forestalling other appeals to statutory language. We might, for example, consider only public health, say, not costs; or, we might by fiat act as though there are thresholds, or as though there are no thresholds; or, quantification of benefits might be required, or prohibited, by the statute; and so forth. Sometimes a battle lost in the agency can be won in a court, perhaps on appeal; or vice versa. Notice, too, that risk analysis, with its drive for consistency and comprehensiveness, will find its strongest advocates in the more rather than the less ambitious "decision-makers"; hence officials who are striving for Administration-wide regulatory reform are likely to favor a framework of risk analysis more than the Agency Administrator whose sense of his mission is narrower. This difference may be especially sharp if implementation of an Administration-wide framework is to be highly centralized.

CONCLUSION

It behooves each of us, as a scientific/technical expert in (perhaps) one small part of the larger enterprise called risk analysis, to recognize in some detail how our work fits into the whole and to make sure that our products are not only sound and appropriate, but that they are soundly and appropriately used (Landy, 1981).

In our opinion, it would be a disaster for major decisions about the control of risks to be left in the hands of technical experts (in science or any other discipline). It would also be, in our opinion, a disaster for the technical experts in science to stop short at the point when they deliver a risk analysis to persons with quite different kinds of backgrounds--persons who have their own agendas and may see science as a tool to be used, or misused, or abused in advancing other interests. Risk analysis, broad or narrow, matters. Sometimes it matters a lot. Its ultimate impact is likely to be most beneficial, we believe, when scientists acting as scientists remain involved throughout the entire process.

What are we doing when we think we are doing risk analysis? We are joining a debate that we should not wish to see concluded, and each of us joins as citizen, professional, and somebody's employee simultaneously. To abandon any of those identities is to impoverish our thinking and to diminish our contributions to the public good.

ACKNOWLEDGMENTS

We gratefully acknowledge the support of the Alfred P. Sloan Foundation and the Mobil Corporation for part of the work reported here.

REFERENCES

Beer, S. H., 1977, Political overload and federalism, Polity, 10:5.

Brown, L. D., 1983, "New Policies, New Politics: Government's Response to Government's Growth," Brookings Institute, Washington, D. C.

Fischhoff, B., Lichtenstein, S., Slovic, P., Derby, S. L., and Keeney, R. L., 1981, "Acceptable Risk," Cambridge University Press, Cambridge, U. K.

Gallie, W. B., 1955-56, Essentially contested concepts, Proc. Aristotelian Soc., 56:167.

Graham, J. D., 1982, Some explanations for disparities in life-saving investments, Policy Stud. Rev., 1(4):692.

Landy, M. K., 1981, Policy analysis as a vocation, World Politics, 33:468.
Melnick, R. S., 1983, "Regulation and the Courts: The Case of the Clean Air Act," Brookings Institute, Washington, D. C.
Ronge, V., 1982, Risks and the waning of compromise in politics, in: "The Risk Analysis Controversy," pp.115-25, H. C. Kunreuther and E. V. Ley, eds., Springer-Verlay, New York.
Stewart, R. B., 1975, The reformation of American administrative law, Harvard Law Rev., 88:1669.
Whittemore, A. S., 1983, Facts and values in risk analysis for environmental toxicants, Risk Anal., 3:23.

DISCUSSION

Ricci: I am intrigued by your attempt, if I understood it correctly, of mixing within risk analysis, private liability, public liability, constitutional law, economics, redistribution of welfare economics and the like. Is that the objective of risk analysis?

Bailar: As we define it, risk analysis happens to involve all these things. Of course, one could redefine risk analysis, so that all these other things would drop out, but we feel that to limit the definition in that way would not be appropriate and productive. We prefer to keep the broader definition.

Ricci: My question then is, how can anyone person or any group such as this keep track of constitutional law, keep track of the law, of cancer theories, statistics, and so on? It is a "definition" which I do not find satisfying.

Bailar: No one person can retain a full and deep knowledge of all those things at once; even if one started with a good basis, one could not keep up with the developments which occur rapidly in all those fields. On the other hand, remember that we are talking about an applied discipline. There are decisions to be made and somebody is going to make them. I think there is a substantial question now as to whether we, as scientists and technicians, will be involved. There is a move on foot to freeze into institutional arrangements and practices the break that we now see between what we call "science" and all these later steps of integration. I am arguing against that kind of arrangement for managing important public affairs. The fact that no one person can keep track of all this is unfortunate. Perhaps, however, if we as scientists and technicians remain involved, we can at least see that there is adequate attention given to our own area of contribution.

Ricci: It seems to me that for any eventual theory we think of, such as negligence, strict liability, social issues, criminal law, civil procedures, or whatever, there is a point where science stops and policy begins. In my view, there is a *de facto* barrier between the two which cannot be cracked. Similarly, for welfare economics, we do not have a general welfare economic function. My question is where can the scientist provide answers?

Bailar: I agree with everything you say, except what I think is the implied conclusion, that we should acknowledge the division and simply step out of the later stages of decision making. The alternative is to leave big decisions about risk to the politicians, the lawyers, the economists, the political scientists, the bureaucrats. And are we satisfied with the way they handle things?

Land: I think that for us, it is a question of providing information in such a way that it can be used. Also we must clearly distinguish between the parts of our information that are data, and the parts that are assumptions, and in some way express the uncertainty about assumptions. It seems to me that this is a big enough problem in itself, perhaps even overwhelming.

Bailar: It is a big problem and perhaps overwhelming, and I am not suggesting that every scientist or technician should go beyond that. On the other hand, if none of us do, we will find that our information is misused, abused, ignored, twisted, and that we are not making an appropriate contribution to major public decisions. I am not arguing that we go all the way in either direction, but rather that we keep a scientific presence in these various areas clear to the end of the decision-making track.

Bridges: The moral here is that the man or woman who is prepared to step outside beyond their strict area of expertise into these other areas is the person who will have influence. As a profession, it is our job to insure that those people are sound people. The profession should take this responsbibility seriously, because the people who tend to make this jump are not necessarily the best people to do so.

Bailar: You have said it well. We want to make sure that the people who do move into these areas carry with them a very sound scientific and technical background that they are among the best of us.

Land: When I said that giving information and separating the data from the assumptions was enough, I did not intend to suggest that we should refuse to commit ourselves. I think that assumptions are assumptions, and by pointing out what uncertainties there are in this field we are making a very important contribution. The things that I consider disreputable in public controversies relating to science which I consider disreputable always relate to assumptions presented as facts.

THE CONTROL OF MUTAGENESIS AND CELL DIFFERENTIATION IN CULTURED HUMAN AND RODENT CELLS BY CHEMICALS THAT INITIATE OR PROMOTE TUMOR FORMATION

E. Huberman and C. A. Jones

Division of Biological and Medical Research
Argonne National Laboratory
Argonne, Illinois

INTRODUCTION

Clinical observations, retrospective epidemiological surveys, and studies with experimental animals have provided evidence that chemicals in our environment, including some produced as by-products from our energy generation processes, are responsible for a significant proportion of human cancers (Boyland, 1967; Higginson, 1969; Higginson and Muir, 1979; Maugh, 1979). Furthermore, these studies suggest that certain human cancers result from the interaction of multiple factors in a multistep process (Berenblum, 1969; Van Duuren, 1969; Boutwell, 1974; Diamond et al., 1980b). From an operational point of view, the carcinogenesis process can be divided into three sequential stages: initiation, promotion, and progression. The first two involve the steps that lead to the transformation of a normal into a malignant cell, whereas progression covers processes whereby a transformed cell develops into a malignant tumor.

To understand chemical carcinogenesis, we need to know more about the basic events underlying these processes. Tumor initiation and promotion, in particular, are known to be influenced by environmental chemicals. Several studies have shown that after metabolic activation many chemical carcinogens that act as tumor initiators can bind to cellular macromolecules--including DNA--and can induce mutations in various cells (Brookes and Lawley, 1964; Heidelberger, 1970; Kuroki and Heidelberger, 1971; Ames et al., 1973; Huberman and Sachs, 1976, 1977). Tumor initiation appears to involve a "mutation-like event" in genes that control normal cell growth and differentiation (Knudsen et al., 1975; Bouck and DiMayorca, 1976; Huberman, 1978). On the other hand, chemicals that promote tumor formation usually do not bind to DNA and are devoid of mutagenic activity. These agents, however, affect a number of cellular events (Cohen et al., 1977; Rovera et al., 1977; Yamasaki et al., 1977; Weinstein et al., 1978; Slaga et al., 1980) including, in certain cells, the induction of cell differentiation (Huberman and Callaham, 1979; Lotem and Sachs, 1979; Rovera et al., 1979; Huberman et al., 1982). This last property led us to suggest that in promotion, expression of "mutated tumor genes" occurs in a process similar to gene expression during normal cell differentiation. To examine this idea further and to investigate questions related to tumor initiation, we are studying the ability of tumor initiators and promoters to induce mutagenesis and cell differentiation in a variety of mammalian cell systems.

MUTAGENESIS BY CHEMICAL CARCINOGENS IN A HUMAN CELL-MEDIATED ASSAY

Mammalian cell-mediated mutagenesis assays are useful for mechanistic studies with chemical carcinogens (initiators) (Huberman and Sachs, 1974, 1976; Jones and Huberman, 1980; Bradley et al., 1981; Jones et al., 1981; Langenbach et al., 1981; Huberman et al., 1984). In this assay, mutagenesis is detected in a standard mammalian target cell in the presence of cells from another tissue, which metabolically activates the carcinogen being tested. Results from such cell-mediated mutagenicity tests correlate with carcinogenicity data more closely in regard to potency and cell specificity than do results from similar assays in which subcellular fractions are used for carcinogen activation (Newbold et al., 1977; Selkirk, 1977; Bigger et al., 1980). Originally, we used primary rodent cells to activate carcinogens and a standard type of rodent cell to determine the genetic damage. However, human cells may respond to environmental chemicals differently than rodent cells because of differences in cell permeability, carcinogen metabolism, or DNA repair. Thus, the most useful results are provided by assays in which human cells--ideally, epithelial cells--are used since most human malignancies are carcinomas.

To establish this type of assay, we have isolated a highly mutable clonal cell population from a human teratoma cell line (Zeuthen et al., 1980). These cells, designated P3, exhibit an epithelial cell morphology and a stable diploid karyotype with 46(XX) chromosomes, including a translocation between chromosomes 15 and 20. To characterize their response to chemical mutagens, we analyzed the ability of the affected cells to acquire resistance to 6-thioguanine (TG), a phenotypic change associated with a mutation at the hypoxanthine-guanine phosphoribosyl transferase (HGPRT) locus (Caskey and Kruh, 1979). Analysis of six P3 TG-resistant mutants, induced by either _N_-methyl-_N_′-nitro-

N-nitrosoguanidine (MNNG) or benzo(a)pyrene (B[a]P), revealed that acquisition of TG resistance was associated with a more than 10-fold reduction in HGPRT activity (Table 1).

We also determined the conditions under which such mutants can be efficiently selected. After treatment with the mutagen, the P3 cells required a period of 7 days to eliminate the functional wild-type enzyme molecules via cell division, dilution, and/or degradation (van Zeeland et al., 1972; van Zeeland and Simons, 1976; O'Neill et al., 1977; Bradley et al., 1981). Metabolic cooperation, which occurs in dense cultures during the selection process with TG (Subak-Sharpe et al., 1969; Trosko et al., 1977; Yotti et al., 1979), is minimized by seeding the P3 cells at a density of 2×10^4 cells/60 mm Petri dish in growth medium containing 40 μg TG.

Table 1. Susceptibility to the Cytotoxic Effect of TG and HGPRT Activity in Parental P3 and Six TG-Resistant Mutant Cell Types

Cell Type	Relative Percentage of Colony-Forming Cells after Treatment with 40-μM TG	Specific Activity of HGPRT (pmol inosine-5-monophosphate/min/μg protein)
P3	< 0.03	4.6 ± 1.0
NG-1	94	0.2 ± 0.1
NG-2	100	0.2 ± 0.1
NG-3	94	0.3 ± 0.1
BP-1	97	0.1 ± 0.1
BP-2	108	0.2 ± 0.1
BP-3	95	0.2 ± 0.1

NG-1,-2,-3 cells are the progeny of TG-resistant colonies isolated from P3 cells after treatment with 7 μM MNNG and BP-1,-2,-3 are progeny of P3 cells treated with 4 μM B(a)P activated in a cell-mediated assay with human breast carcinoma BJ cells. The cells were used in the assays after the mutant cells were subcultured for 4-6 weeks in the absence of TG. The cloning efficiency of the P3 cells was 69%, whereas the cloning efficiency of the mutant cells varied from 45-75%. These cloning efficiencies were considered as 100% for the relative percentage of colony-forming cells. The HGPRT activities are the mean ± SD of seven measurements (based on Huberman et al., 1984).

Using these conditions, we were able to test effectively the mutagenic response to a series of different polycyclic aromatic hydrocarbons (PAH). Like other types of cells used in mutagenesis studies, the P3 cells cannot activate many proximate mutagens/carcinogens (Huberman and Sachs, 1974; Bigger et al., 1980). To detect mutagenic activity, P3 cells have to be supplemented with an exogenous metabolic activation system. We therefore cocultivated the P3 cells (Huberman and Sachs, 1974; Harris et al., 1978; Aust et al., 1980; Diamond et al., 1980a; Gould et al., 1982) with human breast carcinoma BJ cells. Preliminary data indicated that the BJ cells can metabolize B(a)P, a PAH prototype, into water-soluble products (unpublished results). In this human cell-mediated assay, we demonstrated that 7,12-dimethylbenz(a)anthracene, 3-methylcholanthrene, B(a)P, and chrysene induced TG-resistant mutants--with the degree of induction related to the carcinogenic potency of the PAH (IARC, 1973). Pyrene and benzo(e)pyrene, which are not carcinogenic, were either inactive or exhibited a limited induction of TG resistance (Table 2). Also P3 cells can detect the mutagenic response of B(a)P in an assay when Syrian hamster embryo cells are used for activation. Dimethylnitrosamine becomes mutagenic for P3 cells after metabolism by rat liver hepatocytes (Huberman and Sachs, 1974; Langenbach et al., 1978; Jones and Huberman, 1980).

Our results indicate that human epithelial teratoma P3 cells are useful in detecting mutations induced at the HGPRT locus by chemical carcinogens activated in a cell-mediated assay with either human or rodent cells. Furthermore, there is a relationship between the degree of mutagenesis of PAH and their ability to induce tumor formation in experimental animals.

CELL-SPECIFIC ACTIVATION OF BENZO(A)PYRENE TO MUTAGENIC METABOLITES BY FIBROBLASTS AND HEPATOCYTES

Prediction of human risk from chemical carcinogens is complicated because carcinogens act only in specific organs or in certain species. In different species, a given chemical can affect a different spectrum of organs (Miller et al., 1964; Lijinsky, 1983) or even different types of cells within an organ (Rice and Frith, 1983; Lewis and Swenberg, 1980). Our understanding of the critical determinants involved is very limited.

For our studies into the mechanisms underlying organ specificity, we chose the ubiquitous environmental carcinogen, B(a)P, which exhibits organ specificity *in vivo*. In rodents, fibrosarcomas, which originate from fibroblasts, are a predominant class of neoplasms caused by B(a)P; hepatocarcinomas are rarely induced by B(a)P except under unusual circumstances such as treatment of newborn animals with immature and rapidly dividing

Table 2. Induction of TG-Resistant P3 Mutants by Carcinogenic PAH Activated in a Human Cell-Mediated Assay

PAH[a]	Concentration (μM)	No. of TG-Resistant Mutants/10^6 CFC[b]
Control	--	2 ± 2
Pyrene	4	2 ± 2
	12	5 ± 2
B(e)P	4	3 ± 2
	12	7 ± 3
Chrysene	4	12 ± 4
	12	24 ± 5
B(a)P	0.4	47 ± 8
	1.3	103 ± 32
	4.0	124 ± 60
MCA	0.15	45 ± 12
	0.4	188 ± 25
	1.3	196 ± 30
DMBA	0.05	56 ± 21
	0.15	106 ± 18
	0.4	211 ± 26

[a]Two days after treatment with DMBA, B(a)P, and MCA, there were 0.3-1.6 x 10^6 cells/Petri dish. The cloning efficiencies of these treated cells varied from 7-48%. In all other cases there were 1.4-1.8 x 10^6 cells/Petri dish and the cloning efficiencies of these cells varied from 31-52%. The results are the mean ± SD of 4-6 measurements.
[b]Colony-forming cells.
(Table from Huberman et al., 1984).

hepatocytes, partial hepatectomy, or disruption of normal liver function by treatment with enzyme inducers.

Using rodent fibroblasts and hepatocytes to metabolize B(a)P, we examined some of the factors contributing to the expression of differing cell susceptibility. In these experiments, B(a)P was activated into a mutagen for Chinese hamster V79 cells by fibroblasts but not by hepatocytes (Table 3). Analysis of the profiles of metabolites and DNA adducts obtained following incubation of hepatocytes and fibroblasts with [^{3}H]B(a)P indicated that cell

Table 3. Induction of Ouabain-Resistant Mutants by B(a)P in V79 Cells in the Fibroblast-Mediated and Hepatocyte-Mediated Assays

	Fibroblast-mediated Assay		Hepatocyte-mediated Assay	
Carcinogen (μM)	Mutants/10^6 Survivors	% Cell Survival	Mutants/10^6 Survivors	% Cell Survival
Control	1	100	1	100
BP				
0.4	9	85	1	75
4	38	58	1	70
12	68	43	1	68
40	74	41	2	69

V79 cells were cocultivated with either primary hepatocytes or fibroblasts and treated with the carcinogen for 18 h. After trypsinization, the V79 cells were seeded for determination of cell survival and the frequency of ouabain-resistant mutants. The average cloning efficiency of the untreated controls, 45 ± 10%, was normalized as 100% for expression of the percentage of cells surviving after treatment. The spontaneous mutation frequency is 1×10^{-6}. Cell survival relative to survival of untreated cells is shown in parentheses (table from Jones et al., 1983).

specificity could be explained by a difference in the metabolism of B(a)P in the two types of cells. In hepatocytes, metabolism occurred principally in the K-region of the molecule, resulting in the formation of benzo(a)pyrene-4,5-dihydrodiol (BP-4,5-diol), phenols, and quinones, a major portion of which underwent subsequent metabolism to water-soluble conjugates (Fig. 1). Only small amounts of BP-7,8-diol, the precursor of BP-diol epoxide (BPDE), the ultimate carcinogenic and mutagenic metabolite of BP, were generated, and these precursors were rapidly converted to nonmutagenic triols and tetraols. Thus, no appreciable levels of BP-7,8-diol remained within the hepatocytes to allow the generation of significant amounts of BPDE. Metabolism of B(a)P in the fibroblasts, on the other hand, was directed primarily toward the formation of BP-7,8-diol and hence BPDE.

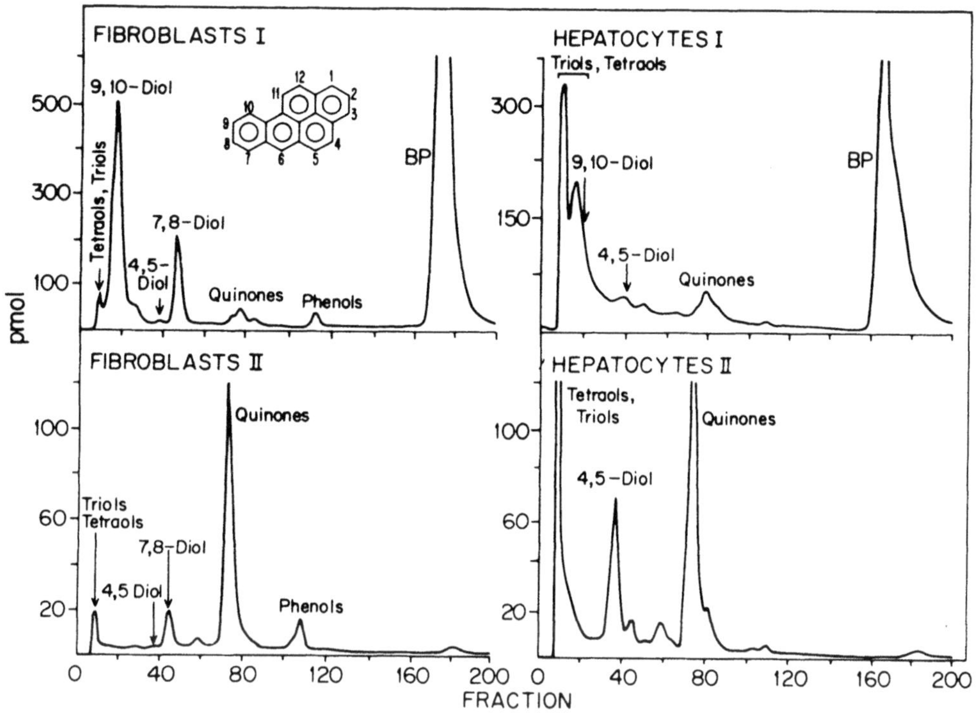

Fig. 1. HPLC chromatograms of the organic-extractable metabolites (I) and water-soluble conjugates (II) of B(a)P formed by fibroblasts and hepatocytes after incubation for 18 h with 4 μM [^{3}H]B(a)P (from Jones et al., 1983).

Analysis of the DNA adducts in fibroblasts showed that the major adduct was derived from DNA reacting with BPDE (Fig. 2). More specifically, the adducts resulted from an interaction between BPDE and the deoxyguanosine bases. Although the biological consequence of B(a)P-DNA interaction, i.e., induction of gene mutations, was determined in the cocultured V79 cells and not in the fibroblasts, it has been demonstrated that the nature and relative proportion of DNA adducts are similar in both types of cells indicating that the transfer of mutagenic B(a)P intermediates from metabolizing cells to target cells is an efficient process (Sebti et al., 1982).

The extent of B(a)P-DNA interaction in the hepatocytes was significantly higher than in the fibroblasts, indicating that B(a)P was metabolized to intermediates that can interact with DNA but do not result in mutagenic lesions. The major DNA adducts in the hepatocytes were hydrophilic deoxyribonucleoside derivatives (Jones et al., 1983). This material has not been identified, but has been observed in other studies, including binding studies with microsomes and perfused liver (King et al., 1975; Kahl et al.,

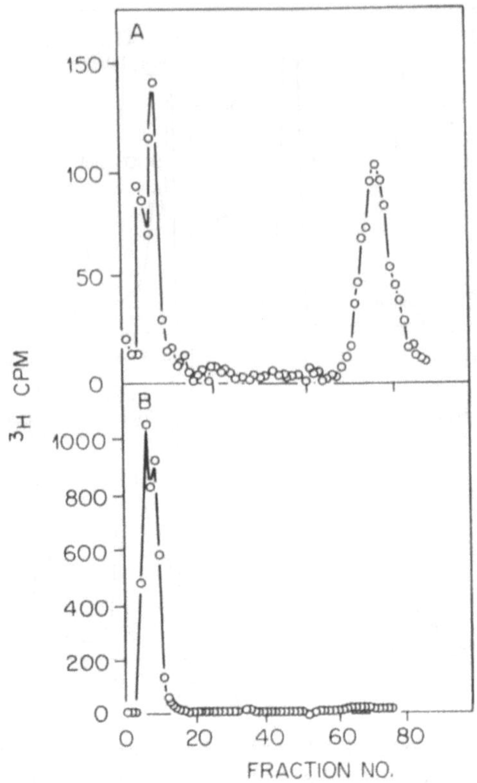

Fig. 2. Sephadex LH-20 column chromatography of enzymatic digests of DNA isolated from cells incubated with 4μM [^{3}H]B(a)P. The column was eluted with water until fraction 40, when the solvent was changed to methanol. (A) Fibroblast DNA. The peak around fraction 80 corresponds to the BPDE adduct. (B) Hepatocyte DNA (from Jones et al., 1983).

1979) and whole-animal experiments (Eastman et al., 1978; Boroujerdi et al., 1981). Only trace levels of BPDE-DNA adducts were detected in the hepatocytes following incubations with B(a)P. Thus, it seems that in the hepatocytes, B(a)P was metabolized via pathways other than those leading to high levels of the mutagenic BPDE.

Our results indicate that some of the factors contributing to the cell specificity of chemical carcinogens can be investigated in specific tissue/cell-mediated mutagenesis assays in combination with metabolism studies in primary cell cultures. In particular, the low susceptibility of the liver to B(a)P-induced carcinogenesis may be explained by the preferential metabolism of B(a)P via routes other than those leading to formation of significant amounts of BPDE; thus critical binding of this carcinogenic B(a)P intermediate to hepatocyte DNA does not occur under normal circumstances.

THE EXPRESSION OF CELL DIFFERENTIATION MARKERS IN CULTURED HUMAN CELLS AFTER TREATMENT WITH TUMOR-PROMOTING AGENTS

Many studies have demonstrated that certain chemicals (e.g., phorbol diesters, teleocidin) can promote the formation of tumors on mouse skin previously treated with a low dose of an initiator such as 7,12-dimethylbenz(a)anthracene (Berenblum, 1969; Boutwell, 1974; Slaga et al., 1978; Fujiki et al., 1982). Unlike tumor initiators, the chemicals that promote tumor formation are devoid of mutagenic activity. These chemicals may exert their promotional effect by causing the expression of the "mutated tumor genes" in a process similar to the process of gene expression during cell differentiation. Indeed, phorbol diesters and teleocidin act as inducers of cell differentiation in a number of cell types (Miao et al., 1978; Huberman and Callaham, 1979; Huberman et al., 1979; Kasukabe et al., 1979; Lotem and Sachs, 1979; Rovera et al., 1979; Nagasawa and Mak, 1980; Koeffler, 1981; Pahlman et al., 1981; Huberman et al., 1982). In our cell differentiation studies with tumor promoters, we used three different types of cells: HO melanoma cells (Huberman et al., 1979), the promyelocytic HL-60 (Huberman and Callaham, 1979), and T lymphoid CEM leukemia cells (Ryffel et al., 1982). These cells display useful markers of cell differentiation.

The prototype phorbol diester phorbol-12-myristate-13-acetate (PMA) at doses as low as 10^{-10} to 10^{-9} M induces a cell differentiation in the HO melanoma cells characterized by an inhibition of cell growth, increased synthesis of melanin, and induction of dendrite-like structures (Huberman et al., 1979). We demonstrated a relationship between the tumor-promoting activity of a series of phorbol diesters *in vivo* and the degree to which these agents induce differentiation *in vitro*. A similar relationship was demonstrated with the HL-60 cells in which phorbol diesters and teleocidin induced a macrophage cell differentiation, characterized by an inhibition of cell growth, appearance of morphologically mature cells, increased phagocytic capability, changes in reactivity to monoclonal antibodies and in the activity of certain enzymes (Huberman et al., 1982; Murao et al., 1983).

In the CEM cells, PMA causes the expression of a phenotype that resembles that of a suppressor T lymphocyte (Ryffel et al., 1982). The new phenotype is characterized in part by the appearance of a specific antigenic pattern. More specifically, the cells exhibit increased reactivity with OKT3 and OKT8 monoclonal antibodies, which characterize mature suppressor and cytotoxic T lymphocytes, respectively (Reinherz et al., 1979, 1980; Reinherz and Schlossman, 1981; Schroff et al., 1982). PMA also reduces the reactivity of the treated cells with the OKT4 monoclonal antibody, which detects inducer/helper T lymphocytes, and with the OKT6 antibody, which detects immature "common" thymocytes (Reinherz et al., 1979, 1980; Reinherz and Schlossman, 1981). The treated CEM cells, in common with mature suppressor T lymphocytes, also suppressed [^{3}H]thymidine incorporation into phytohemaglutinin-activated peripheral blood lymphocytes (Rose and Friedman, 1980) but, unlike cytotoxic lymphocytes, were unable to induce a cytotoxic response in a number of human and rodent target cells.

The ability of PMA to induce differentiation of CEM cells into suppressor cells raises the possibility that tumor promotion in the mouse skin may also involve suppression of the immunity against "initiator"-induced papillomas and carcinomas. Such a suppression could result from an increase in the number of mature suppressor T lymphocytes in the promoted animal.

Despite enormous scientific effort, the mechanism by which these chemicals promote tumorigenesis and alter differentiation processes is poorly understood. Recently it was found that phorbol diesters and teleocidin bind to cellular receptors (Driedger and Blumberg, 1980; Solanki et al., 1981; Umezawa et al., 1981) that seem to be associated with a specific protein kinase (Cooper et al., 1982). We have reported that in HL-60 cells this binding reaches a maximum at 20 minutes and is followed by a diminished binding, a "down regulation" (Fox and Das, 1979; Solanki et al., 1981). Decreased binding of phorbol diester to the cellular receptors did not occur in an HL-60 cell variant that is resistant to induction of cell differentiation by PMA and teleocidin; however, total binding of the tumor promoter was equivalent to that in HL-60 cells (Solanki et al., 1981). On the basis of these results, we suggested that binding of phorbol diesters to cellular receptors and the down regulation of binding are required for the initiation of the biological activity of the phorbol diesters in the HL-60 cells, and perhaps other types of cells. The interaction of these agents with the specific receptors may activate protein kinase and consequently phosphorylate specific proteins. Indeed, treatment of HL-60 cells with PMA for less than 30 minutes resulted in specific changes in the pattern of protein phosphorylation. A number of these changes were absent or less pronounced in R-80 cells, which derived from an HL-60 cell variant that is resistant to PMA-induced cell differentiation (unpublished results) but not to other inducers, such as dimethylsulfoxide or retinoic acid. Perhaps some of these proteins are the ones that are associated with the induction of cell differentiation and expression of the "mutated tumor genes."

Certainly, cultured human cells possessing clear markers of cell differentiation offer a useful tool for studying the mode of action of some types of tumor promoters.

THE INDUCTION OF A MACROPHAGE CELL DIFFERENTIATION IN HUMAN PROMYELOCYTIC (HL-60) LEUKEMIA CELLS BY 1,25-DIHYDROXYVITAMIN D_3, THE BIOLOGICALLY ACTIVE METABOLITE OF VITAMIN D_3

It was reported that 1,25-dihydroxyvitamin D_3 (1,25-$[OH]_2D_3$), the biologically active metabolite of vitamin D_3 (Haussler and McCain, 1977), at low doses induced in HL-60 cells a number of differentiation markers (Miyaura et al., 1981; Tanaka et al., 1982). The biological activity of 1,25-$(OH)_2D_3$, as for both PMA and teleocidin (Dunphy et al., 1980; Horowitz et al., 1981; Solanki et al., 1981; Umezawa et al., 1981), requires binding to specific cellular receptors (Colston et al., 1980, 1981; Frampton et al., 1982; Manolagas and Deftos, 1980). These studies raised the possibility that 1,25-$(OH)_2D_3$ may induce differentiation processes in the HL-60 cells via a mechanism similar to the action of PMA and teleocidin. To examine this possibility, we compared the ability of 1,25-$(OH)_2D_3$ and PMA to induce cell differentiation in HL-60 cells and in the R-80 cell variant that is resistant to PMA-induced cell differentiation (Major et al., 1981; Huberman et al., 1981, 1982). In addition, we investigated the ability of 1,25-$(OH)_2D_3$ to inhibit specific phorbol diester receptor binding (Umezawa et al., 1981) as a way to establish whether 1,25-$(OH)_2D_3$, like teleocidin, shares similar binding sites with phorbol diesters.

Our results show that treatment of HL-60 cells with 1,25-$(OH)_2D_3$ caused an increase in lysozyme and nonspecific esterase activities, increased reactivity with the myeloid-specific OKM1 monoclonal antibody, and an increase in the fraction of morphologically mature cells; the inductions of these markers were both time and dose dependent (Fig. 3). The increases in nonspecific esterase activity and in the synthesis of a number of monocyte/macrophage protein markers and the absence of a large fraction of cells with banded or segmented nuclei (characteristic of granulocytes) indicate that 1,25-$(OH)_2D_3$ treatment of HL-60 results in a cell type that more closely resembles that of a monocyte/macrophage than a granulocyte. These data refute the previous suggestions (Miyaura et al., 1981; Tanaka et al., 1982) that a granulocyte phenotype is produced.

An interesting result was that the R-80 cells were relatively resistant to the differentiation induced by both PMA and 1,25-$(OH)_2D_3$. These, and the previous results, suggest that the two inducers may affect a similar process that leads to the monocyte/macrophage-like phenotype. This process may involve a signal that follows the binding of these inducers to their specific receptors, since 1,25-$(OH)_2D_3$ did not compete for the phorbol diester binding sites (Fig. 4). Various events including "down modulation" of specific binding (Manolagas and Deftos, 1980; Solanki et al., 1981), alterations in phospholipid (Cabot et al., transformation of cells pretreated (initiated) with chemical carcinogens. We used the hamster embryo cell transformation assay (Berwald and Sachs, 1965; Huberman and Sachs, 1966) with a protocol designed to evaluate the activity of presumptive tumor-promoting agents (Rivedal and Sanner, 1982).

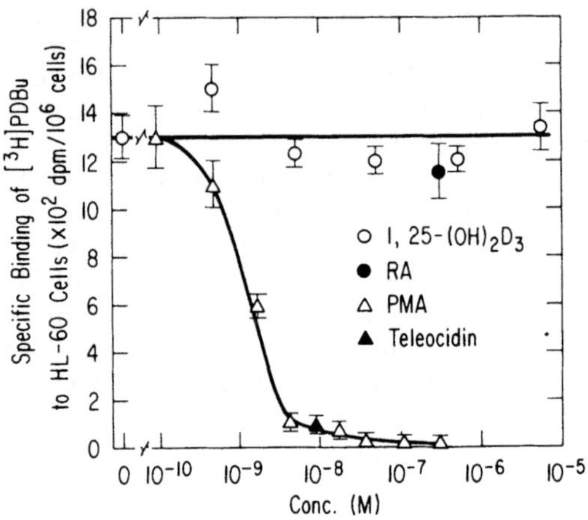

Fig. 4. Inhibition of the specific binding of (^{3}H)PDBu to intact HL-60 cells by (O) 1,25-(OH)$_2$D$_3$, (●) RA, (Δ) PMA, or (▲) teleocidin. Each point represents the mean ± S.E. for three separate experiments done in triplicate. Control value was 13.2 ± 0.8 x 10^2 dpm/10^6 cells (from Murao et al., 1983).

The transformed phenotype of the Syrian hamster embryo cell is characterized by defined changes in cell organization within a colony (Berwald and Sachs, 1965; Borek and Sachs, 1966; Huberman and Sachs, 1966; DiPaolo et al., 1971). These changes result from alterations in the growth pattern of the transformed cells. Such cells upon isolation, propagation, and inoculation into appropriate hosts can acquire an increased malignant potential (Berwald and Sachs, 1965; Kuroki and Sato, 1968; DiPaolo et al., 1969b, 1971; Barrett et al., 1979). This assay is thus a useful tool for

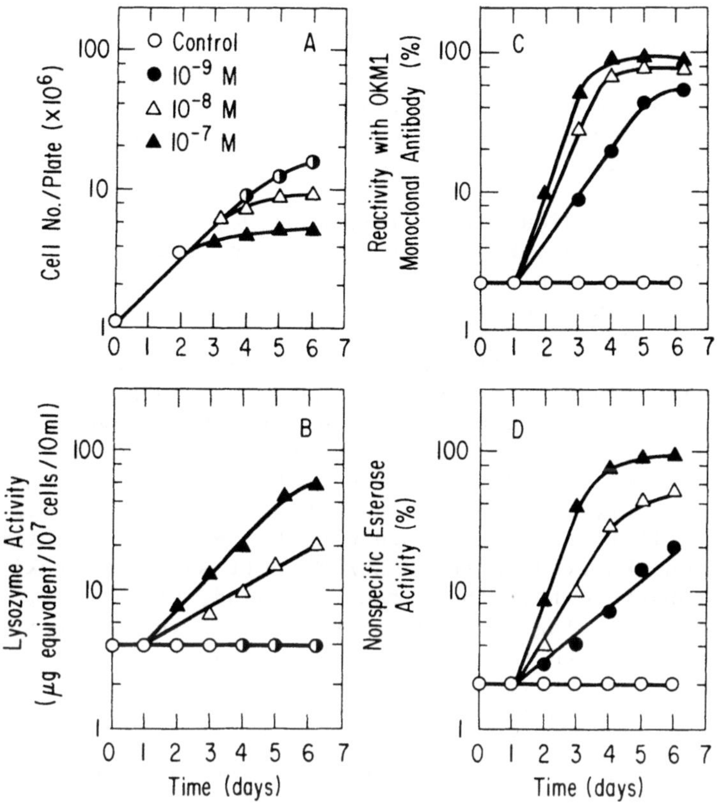

Fig. 3. Cell growth and induction of markers of cell differentiation in HL-60 cells at various times after treatment with different concentrations of 1,25-$(OH)_2D_3$: (A) cell numbers; (B) lysozyme activity; (C) reactivity with OKM1 monoclonal antibody; and (D) nonspecific esterase activity. O, control; ●, 10^{-9} M; Δ, 10^{-8} M; and ▲, 10^{-7} M (from Murao et al., 1983).

1980; Hoffman and Huberman, 1982; Wasserman et al., 1982) and calcium metabolism (Haussler and McCain, 1977; Slaga et al., 1978; Wasserman et al., 1982), caused by both inducers, may perhaps be the common signal(s) that leads to the similar, although not identical, phenotype.

ENHANCEMENT OF CHEMICALLY INDUCED CELL TRANSFORMATION IN HAMSTER EMBRYO CELLS BY 1,25-DIHYDROXYVITAMIN D_3

In view of the similarities in the biological activities of 1,25-$(OH)_2D_3$ and PMA in HL-60 cells, we decided to determine whether 1,25-$(OH)_2D_3$, like PMA, could enhance (promote) the studying the fundamental mechanism(s) underlying the neoplastic process, including tumor initiation and tumor promotion. The assay is sensitive to diverse classes of carcinogens (both tumor initiators and tumor promoters), and can also be used to identify new presumptive carcinogens (DiPaolo et al., 1969a; Pienta et al., 1977; Heidelberger et al., 1983).

Our results indicate that 1,25-$(OH)_2D_3$, in common with some tumor promoters (e.g., PMA), will not effectively induce cell transformation by itself. However, when administered subsequent to known carcinogens/tumor initiators, it enhances the frequency of cell transformation. This enhancing effect has been demonstrated with the polycyclic aromatic hydrocarbons B(a)P and B(e)P, which require metabolic activation (Huberman, 1978; MacLeod et al., 1980), as well as with BPDE (the ultimate mutagenic and carcinogenic form of B(a)P [Huberman et al., 1976; Mager et al., 1977; Slaga et al., 1977]) and MNNG, both of which are themselves chemically reactive (Table 4). An enhancing effect of 1,25-$(OH)_2D_3$ on cell transformation induced by 3-methylcholanthrene, has also been observed in the BALB/3T3 cell system (Kuroki et al., 1983). We observed that the parent compound, vitamin D_3, was also effective in enhancing cell transformation in the hamster embryo cell system, although to a lesser degree than 1,25-$(OH)_2D_3$. Thus a fraction of the embryonic hamster cells may have the capacity to convert vitamin D_3 to its reactive metabolite.

Although 1,25-$(OH)_2D_3$ and PMA share similar biological properties, including enhancement (promotion) of carcinogen/tumor-initiator-induced cell transformation and induction of cell differentiation in human promyelocytic HL-60 leukemia cells (Murao et al., 1983; Tanaka et al., 1983; McCarthy et al., 1983), they differ in their mode of action presumably because they do not have the same receptors for their biological activity (Murao et al., 1983). In this context, 1,25-$(OH)_2D_3$, unlike PMA, does not promote papilloma formation in the mouse. Furthermore, it diminishes the effect of PMA in inducing such tumors (Wood et al., 1983). Our cell transformation studies in which we used a number of cell pools, each deriving from a different pregnant hamster, are also in agreement with this notion (Table 5). Thus the high or low responsiveness of a cell pool to the enhancing (promoting) effect of 1,25-$(OH)_2D_3$ was not necessarily correlated with a similar responsiveness to PMA. These two agents which share similar biological effects in some cell types, such as enhancing (promoting) cell transformation in rodent fibroblasts and inducing cell differentiation in human leukemia cells, may differ in their biological effects in other cell types, including mouse skin epithelial cells.

Table 4. Enhancement of Cell Transformation by 1,25-$(OH)_2D_3$ in Cells Pretreated With Either B(a)P, B(e)P, BPDE, or MNNG[a]

Duration of Treatment 0-3 days (Stage 1)	3-7 days (Stage 2)	Total no. colonies	Cloning efficiency(%)	Total no. transf. colonies	Trans. freq. (%)
-	-	1728	36	1	0.06
1,25-$(OH)_2D_3$	-	1873	39	0	<0.05
-	1,25-$(OH)_2D_3$	1920	40	0	<0.05
1,25-$(OH)_2D_3$[a]	1,25-$(OH)_2D_3$	916	40	0	<0.05
B(a)P (1)[b]	-	1576	33	7	0.4
-	B(a)P (1)	944	20	6	0.6
B(a)P (1)[a]	B(a)P (1)	1133	24	7	0.6
B(a)P (1)	1,25-$(OH)_2D_3$	1080	23	42	3.9
1,25-$(OH)_2D_3$	B(a)P (1)	1122	23	13	1.2
1,25-$(OH)_2D_3$[a] + B(a)P (1)	1,25-$(OH)_2D_3$ + B(a)P (1)	832	17	17	2.0
B(e)P (3)	-	1284	27	0	<0.08
B(e)P (3)	1,25-$(OH)_2D_3$	1459	30	5	0.3
Pyrene (3)	-	1776	31	0	<0.06
Pyrene (3)	1,25-$(OH)_2D_3$	1670	35	0	<0.06
BPDE (0.1)	-	1078	23	9	0.8
BPDE (0.1)	1,25-$(OH)_2D_3$	994	21	28	2.8
MNNG (0.3)	-	1818	38	3	0.2
MNNG (0.3)	1,25-$(OH)_2D_3$	1794	37	24	1.3

[a]Cultures treated for 7 days with the chemical without medium change.
[b]The numbers in the parentheses represent the concentrations of the chemicals in µg/mL. The concentration of 1,25-$(OH)_2D_3$ was 0.1 µg/mL.

In summary, we have discussed in the present report a number of studies in rodent and human cell culture systems that have helped to expand to some degree our comprehension of the cellular processes involved in tumor initiation and promotion.

Table 5. Enhancement by 1,25-$(OH)_2D_3$ of B(a)P-Induced Cell Transformation in Embryo Cell Pools Derived From Different Pregnant Hamsters

Duration of Treatment		Transformation Frequency %					
		Cryopreserved Cell Pools					Noncryopreserved Cell Pool
0-3 days	3-7 days	A	B	C	D	E	
-	-	<0.07	<0.10	<0.07	<0.06	<0.05	<0.06
-	1,25-$(OH)_2D_3$	<0.09	<0.12	<0.07	<0.06	<0.08	0.06
-	PMA	<0.09	<0.08	<0.09	<0.06	<0.12	<0.05
B(a)P	-	0.89	0.20	0.16	0.14	0.08	0.09
B(a)P	1,25-$(OH)_2D_3$	4.50	2.37	0.40	0.26	0.10	0.49
B(a)P	PMA	4.65	1.29	2.62	4.79	0.17	0.07

For each determination 700-2200 colonies were scored. The cloning efficiency of the control cells or cells treated with either 1,25-$(OH)_2D_3$ or PMA ranged from 23-47%. B(a)P treatment reduced these frequencies to 10-28%. The concentration of B(a)P was 1 μg/mL; the concentrations of PMA and 1,25-$(OH)_2D_3$ were 0.5 and 0.1 μg/mL, respectively.

REFERENCES

Ames, B. N., Durston, W. E., Yamasaki, E., and Lee, F. D., 1973, Carcinogens are mutagens: A simple test system combining liver homogenates for activation and bacteria for detection, Proc. Natl. Acad. Sci. U.S.A., 70:2281.

Aust, A. E., Falahee, K. J., Maher, V. M., and McCormick, J. J., 1980, Human cell-mediated benzo(a)pyrene cytotoxicity and mutagenicity in human diploid fibroblasts, Cancer Res., 40:4070.

Barrett, J. C., Crawford, B. D., Mixter, L. O., Schectman, L. M., Ts'o, P. O. P., and Pollack, R., 1979, Correlation of in vitro growth properties and tumorigenicity of Syrian hamster cell lines, Cancer Res., 39:1504.

Berenblum, I., 1969, A re-evaluation of the concept of cocarcinogenesis, Prog. Exp. Tumor Res., 11:21.

Berwald, Y., and Sachs, L., 1965, In vitro transformation of normal cells to tumor cells by carcinogenic hydrocarbons, J. Natl. Cancer Inst., 35:641.

Bigger, C. A. H., Tomaszewski, J. E., Dipple, A., and Lakes, R.S., 1980, Limitations of metabolic activation systems used with in vitro tests for carcinogenesis, Science, 209:503.

Borek, C., and Sachs, L., 1966, In vitro cell transformation by X-irradiation, Nature (London), 210:276.

Boroujerdi, M., Kung, H., Wilson, A. E. G., and Anderson, M. W., 1981, Metabolism and DNA binding of benzo(a)pyrene in vivo in the rat, Cancer Res., 41:951.

Bouck, N., and DiMayorca, G., 1976, Somatic mutation as the basis for malignant transformation of BKH cells by chemical carcinogens, Nature (London), 264:360.

Boutwell, R. K., 1974, Function and mechanisms of promoters of carcinogenesis, CRC Crit. Rev. Toxicol., 2:419.

Boyland, E., 1967, The correlation of experimental carcinogenesis and cancer in man, Progr. Exp. Tumor Res., 11:222.

Bradley, M. O., Bhuyan, B., Francis, M. C., Langenbach, R., Peterson, A., and Huberman, E., 1981, Mutagenesis by chemical agents in V-79 Chinese hamster cells: A review and analysis of the literature: A report of the Gene-Tox Program, Mutat. Res., 87:81.

Brookes, P., and Lawley, P. D., 1964, Evidence for the binding of polynuclear aromatic hydrocarbons to the nucleic acids of mouse skin: Relation between carcinogenic power of hydrocarbons and their binding to deoxyribonucleic acid, Nature (London), 202:781.

Cabot, M. C., Welsh, C. J., Callaham, M. F., and Huberman, E., 1980, Alterations in lipid metabolism induced by 12-O-tetradecanoylphorbol-13-acetate in differentiating human myeloid leukemia cells, Cancer Res., 40:3674.

Caskey, C. T., and Kruh, G. D., 1979, The HPRT locus, Cell, 16:1.

Cohen, R., Pacifici, M., Rubinstein, N., Beneni, Y., and Holtzer, H., 1977, Effect of a tumor promoter on myogenesis, Nature (London), 266:538.

Colston, K., Colston, M. J., and Feldoman, D., 1981, 1,25-Dihydroxyvitamin D_3 and malignant melanoma: the presence of receptors and inhibition of cell growth in culture, Endocrinology, 108:1083.

Colston, K., Hirst, M., and Feldoman, D., 1980, Organ distribution of the cytoplasmic 1α,25-dihydroxycholecalciferol receptor in various mouse tissues, Endocrinology, 107:1916.

Cooper, R. A., Braunwald, A. D., and Kuo, A. L., 1982, Phorbol ester induction of leukemic cell differentiation is a membrane-mediated process, Proc. Natl. Acad. Sci. U.S.A., 79:2865.

Diamond, L., Kruszewski, F., Aden, D. P., Knowles, B. B., and Baird, W. M., 1980a, Metabolic activation of benzo(a)-pyrene by a human hepatoma cell line, Carcinogenesis, 1:871.

Diamond, L., O'Brien, T. G., and Baird, W. M., 1980b, Tumor promoters and the mechanisms of tumor promotion, Adv. Cancer Res., 32:1.

DiPaolo, J. A., Nelson, R. L., and Donovan, P. J., 1969a, Sarcoma producing cell clones derived from clones transformed in vitro by benzo(a)pyrene, Science, 165:917.

DiPaolo, J. A., Nelson, R. L., and Donovan, P. J., 1969b, Quantitative studies of in vitro transformation by chemical carcinogens, J. Natl. Cancer Inst., 42:867.

DiPaolo, J. A., Nelson, R. L., and Donovan, P. J., 1971, Morphological, oncogenic, and karyological characteristics of Syrian hamster embryo cells transformed in vitro by carcinogenic polycyclic hydrocarbons, Cancer Res., 31:1118.

Driedger, P. E., and Blumberg, P. M., 1980, Specific binding of phorbol ester tumor promoters, Proc. Natl. Acad. Sci. U.S.A., 77:576.

Dunphy, W. G., Delclos, K.B., and Blumberg, P. M., 1980, Characterization of specific binding of [^{3}H]phorbol-12,13-dibutyrate and [^{3}H]phorbol-12-myristate-13-acetate to mouse brain, Cancer Res., 40:3635.

Eastman, A., Sweetenham, J., and Bresnick, E., 1978, Comparison of in vitro and in vivo binding of polycyclic hydrocarbons to DNA, Chem. Biol. Interact., 23:345.

Fox, C. F., and Das, M., 1979, Internalization and processing of the EFG receptor in the induction of DNA synthesis in cultured fibroblasts: The endocytic activation hypothesis, J. Supramol. Struct., 10:199.

Frampton, R. J., Suva, 1. J., Eisman, J. A., Findlay, D. M., Moore, G. E., Moseley, J. M., and Martin, T. J., 1982, Presence of 1,25-dihydroxyvitamin D_3 receptors in established human cancer cell lines in culture, Cancer Res., 42:1116.

Fujiki, H., Suganuma, M., Matsukura, N., Sugimura, T., and Takayama, S., 1982, Teleocidin from Streptomyces is a potent promoter of mouse skin carcinogenesis, Carcinogenesis, 3:895.

Gould, M. N., Cathers, L. E., and Moore, C. J., 1982, Human breast cell-mediated mutagenesis of mammalian cells by polycyclic aromatic hydrocarbons, Cancer Res., 42:4619.

Harris, C. C., Hsu, I. C., Stoner, G. D., Trump, B. F., and Selkirk, J. K., 1978, Human pulmonary alveolar macrophages metabolise benzo(a)pyrene to proximate and ultimate mutagens, Nature (London), 272:633.

Haussler, M. R., and McCain, T. A., 1977, Basic and clinical concepts related to vitamin D metabolism and action, N. Engl. J. Med., 297:974.

Heidelberger, C., 1970, Chemical carcinogenesis, Ann. Rev. Biochem., 44:78.

Heidelberger, C., Freeman, A. E., Pienta, R. J., Sivak, A., Bertram, J. S., Casto, B. C., Dunkel, V. C., Francis, M. W., Kakunaga, T., Little, J. B., and Schechtman, L. M., 1983, Cell transformation by chemical agents--a review and analysis of the literature, Mutat. Res., 114:283.

Higginson, J., 1969, Present trends in cancer epidemiology, Proc. Can. Cancer Res. Conf., 8:40.

Higginson, J., and Muir, C. S., 1979, Environmental carcinogenesis; misconceptions and limitations to cancer control, J. Natl. Cancer Inst., 63:1291.

Hoffman, D. R., and Huberman, E., 1982, The control of phospholipid methylation by phorbol diesters in differentiating human myeloid HL-60 leukemia cells, Carcinogenesis, 8:875.

Horowitz, A. D., Greenbaum, E., and Weinstein, I. B., 1981, Identification of receptors for phorbol ester tumor promoters in intact mammalian cells and of an inhibitor of receptor binding in biological fluids, Proc. Natl. Acad. Sci. U.S.A., 78:2315.

Huberman, E., 1978, Mutagenesis and cell transformation of mammalian cells in culture by chemical carcinogens, J. Environ. Pathol. Toxicol., 2:29.

Huberman, E., Braslawsky, G. R., Callaham, M. F., and Fujiki, H., 1982, Induction of differentiation of human promyelocytic leukemia (HL-60 cell) cells by teleocidin and phorbol-12-myristate-13-acetate, Carcinogenesis, 3:111.

Huberman, E., and Callaham, M. F., 1979, Induction of terminal differentiation in human promyelocytic leukemia cells by tumor-promoting agents, Proc. Natl. Acad. Sci. U.S.A., 76:1293.

Huberman, E., Heckman, C., and Langenbach, R., 1979, Stimulation of differentiated functions in human melanoma cells by tumor-promoting agents and dimethylsulfoxide, Cancer Res., 39:2618.

Huberman, E., McKeown, C. K., Jones, C. A., Hoffman, D. R., and Murao, S.-I., 1984, Induction of mutations by chemical agents at the hypoxanthine-guanine phosphoribosyl transferase locus in human epithelial teratoma cells, Mutat. Res., 130:127.

Huberman, E., and Sachs, L., 1966, Cell susceptibility to transformation and cytotoxicity to the carcinogenic hydrocarbon benzo(a)pyrene, Proc. Natl. Acad. Sci. U.S.A., 56:1123.

Huberman, E., and Sachs, L., 1974, Cell-mediated mutagenesis of mammalian cells with chemical carcinogens, Int. J. Cancer, 13:326.

Huberman, E., and Sachs, L., 1976, Mutability of different genetic loci in mammalian cells by metabolically activated polycyclic hydrocarbons, Proc. Natl. Acad. Sci. U.S.A., 73:88.

Huberman, E., and Sachs, L., 1977, DNA binding and its relationship to carcinogenesis by different polycyclic hydrocarbons, Int. J. Cancer, 19:122.

Huberman, E., Sachs, L., Yang, S. K., and Gelboin, H. V., 1976, Identification of mutagenic metabolites of benzo(a)pyrene in mammalian cells, Proc. Natl. Acad. Sci. U.S.A., 73:607.

Huberman, E., Weeks, C., Herrmann, A., Callaham, M. F., and Slaga, T. J., 1981, Alterations in polyamine levels induced by phorbol diesters and other agents that promote differentiation in human promyelocytic leukemia cells, Proc. Natl. Acad. Sci. U.S.A., 78:1062.

IARC, 1973, International Agency for Research on Cancer Monographs, Evaluation of Carcinogenic Risk of Chemicals to Man, Vol. 3, Lyon, France.

Jones, C. A., and Huberman, E., 1980, A sensitive hepatocyte-mediated assay for the metabolism of nitrosamines to mutagens for mammalian cells, Cancer Res., 40:406.

Jones, C. A., Marlino, P. J., Lijinsky, W., and Huberman, E., 1981, The relationship between the carcinogenicity and mutagenicity of nitrosamines in a hepatocyte-mediated mutagenicity assay, Carcinogenesis, 2:1075.

Jones, C. A., Santella, R. M., Huberman, E., Selkirk, J. K., and Grunberger, D., 1983, Cell specific activation of benzo(a)pyrene by fibroblasts and hepatocytes, Carcinogenesis, 4:1351.

Kahl, G. F., Klaus, E., Legraverend, C., Nebert, D. W., and Pelkonen, O., 1979, Formation of benzo(a)pyrene metabolite-nucleoside adducts in isolated perfused rat and mouse liver and in mouse lung slices, Biochem. Pharmacol., 28:1051.

Kasukabe, T., Honma, Y., and Hozumi, M., 1979, Induction of lysosomal enzyme activities with glucocorticoids during differentiation of cultured mouse myeloid leukemia cells, Gann, 70:119.

King, H. W. S., Thompson, M. H., and Brookes, P., 1975, The benzo(a)pyrene deoxyribonucleoside products isolated from DNA after metabolism of benzo(a)pyrene by rat liver microsomes in the presence of DNA, Cancer Res., 34:1263.

Knudsen, A. G., Jr., Hetchcote, H. W., and Brown, B. W., 1975, Mutation and childhood cancer: A probabilistic model for the incidence of retinoblastoma, Proc. Natl. Acad. Sci. U.S.A., 72:166.

Koeffler, H. P., 1981, Human myelogenous leukemia: enhanced clonal proliferation in the presence of phorbol diesters, Blood, 57:256.

Kuroki, T., and Heidelberger, C., 1971, The binding of polycyclic aromatic hydrocarbons to the DNA, RNA and proteins of transformable cells in culture, Cancer Res., 31:2168.

Kuroki, T., Sasaki, K., Chida, K., Abe, E., and Suda, T., 1983, 1,25-Dihydroxyvitamin D_3 markedly enhances chemically-induced transformation in BALB 3T3 cells, Gann, 74:611.

Kuroki, T., and Sato, H., 1968, Transformation and neoplastic development in vitro of hamster embryonic cells by 4-nitroquinoline-1-oxide and its derivatives, J. Natl. Cancer Inst., 41:53.

Langenbach, R., Freed, H. J., and Huberman, E., 1978, Liver cell-mediated mutagenesis of mammalian cells by liver carcinogens, Proc. Natl. Acad. Sci. U.S.A., 75:2864.

Langenbach, R., Nesnow, S., Tompa, A., Gingell, R., and Kuszynski, C., 1981, Lung and liver cell-mediated mutagenesis systems: Specificities in the activation of chemical carcinogens, Carcinogenesis, 2:851.

Lewis, J. G., and Swenburg, J. A., 1980, Differential repair of O^6-methylguanine in DNA of rat hepatocytes and nonparenchymal cells, Nature (London), 288:185.

Lijinsky, W., 1983, Species specificity in nitrosamine carcinogenesis, in "Organ and Species Specificity in Chemical Carcinogenesis," pp. 63-75, R. Langenbach, S. Nesnow, and J. M. Rice, eds., Plenum Press, New York.

Lotem, J., and Sachs, L., 1979. Regulation of normal differentiation in mouse and human myeloid leukemic cells by phorbol esters and the mechanism of tumor promotion, Proc. Natl. Acad. Sci. U.S.A., 76:5158.

MacLeod, M. C., Levin, W., Conney, A. M., Lehr, R. E., Mansfield, B. V., Jerina, D. M., and Selkirk, J. K., 1980, Metabolism of benzo(e)pyrene by rat liver microsomal enzymes, Carcinogenesis, 1:165.

Mager, R., Huberman, E., Yang, S. K., Gelboin, H. V., and Sachs, L., 1977, Transformation of normal hamster cells by benzo(a)pyrene and diol-epoxide, Int. J. Cancer, 19:814.

Major, P. P., Griffin, J. D., Minden, M., and Kufe, D. W., 1981, A blast subclone of the HL-60 human promyeloycytic cell line, Leukemia Res., 5:429.

Manolagas, S. C., and Deftos, L. J., 1980, Studies of the internalization of vitamin D_3 metabolites by cultured osteogenic sarcoma cells and their application to a non-chromatographic cytoreceptor assay for 1,25-dihydroxy-vitamin D_3, Biochem. Biophys. Res. Commun., 95:596.

Maugh, T. H., 1979, Cancer and environment: Higginson speaks out, Science, 205:1363.

McCarthy, D., SanMiguel, J. F., Freake, H. C., Green, P. M., Zola, H., Catovsky, D., and Goldman, J. M., 1983, 1,25-dihydroxy-vitamin D_3 inhibits proliferation of human promyelocytic leukemia (HL60) cells and induces monocyte-macrophage differentiation in HL60 and normal human bone marrow cells, Leukemia Res., 7:51.

Miao, R. M., Friedstell, A. H., and Fodge, D. W., 1978, Opposing effects of tumour promoters on erythroid differentiation, Nature (London), 274:271.

Miller, E. C., Miller, J. A., and Enomoto, M., 1964, The comparative carcinogenicities of 2-acetylaminofluorene and its N-hydroxy metabolite in mice, hamsters and guinea pigs, Cancer Res., 24:2018.

Miyaura, C., Abe, E., Kuribayashi, T., Tanaka, H., Konno, K., Nishii, Y., and Suda, T., 1981, 1,25-dihydroxyvitamin D_3 induces differentiation of human myeloid leukemia cells, Biochem. Biophys. Res. Commun., 102:937.

Murao, S.-I., Gemmell, M. A., Callaham, M. F., Anderson, N. L., and Huberman, E., 1983, Control of macrophage cell differentiation in human promyelocytic HL-60 leukemia cells by 1,25-dihydroxyvitamin D_3 and phorbol-12-myristate-13-acetate, Cancer Res., 43:4989.

Nagasawa, K., and Mak, T. W., 1980, Phorbol esters induce differentiation in human malignant T lymphoblasts, Proc. Natl. Acad. Sci. U.S.A., 77:2964.

Newbold, R. F., Wigley, C. B., Thompson, M. H., and Brookes, P., 1977, Cell-mediated mutagenesis in cultured Chinese hamster cells by carcinogenic hydrocarbons. Nature and extent of the associated hydrocarbon-DNA reaction, Mutat. Res., 43:101.

O'Neill, J. P., Brimmer, P.A., Machanoff, R., Hirsch, G. P., and Hsie, A. W., 1977, A quantitative assay of mutation induction at the hypoxanthine-guanine phosphoribosyl transferase in Chinese hamster ovary cells (CHO/HGPRT system): Development and definition of the system, Mutat. Res., 45:91.

Pahlman, S., Odelstad, l., Larsson, E., Grottle, G., and Nillson, K., 1981, Phenotypic changes of human neuroblastoma cells in culture induced by 12-O-tetradecanoyl-phorbol-13-acetate, Int. J. Cancer, 28:583.

Pienta, R. J., Poiley, J. A., and Lebherz, W. B., 1977, Morphological transformation of early passage golden Syrian hamster embryo cells derived from cryopreserved primary cultures as a reliable in vitro bioassay for identifying diverse carcinogens, Int. J. Cancer, 19:641.

Reinherz, E. L., Kung, P. C., Goldstein, G., Levey, R. H., and Schlossman, S. F., 1980, Discrete stages of human intrathymic differentiation: Analysis of normal thymocytes and leukemic lymphoblasts of T-cell lineage, Proc. Natl. Acad. Sci. U.S.A., 77:1588.

Reinherz, E. L., Kung, P. C., Goldstein, G., and Schlossman, S. F., 1979, Separation of functional subsets of human T cells by a monoclonal antibody, Proc. Natl. Acad. Sci. U.S.A., 76:4061.

Reinherz, E. L., and Schlossman, S. F., 1981, Derivation of human T-cell leukemias, Cancer Res., 41:4767.

Rice, J. M., and Frith, C. M., 1983, The nature of organ specificity in chemical carcinogenesis, in "Organ and Species Specificity in Chemical Carcinogenesis," pp. 1-22, R. Langenbach, S. Nesnow, and J. M. Rice, eds., Plenum Press, New York.

Rivedal, E., and Sanner, T., 1982, Promotional effect of different phorbol esters on morphological transformation of hamster embryo cells, Cancer Lett., 17:1.

Rose, N. R., and Friedman, H., 1980, Manual of Clinical Immunology American Society for Microbiology, Washington.

Rovera, G., O'Brien, T. J., and Diamond, L., 1977, Tumor promoters inhibit spontaneous differentiation of friend erythroleukemia cells in culture, Proc. Natl. Acad. Sci. U.S.A., 74:2894.

Rovera, G., Santoli, D., and Damski, C., 1979, Human promyelocytic leukemia cells in culture differentiate into macrophage-like cells when treated with a phorbol diester, Proc. Natl. Acad. Sci. U.S.A., 76:2779.

Ryffel, B., Henning, C. B., and Huberman, E., 1982, Differentiation of human T lymphoid leukemia cells into cells that have a suppressor phenotype is induced by phorbol-12-myristate-13-acetate, Proc. Natl. Acad. Sci. U.S.A., 79:7336.

Schroff, R. W., Foon, K. A., Billing, R. J., and Fahey, J. L., 1982, Immunologic classification of lymphocytic leukemias based on monoclonal antibody-defined cell surface antigens, Blood, 59:207.

Sebti, S. M., Baird, W. M., Knowles, B. B., and Diamond, L., 1982, Benzo(a)pyrene-DNA adduct formation in target cells in a cell-mediated mutation assay, Carcinogenesis, 3:1317.

Selkirk, J. K., 1977, Benzo(a)pyrene carcinogenesis: A biochemical selection mechanism, J. Toxicol. Environ. Health, 2:1245.

Slaga, T. J., Bracken, W. M., Vaije, A., Levin, W., Yagi, H., Jerina, D. M., and Conney, A. H., 1977, Comparison of the tumor-initiating activities of benzo(a)pyrene arene oxides and diol-epoxides, Cancer Res., 37:4130.

Slaga, T. J., Fischer, S. M., Nelson, K., and Gleason, E. L., 1980, Studies on the mechanism of skin tumor promotion. Evidence for several stages in promotion, Proc. Natl. Acad. Sci. U.S.A., 77:3659.

Slaga, T. J., Sivak, A., and Boutwell, R. K., (eds.), 1978, "Carcinogenesis--a Comprehensive Survey," Raven Press, New York.

Solanki, V., Slaga, T. J., Callaham, M., and Huberman, E., 1981, The down regulation of specific binding of [20-^{3}H]phorbol-12,13-dibutyrate and phorbol ester-induced differentiation of human promyelocytic leukemia cells, Proc. Natl. Acad. Sci. U.S.A., 78:1722.

Subak-Sharpe, J. H., Burk, R. R., and Pitts, J. D., 1969, Metabolic cooperation between biochemically marked mammalian cells in tissue culture, J. Cell Sci., 4:353.

Tanaka, H., Abe, E., Miyaura, C., Kuribayashi, T., Konno, K., Nishii, Y., and Suda, T., 1982, 1α,25-Dihydroxycholecalciferol and a human myeloid leukaemia cell line (HL-60), Biochem. J., 204:713.

Tanaka, H., Abe, E., Miyaura, C., Shiina, Y., and Suda, T., 1983, 1,25-Dihydroxyvitamin D_3 induces differentiation of human promyelocytic leukemia cells (HL-60) into monocyte-macrophages, but not into granulocytes, Biochem. Biophys. Res. Commun., 117:86.

Trosko, J. E., Chang, C. C., Yotti, L. P., and Chu, E. H. Y., 1977, Effect of phorbol myristate acetate on the recovery of spontaneous and ultraviolet light-induced 6-thioguanine and ouabain-resistant Chinese hamster cells, Cancer Res., 37:188.

Umezawa, K., Weinstein, I. B., Horowitz, A., Fujiki, H., Matsushima, T., and Sugimura, T., 1981, Similarity of teleocidin B and phorbol ester tumour promoters in effects on membrane receptors, Nature (London), 290:411.

Van Duuren, B. L., 1969, Tumor-promoting agents in two stage carcinogenesis, Prog. Exp. Tumor Res., 11:31.

van Zeeland, A. A., and Simons, J. W. I. M., 1976, Linear dose-response relationship after prolonged expression times in V-79 Chinese hamster cells, Mutat. Res., 35:129.

van Zeeland, A. A., van Diggelen, M. C. E., and Simons, J. W. I. M., 1972, The role of metabolic cooperation in selection of hypoxanthine-guanine-phosphoribosyl-transferase (HGPRT)-deficient mutants from diploid mammalian cell strains, Mutat. Res., 14:355.

Wasserman, R. H., Brindak, M. E., Meyer, S. A., and Fullmer, C. S., 1982, Evidence for multiple effects of vitamin D_3 repletion to 1,25-dihydroxyvitamin D_3, Proc. Natl. Acad. Sci. U.S.A., 79:7939.

Weinstein, I. B., Wigler, M., Fisher, P. B., Sisskin, E., and Pietropaolo, C., 1978, Cell culture studies on biological effects of tumor promoters, in: "Mechanisms of Tumor Promotion and Cocarcinogenesis," pp. 313-333, T. J. Slaga, A. Sivak, and R. K. Boutwell, eds., Raven Press, New York.

Wood, A. W., Chang, R. L., Huang, M.-T., Uskokovic, M., and Conney, A. H., 1983, 1,25-Dihydroxyvitamin D_3 inhibits phorbol ester-dependent chemical carcinogenesis in mouse skin, Biochem. Biophys. Res. Commun., 116:605.

Yamasaki, H., Fibach, E., Nudel, U., Weinstein, I. B., Rifkind, R. A., and Marks, P. A., 1977, Tumor promoters inhibit spontaneous and induced differentiation of murine erythroleukemia cells in culture, Proc. Natl. Acad. Sci. U.S.A., 74:3451.

Yotti, L. P., Chang, C. C., and Trosko, J. E., 1979, Elimination of metabolic cooperation in Chinese hamster cells by a tumor promoter, Science, 206:1089.

Zeuthen, J., Norgaard, J. O. R., Avner, P., Fellous, M., Wariovaara, J., Vaheri, A., Rosen, A., and Giovanella, B. C., 1980, Characterization of a human ovarian teratocarcinoma-derived cell line, Int. J. Cancer, 25:19.

THE ROLE OF MECHANISTIC DATA IN DOSE-RESPONSE MODELING

Thomas B. Starr

Department of Epidemiology
Chemical Industry Institute of Toxicology
Research Triangle Park, NC 27709

INTRODUCTION

A critical issue in carcinogenic dose-response modeling involves the functional form of the relationship between the two distinct measures of exposure denoted by the terms administered dose and delivered dose. Administered dose refers to the external measures of exposure that are directly controlled in laboratory studies of toxicity. For inhalation studies, it typically denotes the concentration of a test chemical in the inhalation chamber air. In contrast, delivered dose refers to internal measures of exposure, such as the quantity or concentration of the biologically active form of a test chemical that is present in specific target tissue components. It is this latter measure, the delivered dose, that is presumed to be the direct causative variable in mechanistic descriptions of the carcinogenic process at the cellular and molecular levels.

When data regarding delivered dose are absent, as is usually the case, it has been customary to assume that the delivered dose/administered dose relationship is linear. In such cases the administered dose is thus implicitly taken to be a proxy measure of the unknown delivered dose, differing from it at most by some constant scaling factor. It must be kept in mind, however, that the true delivered dose/administered dose relationship reflects the entire spectrum of biological responses to *in vivo* exposure, ranging from physiologic responses of the whole animal to intra-cellular biochemical responses in target tissues. Given the remarkable complexity of these multi-level, multi-site responses, a linear delivered dose/administered dose relationship appears more likely to be the occasional exception rather than the general rule. This is especially important because low-dose extrapolations that assume linearity are known to provide risk estimates that are either excessively conservative, i.e., too high, or anti-conservative (too low) when the true relationship is nonlinear (Hoel et al., 1983).

Mechanistic studies that identify important factors in the delivered dose/administered dose relationship can therefore provide information that is of great value to dose-response modeling. For low-dose extrapolation in particular, the crux of the matter is whether or not measures of exposure in target tissue depend nonlinearly on administered dose. If the data collected in mechanistic studies do not contradict the assumption of linearity, then they imply only that dose-response models that employ administered dose as the independent variable are not in obvious error. On the other hand, if the data provide significant evidence of nonlinearity, then models based on administered dose are certain to yield biased estimates of low-dose risk; these models must be re-estimated with a representative measure of delivered dose as the independent variable if this bias is to be corrected. This point is illustrated with two compounds, vinyl chloride and formaldehyde, for which data on delivered dose are available and for which the delivered dose/administered dose relationship appears to be significantly nonlinear.

VINYL CHLORIDE

Maltoni and Lefemine (1975) reported the incidence at death of hepatic angiosarcomas among Sprague-Dawley rats exposed to vinyl chloride by inhalation for 4 hours/day, 5 days/week, for up to 52 weeks. These data are displayed versus the administered dose, i.e., airborne vinyl chloride concentration (Fig. 1). The tumor response is unusual in that it is very steep at the lower administered doses, levels off at about 2000 ppm, and finally appears to fall off at still higher concentrations.

The multistage model is commonly employed by regulatory agencies for dose-response modeling because it is ordinarily conservative, i.e., it tends to overestimate risk at low doses, even in the observable response range. If one fits the multistage model to the vinyl chloride data, its most conservative form, the one-stage model, is selected by the fitting process, but the quality of the fit is remarkably poor. Specifically, the tumor incidence at low airborne vinyl chloride concentrations is seriously underestimated, while the incidence at high vinyl chloride concentrations is seriously overestimated. The family of multistage models simply fails to provide an adequate characterization of the apparent negative curvature of tumor incidence versus administered

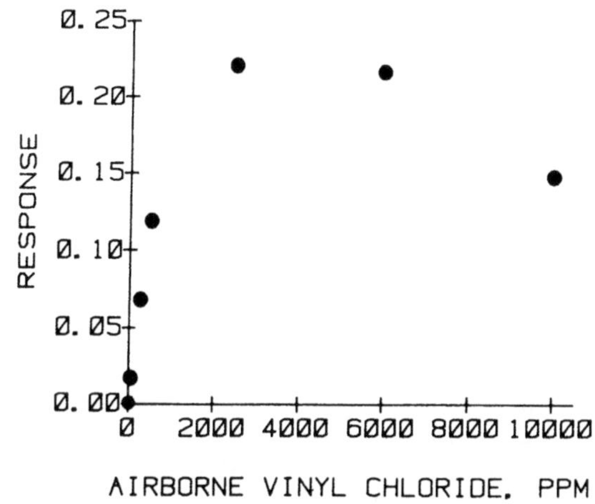

Fig. 1. Incidence of hepatic angiosarcomas (RESPONSE) among Sprague-Dawley rats versus airborne vinyl chloride concentration.

dose, and the ensuing low-dose risk estimates are demonstrably anti-conservative (too low) in the observable response range.

This problem has been resolved by modeling the incidence data with a direct measure of delivered dose as the independent variable (Reitz et al., 1980). It is now known that vinyl chloride must be metabolically activated before it can produce hepatic angiosarcoma (Bolt et al., 1975), and further, that the biotransformation of vinyl chloride in rats proceeds in accordance with nonlinear Michaelis-Menten, i.e., saturable, kinetics (Watanabe et al., 1976). These kinetics imply the existence of an upper limit on the absolute rate of production of the active metabolite responsible for tumor induction. Thus, although linear proportionality between administered and delivered doses may hold at low airborne vinyl chloride concentrations, it breaks down at sufficiently high concentrations.

The nonlinear dependence of the amount of vinyl chloride metabolized per 4-hour period upon the airborne vinyl chloride concentration is illustrated by the lower curve in Fig. 2 using the relationship established by Gehring et al. (1978). For comparison, the upper curve in Fig. 2 depicts the linear relationship implied by the assumption of first-order kinetics. It is apparent that in order to produce a unit change in the amount metabolized, a much larger change in the administered dose is required at high airborne concentrations than at low concentrations. The net effect is to stretch nonlinearly the high end of the administered dose scale relative to that of the delivered dose.

If the hepatic angiosarcoma incidence data of Fig. 1 are re-plotted against the metabolized dose given by the nonlinear curve in Fig. 2, the result is the much "better behaved" dose-response depicted in Fig. 3. It is primarily linear, with a slight amount of upward (positive) curvature in the intermediate delivered dose range. Van Ryzin and Rai (1980) have described the dramatically improved fit of the multistage model to these data when this delivered dose measure is employed as the independent variable. Most importantly, the multistage model's risk estimates at the low end of the observable response range are no longer

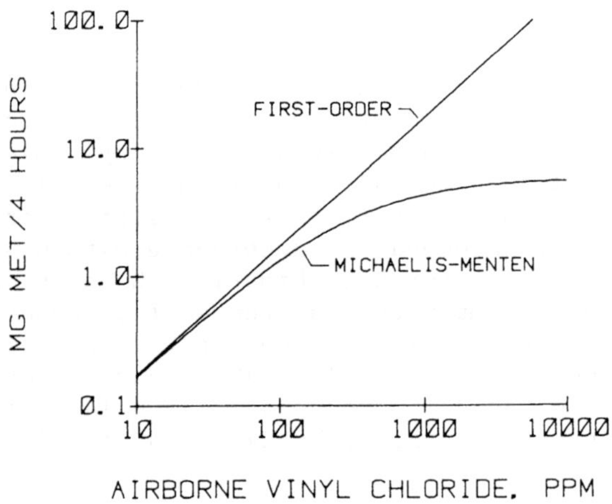

Fig. 2. Relationship between airborne vinyl chloride concentration and the amount of vinyl chloride metabolized (MG MET/4 HOURS).

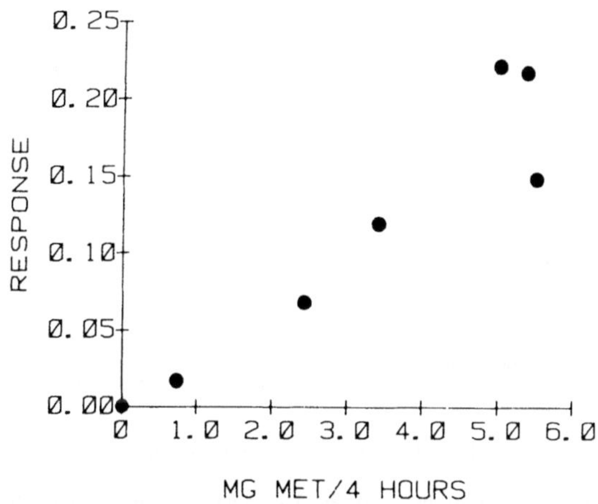

Fig. 3. Incidence of hepatic angiosarcomas (RESPONSE) among Sprague-Dawley rats verus amount of vinyl chloride metabolized (MG MET/4 HOURS).

clearly anti-conservative, but rather are consistent with the observed tumor incidence.

FORMALDEHYDE

Chronic inhalation exposure of Fischer-344 rats to sufficiently high airborne formaldehyde concentrations has been shown to induce a high incidence of nasal cavity squamous cell carcinoma, a tumor which is otherwise extremely rare in rodents (Kerns et al., 1983). The observed tumor response is plotted versus the administered dose, i.e., airborne concentration, in Fig. 4. Its shape contrasts dramatically with that for vinyl chloride (Fig. 1), being essentially flat to 6 ppm, and then rising very steeply to over 50% incidence at 14.3 ppm.

Studies conducted at the Chemical Industry Institute of Toxicology have sought to provide a mechanistic explanation for this extreme nonlinearity. We have identified several biological responses to formaldehyde exposure that appear to be important

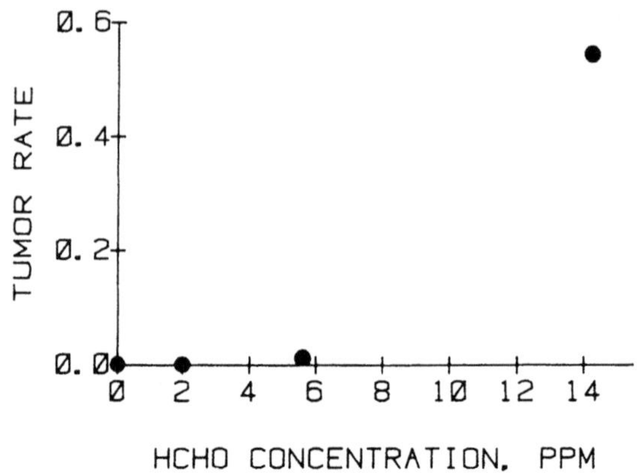

Fig. 4. Incidence of squamous cell carcinomas (TUMOR RATE) among Fischer-344 rats versus airborne formaldehyde (HCHO) concentration.

determinants of the dose delivered to target tissues in the rodent nasal cavity during in vivo inhalation exposure. These responses include the respiratory depression response to sensory irritation, the inhibition of mucociliary clearance, the cellular proliferative response to cytotoxicity, and metabolic incorporation and covalent binding of formaldehyde to intracellular macromolecules in target tissues. In the following section, results from these mechanistic studies are reviewed, focusing on the issue of nonlinearity and its implications for dose-response modeling and low-dose risk extrapolation.

Respiratory Depression Response to Sensory Irritation

Inhaled gaseous formaldehyde is a potent sensory irritant, capable of stimulating a variety of respiratory tract receptors, particularly those in the nasal cavity associated with the trigeminal nerve. An important response to the sensory irritation is a reduction in the amount of air that animals inhale per unit time. This response serves to minimize the uptake and deposition of formaldehyde, but its magnitude is dependent upon the airborne concentration. For example, rats exposed to 15 ppm formaldehyde for 6 hours inhale only twice the amount of formaldehyde per unit time as do rats similarly exposed to 6 ppm, even though 15 ppm is two and one-half times larger a concentration than 6 ppm (Swenberg et al., 1983a). This mildly nonlinear relationship between inspired formaldehyde and ambient air concentration is due to the larger depression in minute volume induced in rats by 15 ppm (about 20%) relative to that induced by 6 ppm (about 10%).

One consequence of this effect is that in order to produce a unit change in the rate at which formaldehyde is inhaled by rats, a larger change in the administered dose is required at high airborne concentrations than at low concentrations. Thus, as with vinyl chloride metabolism, the net effect is to stretch nonlinearly the high end of the administered dose scale relative to that of the inhaled dose. Consequently, the precipitously steep rise in squamous cell carcinoma incidence from 1% among rats exposed to 5.6 ppm formaldehyde to nearly 50% among rats exposed to 14.3 ppm (Fig. 4) becomes steeper yet, i.e., more severely nonlinear, when the amount of formaldehyde inhaled per unit time is used as the measure of exposure.

If minute volume depression were the only factor affecting the delivered formaldehyde dose, low-dose risk estimates based on inhaled dose would actually be increased relative to estimates based upon the linearity assumption, although the increase is less than a few percent, because the respiratory depression response is manifest in rats primarily at concentrations higher than 15 ppm. However, minute volume depression is not the only response of rodents to formaldehyde exposure.

Inhibition of Mucociliary Clearance

Under normal conditions, a flowing and continously replenished layer of mucus covers those areas of nasal respiratory epithelium in which squamous cell carcinomas were induced by chronic exposure to formaldehyde. This mucus layer consists primarily of water (95%) and mucus glycoproteins (0.5-1%). It is propelled across underlying tissues by the synchronized beating of ciliated cells, flowing ultimately to the rear of the nasal cavity where it is subsequently swallowed. The primary function of this complex apparatus is to warm and humidify inhaled air, but it also protects the nasal passages and lower airways from inhaled particulate material by trapping particles in the mucus layer until they, and the mucus itself, are swallowed.

The mucociliary clearance system may also have a significant capacity to protect underlying epithelial cells from low-level exposures to toxic gases. However, since formaldehyde is a highly reactive chemical that binds readily to proteins, it is reasonable to suppose that at least some of the inhaled formaldehyde that is deposited on the mucus surface becomes bound to mucus glycoproteins. At low formaldehyde concentrations, only a fraction of the amount deposited per unit time would be free to diffuse through the layer to underlying tissue, while most of the glycoprotein-bound formaldehyde would be removed from the nasal cavity by normal mucociliary clearance.

It has been demonstrated, however, that exposure to sufficiently high airborne formaldehyde concentrations impairs the functioning of the mucociliary apparatus in the rat nasal cavity, with sequential induction of reduced mucus flow rate, cessation of flow, and finally ciliastasis (Morgan et al., 1983). For example, exposure of rats to 15 ppm formaldehyde, 6 hours per day, for 1-9 days produced distinct areas of ciliastasis on the naso- and maxilloturbinates, with progression of this effect more deeply into the nasal cavity with inceasing days of exposure (Fig. 5). Similar exposure to 6 ppm formaldehyde produced only focal effects on mucociliary activity in anterior portions of the naso- and maxilloturbinates. In contrast, with 2 ppm small areas of ciliastasis were apparent only in some rats, and exposures to 0.5 ppm formaldehyde yielded no evidence at all of impaired mucociliary function.

At airborne formaldehyde concentrations high enough to impair this clearance mechanism, the fraction of inhaled formaldehyde that would normally be removed from the nasal cavity remains instead at the site of deposition, where it may then diffuse into underlying tissues. A unit change in the rate of formaldehyde penetration is thus likely to be achieved with a smaller change in airborne formaldehyde concentration at clearance-impairing concentrations than at lower concentrations which leave clearance unaffected. The net effect of impairment is therefore to compress nonlinearly the high end of the administered dose scale relative to that of the rate at which formaldehyde penetrates to underlying tissue.

Restorative Cell Proliferation in Response to Acute Cytotoxicity

Acute cytotoxicity is also induced in the nasal turbinates of rats by a single 6-hour exposure to 15 ppm formaldehyde, with additional exposures producing more extensive cytotoxicity. A prominent defensive response to this cell-killing is restorative cell proliferation and hyperplasia, which can be detected via the incorporation of tritiated thymidine in DNA during cell replication. For example, when the nasal passages of rats were examined after 3 days of exposure to 0.5 or 2 ppm formaldehyde, the percent of respiratory epithelial cells that were radiolabeled was no different from that observed in control rats, but similar exposure to 6 ppm yielded a dramatic 20-fold increase in this percentage (Swenberg et al., 1983b). This increased proliferation enhances

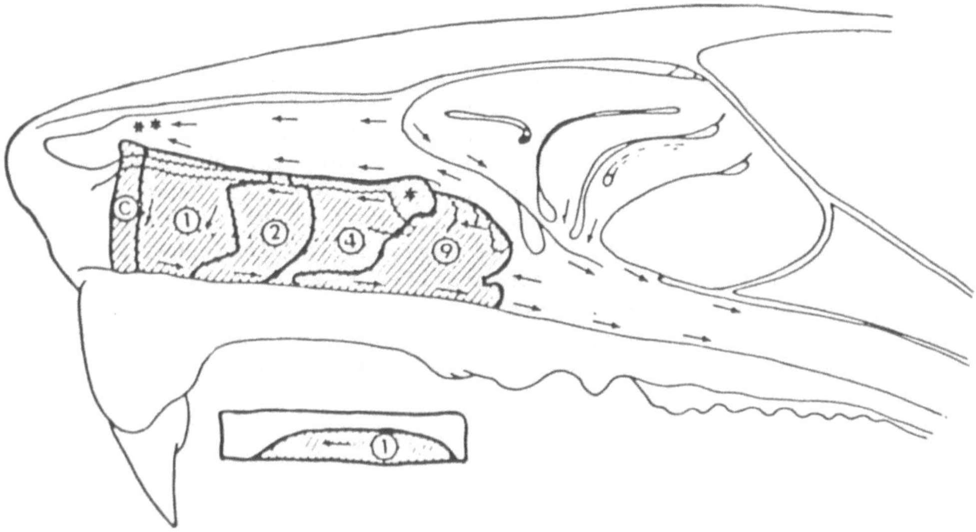

Fig. 5. Diagram of a rat head opened adjacent to the midline with the nasal septum removed to demonstrate the outline of the turbinates, part of the lateral wall and nasopharynx. The inset represents the left nasoturbinate which is placed to reveal its lateral aspect. The general direction of mucus flow is indicated by the arrows. Progressive extension of the areas of formaldehyde-induced mucostasis and partial to complete ciliastasis with increasing days of exposure (15 ppm; 1, 2, 4, or 9 days, 6 hours per day) is indicated by the cross-hatching. C indicates the anterior areas of respiratory epithelium in which mucus flow and ciliary activity are generally not apparent in controls. The medial nasoturbinate (**) clears via the nasal septum and flow from the lateral wall (*) clears via the dorsal or anterior aspect of the maxilloturbinates. From Swenberg et al. (1983a).

the likelihood of irreversible genotoxic events by increasing the number of single-stranded DNA sites at which formaldehyde may covalently bind, and also by decreasing the amount of time available for the repair of such lesions.

There is no question that cytotoxicity and subsequent cell proliferation are indicative of the penetration of formaldehyde to target cells. However, several problems exist with the use of proliferation itself as a quantitative measure of the delivered formaldehyde dose. First, when exposure was increased from 6 to 15 ppm formaldehyde, the percentage of labeled cells actually decreased, although it remained well above control levels as is shown in Fig. 6. One can speculate that the toxicity at this high concentration was so acute as to interfere with the proliferative process. Second, no increase in proliferation above control levels was detected at 0.5 or 2 ppm. This could mean that a threshold formaldehyde concentration exists for the induction of proliferation, but it could also be explained by inadequate experimental sensitivity. Third, since proliferation is a biological process, it could be argued that it is simply a nonlinear biological response to a linear delivered dose. For these reasons, the cellular proliferative response to formaldehyde exposure provides a less than ideal quantitative measure of the delivered dose.

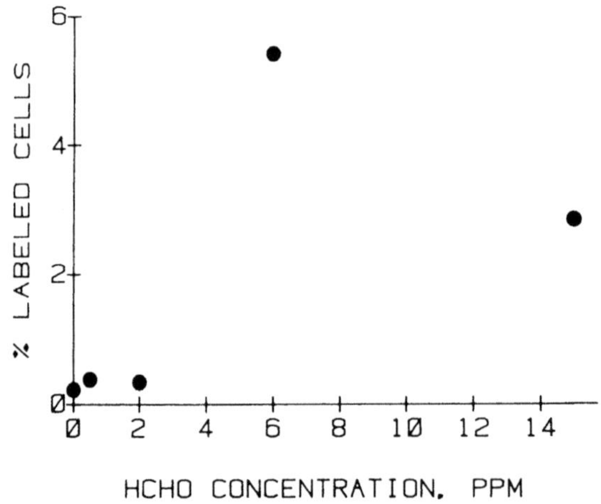

Fig. 6. Percent respiratory epithelial cells labeled with [^{3}H]-thymidine versus airborne formaldehyde (HCHO) concentration.

Metabolic Incorporation and Covalent Binding in Target Tissues

While the data regarding respiratory depression, mucociliary clearance, and restorative proliferation indicate that the delivered dose in target tissues is likely to depend nonlinearly on airborne formaldehyde concentration, they provide only indirect evidence regarding this issue. Recent biochemical studies have provided important information on the disposition of formaldehyde in rat nasal cavity tissues, including the first direct quantitative measurements of the amount of formaldehyde that reaches the DNA of target tissues. These studies are of critical importance for several reasons. First, they provide quantitative data which demonstrate that the delivered dose/administered dose relationship is nonlinear, as would be anticipated from consideration of the observed spectrum of effects of formaldehyde on minute volume, mucociliary clearance, and cell proliferation. Second, these studies also provide evidence that metabolic incorporation, a process by which delivered formaldehyde is detoxified, is less efficient at high airborne concentrations than at low concentrations. Thus, a removal pathway that provides protection from formaldehyde toxicity at low airborne concentrations appears to be inhibited at concentrations greater than 2 ppm. Third, the data for covalent binding of formaldehyde to target tissue DNA are in a form that makes it possible to reanalyze easily the chronic bioassay tumor incidence with this delivered formaldehyde dose as the measure of exposure.

Early studies in this series provided the first evidence that inhaled formaldehyde interacts directly with respiratory mucosal DNA (Casanova-Schmitz and Heck, 1983). Exposure of rats to formaldehyde concentrations of 6 ppm and greater resulted in statistically significant increases in the percent of respiratory mucosal DNA that appeared in an interfacial layer when tissue homogenates were extracted with a strongly denaturing aqueous immiscible organic solvent mixture. This interfacial DNA could not be extracted from proteins without enzymatic digestion. These data, plotted versus airborne formaldehyde concentration in Fig. 7, provide presumptive evidence for the formation of DNA-protein crosslinks, but only at formaldehyde concentrations of 6 ppm and greater. The measurements at 0.3 and 2 ppm were not significantly different from control levels. Furthermore, the data for all concentrations were not inconsistent with the hypothesis of a purely linear relationship between this measure of delivered dose and the airborne formaldehyde concentration, as is apparent from the close agreement between the observations and the best-fitting straight line that is also displayed in Fig. 7.

Subsequent studies were undertaken to investigate the mechanisms of labeling of respiratory mucosal DNA following inhalation exposure to [^{14}C]- and [^{3}H]-labeled formaldehyde (Casanova-

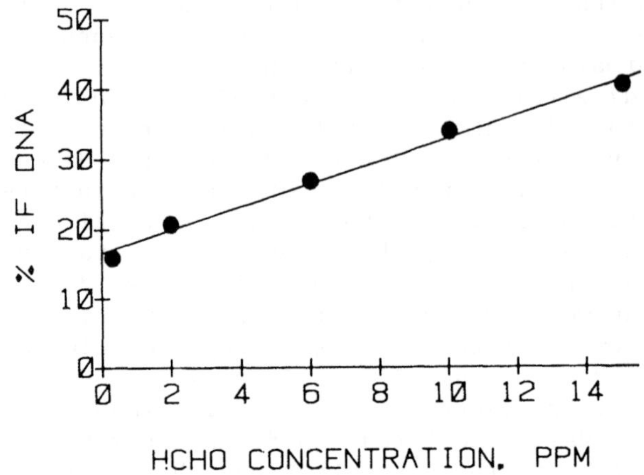

Fig. 7. Percent respiratory mucosal DNA appearing in the interfacial fraction (% IF DNA) when tissue homogenates were extracted with a strongly denaturing aqueous immiscible organic solvent mixture plotted versus airborne formaldehyde (HCHO) concentration.

Schmitz et al., 1984). These studies demonstrated that the [^{3}H] to [^{14}C] isotope ratio in the interfacial DNA fraction increased with increasing airborne formaldehyde concentration, as is shown in Fig. 8. They thus provided additional evidence for the covalent binding of formaldehyde to target tissue DNA. However, as with the interfacial DNA fraction, these data are not inconsistent with the hypothesis of a linear relationship between the isotope ratio in interfacial DNA and airborne formaldehyde concentration.

Very recently, however, it was discovered that when the data for interfacial DNA fraction were analyzed in combination with similar measurements for the aqueous DNA fraction, then it was possible to calculate the actual amount of formaldehyde that binds covalently to respiratory mucosal DNA (Casanova-Schmitz et al., 1984). These observations, expressed as nanomoles of covalently bound formaldehyde per milligram of respiratory mucosal DNA, are displayed in Fig. 9. Covalent binding increases gradually between 0.3 and 2 ppm, very steeply between 2 and 6 ppm, and less

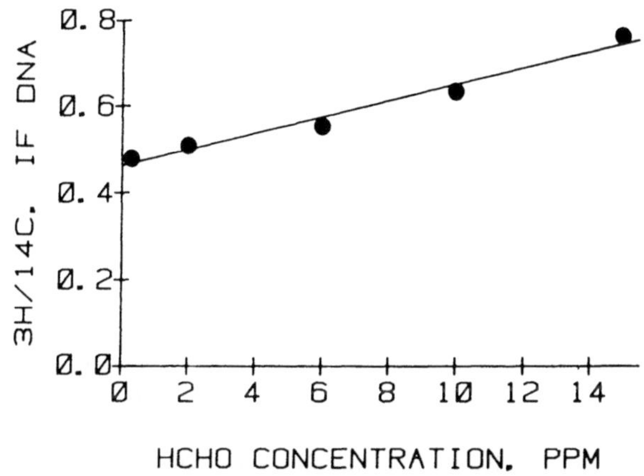

Fig. 8. [^{3}H] to [^{14}C] isotope ratio in the interfacial fraction (IF) of respiratory mucosal DNA versus airborne formaldehyde (HCHO) concentration.

steeply again at higher concentrations. This nonlinear relationship implies that in order to produce a unit change in the amount of formaldehyde covalently bound to DNA, a smaller change in the administered dose is required at airborne concentrations above 2 ppm than at concentrations below 2 ppm. The net effect is to compress nonlinearly the high end of the administered dose scale relative that of the dose delivered to respiratory mucosal DNA.

The straight line in Fig. 9 shows the result to be expected if covalent binding were a linear function of concentration that passed from the origin through the observation at 6 ppm. The binding measurement at 2 ppm is of critical importance. First, it is significantly different from zero, so binding was indeed detected. Second, the amount of binding was also significantly less, by about a factor of three, than would be predicted by linear extrapolation from the observations at higher concentrations. The disparity between the observations at 0.3 and 2 ppm and the assumption of linearity is even more apparent when these data are shown on a log-log plot which expands the lower end of both dose measure scales (Fig. 10). The upper straight line in

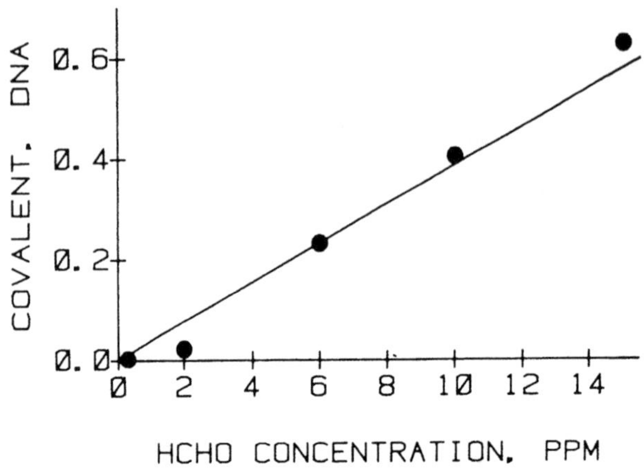

Fig. 9. Amount (NM/MG DNA) of formaldehyde covalently bound to respiratory mucosal DNA versus airborne formaldehyde (HCHO) concentration.

this figure is the same as the one in Fig. 9, passing from zero through the observation at 6 ppm. The observations at 0.3 and 2 ppm clearly fall well below the predictions based upon linearity.

Implications for Dose-Response Modeling and Low-Dose Risk Estimation

In order to assess the significance of these findings for dose-response modeling and risk estimation, the tumor incidence observed in the chronic bioassay has been reanalyzed using both covalent binding and airborne formaldehyde concentration as measures of exposure (Starr and Buck, 1984). Four commonly used dose-response models, the probit, logit, Weibull, and multistage, were employed. Model parameters were estimated using standard maximum likelihood techniques. Maximum likelihood estimates and their upper 95% confidence bounds were calculated for airborne formaldehyde concentrations ranging from 0.1 to 1 ppm.

It was assumed that the delivered dose/administered dose relationship was linear in this range, and given by the lower straight line shown in Fig. 10, which passes from zero through

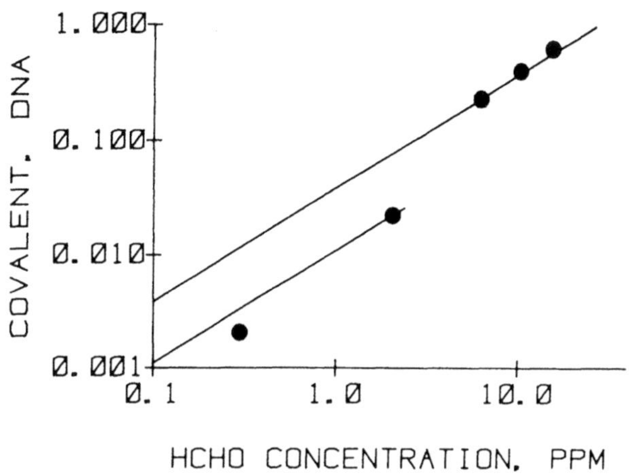

Fig. 10. Amount (NM/MG DNA) of formaldehyde covalently bound to respiratory mucosal DNA versus airborne formaldehyde (HCHO) concentration with both axes logarithmic. The upper line passes from the origin through the observation at 6 ppm. The lower line passes from the origin through the observation at 2 ppm.

the amount of covalent binding observed at 2 ppm. Since the binding observed at 0.3 ppm falls slightly below this line, this low-dose linearity assumption is likely to overestimate the amount of covalent binding that actually occurs at concentrations below 2 ppm.

The maximum likelihood estimates of risk that were obtained at 1 ppm are presented in Table 1 with powers of 10 indicated in parentheses. These estimates ranged over 16 orders of magnitude across models, with the probit model being the least conservative and the multistage model the most conservative. More importantly, however, the estimates obtained with covalent binding as the measure of exposure are unilaterally lower than the corresponding estimates obtained with airborne formaldehyde concentration. Risk reduction factors, i.e., the ratios of administered dose based risk estimates to corresponding delivered dose based estimates appear in the third row. These reduction factors ranged from a low of 53 for the multistage model to more than 8 orders of magnitude for the probit model.

Estimates obtained at 0.1 ppm are shown in Table 2. Again, there is a unilateral reduction when the risk estimates are derived with covalent binding as the measure of exposure. The only exception is for the probit model estimates, which were both too small, less than 1×10^{-26}, to be calculated with any real precision. The calculable risk reduction factors ranged from a low of 35 for the Weibull model to over 53 for the multistage.

Table 1. Maximum Likelihood Risk Estimates Associated with Exposure to 1 ppm Formaldehyde

Dose Measure	Model			
	Probit	Logit	Weibull	Multistage
Administered	2.65(-11)	2.87(- 6)	5.94(- 6)	2.51(- 4)
Delivered	4.00(-20)	2.15(- 8)	7.13(- 8)	4.70(- 6)
Reduction Factor	6.60(+ 8)	133.0	83.4	53.4

Table 2. Maximum Likelihood Risk Estimates Associated with Exposure to 0.1 ppm Formaldehyde

Dose Measure	Model			
	Probit	Logit	Weibull	Multistage
Administered	< 1.00(-26)	3.92(-11)	2.20(-10)	2.51(- 7)
Delivered	< 1.00(-26)	7.40(-13)	6.20(-12)	4.70(- 9)
Reduction Factor	NC	53.0	35.4	53.4

NC: not calculable

Similar results were obtained for upper 95% confidence bounds on these risk estimates. Table 3 shows these upper bounds for 1 ppm formaldehyde. Again, the range of estimates across models is enormous, over 16 orders of magnitude. However, there is also a unilateral reduction in the bounds when they are derived with covalent binding as the measure of exposure. Upper confidence-bound reduction factors, shown in the third row, range from a low of 2.9 for the multistage to more than 8 orders of magnitude for the probit.

Finally, Table 4 displays the upper bound results obtained for 0.1 ppm. Both probit model bounds were too small to be cal-

Table 3. Upper 95% Confidence Bounds on Risk Estimates Associated with Exposure to 1 ppm Formaldehyde

Dose Measure	Model			
	Probit	Logit	Weibull	Multistage
Administered	2.58(-10)	1.24(- 5)	2.54(- 5)	1.80(- 3)
Delivered	7.09(-19)	1.22(- 7)	3.98(- 7)	6.24(- 4)
Reduction Factor	3.63(+ 8)	101.8	63.8	2.9

Table 4. Upper 95% Confidence Bounds on Risk Estimates Associated with Exposure to 0.1 ppm Formaldehyde

Dose Measure	Model			
	Probit	Logit	Weibull	Multistage
Administered	< 1.00(-26)	2.84(-10)	1.57(- 9)	1.56(- 4)
Delivered	< 1.00(-26)	6.19(-12)	5.12(-11)	6.19(- 5)
Reduction Factor	NC	46.0	30.6	2.5

NC: not calculable

culated meaningfully. The other model bounds ranged over more than 8 orders of magnitude. For these models, however, there was again a unilateral reduction in the estimates derived with covalent binding. The calculable upper-bound reduction factors ranged from a low of 2.5 for the multistage model to about 46 for the logit model.

SUMMARY

The results from employing metabolized dose in the case of vinyl chloride demonstrate that the delivered dose concept can radically alter the shape of the dose-response curve in the observable response range. Simultaneously, it can eliminate the obvious bias in low-dose risk estimates that arises from use of an inappropriate dose measure, namely, the administered airborne vinyl chloride concentration.

For formaldehyde, the data regarding covalent binding to respiratory mucosal DNA provide the best measure of target tissue exposure that is currently available. The results obtained with this dose measure demonstrate that incorporation of the delivered dose concept into any of the commonly used low-dose extrapolation procedures leads to a unilateral reduction in estimated cancer risk associated with the exposure of rats to low airborne formaldehyde concentrations. These findings strongly suggest that low-dose risk estimates obtained with the administered formaldehyde dose as the measure of exposure are biased conservatively, and that regulatory assessments of formaldehyde should consider the delivered formaldehyde dose to be the valid and relevant measure of exposure.

Additional research is, of course, required to refine and elaborate the delivered dose concept, especially for human exposures. Nevertheless, this concept provides a convenient vehicle for incorporating mechanistic information directly into the quantitative risk assessment process in a meaningful and relevant manner. Clearly, risk estimates based on delivered dose reflect what is known of the underlying biological reality more faithfully than estimates that are based solely on unverified assumptions and the findings in chronic bioassays. Use of the delivered dose concept should be strongly encouraged, since it can help place quantitative risk assessment on a sound and scientifically defensible footing.

REFERENCES

Bolt, H. M., Kappus, H., Kaufmann, R., Appel, K. E., Buchter, A., and Bolt, W., 1975, Metabolism of [^{14}C]-vinyl chloride *in vitro* and *in vivo*, *Inserm.*, 52:151.

Casanova-Schmitz, M., and Heck, H. d'A., 1983, Effects of formaldehyde exposure on the extractability of DNA from proteins in the rat nasal mucosa, *Toxicol. Appl. Pharmacol.*, 70:121.

Casanova-Schmitz, M., Starr, T. B., and Heck, H. d'A., 1984, Differentiation between metabolic incorporation and covalent binding in the labeling of macromolecules in the rat nasal mucosa and bone marrow by inhaled [^{14}C]- and [^{3}H]formaldehyde, *Toxicol. Appl. Pharmacol.*, 76:26.

Gehring, P. J., Watanabe, P. G., and Park, C. N., 1978, Resolution of dose-response toxicity data for chemicals requiring metabolic activation. Example: vinyl chloride, *Toxicol. Appl. Pharmacol.*, 44:581.

Hoel, D. G., Kaplan, N. L., and Anderson, M. W., 1983, Implication of nonlinear kinetics on risk estimation in carcinogenesis, *Science*, 219:1032.

Kerns, W. D., Pavkov, K. L., Donofrio, D. J., and Swenberg, J. A., 1983, Carcinogenicity of formaldehyde in rats and mice after long-term inhalation exposure, *Cancer Res.*, 43:4382.

Maltoni, C., and Lefemine, G., 1975, Carcinogenicity assays of vinyl chloride: Current results, *Ann. N.Y. Acad. Sci.*, 246:195.

Morgan, K. T., Patterson, D. L., and Gross, E. A., 1983, Formaldehyde and the nasal mucociliary apparatus, *in*: "Formaldehyde: Toxicology, Epidemiology, Mechanisms," pp. 193-209, J. J. Clary, J. E. Gibson, and R. S. Waritz, eds., Marcel Dekker, New York.

Reitz, R. H., Quast, J. F., Schumann, A. M., Watanabe, P. G., and Gehring, P. J., 1980, Non-linear pharmacokinetic parameters need to be considered in high dose/low dose extrapolation, *in*: "Quantitative Aspects of Risk Assessment in Chemical Carcinogenesis," *Arch. Toxicol.*, Suppl. 3:79.

Starr, T. B., and Buck, R. D., 1984, The importance of delivered dose in estimating low-dose cancer risk from inhalation exposure to formaldehyde, *Fundam. Appl. Toxicol.*, 4:740.

Swenberg, J. A., Barrow, C. S., Boreiko, C. J., Heck, H. d'A., Levine, R. J., Morgan, K. T., and Starr, T. B., 1983a, Nonlinear biological responses to formaldehyde and their implications for carcinogenic risk assessment, Carcinogenesis, 4:945.

Swenberg, J. A., Gross, E. A., Randall, H. W., and Barrow, C. S., 1983b, The effect of formaldehyde exposure on cytotoxicity and cell proliferation, in: "Formaldehyde: Toxicology, Epidemiology, Mechanisms," pp. 225-236, J. J. Clary, J. E. Gibson, and R. S. Waritz, eds., Marcel Dekker, New York.

Van Ryzin, J., and Rai, K., 1980, The use of quantal response data to make predictions, in: "The Scientific Basis of Toxicity Assessment," pp. 273-290, H. R. Witschi, ed., Elsevier/North Holland, New York.

Watanabe, P. G., Hefner Jr., R. E., and Gehring, P. J., 1976, Fate of [^{14}C]-vinyl chloride following inhalation exposure in rats, Toxicol. Appl. Pharmacol., 37:49.

DISCUSSION

Portier: You said you would discuss the goodness-of-fit. Did the use of an effective dose improve the fit?

Starr: It looked a little bit better, but not enough to make a statistically significant difference. Furthermore, since you cannot categorically reject the models when you begin, you certainly cannot reject them when they do a better job. With respect to "goodness-of-fit", however, I do have some problems. You know that when you assess the fit of the tolerance distribution models, you must make assumptions about the asymptotic normality of the parameter estimates. That is equivalent to assuming that the sample size for the bioassay is big. It is sufficiently big that the estimates of the parameter are approximately normally distributed about their true values. We have done some work to indicate that that is not true for this data set for any of the tolerance distribution models. We get multimodal distributions of parameter estimates, i.e., cumulative distributions that are decidedly not normal, so the goodness-of-fit test, as based on chi-square with one degree of freedom is not valid. Futhermore, if you use the three degrees of freedom that are there (there is no degree of freedom for the control group because there is no response there), you have not got any degrees of freedom left if you estimate an alpha, a beta, and an additive or independent background parameter, so goodness-of-fit cannot be assessed.

On the other hand, the point should be made that when a model does not fit the data, it is usually symptomatic of something other than a poor model in just that respect. It is an indication, perhaps, of non-linearity in some biological process; that is the point I made with respect to Dr. Land's talk. The flattening-out of a dose-response curve is not accomodated by these models, so we look elsewhere for an explanation.

Gough: In the last few years, some attention has been directed to time-to-tumor models, but they have not been mentioned today. Are they so much less reliable or useful that they should be ignored?

Starr: My impression of them is that normally there is not sufficient information routinely collected in bioassays to implement them. One of the critical factors is, did the pathologist specify the cause of death? I think, in effect, it has been shown that the risk estimates are not very different if you make lifetime exposures and try to assess the effects over a long period of exposure. Taking into account time-to-tumor does not appear to make much difference.

Hughes: I wanted to bring to your attention the result of a workshop which was held about two years ago by the Society of Toxicology and the National Center for Toxicologic Research. At that meeting we attempted to get at some consensus of understanding of what was happening with the ED_{01} Study at low dose-response exposures. I want to illustrate to you the impact time-to-tumor on the shape of the response curve. The effect of time-to-tumor was evaluated in the ED_{01} study via interim sacrifice at 9, 19, 16, 18, 24 and 34 months. The accompanying 3-dimensional projections of the bladder and liver responses to 2-AAF are updates of the figure presented in ED_{01} Workshop publication (D. Hughes et al., Fundamental and Applied Toxicology, 3, May/June 1983).

Painter: You gave two sets of DNA adduct data where a statistical difference between treated and controls could not be shown. However, it was found that if the one nonstatistically different datum was divided by the other nonstatistically different datum, one could find a statistical difference from controls. Can you tell us how this was done?

Starr: That is a good question, and it is not entirely a simple matter to explain. The observations of isotope ratios in aqueous and interfacial fractions were made simultaneously on the same animals. If we had not paired the observations, we would not have been able to subtract them and declare the difference statistically significant. The null hypothesis requires that the aqueous and interfacial fractions have the same isotope ratio in the same animal, and when we do a paired analysis, we find a significant difference at 2 parts per million. It is not a simple point to clarify for non-statisticians, and it has been a bone of contention between us and several regulatory agencies. However, a paper has been accepted by Toxicology and Applied Pharmacology and will appear soon. It includes an appendix that gives details of exactly what statistical procedures we used and performs the calculation with raw data for the 2 ppm group, which is the most important one.

Peto: I want to comment on Dr. Hughes' data for the AAF and ED_{01} study. I think the dose-response relationship that you obtained is rather unusual in the flatness of the apparent response at very low doses. We have just done a large nitrosamine study, the results of which will appear in the Proceedings of the Banff Conference on Nitrosamines. What we found is an approximately cubic dose-response relationship at high doses, which changed into a linear dose-response relationship (i.e., one of siple proportionality) as we moved down to lower doses. It appears as though the risk is related to the cube of the total effective dose, where the "total effective dose" is defined to equal the applied dose plus a small background dose. The small background dose makes a negligible contribution at high doses, so the relationship appears cubic, but the fact that there is a background dose means that the relationship appears linear at low doses. I think that the apparent completely null effects that you observed may be an exception rather than a rule. Perhaps if you could plot the total risk against the total amount reaching the bladder, you would not get that kind of shape. For example, for the liver, you would not get absolute apparent abolition of the risk. The key thing in the downwards extrapolation of all dose-response relationships is the acceptance that in many cases, there may be some background (Peto 1978a). This conclusion even applies to the probit model. If instead of plotting a probit relationship to the applied dose, you add some background dose, where the background is itself a parameter to be estimated, then you would get linearity at low doses from any model. It does not matter whether the mathematical formula of the model is cubic, probit, or linear. As long as you do not, by assumption, exclude background, you will get linearity at least of the upper confidence limits of risk with respect to dose at low doses. It is, to my mind, a very strong and very implausible biological assumption that statisticians make when, in their mathematicl formulae, they say, "I know for certain that there is nothing else like this carcinogen in the whole of the universe". It is implicit in the fitting of risk against the probit of dose, or of risk against the square of dose, with no background.

Starr: No, I disagree. This has reminded me of a point about goodness-of-fit that I wanted to make before. We have become very uncomfortable about using asymptotic normality to estimate upper bounds on risk from the probit, logit and other tolerance distribution models because for sample sizes comparable to this rather large bioassay, the parameter estimates are not normal. Their distribution estimated by simulation are decidedly not normal, so we cannot use asymptotic normality. So we looked at the results of an analysis similar to the one made by Crump (1977) for the multi-stage models (whose parameter estimates are also not normal when the true values are close to the edges of the feasible space). We tried to estimate an additive background upper bound with their constrained approach,--to force the additive background to be big, and re-optimize the other parameters. When we got to 100,000 parts per million, the log-likelihood had not changed by more than a tenth of one percent. With just two parameters, the tolerance distribution models are flexible enough to describe most any data, and when you estimate an additive parameter as well, you have given them license to kill. You cannot get away from a near-optimal description of the data. there is simply not enough data in the bioassay to tell you anything about the upper bound on an additive background component. I can concede low-dose additivity in principle, but I find it very difficult to believe that any bioassay like this one, which was well-done, tells me anything about the slope of the dose-response at very low doses.

Peto: I think you are misunderstanding my question. What I am saying, and what I have explained at more length in my Environmental Health Perspectives paper (Peto, 1978b), is that you cannot by any kind of complicated mathematical model whatsoever justifiably prove that linearity is excluded by any conceivable set of data; some of your models were doing this. You were putting dose-reduction factors up in the millions.

Starr: Yes, but that was at one part per million. I think you are talking about additive background levels on the order of hundredths of a part per million or less, -- parts per trillion. I concede that there the dose-response could be linear. However, I do not believe that at one part per million there is an additive background component that is of any consequence.

Schneiderman: I also have a question about linearity. Do we expect linearity against dose or against logarithm of dose? Your plots are all against the logarithm of dose. Is that the appropriate measure?

Starr: Yes, when the risk is plotted on the same logarithmic scale. Linear functions without an interceptive linear on a log/log plot as well as on a linear/linear plot.

REFERENCES

Crump, K.S., Guess, H.A., and Deal, K.L., 1977, Confidence intervals and test of hypotheses concerning dose/response relations inferred from animal carcinogenicity data, Biometrics, 33:437.

Peto, R., 1978a, Control of industrially induced cancers, Env. Health Persp., 22:153.

Peto, R., 1978b, Carcinogenic effects of chronic exposure to low levels of toxic substances, Env. Health Persp., 22:155.

STATISTICAL ASPECTS OF THE ESTIMATION OF HUMAN RISKS

Charles C. Brown

Biostatistics Branch
National Cancer Institute
Bethesda, Maryland 20205

Quantitative risk assessment requires extrapolation from results of experimental assays conducted at high dose levels to predicted effects at lower dose levels which correspond to human exposures. The meaning of this high to low dose extrapolation within an animal species will be discussed, along with its inherent limitations. A number of commonly used mathematical models of dose response necessary for this extrapolation will be discussed and I will comment on the limitations in their ability to provide precise quantitative low-dose risk estimates. These constraints include: the existence of thresholds; incorporation of background, or spontaneous responses; and modification of the dose response by pharmacokinetic processes.

As the serious long-range health hazards of environmental toxicants have become recognized, the need has arisen to quantitatively estimate the effects upon humans exposed to low levels of these toxic agents. Often inherent in this estimation procedure is the necessity to extrapolate evidence observed under one set of conditions in one population or biological system to arrive at an estimate of the effects expected in another population under a different set of conditions. Since this extrapolation from one situation to another is often based on incomplete knowledge, derived risk estimates involve a number of uncertainties. The purpose of this report is to describe the major sources of these uncertainties and to demonstrate their potential magnitude. The discussion will be limited to those uncertainties involved in extrapolation from high to low doses; extrapolation from one species to another (e.g., animals to humans) are another major source of risk estimation uncertainty but will not be discussed here.

By necessity, the potentially deleterious effects of chemical compounds often must be tested in laboratory animals which are exposed to levels higher than those for which the risk estimation is to be made. Some consideration has been given to the possibility of conducting extremely large experiments at very low dose levels. Use of large numbers of experimental subjects reduces statistical error so that very small effects can be adequately quantified. However, as Schneiderman et al. (1975) remark, "purely logistical problems might guarantee failure." Therefore, to obtain reliably measureable effects, the experimental information must be based on levels of exposure high enough to detect positive results. Since large segments of the human populations are often exposed to much lower levels, these high exposure level data must be extrapolated to lower levels of exposure.

The high to low dose extrapolation problem is conceptually straight forward. The probability of a toxic response is modeled by a dose-response function P(D) which represents the probability of a response when exposed to D units of the toxic agent. A general mathematical model is chosen to describe this functional relationship, its unknown parameters are estimated from the available data, and this estimated dose-response function P(D) is then used either to: (1) estimate the response at a particular low dose level; or (2) estimate that dose level corresponding to a desired low level of response (this dose estimate is commonly known as the virtually safe dose, VSD). One of the major difficulties inherent in this high to low dose extrapolation problem is that the estimates of risk at low doses, and correspondingly, the estimates of VSD's for low response levels, are highly dependent upon the mathematical form assumed for the underlying dose response.

Mathematical models of dose response can be categorized into three classes: (1) tolerance distribution models; (2) models based on "hit" theory; and (3) models based on quantitative theories of carcinogenesis. Tolerance distribution models are based on the following assumptions that individuals in the population have their own "tolerance" for exposure to the toxic agent, such that no response will occur if the level of exposure is below their tolerance and response will occur if the level is above their tolerance; individuals in the population have different tolerances. This class of models assumes a deterministic process relating exposure to response, and that the dose-response relationship is produced by the distribution of tolerances within the population. Small population variation leads to a steep dose response while large variation produces a shallow dose response. Examples of models in this class are the log normal, or probit, model (Gaddum, 1933; Bliss, 1935; Finney, 1971) and the log logistic model (Worcester and Wilson, 1943; Berkson, 1944).

Another class of models are those which are based on the "hit theory" for the interaction of toxic molecules with susceptible biological targets (Turner, 1975). In general, this theory rests upon a number of postulates, which include: the organism has some number, M, of "critical targets" (usually assumed to be infinitely large); the organism responds if m or more of these critical targets are "destroyed"; a critical target is destroyed if it is "hit" by k or more toxic particles; and, the probability of a hit in the low dose region is proportional to the dose level of the toxic agent. Some commonly used cases of this general theory are the single-hit model, where the subject responds if a single critical target is destroyed by a single hit, and the multihit model, where the subject responds if a single critical target is destroyed by k hits. For a discussion of the single-hit model as applied to the high to low dose extrapolation problem, see National Academy of Sciences (1980). The Report of the Scientific Committee of the Food Safety Council (1978) and Rai and Van Ryzin (1979) discuss the application of the multihit model for dose-response extrapolation.

Other mechanistic models have also been derived from quantitative theories of carcinogenesis. The multistage carcinogenesis theory which assumes that a single cell can generate a malignant tumor only after it has undergone a certain number of heritable changes leads to the multistage model (Armitage and Doll, 1961). The use of this model for extrapolation has been described by Brown (1978) and Guess and Crump (1978). The multicell carcinogenesis theory of Fisher and Holloman (1951) which assumes that a tumor arises from a clone of cells each of which has undergone a single cellular change leads to a dose-response function having extrapolation characteristics similar to the multihit model. This model has also been termed the Weibull model and Van Ryzin (1980) discusses its application.

It might be thought that sufficient information for selection of one particular model over the others would be provided by the observed dose response. However, this is often not the case, as many dose-response models appear similar to one another over the range of observable response rates. Tables 1 and 2 compare the dose-response relationships of the more commonly used models; Table 1 compares the log normal, log logistic and single-hit models; Table 2 compares the multihit, Weibull and multistage models. If the estimated dose response is to be used to predict the expected response rate from an exposure level within the range of observable rates, then the models within each of the two sets will give similar results. However, extrapolation to exposure levels expected to give very low response rates is highly dependent upon the choice of model, as shown in the lower portions of Tables 1 and 2. These tables extend the dose response in the upper portions to much lower dose levels. The further one extrapolates from the observable response range, the more divergent the models become. At a dose level which is 1/1000 of the dose giving a 50% response, the single-hit model gives an estimated response rate 200 times that of the log normal model, and the multistage model gives an estimated response rate over 200 times that of the multihit model.

Table 1. Comparison of Dose-Response Relationships over Range of Observable and Extrapolated Response Rates; Log normal, Log logistic, Single-hit Models

Dose Level	Percent Responders		
	Log normal	Log logistic	Single hit
16	98%	96%	100%
8	93	92	99
4	84	84	94
2	69	70	75
1	50	50	50
1/2	31	30	29
1/4	16	16	16
1/8	7	8	8
1/16	2	4	4
1/100	5×10^{-2}	4×10^{-1}	7×10^{-1}
1/1000	4×10^{-4}	3×10^{-2}	7×10^{-2}
1/1000	1×10^{-7}	2×10^{-3}	7×10^{-3}

Table 2. Comparison of Dose-Response Relationships over Range of Observable and Extrapolated Rates; Multihit, Weibull, and Multistage Models

Dose Level	Percent Responders		
	Multihit	Weibull	Multistage
4	99%	99%	100%
3	96	97	98
2	85	85	85
1	50	49	46
0.75	36	35	33
0.50	21	21	21
0.25	7	8	9
0.01	1×10^{-2}	7×10^{-2}	3×10^{-1}
0.001	1×10^{-4}	2×10^{-3}	3×10^{-2}
0.000	1×10^{-6}	7×10^{-5}	3×10^{-3}

Table 3 provides a comparison of the behavior of these models when applied to the dose response of liver hepatomas in mice exposed to various levels of DDT (Tomatis et al., 1972). This example shows that each of the six dose-response models fit the observed data nearly equally well. Therefore, the data cannot discriminate among these models. Based on the goodness-of-fit statistics, the Weibull model fits the best (P = 0.22), but not significantly better than any of the other models. However, there is a substantial difference among the VSD's estimated from these models; the log normal model estimates a VSD over 3000 times as large as the single-hit model, leaving the true VSD open to wide speculation. The fact that an experimental study conducted at exposure levels high enough to give measurable response rates cannot clearly discriminate among these various models, along with the fact that those models show substantial divergence at low exposure levels present one of the major difficulties for the problem of low dose extrapolation.

One would naturally think that since many experimental dose-response studies are conducted with a limited number of animals at each dose level (usually on the order of 100 or fewer) over a range of response rates on the order of 10% - 90%, this problem of wide variation in the VSD might be reduced by testing more animals and using lower dose levels. However, that this will not necessarily be the case is demonstrated by the "megamouse" study of dietary exposure of 30 ppm - 150 ppm 2-acetylaminofluorene (2-AAF).

Table 3. Comparison of Virtually Safe Doses (VSD) Leading to an Excess Risk of 10^{-6} for Various Dose-Response Extrapolation Models

Extrapolation Model	VSD* (ppm DDT in daily diet)	Goodness-of-fit Statistic of Model to Observed Data χ^2	(d.f.)	P-value
Log normal	6.8×10^{-1}	3.93	(2)	0.14
Weibull	5.0×10^{-2}	3.01	(2)	0.22
Multihit	1.3×10^{-2}	3.31	(2)	0.19
Log logistic	6.6×10^{-3}	3.45	(2)	0.18
Multistage	2.5×10^{-4}		-------**	
Single hit	2.1×10^{-4}	5.10	(3)	0.16

* 97.5% lower confidence limit on VSD

** no goodness-of-fit statistic since the number of parameters equals the number of data points

Table 4. Virtually Safe Doses (VSD) for 2-AFF Based on Multi-Stage and Log Normal Models applied to Different Dose Level Combinations

Dose Levels Used (ppm)	Bladder Neoplasms Multistage	Bladder Neoplasms Log normal	Liver Neoplasms Multistage	Liver Neoplasms Log normal
	VSD (ppm)*			
0, 60 - 150	3.07×10^{-2}	34.4	3.50×10^{-4}	4.84×10^{-1}
0, 45 - 150	4.12×10^{-2}	34.5	3.82×10^{-4}	5.27×10^{-1}
0, 35 - 150	4.48×10^{-2}	34.5	3.99×10^{-4}	4.03×10^{-1}
0, 30 - 150	3.63×10^{-2}	34.5	4.32×10^{-4}	3.84×10^{-1}

* 97.5% lower confidence limit on VSD leading to an excess risk of 10^{-6}

The occurrence of bladder and liver neoplasms after 24 months on study were examined to see if the addition of data at dose levels giving low response rates would lead to a reduction in the variation of the VSD estimates (Farmer et al., 1980). Two extrapolation models, the multistage and the log normal, were applied to these data in a series of calculations. In each case, both models fit the observed data very well. First, the VSD's leading to an excess risk of 10^{-6} are estimated using the controls and the four highest dose groups, 60 - 150 ppm, then the VSD's are reestimated by adding the next lower dose, one at a time. These VSD estimates are shown in Table 4.

These results show that the inclusion of additional low dose data has little effect on the VSD estimates. For bladder neoplasms, which show a highly curvilinear dose response, addition of data from 3 low doses increases the lower confidence limit on the VSD from the multistage model only 18% (from 3.07×10^{-2} to 3.63×10^{-2}), while that based on the log normal model is hardly changed. For liver neoplasms, which show a nearly linear dose-response, the additional data increases the lower confidence limit on the VSD only 23% for the multistage model and decreases the VSD confidence limit for the log normal model. The differences in the VSD estimates from these two extrapolation models is little affected by these additional low dose data: for bladder neoplasms, the additional data decreases the difference from a log normal/multistage ratio of 1120 ($34.4/3.07 \times 10^{-2}$) to 950 ($34.5/3.63 \times 10^{-2}$); for liver neoplasms, this ratio is reduced from 1380 to 890. Therefore, these additional low-dose data, based on substantial numbers of animals have little effect on the VSD estimates for a particular extrapolation model, and, more importantly, have little effect on reducing the variation in VSD estimates between different models.

Each of these mathematical dose-response models have assumed that the responses of the subjects are due solely to the applied stimuli. However, many studies show clear evidence that responses can occur even at a zero dose. Thus, any mathematical dose-response function should properly allow for this natural, or "background," responsiveness. Two methods have been proposed to incorporate the background response. The first is commonly known as 'Abbott's correction' which is based on the assumption of an independent action between the stimulus and the background (Abbott, 1925). The second method assumes that the dose acts mechanistically in the same manner as the background environment, such that the dose of the toxic agent can be considered as being "added" to a background dose (Albert and Altschuler, 1973).

It is often difficult to discriminate between the independent and additivity assumption on the basis of dose-response data (Brown, 1983). Hoel (1980) compares low dose risk extrapolations based on the two assumptions applied to a log normal dose-response model (Table 5). This table clearly shows the low-dose linearity of the additive assumption, and the substantial difference between the additive and independence assumptions at low-dose levels. Hoel also examined models which incorporate a mixture of independent and additive background response, and found that low-dose linearity prevails except when the background mechanism is totally independent of the dose-induced mechanism.

Table 5. Excess Risk P(D)-P(0) for Log Normal Dose-Response Model assuming Independent and Additive Background

Dose (D)	Type of Background: Independent	Additive
10^{0}	4.0×10^{-1}	4.0×10^{-1}
10^{-1}	1.5×10^{-2}	5.2×10^{-2}
10^{-2}	1.6×10^{-5}	5.2×10^{-3}
10^{-3}	3.8×10^{-10}	5.1×10^{-4}
10^{-4}	1.8×10^{-16}	5.1×10^{-5}

Another source of uncertainty in risk extrapolation lies with the measure of dose which is used in these mathematical models of dose response. Commonly used is the "environmental" dose, the level of the toxic agent to which humans or experimental animals are exposed. However, quantitative theories of carcinogenesis lead to mathematical models which are functions of the "effective" dose, the level of the "active" toxic agent at its site of action in the body. The many biochemical interactions between the foreign substance and components of the body can substantially modify this dose-response relationship (Hoel et al., 1983).

A critical problem in the application of pharmacokinetic principles to risk extrapolation is the potential change in metabolism or other biochemical reactions as external exposure levels of the toxic agent decrease. Linear pharmacokinetic models are often used. However, there are numerous examples of nonlinear behavior in the dose range studied, and these nonlinear kinetics pose significant problems for quantitative extrapolation from high to low doses if the kinetic parameters are not measured. Linear kinetics assume that the reaction rate per unit time of a chemical reaction is proportional to the concentration of the substance being acted upon. Nonlinear kinetics is most often described in the form of a Michaelis-Menten expression which for low concentrations is linear, while for high concentrations is often referred to as "saturable" kinetics.

If all processes are linear, then the concentration rate of the toxic substance at its site of action ("effective dose") will be proportional to the external exposure rate ("administered dose"). However, saturation phenomena may produce different results depending upon the processes affected; if elimination and/or detoxification pathways are saturable, then the effective dose will increase more rapidly with the administered dose than linear kinetics would suggest. If the distribution and/or activation pathways are saturable, then the effective dose will increase less rapidly with the administered dose. These simplified pharmacokinetic models may provide more realistic explanations of observed nonlinear dose-response relationships than other dose-response models currently in use.

Pharmacokinetic models involving nonlinear kinetics of the Michaelis-Menten form have the important extrapolation characteristic of being linear at low dose levels. This low dose linearity contrasts with the low dose nonlinearity of the multihit and Weibull models. Each model, pharmacokinetic, multihit, and Weibull, has the desirable ability to describe either convex (upward curvature) or concave (downward curvature) dose-response relationships. Other models, such as the log normal or multistage, are not consistent with concave relationships. However, the pharmacokinetic model differs from the multihit and Weibull in that it does not assume the nonlinear behavior observed at high dose levels will necessarily correspond to the same nonlinear behavior at low dose levels.

Brown (1984) gives two examples of how nonlinear pharmacokinetics may modify a dose-response relationship; one example is based on vinyl-chloride induced liver angiosarcomas in rats where the metabolic activation of vinyl chloride shows nonlinear kinetics; the second example is based on urethane-induced lung adenomas in mice where the excretion of urethane shows nonlinear kinetics. In both cases the dose response is nearly linear when "effective" dose is used, and substantially nonlinear when "environmental" dose is used.

Other sources of uncertainty in high to low dose extrapolation include: (1) the possible existence of thresholds; (2) heterogeneity of sensitivity to the toxic agent among members of the exposed population; and (3) different exposure patterns for individuals in the exposed population. The mathematical models discussed earlier do not assume a threshold value below which no exposure-induced response is possible. The assumption of a single threshold for the entire exposed population can be incorporated into these models in order to estimate a clearly safe level of exposure for all members of the population. However, as shown by Brown (1976), its estimation is subject to large statistical uncertainties. Heterogeneity among individuals in their sensitivity to the toxic agent and with regard to concomitant exposure to other environmental risk factors could lead to modification of the dose response. A large degree of heterogeneity will lead to a shallow dose-response relationship which implies larger low dose risks than for a steep dose response.

For situations of long-term chronic exposure to a toxic agent, the relationship of risk to the rate and duration of exposure is often of importance when estimating risk for different exposure situations. The multistage theory of carcinogenesis predicts that cancer risk is dependent upon the dose rate and duration of exposure, but not necessarily leading to a relationship with total dose, the product of rate and duration. Whittemore (1977) and Day and Brown (1980) discuss these multistage theories and indicate that the risk of cancer is likely to be the product of two different functions of dose rate and duration. Besides being a function of both dose rate and duration, the multistage theory also predicts that cancer risk may be a function of the age at which exposure first begins and the amount of time following cessation of exposure. For example, exposure at a young age to a carcinogen affecting an early stage (i.e., an initiator) is predicted by the theory to have a greater effect on future cancer risk than the same exposure at a later age. The converse is predicted for exposure to a carcinogen which affects a late stage in the process. Therefore, the multistage theory of carcinogenesis predicts another level of complexity in extrapolating cancer risk from one exposure situation to another, since a limited duration (e.g., 10 years) of exposure to the same dose rate will not necessarily produce the same excess cancer risk in two otherwise identical individuals whose exposure period is during different ages of their life.

Summary and Conclusions

The preceding paper has discussed the general problem of high dose to low dose extrapolation within a single animal species. The purpose of this extrapolation is to estimate the effects of low level exposure to toxic agents known to be associated with undesired effects at high dose levels. Mathematical models of dose response are necessary for this extrapolation process since the low dose effects, expected to be on the order of response rates of 10^{-6}, are too small to be accurately measured with limited study sample sizes. A number of mathematical dose-response models have been proposed for extrapolation purposes; we have shown how similar they can appear to one another in the range of observable response rates, yet how different they become at lower, unobservable response rates, the region of primary interest. This is the single, most important limitation of this extrapolation methodology. An estimate of risk at a particular low dose, or an estimate of the dose leading to a particular level of risk is highly dependent upon the mathematical form of the presumed dose response and differences of 3 - 4 orders of magnitude are not uncommon. Therefore, all these sources of uncertainty, (1) dose-response model, (2) pharmacokinetic behavior of the toxic agent, (3) thresholds, (4) heterogeneity, and (5) patterns of exposure, lead to substantial uncertainties in estimates of risk based on high to low dose extrapolations.

References

Abbott, W. S., 1925, A method of computing the effectiveness of an insecticide, J. Econ. Entomol., 18:265.

Albert, R., and Altschuler, B., 1973, Considerations relating to the formulation of limits for unavoidable population exposures to environmental carcinogens, in: "Radionuclide Carcinogenesis," p. 233, J. Ballou, R. Busch, D. Mahlum and C. Sanders, eds., AES Symposium Series CONF-720505 NIIS, Springfield, Virginia.

Armitage, P., and Doll, R., 1961, Stochastic models for carcinogenesis, in: "Proceedings of the Fourth Berkeley Symposium on Mathematical Statistics and Probability, Vol. 4," p. 19, J. Neyman, ed., University of California Press, Berkeley and Los Angeles, California.

Berkson, J., 1944, The application of the logistic function to bioassay, J. Am. Stat. Assoc., 39:134.

Bliss, C. I., 1935, The calculation of the dosage-mortality curve, Ann. Appl. Biol., 22:134.

Brown, C. C., 1976, Mathematical aspects of dose-response studies in carcinogenesis - the concept of thresholds, Oncology, 33:62.

Brown, C. C., 1978, Statistical aspects of extrapolation of dichotomous dose response data, J. Natl. Cancer Inst., 60:101.

Brown, C. C., 1983, Learning about toxicity in humans from studies on animals, CHEMTECH, 13:350.

Brown, C. C., 1984, High to low dose extrapolation of experimental animal carcinogenesis studies, in: "Proceedings of the Twenty-ninth Conference on the Design of Experiments in Army Research, Development and Testing," U.S. Army Research Office, Research Triangle Park, North Carolina (in press).

Day, N. E., and Brown, C. C., 1980, Multistage models and primary prevention of cancer, J. Natl. Cancer Inst., 64:977.

Farmer, J. H., Kodell, R. L., Greenman, D. L., and Shaw, G. W., 1980, Dose and time response models for the incidence of bladder and liver neoplasms in mice fed 2-acetylamino-fluorene continuously, J. Environ. Pathol. Toxicol., 3:55.

Finney, D. J., 1971, "Probit Analysis," Cambridge University Press, London.

Fisher, J. C., and Holloman, J. H., 1951, A new hypothesis for the origin of cancer foci, Cancer, 4:916.

Gaddum, J. H., 1933, Methods and biological assay depending on a quantal response, in: "Medical Research Council, Special Report Series No. 183," London.

Guess, H. A., and Crump, K. S., 1978, Best-estimate low-dose extrapolation of carcinogenicity data, Environ. Health Perspect., 22:149.

Hoel, D. G., 1980, Incorporation of background response in dose-response models, Fed. Proceed., 39:67.

Hoel, D. G., Kaplan, N. L., and Anderson, M. W., 1983, Implications of nonlinear kinetics on risk estimation in carcinogenesis, Science, 219:1032.

National Academy of Sciences, 1980, "The Effects on Populations of Exposure to Low Levels of Ionizing Radiation," National Academy of Sciences, Washington, D.C.

Rai, K. and Van Ryzin, J., 1979, Risk assessment of toxic environmental substances using a generalized multi-hit dose response model, in: "Energy and Health," p. 99, N. Breslow and A. Whittemore, eds., SIAM, Philadelphia, Pennsylvania.

Schneiderman, M. A., Mantel, N., and Brown, C. C., 1975, From mouse to man -- or how to get from the laboratory to Park Avenue and 59th Street, Ann. N.Y. Acad. Sci., 246:237.

Scientific Committee, Food Safety Council, 1978, Proposed system for food safety assessment, Food and Cosmet. Toxicol., 16, Supplement 2:1.

Tomatis, L., Turusov, V., Day, N., and Charles, R. T., 1972, The effects of long term exposure to DDT on CF-1 mice, Int. J. Cancer, 10:489.

Turner, M., 1975, Some classes of hit theory models, Math. Biosci., 23:219.

Van Ryzin, J., 1980, Quantitative risk assessment, J. Occup. Med., 22:321.

Whittemore, A. S., 1977, The age distribution of human cancers for carcinogenic exposures of varying intensity, Am. J. Epidemiol., 106:418.

Worcester, J. and Wilson, E. B., 1943, The determination of LD50 and its sampling error in bioassay, Proc. Natl. Acad. Sci. USA, 29:79.

DISCUSSION

Ricci: I suggest we should separate dose-response models on either a biological basis (for example, multistage v. hit models) or a deterministic basis (such as tolerance distribution). Results should not be shown in the same graph or table, as these classes of models differ fundamentally. In other words, let us not mix oranges with bananas.

Brown: I do not know enough cancer biology in order to eliminate from any consideration tolerance distribution models and I would argue that I do not think that anybody in this room can eliminate such models. I can visualize a tolerance-type model might be applicable when the carcinogenic process for a particular chemical works in some indirect manner. There may be other situations and I would be afraid to eliminate any of these models from consideration.

Brown, S.: What is the source of convexity in single-hit models at low dose?

Brown: None of these models are strictly linear at low dose. The single-hit model is convex at low doses, but it is very closely linear. I don't know how to express this without going into epsilons and deltas and complicated mathematics. But it is not strictly linear at low doses, none of these models are; it is very, very, very close.

Portier: I have two comments and one point that I would like you to expand upon. My first comment is that when you talked about additional low-dose information, the fact that it did not change the point estimate of virtually safe dose is not really a justification for not including additional low doses. In fact, what you probably wanted to look at were variances, or not the estimates themselves. Clearly, the mathematics would support the use of only two doses for estimation. But when dealing with small numbers of animals and trying to minimize the variability of the estimates, then those low doses may play an important role. The second point pertains to the question about whether we should combine tolerance distribution models with multistage models. I would say that even if we only look at the subset of the multistage model, the Weibull model, there will still be order of magnitude differences. Even in that small class of models, supposedly of the same type, there are very large discrepancies. The question I would like you to expand upon is, do you think that models which have a significant lack-of-fit should be excluded when we have models with an acceptable fit?

Brown: Let me comment on your first comment, that had to do with the variability in comparing the additional information at low doses. Those numbers that I presented in Table 4 are confidence limits on virtually safe doses. So they do incorporate this variability. I agree with your second comment completely, and I think with just the multistage model and the Weibull model there is this large level of uncertainty. The third comment concerned what to do when a model does not fit. Well, if you believe as I do that goodness of fit is meaningless because what happens at the high dose is not necessarily indicative of what to expect at the low dose, then there is no sense in using goodness of fit. Upper bounds are about all you can do right now with the current state of knowledge about mechanisms of cancer. As far as I am concerned, all you need is to have one dose in a bioassay, fit a conservative model--just draw a straight line basically between the one dose and the background--and that will give you an upper bound.

Portier: I agree, and that was what I was hoping you would say. I would again call upon the biologist and toxicologist, and stress the point that they need to decide what model is appropriate, not the mathematician. I think it is a biological issue; we can fit any number of models.

Bailar: Several biological mechanisms are thought to cause dose-response curves for carcinogens to bend downward as doses increase; these include heterogeneity in the target population, saturation of activating mechanisms, cell killing or sterilization, and non-independent competing risks. When these effects can be demonstrated at high and moderate doses, do we have reason to think that they do not occur at low doses? If not, where does that leave us with respect to using the one-hit model as a "conservative" tool for assessing risks?

Brown: The only way I know of to get a dose-response curve that is concave at low doses, and I emphasize the words "low doses," is to have a situation which the population at risk is very heterogeneous. For example, if there is a population at very, very high risk and a population at low or moderate risk, that situation can generate this kind of dose-response. But this is the only way I know of, and to me it is a little far-fetched. At high doses, that is a very different situation--we are concerned here about the action of low doses.

Starr: I take issue with the notion that upper bound is an upper bound for humans. It may very well be an upper bound for rats and mice under the conditions of a bioassay, but I really do not think we know that it is an upper bound for any other species under any other conditions.

Brown: I was hoping to get the sense of that across, by talking about all the different sources of uncertainty. You can bound the model uncertainty with the linear-at-low-dose models but I do not know about the other sources of uncertainty that I mentioned. For example, the homogeneity versus the heterogeneity would compromise the upper bound when extrapolating from animals to humans. Another example is the source of uncertainty involved in different exposure patterns. I was trying to suggest that the upper bound or linear-at-low-dose models is simply an upper bound on the dose-response in that particular species. What that has to do when extrapolating across species is completely unknown.

Starr: I would like to make one other comment. I disagree with you with respect to the goodness-of-fit criteria. If a model fails to describe the data in the observable response range, that usually is symptomatic of some problem with the dose-response that has been observed. It may be leveling off as a function of exposure, it may be going up too steeply as a function of exposure, or it may very well be an indication of nonlinear pharmacokinetics.

Haynes: I think it is a bit hazardous to rule out of court the idea of the existence of population heterogeneity, and perhaps generate a concave curve at low doses. When you consider the large number of genes that might confer DNA repair deficiency (there is about 100 in yeast, and we do not know how many in humans) and if you consider that individuals heterozygous for these genes might be at a risk of cancer, then there may be small numbers of individuals in the population whose "probabilistic risks" would be greater than expected on the basis of linear extrapolation.

Brown: More reason for the biologists to help us statisticians out with the problems we have!

Gough: Five years ago David Gaylor (National Center for Toxicologic Research) gave a seminar at the Office of Technology Assessment. He discussed the ED_{01} ("mega-mouse") study (Gaylor, 1980), and expressed all the reservations and caveats that you did today. He also asked for help from biologists in model construction. Has anything changed in the last five years?

Brown: Not as far as I am concerned. I honestly do not think that it will change in my lifetime. I am a real pessimist, I'm afraid.

Land: I question the use of a particular extrapolation rule "because it is conservative." This isn't always a bad idea because sometimes it can lead to a simple answer. Thus, if an estimated excess risk is negligible according to a "conservative" rule, it is easy to conclude that the exposure in question is acceptable. Similarly, if an unacceptable excess is estimated using a model that is "conservative" in the opposite direction, again the decision is easy. But if neither of these events obtains, should the ultimate decision about acceptibility of the exposure depend upon the preference of the statistician? There may, for example, be significant penalties associated with avoiding exposure.

Brown: I think you are bringing up the point that there is a lot of politics involved in risk assessment: you are not saying those words, but I am interpreting it that way. There have to be people making decisions about the scientific knowledge that they have, which is often very limited, and what one can do with that scientific knowledge. One cannot simply plug these bioassay results into a computer and come out with an answer without any thought going into it.

Trosko: To illustrate how the biology of carcinogenesis, using the initiation/promotion model, should help understand mathematical models, I wish to convey the observations related to polybrominated biphenyl as a tumor promoter.

My colleagues and I have initiated rats with DEN. We then exposed the animals to either 2,4,5,2',4',5'-hexabromobiphenyl (which is non-cytotoxic) or 3,4,5,3',4',5'-hexabromobiphenyl (a cytotoxic chemical). The 2,4,5,2',4',5'-HBB was a powerful tumor promoter, whereas, at non-cytotoxic levels, 3,4,5,3'4'5'-HBB was not a tumor promoter. However, at slightly cytotoxic levels, the 3,4,5,3'4',5'-HBB was a weak/moderate promoter. At higher doses of 3,4,5,3',4',5'-HBB, because of the greater cytotoxicity, was not as effective. A possible explanation is that, since the number of initiated cells was the same in all animals, the 2,4,5,2',4',5'-HBB promoted "all" the potential initiated cells by a non-cytotoxic mechanism. The 3,4,5,3',4',5'-HBB, at low cytotoxicity, promoted some of the initiated cells by "compensatory" hyperplasia (that is, forcing the surviving initiated cells to regenerate the necrotic tissue). At higher cytotoxicity, more of the initiated cells are killed, and cannot contribute to the compensatory hyperplasia.

REFERENCES

Gaylor, D.W., 1980, The ED_{01} Study: Summary and Conclusions, J. Environ. Path. Toxicol. 3:179.

THE ISOLATION AND CHARACTERIZATION OF THE BLYM-1 TRANSFORMING GENE

Alan Diamond[1], Joan M. Devine[1], Mary-Ann Lane[2] and Geoffrey M. Cooper[1]

[1]Laboratory of Molecular Carcinogenesis
Harvard Medical School
Boston, Massachusetts 02115

[2]Laboratory of Molecular Immunobiology
Dana-Farber Cancer Institute and Dept. of Pathology
Harvard Medical School
Boston, Massachusetts 02115

INTRODUCTION

The study of the genetic and molecular events which are involved in tumor development has been stimulated by the observation of discrete genes with oncogenic potential. Such genes have been detected in the genomes of acutely transforming retroviruses which induce tumors in susceptible hosts within a relatively short latent period (reviewed by Cooper, 1982). The mutation or deletion of such genes from the retroviral genome results in their loss of tumorigenicity. In addition, the ability of subgenomic fragments of retroviral DNA to induce the morphological transformation of recipient cells by transfection has provided further evidence for their role in neoplastic diseases (Anderson et al., 1979; Blair et al., 1980; Chang et al., 1980; Copeland et al., 1980; Barbacid, 1981). Sequences homologous to retroviral transforming genes have been detected in normal cellular DNA, suggesting that these sequences were acquired during the evolution of the viruses (Bishop, 1981). The first evidence that cellular DNA from neoplasms could efficiently transform recipient cells was reported by Shih et al. in 1979 who showed that the high molecular weight DNA of chemically transformed mouse fibroblasts could transform other mouse cells by transfection. Since this observation, the DNAs of approximately 50% of the tumors and tumor cell lines tested were capable of transforming mouse cells (Cooper, 1982). Many of a wide variety of neoplasms tested contained an activated transforming gene homologous to the ras gene, the transforming sequence of the Kirsten and Harvey murine sarcoma viruses (Cooper and Lane, 1984). In contrast, activation of other transforming genes, such as Blym-1, occurs consistently in tumors of a specific cell type. Blym-1 has been shown to be activated in B-cell lymphomas of both chicken and man (Goubin et al., 1983; Diamond et al., 1983).

THE BLYM-1 TRANSFORMING GENE ACTIVATED IN CHICKEN BURSAL LYMPHOMAS

Chicken bursal lymphomas are B-cell neoplasms which arise after a long latent period following infection with avian lymphoid leukosis virus (LLV). These viruses are retroviruses that do not contain their own transforming gene. In approximately 90% of LLV-induced lymphomas, the LLV genome integrates in the vicinity of the chicken c-myc gene resulting in enhanced c-myc expression (Hayward et al., 1981). High molecular weight DNA of LLV-induced lymphomas induces the transformation of mouse NIH 3T3 cells although the transforming gene detected by this assay was not linked to either LLV or c-myc sequences (Cooper and Neiman, 1980; 1981). These results indicated the activation of a second cellular gene in these lymphomas which was detectable by transfection.

The activated transforming gene detected by transfection of chicken bursal lymphoma DNA was isolated by sib-selection (Goubin et al., 1983). A recombinant phage library was generated from the DNA of NIH 3T3 cells transformed by the DNA of the RP9 chicken bursal lypmhoma cell line, and pools of 20,000 phage were screened by transfection to determine which phage contained the activated transforming gene. Positive pools were subdivided into smaller pools and assayed by transfection. This process was continued until a single biologically active phage was obtained. The gene contained within this phage was designated chicken Blym-1 and its complete nucleotide sequence was determined (Goubin et al., 1983). The chicken Blym-1 gene codes for a small protein consisting of 65 amino acids that shows significant homology to the transferrin family of proteins. Transferrins are a group of large iron binding proteins that have evolved apparently by gene duplication resulting in homologous amino- and carboxy-terminal halves (Macgilliay et al., 1977); chicken Blym-1 shows homology to an amino-terminal region within each half (Goubin et al., 1983). Transferrins show a distinct pattern of amino-acid conservation in this region and this pattern is also conserved in chicken Blym-1. This pattern of homology is also maintained by the human Blym-1 gene (as we will discuss).

Hybridization experiments revealed that chicken Blym-1 hybridized to a small gene family of normal and lymphoma chicken DNA as well as to human DNA (Goubin et al., 1983). The presence of this gene family in human DNA indicated that it is well conserved throughout vertebrate evolution and led us to speculate that sequences homologous to chicken Blym-1 were activated in human lymphomas. We examined Burkitt's lymphoma as it represents a B-cell neoplasm at the same stage at differentiation as chicken bursal lymphomas.

THE BLYM-1 TRANSFORMING GENE OF HUMAN BURKITT'S LYMPHOMA

In order to assess the transforming activity of Burkitt's lymphoma DNA, DNA was obtained from six Burkitt's lymphoma cell lines (Raji, BJAB, Namalwa, CW678, EW 36 and MC116) and analyzed by transfection (Diamond et al., 1983). All six DNAs transformed mouse NIH 3T3 cells with efficiences ranging from 0.17 to 1.0 foci per µg of DNA. DNA from transformants was, in turn, capable of inducing foci in secondary rounds of transfection. Restriction enzyme sensitivity studies were performed to determine if the detected transforming gene was the same in all six cell lines. The transforming activity of the DNA of each of the six cell lines was inactivated by digestion with the restriction endonuclease Bam H1 but not by digestion with Hind III, Xho I or Eco R1, providing evidence that in each case the same gene was activated.

The transforming gene activated in Burkitt's lymphomas was isolated from a recombinant phage library prepared from the DNA of the CW678 cell line using chicken Blym-1 as a hybridization probe (Diamond et al., 1983). This lambda clone transformed mouse cells with an efficiency of 2-3 x 10^4 transformants per pmole of DNA. Furthermore, restriction enzyme sensitivity studies revealed the same pattern of sensitivity as that observed for total Burkitt's lymphoma DNA indicating that the transforming gene (human Blym-1) isolated from a recombinant phage library was the same gene detected by transfection of genomic DNA.

The biologically active region of the lambda clone containing the human Blym-1 gene was localized to a 0.95 kb fragment (Diamond et al., 1983). Southern blot hybridization studies indicated that this fragment also contained the chicken Blym-1 homologous sequences. Thus, the biologically active transforming gene and the chicken Blym-1 homologous sequences reside on the same small fragment of DNA. Further hybridization experiments using this Eco R1 fragment revealed that human Blym-1 was not homologous to any of the viral transforming genes myc, ras^H, ras^K, fms, erb, myb, abl, src, sis, fes, mos or rel. Similarly, human Blym-1 is not homologous to the Tlym transforming gene isolated from mouse T-cell lymphomas (Lane et al., 1984).

In order to obtain information about the genomic organization of human Blym-1 in normal and tumor DNA, the 0.95 kb Eco R1 fragment was hybridized to normal human fibroblast DNA and to the DNA of six Burkitts lymphoma cell lines (Diamond et al., 1983). In each case, human Blym-1 hybridized to a single 5.0 kb Hind III fragment indicating that activation of this gene is probably not accompanied by gross gene rearrangements. In addition, the similar staining intensity of the 5.0 kb Hind III band in normal and tumor DNA suggested that activation is not a result of gene amplification. The detection of this 5.0 kb Hind III fragment by hybridization of human Blym-1 to the DNA of mouse cells transformed by Burkitt's lymphoma DNA but not to normal mouse DNA showed that transformants had indeed acquired the human Blym-1 gene.

As a first step in determining the mechanism of activation of human Blym-1, the biologically active 0.95 kb Eco R1 fragment was subcloned into pBR322 and the entire nucleotide sequence determined (Diamond et al., 1984). The sequence contained signals associated with eucaryotic promoters followed by an open reading frame interrupted by a single intervening sequence. The reading frame was terminated by the UGA stop codon and was followed by the polyadenylation signal AATAAA. The open reading frame could code for a small protein of 58 amino acids that showed 33% homology to chicken Blym-1. The predicted human Blym-1 gene product showed significant homology ($p<.005$) to the same amino-terminal region of transferrins as did chicken Blym-1 suggesting that the observed amino acid homology among chicken and human Blym-1 and transferrins reflects some common functional property. Transferrins have been implicated as having growth-related properties sometimes independent of their ability to transport iron (Haynes et at., 1981; Sutherland et al., 1981; Trowbridge and Domingo, 1981; Trowbridge and Omary, 1981; Trowbridge et al., 1982; Trowbridge and Lopen, 1982; Taetle et al., 1983). It is therefore intriguing that transforming genes such as the Blym-1 genes show homology to this family of proteins.

The observed nucleotide homology between the human and chicken Blym-1 genes was only 50-60% with several regions of approximately 20 nucleotides in length showing 70% homology (Diamond et al., 1984). Chicken Blym-1 hybridized more strongly to sequences in human DNA other than human Blym-1 (Goubin et al., 1983) which implies that, although human and chicken Blym-1 belong to the same gene family, they represent rather divergent members.

THE CHROMOSOMAL LOCATION OF HUMAN BLYM-1

Chromosomal abnormalities have been associated with a variety of different neoplasms (Klein, 1981; Rowley, 1982), among the most consistent of which is the translocation between chromosome 8 and either 14, 2 or 22 observed in Burkitt's lymphomas (Manolov and Manolova, 1972). The chromosomal location of human Blym-1 was therefore looked for by in situ hybridization to human metaphase chromosomes to assess whether these observed chromosomal translocations may have a direct effect on the human Blym-1 gene (Morton et al., 1984). Human Blym-1 was assigned to chromosome 1 at position 1p32 and thus is at a locus that is not associated with the typical Burkitt's lymphoma translocations.

A number of investigators have determined that the human c-myc gene resides on chromosome 8 and is translocated to a recipient chromosome in Burkitt's lymphomas implicating c-myc in the disease process (Dalla-Favera et al., 1982; Neel et al., 1982; Taub et al., Marcu et al., 1983). It appears, therefore, that in Burkitt's lymphoma, as in chicken bursal lymphomas, both the myc and Blym-1 genes are involved in tumorigenesis.

SUMMARY

There are a number of similarities between chicken bursal lymphomas and human Burkitt's lymphomas. Both lymphomas are associated with viral infection, by LLV in bursal lymphomas and Epstein-Barr virus in Burkitt's lymphoma. Avian lymphoid leukosis virus integration is associated with enhanced c-myc expression, while the role EBV plays in tumorigenesis remains unclear. In Burkitt's lymphoma, however, c-myc activation does occur as a result of specific chromosomal translocations involving the human c-myc locus. Furthermore, the activated transforming genes detected by transfection of both bursal lymphoma and Burkitt's lymphoma DNAs are homologous members of the Blym family of genes. These similarities between chicken and human lymphomas provide evidence that viral involvement and oncogene activation are significant in tumor development and suggest they are involved in the multi-step progression to the neoplastic phenotype.

The function of the Blym genes remains to be determined. Although the chicken and human Blym genes are only distantly related, they have maintained their homology to the amino-terminal regions of transferrins. This fact may reflect some functional constraint on the evolution of these genes. It is therefore possible that transforming genes such as Blym may function via a transferrin-related mechanism.

ACKNOWLEDGEMENTS

These studies were supported by National Institutes of Health grant CA 28946, National Institutes of Health Fellowship CA 07250, and a Damon Runyon-Walter Winchell Cancer Fund fellowship to A.D., a travel fellowship from the Imperial Cancer Research Fund to J.M.D. and faculty award from the American Cancer Society to G.M.C.

REFERENCES

Andersson, P., Goldfarb, M.P. and Weinberg, R.A., 1979, A defined subgenomic fragment of in vitro synthesized Moloney-sarcoma virus DNA can induce cell transformation upon transfection, Cell, 16:63.

Barbacid, M., 1981, Cellular transformation by subgenomic feline sarcoma virus DNA, J. Virol., 37:518

Bishop, J.M., 1981, Enemies within: The genesis of retrovirus oncogenes, Cell, 23:5.

Blair, D.G., McClements, W.L., Oskarsson, M.K., Fischinger, P.J., and Wande Woude, G.F., 1980, Biological activity of cloned Moloney sarcoma virus DNA: Terminally redundant sequences may enhance transformation efficiency, Proc. Natl. Acad. Sci. USA, 77:3504.

Chang, E.H., Maryak, J.M., Wei, C.-M., Shih, T.Y., Shober, R., Cheung, H.L., Ellis, R.W., Hager, G.L., Scolnick, E.M., and Lowy, D.R., 1980, Functional organization of the Harvey murine sarcoma virus genome, J. Virol., 35:76.

Cooper, G.M., 1982, Cellular transforming genes, Science, 217:801.

Cooper, G.M., and Lane, M.-A., 1984, Cellular transforming genes and oncogenesis, Biochem. Biophys. Acta, in press.

Cooper, G.M., and Neiman, P.E., 1980, Transforming genes of neoplasms induced by avian lymphoid leukosis viruses, Nature (London), 287:656.

Cooper, G.M., and Neiman, P.E., 1981, Two distinct candidate transforming genes of lymphoid leukosis virus-induced neoplasms, Nature (London), 292:857.

Copeland, N.G., Zelentz, A.D., and Cooper, G.M., 1980, Transformation by subgenomic fragments of Rous sarcoma virus DNA, Cell, 19:863.

Dalla-Favera, R., Gregni, M., Erikson, J., Patterson, D., Gallo, R.C., and Croce, C.M., 1982, Human c-myc onc gene is located on the region of chromosome 8 that is translocated in Burkitt lymphoma cells, Proc. Natl. Acad. Sci. USA, 79:7824.

Diamond, A., Cooper, G.M., Ritz, J., and Lane, M.-A., Identification and molecular cloning of the human Blym transforming gene activated in Burkitt's lymphomas, Nature (London), 305:112.

Diamond, A., Devine, J.M., and Cooper, G.M., 1984, Nucleotide sequence of a human Blym transforming gene activated in a Burkitt's lymphoma, Science, 225:516.

Goubin, G., Goldman, D.S., Luce, J., Neiman, P.E., and Cooper, G.M., 1983, Molecular cloning and nucleotide sequence of a transforming gene detected by transfection of chicken B-cell lymphoma DNA, Nature (London), 302:114.

Haynes, B.F., Hemler, M., Cotner, T., Mann, D.L., Eisenbarth, G.S., Strominger, J., and Fauci, A.S., 1981, Characterization of a monoclonal antibody (5E9) that defines a human cell surface antigen of cell activation, J. Immunol., 127:347.

Hayward, W.S., Neel, B.G., and Astrin, S.M., 1981, Activation of a cellular onc gene by promoter insertion in ALV-induced lymphoid leukosis, Nature (London), 290:475.

Klein, G., 1981, The role of gene dosage and genetic transpositions in carcinogenesis, Nature (London), 294:313.

Lane, M.-A., Sainten, A., Doherty, K.M., and Cooper, G.M., 1984, Isolation and characterization of a stage-specific transforming gene, Tlym-I, from T-cell lymphomas, Proc. Natl. Acad. Sci. USA, 81:2227.

Macgilliray, R.T.A., Mendez, E., and Brew, K., 1977, in: "Proteins of Iron Metabolism," E.B. Brown, P. Aisen, J. Fielding, and R.R. Crichton, eds., pp. 133-141, Grune and Stratton, New York.

Manolov, G., and Manolova, Y., 1972, Marker band in one chromosome 14 from Burkitt lymphomas, Nature (London), 237:33.

Marcu, K.B., Harris, L.J., Stanton, L.W., Erikson, J., Watt, R., and Croce, C.M., 1983, Transcriptionally active c-myc oncogene is contained within NIARD, a DNA sequence associated with chromosome translocations in B-cell neoplasia, Proc. Natl. Acad. Sci. USA, 80:519.

Morton, C.C., Taub, R., Diamond, A., Lane, M.-A., Cooper, G.M., and Leder, P., 1984, Mapping of the human Blym-1 transforming gene activated in Burkitt lymphomas to chromosome 1, Science, 223:173.

Neel, B.G., Jhanwar, S.C., Changant, R.S.K., and Hayward, W.S., 1982, Two human c-onc genes are located on the long arm of chromosome 8, Proc. Natl. Acad. Sci. USA, 79:7842.

Rowley, J., 1982, Identification of constant chromosome regions involved in human hematologic malignant disease, Science, 216:749.

Shih, C., Shilo, B.-Z., Goldbarb, M.P., Dannenberg, A., and Weinberg, R.A., 1979, Passage of phenotypes of chemically transformed cells via transfection of DNA and chromatin, Proc. Natl. Acad. Sci. USA, 76:5714.

Sutherland, R., Delia, D., Schneider, C., Newman, R., Kemshead, J., and Greaves, M., 1981, Ubiquitous cell-surface glycoprotein on tumor cells is proliferation-associated receptor for transferrin, Proc. Natl. Acad. Sci. USA, 78:4515.

Taetle, R., Honeysett, J.M., and Trowbridge, I.S., 1983, Effects of anti-transferrin receptor antibodies on growth of normal and malignant myeloid cells, Int. J. Cancer, 32:343.

Taub, R., Kirsch, I., Morton, C., Lenoir, G., Swan, D., Tronick, S., Aaronson, S., and Leder, P., 1982, Translocation of the c-myc gene into the immunoglobulin heavy chain locus in human Burkitt lymphoma and murine plasmacytoma cells, Proc. Natl. Acad. Sci. USA, 79:7837.

Trowbridge, I.S., and Domingo, D.L., 1981, Effect on growth of human tumor cells of anti-transferrin receptor, monoclonal antibody and toxin antibody conjugates, Nature (London), 294:171.

Trowbridge, I.S., Lesly, J., and Schulte, R., 1982, Murine cell surface transferrin receptor: studies with an anti-receptor monoclonal antibody, J. Cell. Physiol., 112:403.

Trowbridge, I.S., and Lopez, F., 1982, Monoclonal antibody to transferrin receptor blocks transferrin binding and inhibits tumor cell growth in vitro, Proc. Natl. Acad. Sci. USA, 79:1175.

Trowbridge, I.S., and Omary, M.B., 1981, Human cell surface glycoprotein related to cell proliferation is the receptor for transferrin, Proc. Natl. Acad. Sci. USA, 78:3039.

DISCUSSION

Sutherland: Do you find Blym activation in Burkitt's lymphoma lines, in which there have been found no chromosomal rearrangements, for example, Raji?

Diamond: It is very infrequent to find Burkitt's lymphomas that do not have translocations; about 90 to 95% have some type of rearrangement involving chromosome 8, and either 14, 2, or 22. Raji is an unusual example of Burkitt's lymphoma in which chromosomal rearrangements have not been seen. However, recent evidence shows a small rearrangement in the c-myc locus.

Evans: Can you make that clearer and distinguish here between chromosomal rearrangements and sequence rearrangements, because Raji has got an 8 to 14 translocation.

Diamond: I agree.

Evans: Raji has a small deletion of c-myc. A further point to make is that very few chromosomal rearrangement sites within the 8 q 24 region are common to all Burkitt lymphomas; there is a whole variety of breakage and exchange sites within that region.

Diamond: The actual point of breakage is always variable in all the Burkitt's lymphomas. The one common denominator is that breakage always occurs in the vicinity of the myc gene. This finding implies that myc is involved in tumorigenesis although the mechanism involved is still unclear.

Evans: You suggested that in many cases virus integration may be one of the firt steps in activation. Did you imply that this is the case for Burkitt's lymphoma?

Diamond: No.

Evans: The question I ask then is, in the case of Burkitt's lymphoma, an important early phenomenon is the uptake of the Epstein-Barr (EB) virus, although it is not integrated specifically into chromosome number 8. What evidence do you have that EB virus activates c-myc?

Diamond: I know of no evidence that EB virus activates c-myc. Burkitt's lymphoma is endemic in Africa, where it is virtually always associated with EB virus infection. In other parts of the world, including the U.S., there is a much lower incidence of the disease where it tends to occur in the absence of EB virus. There are a number of studies which have observed integration into chromosome 1. We feel it is too far distant from Blym to be directly activating it.

Evans: My reason for asking the question was simply that I think it is an oversimplification to consider that you've got two events, i.e. an activation of c-myc and Blyml 1. In Burkitt's lymphoma, there are at least three and maybe more.

Diamond: Certainly; two was a minimum number. We were limited by the techniques and we are just at the beginning of the dissection of the steps. Where we show myc and Blym activation, these may be an early step and a later step, but there may be steps in between, before and after.

Hei: Most of the human tumors you cited are highly anaplastic, malignant tumors. Have you done any study with regard to benign tumors? Are there benign tumors capable of transfecting cells in culture?

Diamond: I think benign tumors tend not to transform efficiently.

Evans: The answer may be yes. If you look at papillomas in rats induced by chemical carcinogens, you can pick up activation of the c-Harvey ras gene, and the tumors are benign. If a promoter is added to the backs of these rats, the papillomas become malignant and the ras gene is still active. DNAs from both the premalignant benign papillomas and from the malignant carcinomas are active in transformation (Balmain _et al_., 1984).

Diamond: I would add that where there are preneoplastic or benign stages, for example in polyps, certainly there has been no activation of the transforming gene.

Huberman: The _ras_ gene family is the one most often mentioned in the transfection experiments. Are these the only type of genes that can be identified in these types of studies?

Diamond: Ras is by far the easiest gene to detect, simply because of the type of foci it gives in transfections. The cells are very obviously transformed and have a strong growth advantage. In contrast, the Blym gene gives a very different type of focus that is extremely difficult to see. The ras gene appears to be activated in a wide variety of different tumors, and my guess is that 20% of any type of tumor would have ras activation.

Huberman: Are there other types of cells, in addition to the 3T3 cells, susceptible to transfection with DNA-containing active oncogenes?

Diamond: There are other cell lines which you can use, but they will present the same problem. There are other approaches based upon the transfection procedure, in which you transform cells, and put them back into nude mice, for example, or into soft agaar, which is a little more stringent. Essentially, we are very limited by having to use the NIH 3T3 cell assay, which is the best we have available.

Sutherland: I might add here, that in my laboratory we are doing DNA transformation of normal human cells. It is not as easy as 3T3, but it works nicely and we are getting some answers. One interesting finding is that the human cells are transformed to at least anchorage-independence by tumor DNAs, which do not transform the 3T3 cells.

Huberman: What kind of genes are activated in these human cells? Are they also ras genes?

Sutherland: One does not know because it is a tumor DNA which does not transform 3T3s. All one can say is that it transforms human cells.

Hei: This is an answer to Dr. Huberman's question. Carmia Borek from our laboratory at Columbia also has done some transformation transfection studies using 3T3 cells and 10T1/2 cells and has succeeded in transforming 3T3 cells and vice versa.

Painter: What was the library used for the human gene? Was it just luck that you got a positive signal in a 140,000 copy library?

Diamond: In the case of the chicken gene isolation, it was NIH 3T3 cells that had been transformed by the bursa lymphoma DNA. In the case of the human gene it was a human Burkitt's lymphoma library; it was an amplified library. There were 15 positives, but only one of the 15 was active in the transfection assay.

Bertram: Have you initiated any studies to look at levels of the gene product of the Blym family?

Diamond: We made synthetic peptides from the nucleotide sequence, and those are now sitting in rabbits.

Bertram: What is the nature of the p21/90 kdalton association?

Diamond: Two interesting facts that have been verified are that P21 is a GTP binding protein, and GTP has no effect on that interaction. But transferin itself does dissociate the complex.

Martner: When you precipitate the transferin receptor with anti-transferin receptor antibodies, you do not bring down p21. Could you comment on this?

Diamond: We are not sure why that is. I could simply be that the antibody affects the tertiary structure of the antibody complex.

Thilly: Have things got to the point where you can consider studying whether Blym is transiently expressed, and that this might be a sufficient action in order to be active against a myc background? One of the things that always interested me in chemical transformation was that we talk about an initiating event, a heritable mutation, for our purposes let us call it a "remembered" event. Yet, on the mouse back, and in the C3H10T1/2 system you then need a period of extended treatment with phorbol esters or chemicals that mimic their behavior. I wondered if the second step was a stoichastic one, which, against a potentiating background, a (transient expression for transient phenomena) would permit a stage change in order to move over to a stable transformed state. Are you anticipating experiments in which you model Blym expression, not necessarily continually expressed, to bring about transformation.

Diamond: Since Blym transforms mouse cells, whereas normal DNA does not, we may conclude that a dominant genetic change has occurred. This essentially answers the question, but, as far as expression of this gene is concerned, when we look at messenger RNA levels in Burkitt's lymphoma cells, there is no difference in the levels of the messenger RNAs for Blym, as compared to that in EBV immortalized cells.

Thilly: This does not seem to geometrically demonstrate that continued expression is required.

Diamond: Right.

REFERENCES

Balmain, A., Ramsden, M., Bowder, G. T., and Smith, J., 1984, Activation of the mouse cellular Harvey-ras gene in chemically induced benign skin papillomas, Nature (London), 307:658.

RETROVIRAL ONCOGENES AND HUMAN NEOPLASIA

E. Premkumar Reddy

Laboratory of Cellular and Molecular Biology
National Cancer Institute, Bethesda, MD 20205

INTRODUCTION

Carcinogenesis is a complex multistep process. For a molecular biologist seeking simpler model systems to gain insight into this neoplastic process, retroviruses have provided an excellent system because of their relatively simple biochemical organization. These viruses fall into two major groups: chronic leukemia viruses and acute transforming viruses. Chronic leukemia viruses cause tumors in susceptible hosts but only after a prolonged latent period of several months. These viruses are replication competant and can be propagated _in vitro_ without transforming their host cells. In contrast, acute transforming viruses induce a variety of tumors including sarcomas, carinomas and hematopoietic tumors with a short latent perod of days to weeks. These viruses are generally replication defective and induce transformation of appropriate assay cells _in vitro_. Because of these properties, acute transforming viruses have been the subject of intensive investigations in the past two decades which have led to several important insights into the mechanisms involved in carcinogenesis. Recent studies on the isolation and characterization of oncogenes from human tumors have indicated further that several of the oncogenes associated with acute transforming viruses play an important role in human malignancies. Thus it appears that the knowledge gained from studies on retroviruses can be applied directly to the understanding of human oncogenesis.

Present Address: Department of Molecular Genetics
Hoffmann-La Roche, Nutley, NJ 07110

STRUCTURE AND ORIGIN OF ACUTE TRANSFORMING VIRUSES

The chronic leukemia viruses contain coding information for three structural genes termed _gag_, _pol_ and _env_ (Baltimore, 1974). The _gag_ gene codes for internal structural proteins while the _pol_ and _env_ genes code for reverse transcriptase and envelope protein respectively. An important structural feature of the retroviral genome is the occurrence of two large terminal repeats (LTRs) of 300-600 base pairs (bps) at both 5' and 3' ends of proviral genome (Shimotohono _et al._, 1980; Dhar _et al._, 1980). These LTRs have been shown to contain signals for RNA transcription (such as the promoter and enhancer sequences) as well as messenger RNA capping and polyadenylation. Chronic leukemia viruses do not possess discrete transforming genes within their genome and the mechanism of transformation by these agents is yet to be resolved.

Acute transforming viruses arose in nature by recombination of leukemia viruses with host cellular sequences during which process portions of the viral genome have been substituted by the cellular sequences (Coffin, 1982). In recent years these cellular sequences have been shown to be essential for the transforming activity of the acute transforming viruses and therefore have been termed oncogenes. To date approximately 20 different oncogenes have been identified in nature and all of them have been found to contain analogs in normal cellular DNA (which have been termed as protooncogenes or c-oncogenes). This implies that the acute transforming viruses gain their transforming ability by virtue of acquiring sequences from their host cell DNAs. This leads us to an important question as to how a gene which otherwise is innocuous in vertebrate cells acquires this ability to induce cancer after recombination with a viral element. Two obvious explanations are a) enhanced expression of these oncogenes occurs as a result of their incorporation into a proviral genome which in turn results in the transformation of the host cells, and b) the process of recombination leads to structural alterations which result in the activation of a normal gene to its malignant counterpart. In an attempt to gain insights into the mechanism of activation of the protooncogenes, we have undertaken the molecular cloning and nucleotide sequence analysis of several of the retroviral transforming genes and their normal cellular analogs. Comparative analysis of three of these genes (_abl_, _myb_, and _ras_) with their normal cellular homologs are discussed in this chapter.

Structural Organization of Abelson Murine Leukemia Virus and a Comparison of v-abl and c-abl Encoded Messenger RNAs

Abelson murine leukemia virus (A-MuLV) is a replication-defective transforming retrovirus that was isolated after inoculation of Moloney murine leukemia virus (M-MuLV) in prednisolone-treated BALB/c mice (Abelson and Rabstein, 1970). A-MuLV induces B-cell lymphomas _in vivo_ (Premkumar _et al._, 1975) and is able to transform both lymphoid and fibroblastic cells in vitro (Rosenberg and Baltimore, 1980). Several lines of evidence indicate that A-MuLV arose by recombination of nondefective helper virus (M-MuLV) and cellular sequences present within the normal mouse genome (Goff et al., 1980). The later sequences termed abl appear to code for the transforming properties of the virus. In an effort to understand the structural organization of this viral genome, we molecularly cloned the A-MuLV genome and determined its complete nucleotide sequence (Reddy et al., 1983). These studies have revealed several important features of its molecular organization. Like many of the transforming viruses, A-MuLV appears to synthesize its transforming protein by means of a gag-abl polyprotein, the amino terminal region of which is composed of helper virus gag gene products. This protein contains 240 amino acids derived from the amino terminus of the gag region followed by 745 amino acids that are specific to the v-abl region. Thus, the transforming gene utilized the helper viral sequences for the initiation of its synthesis.

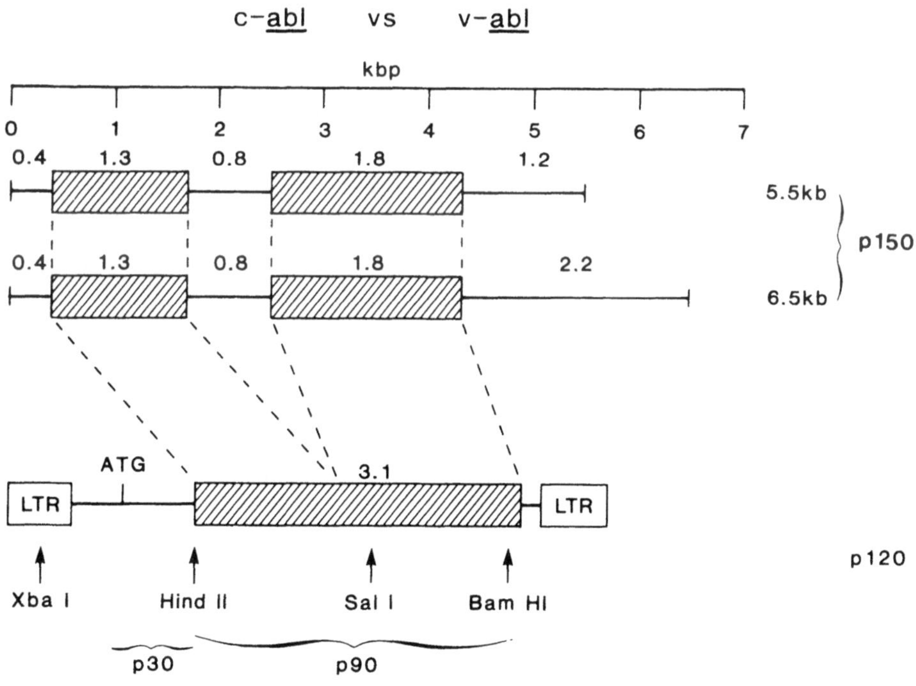

Fig. 1. Comparison of mouse v-abl and c-abl sequences. The portion of the cellular genome incorporated into the virus is indicated by the hatched boxes. The c-abl genome codes for two mRNAs of 5.5 Kb and 6.5 Kb long and the structure of the cDNA clones derived from these mRNAs is shown above.

Like all other transforming genes of retroviruses, DNA sequences homologous to abl (termed c-abl) are found in normal mouse DNA (Goff et al., 1980). Thus the viral oncogene represents a transduced cellular gene. The c-abl gene appears to be functionally active in several lymphoid organs, including thymus, coding for a protein product of 150,000 daltons (Ponticelli et al., 1982). Mouse thymus also synthesizes mRNA species 5.5 and 6.5 kb long that cross hybridize with v-abl sequences (Wang and Baltimore, 1983). In order to compare the structure of v-abl and c-abl encoded mRNAs we isolated several c-DNA clones of the c-abl encoded mRNAs and compared their structure with that of the v-abl gene. Preliminary results indicate that the v-abl genome arose from the c-abl gene as a result of extensive deletional process at both the amino and carboxy terminal ends of the molecule. This is schematically demonstrated in Fig. 1. These structural changes also affect the size and biochemical properties of the protein encoded by the c-abl and v-abl encoded proteins. While the v-abl encodes a polypeptide of approximately 90,000 daltons fused to a 30,000 dalton polypeptide of the gag gene, the c-abl gene has been shown to code for a protein product of 150,000 daltons (Ponticelli, et al., 1982). Thus, it appears that the c-abl gene codes for a protein approximately 60,000 daltons longer than that encoded by the v-abl gene. In addition, the transforming gene product of A-MuLV (p120) has been shown to possess closely associated kinase activity with specificity for tyrosine phosphorylation. On the other hand, the c-abl encoded protein termed NCP150 seemed to lack this property even though the c-abl gene contains all the information present in the v-abl genome (Ponticelli et al., 1982; Wang et al., 1984). These results imply that the phosphotyrosine kinase activity of the c-abl encoded product is highly controlled, and the deletional process that occurred in the genome during recombination between cellular and viral genes might have resulted in the loss of this regulation of the kinase activity (Ponticelli et al., 1982).

Structural Organization of Avian Myeloblastosis Virus and a Comparison of v-myb and c-myb Encoded Messenger RNA

Avian myeloblastosis virus (AMV), like A-MuLV is a replication defective acute transforming virus and has arisen by recombination of a nondefective helper virus and host cellular sequences present within the normal avian genome. This retrovirus causes acute myeloblastic leukemia in chickens and transforms a specific class of hematopoietic cells *in vitro* (Baluda and Goetz, 1961). The AMV provirus was cloned by Souza et al. (1980) and the complete nuceotide sequence of its transforming gene determined by Rushlow et al. (1982). Recently, we have also cloned the reverse transcript of the subgenomic messenger RNA encoded by this viral genome and determined the nucleotide sequence of this cDNA clone (Sanders and Reddy, in preparation). These results indicate that AMV transforming gene is synthesized via a spliced messenger RNA in which the leader sequence derived from the 5' terminus of the genomic RNA, is spliced to the body of v-*myb* encoded sequences. Protein synthesized by the spliced messenger would contain 6 amino acids derived from the *gag* gene of the helper virus and 371 amino acids derived from the v-*myb* sequences. The open reading frame of the v-*myb* region extends into the viral sequences at the 3' end, terminating in the envelope portion of the viral gene. Thus the last 11 amino acids at the carboxy terminal end of the transforming protein are derived from the 3' end of the envelope gene. These results indicated that the transduced transforming gene is incomplete and it utilized viral initiator and terminator codons for the synthesis of its protein.

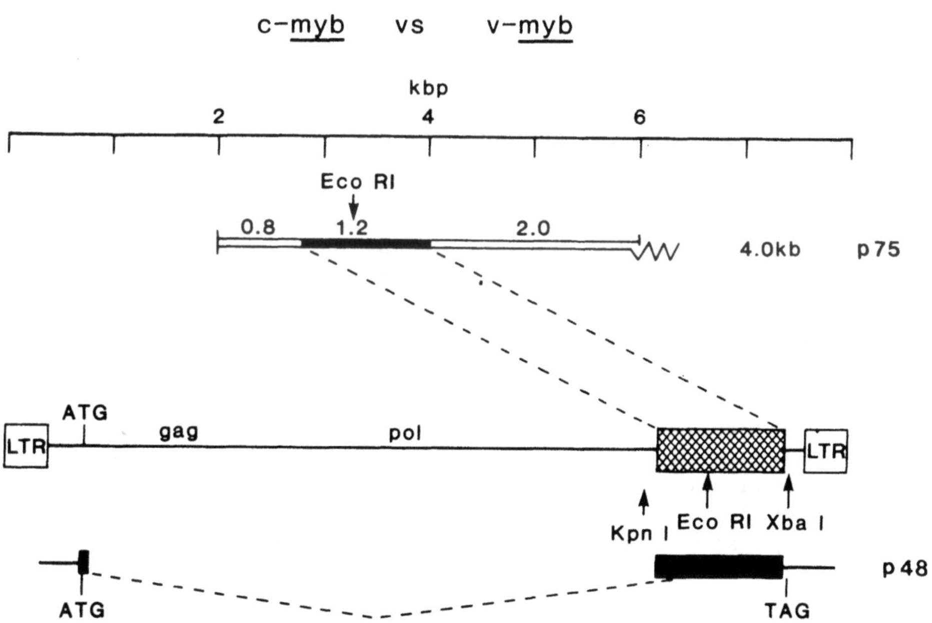

Fig. 2. Comparison of chicken c-*myb* and v-*myb* sequences. The portion of the cellular genome incorporated into the virus is indicated by the solid and/or hatched boxes. The bottom portion shows the structure of the spliced subgenomic mRNA that codes for the transforming protein.

In order to compare the structure of v-myb encoded sequences with c-myb gene product, we resorted to cDNA cloning of its mRNAs The c-myb genome is functionally active in several lymphoid organs including the thymus coding for a mRNA species which is 4.0 Kb long. Comparison of the structure of the two clones revealed that the v-myb gene arose from the c-myb gene as a result of extensive deletional process at both the amino-and carboxy-terminal ends of the molecule. This is schematically demonstrated in Fig. 2. These structural changes also affect the size of the proteins encoded by v-myb and c-myb gene. While v-myb encodes for a protein of approximately 45,000 daltons, the c-myb encoded protein appears to be approximately twice as big (Boyle et al, 1983; Klempnauer et al., 1983). How these deletions affect the biochemical properties of the two proteins is yet to be determined. It is tempting to speculate that the deletions suffered by the myb gene result in the activation of this gene and the availability of the c-myb clones allows this hypothesis to be tested.

Structural Organization of v-bas Oncogene

BALB/c murine sarcoma virus (BALB-MSV) is a replication-defective transforming retrovirus of mouse origin that was isolated from a spontaneously occurring BALB/c mouse tumor (Peters et al., 1974). Recent studies have demonstrated that BALB-MSV arose by recombination between endogenous mouse leukemia virus and a cellular DNA sequence which was termed bas (Anderson et al., 1981a). Molecular cloning and detailed structural analysis of the proviral genome of BALB-MSV has revealed that large portions of helper viral pol and env genes have been deleted as a result of this recombination and have been replaced by a stretch of cellular sequences approximately 0.6 kb long. Cloned bas sequences exhibited a high degree of sequence homology with the transforming gene of Harvey murine sarcoma virus (H-ras) (Anderson et al., 1981). In an effort to understand the structural relationships between the oncogenes of BALB and Harvey MSV and their normal cellular counterparts, we have undertaken primary nucleotide sequence analysis of these genes (Reddy et al., 1985). The nucleotide sequence of the transforming gene of BALB-MSV revealed that it codes for a protein of 189 amino acids with a molecular weight of approximately 21,000 daltons. The sequence analysis also revealed that BALB-MSV has incorporated the entire coding region of the c-bas gene. While sequence analysis of the normal mouse counterpart of v-bas awaits its molecular cloning, it has been possible to compare its sequence with the rat and human cellular homologs whose sequence has been determined (Reddy, 1983; M. Ruta, personal communication). This comparative analysis indicated that viral transforming gene differs from the normal rat cellular counterpart at amino acid positions 12 and 143. Furthermore, this transforming gene shows extensive sequence homology with v-H-ras (Dhar et al., 1982) and v-k-ras (Tsuchida et al, 1982) oncogenes which also have undergone mutation at amino acid position 12. Thus, it appears that v-bas gene has undergone specific mutations at position 12 and 143. The importance of mutations in ras genes became apparent with the discovery and isolation of oncogenes associated with human tumors which not only linked virus-associated oncogenes with human malignancies but also demonstrated that the mechanism of activation of these oncogenes follows a similar pathway.

TRANSFORMING GENES FROM HUMAN TUMOR CELLS

An independent approach for the detection of transforming genes became available by the advent of DNA-mediated gene transfer techniques (Graham and Van der Eb, 1973). Thus, the DNAs from certain naturally occurring human tumors were found to induce malignant transformation of NIH/3T3 cells, a continuous mouse cell line that is contact inhibited and is highly susceptible to DNA transfection (Cooper, 1982). Although only about 10% of human tumor DNAs have been shown to be capable of transforming NIH/3T3 cells in transfection assays, oncogenes have been detected in tumors representative of each of the major forms of human cancer. An oncogene present in T24 and EJ bladder carcinoma cell line was the first to be isolated by molecular cloning techniques (Goldfarb et al., 1982; Pulciani et al., 1982a,b; Shih and Weinberg, 1982). This bladder carcinoma oncogene was found to be closely related to the oncogenes of Harvey and BALB murine sarcoma viruses (Der et al., 1982; Parada et al., 1982; Santos et al., 1982). Use of the cloned DNAs of bladder carcinoma oncogene as sequence probes showed that this oncogene is derived from a sequence of similar structure present in the normal human genome. Unlike the T24 bladder carcinoma oncogene which exhibited a transforming activity of 10^4 focus-forming units per microgram of DNA, its normal cellular analog did not exhibit transforming activity in transfection assays. Moreover, restriction endonuclease mapping of the T24 gene was found to be identical to its normal counterpart. These results implied that the cellular sequences could be activated by subtle genetic alteration within naturally occurring human cancers independent of retroviral involvement. The mechanism of activation of T24 and EJ oncogenes was studied by construction of several recombinants from parts of the T24 oncogene and its normal homolog (Tabin et al., 1982; Reddy et al., 1982). Analysis of such recombinants and a comparison of their sequences yielded the startling result that the genetic change leading to the activation of this oncogene is a point mutation, with thymidine replacing guanine. This substitution results in the incorporation of valine instead of glycine as the 12th amino acid residue of the T24 oncogene encoded protein p21. Thus, a single amino acid substitution (gly-val) seems to be sufficient to confer transforming properties to the gene product of the T24 human bladder carcinoma oncogene.

A Common Genetic Alteration in Human and Retroviral Transforming Genes

The amino acid sequences of Harvey-MSV (Dhar *et al.*, 1982), BALB-MSV (Reddy *et al.*, 1985) and Kirsten MSV (Tsuchida *et al.*, 1982) encoded p21 proteins have been determined by nucleotide sequence analysis of their corresponding onc gene, v-*has*, v-*bas* and v-*kis*, respectively. Comparison of the first 37 amino acids

	1	2	3	4	5	6	7	8	9	10	11	12	13	14	15	16	17	18	19	20	21	22	23	24	25	26	27	28	29	30	31	32	33	34	35	36	37
Human c-H-*ras* :	M	T	E	Y	K	L	V	V	V	G	A	G	G	B	G	K	S	A	L	T	I	Q	L	I	Q	N	H	F	V	D	E	Y	D	P	T	I	E
T24 oncogene :	M	T	E	Y	K	L	V	V	V	G	A	V	G	B	G	K	S	A	L	T	I	Q	L	I	Q	N	H	F	V	D	E	Y	D	P	T	I	E
BALB-MSV :	M	T	E	Y	K	L	V	V	V	G	A	K	G	B	G	K	S	A	L	T	I	Q	L	I	Q	N	H	F	V	D	E	Y	D	P	T	I	E
Harvey-MSV :	M	T	E	Y	K	L	V	V	V	G	A	R	G	B	G	K	S	A	L	T	I	Q	L	I	Q	N	H	F	V	D	E	Y	D	P	T	I	E
Kirsten-MSV :	M	T	E	Y	K	L	V	V	V	G	A	S	G	B	G	K	S	A	L	T	I	Q	L	I	Q	N	H	F	V	D	E	Y	D	P	T	I	Q

Fig. 3. Predicted amino acid sequence of the first exon of the normal human cH-*ras* gene and its transforming allele, the T24 oncogene. Comparison with the first 37 amino acids predicted for the p21 proteins coded for by BALB-MSV, Harvey-MSV and Kirsten-MSV. The letter code for the amino acids is: A, Ala; D, Asp; E, Glu; F. Phe; G, Gly; H, His; I, Ile; K, Lys; L, Leu; M, Met; N, Asn; P, Pro; Q, Gln; R, Arg; S, Ser; T, Thr; V, Val; and Y, Tyr.

of the p21 protein coded for the c-*has* human gene with their retroviral counterparts shows complete identity except at position 12, the same amino acid residue that is altered in the T24 oncogene (figure 3). Whereas arginine is present in Harvey-MSV, lysine was found in BALB-MSV and serine in Kirsten-MSV. It appears therefore that viral oncogenes and the oncogenes isolated from human tumors have a common mechanism of activation.

The studies on the T24/EJ bladder carcinoma oncogene precipitated an intensive search for other oncogenes associated with human tumors. Several human oncogenes were isolated and characterized from human tumors using NIH/3T3 cells as an assay system. Most of the transforming genes isolated using NIH/3T3 system were found to belong to the ras gene family. The ras gene family consists of three members designated H-ras, N-ras and K-ras (Cooper, 1982; Pulciani et al., 1982a,b; Taparowsky et al., 1983; Shimizu et al., 1983; Capon et al., 1983). Although the three genes have different genetic structures, all of them code for structurally homologous proteins of 189 amino acids. All the three genes seem to be activated in human tumors as a result of point mutations. Several of these mutations seem to result in the alteration of the 12th codon as in the T24 bladder carcinoma oncogene. However, some of the activated ras genes were found to contain glycine in the 12th position as it occurs with their normal counterpart. By the approach utilized for the analysis of the T24 oncogene, the genetic alteration responsible for the transforming activity of these genes have been localized to a point mutation in the second exon resulting in the alteration of the 61st codon which codes for glutamine in the normal ras proteins. Thus, the transforming gene HS242 isolated from a human lung carcinoma derived cell line was found to code for leucine instead of glutamine (Yuasa et al., 1983), while an N-ras related gene isolated from a human neuroblastoma (SK-N-SH) was found to code for lysine as its 61st amino acids (Taparowsky et al, 1981). Recent studies by several groups have localized these genetic lesions that activate ras genes to these same specific codons. Table 1 summarizes the various mutations that have so far been identified in human tumor-associated ras genes. Thus, codons 12 and 61 appear to be hot spots for activating members of the ras family of oncogenes. Presumably these mutations alter the site at which p21 interacts with its normal substrate by altering their secondary structure. This is also seen in the altered electrophoretic mobility of these proteins (Cooper, 1982; Tabin et al., 1982; Yuasa et al., 1983). All of these findings argue that qualitative rather than quantitative alterations are probably responsible for the conversion of a normal ras gene into its malignant counterpart.

CONCLUSIONS

Carcinogenesis primarily involves irreversible alteration of cellular growth controls. In view of the large number of structural genes that are present in the mammalian genome, the molecular dissection of the neoplastic process has been a formidable task. However, retroviruses, because of their unique ability to interact with transforming genes, have been instrumental in identifying a small set of genes which are now termed as oncogenes.

Table 1. Mutations in Human Tumor-Associated *ras* Genes

Oncogene	Tumor Source	Point Mutation		Codon
H-*ras*				
T24	Bladder carcinoma	GGC Gly	GTC Val	12
Hs0578	Mammary carcinoma	GGC Gly	GAC Asp	
Hs242	Lung carcinoma	CAG Gln	CTG Leu	61
N-*ras*				
SK-N-SH	Neuroblastoma	CAA Gln	AAA Lys	61
SW-1271	Lung carcinoma	CAA Gln	CGA Arg	61
K-*ras*				
Calu-1	Lung carcinoma	GGT Gly	TGT Cys	12
PR371	Lung carcinoma	GGT Gly	TGT Cys	12
SW480	Colon carcinoma	GGT Gly	GTT Val	12
A1698	Bladder carcinoma	GGT Gly	CGT Arg	12
A2182	Lung carcinoma	GGT Gly	CGT Arg	12
LC10	Squamous cell carcinoma	GGT Gly	CGT Arg	12

The data are derived from Tabin *et al.* (1982); Reddy *et al.* (1982); Kraus *et al.* (1984); Yuasa *et al.* (1983 and 1984); Taparowsky *et al.* (1983); Shimizu *et al.* (1983); Capon *et al.* (1983); Nakano *et al.* (1984); and Santos *et al.* (1984).

Independent of retrovirus involvement, these oncogenes appear to be activated by a variety of mechanisms including point mutations and chromosomal rearrangments. Identification of this small set of genes and deciphering the mechanism of activation of these genes has made the study of neoplastic process more approachable at the molecular level. Insights gained by the strategies such as those outlined above, hopefully will lead in the future for better methods for diagnosis, treatment and prevention of cancer.

Abelson, H.T., and Rabstein, L.S., 1970, Lymposarcoma: Virus induced thymic-independent disease in mice, Cancer Res., 3:2208.

Anderson, P.R., Devare, S.G., Tronick, S.R., Ellis, R.W., Aaronson, S.A., and Scolnick, E.M., 1981, Generation of BALB-MUSV and HA-MUSV by type C virus transduction of homologous transforming genes from different species, Cell, 26:129.

Anderson, P.R., Tronick, S.R., and Aaronson, S.A., 1981a, Structural organization and biological activity of molecular clones of the integrated genome and BALB/c mouse sarcoma virus, J. Virol., 40:431.

Baltimore, D., 1974, Tumor Viruses, Cold Spring Harbor Symp. Quant. Biol., 39:1187.

Baluda, M.A., and Goetz, I.E., 1961, Morphological conversion of cell cultures by avian myeloblastosis virus, Virology, 15:185

Boyle, W.J., Lipsick, J.S., Reddy, E.P., and Baluda, M.A., 1983, Identification of the leukemogenic protein of avian myeloblastosis virus and of its normal cellular homologue, Proc. Natl. Acad. Sci. USA, 80:2834.

Capon, D.J., Seeburg, P.H., McGratti, J.P., Hayflick, J.S., Edman, U., Levinson, A.D., and Goeddel, D.V., 1983, Activation of Ki-ras-2 gene in human colon and lung carcinomas by two different point mutations, Nature (London), 304:507.

Coffin, J., 1982, Structure of the retroviral genome, in: "RNA Tumor Viruses," pp. 261-368, R. Weiss, N. Teich, H. Varmus, and J. Coffin, eds., Cold Spring Harbor Laboratory, New York.

Cooper, G.M., 1982, Cellular transforming genes, Science, 217:801.

Der, C.J., Krontiris, T.G., and Cooper, G.M., 1982, Transforming genes of human bladder and lung carcinoma cell lines are homologous to the ras genes of Harvey and Kirsten sarcoma viruses, Proc. Natl. Acad. Sci., USA, 79:3637.

Dhar, R., Ellis, R.W., Shih, T.Y., Oroszalan, S., Shapiro, B., Maizel, J., Lowy, D., and Scolnick, E., 1982, Nucleotide sequence of the p21 transforming protein of Harvey murine sarcoma virus, Science, 217:934.

Dhar, R., McClements, W.L., Enquist, L.W., and Vande Wonde, G.W., 1980, Nucleotide sequence of integrated Moloney sarcoma provirus long terminal repeats and their host and viral junctions, Proc. Natl. Acad. Sci USA, 77:3937.

Goff, S.P., Gilboa, E., Witte, O.N., and Baltimore, D., 1980, Structure of the Abelson murine leukemia virus genome and the homologous cellular gene: Studies with cloned viral DNA, Cell, 22:777.

Goldfarb, M., Shimizu, K., Perucho, M., and Wigler, M., 1982, Isolation and preliminary characterization of a human transforming gene from T24 bladder carcinoma cells, Nature (London), 296:404.

Graham, F.L., and Van der Eb, A.J., 1973, Transfection of rat cells by DNA of human adenovirus 5, Virology, 52:456.

Klempnauer, K.H., Ramsay, G., Bishop, J.M., Moscovici, M.G., Moscovici, C.J., McGrath, P., and Levinson, A.D., 1983, The product of the retroviral transforming gene v-myb is a truncated version of the protein encoded by the cellular oncogene c-myb, Cell, 33:345.

Kraus, M., Yuasa, Y., and Aaronson, S., 1984, A position 12 activated H-ras oncogene in all HS578T mammary carcinosarcoma cells but not normal mammary cells of the same patient, Proc. Natl. Natl. Acad. Sci. USA, 81:5384.

Nakano, H., Yamamoto, F., Neville, C., Evans, D., Mizano, T., and Perucho, M., 1984, Isolation of transforming sequences of two human lung carcinomas: Structural and functional analysis of the activated c-k-ras oncogenes, Proc. Natl. Acad. Sci. USA, 81:71.

Parada, L.F., Tabin, C.J., Shih, C., and Weinberg, R.A., 1982, Human EJ bladder carcinoma oncogene is a homologue of Harvey sarcoma virus ras gene, Nature (London), 297:474.

Peters, R.L., Rabstein, L.S., Louise, S., Van Vleck, R., Kelloff, G.J., and Huebner, R.J., 1974, Naturally occurring sarcoma virus of the BALB/c Cr mouse, J. Natl. Cancer Inst., 53:1725.

Ponticelli, A.S., Whitlock, C.A., Rosenberg, N., and Witte, O.N., 1982, In vivo tyrosine phosphorylations of the Abelson virus transforming protein are absent in its normal cellular homolog, Cell, 29:953.

Premkumar, E., Potter, M., Singer, P.A., and Sklar, M.D., 1975, Synthesis, surface deposition, and secretion of immunoglobulins by Abelson virus transformed lymphosarcoma cell lines, Cell, 6:149.

Pulciani, S., Santos, E., Lauver, A.V., Long, L.K., Aaronson, S.A., and Barbacid, M., 1982a, Oncogenes in solid human tumors, Nature (London), 300:539.

Pulciani, S., Santos, E., Lauver, A.V., Long, L.K., Robbins, K.C., and Barbacid, M., 1982b, Oncogenes in human tumor cell lines: molecular cloning of a transforming gene from human bladder carcinoma cells, Proc. Natl. Acad. Sci. USA, 79:2845.

Reddy, E.P., 1983, Nucleotide sequence analysis of the T24 human bladder carcinoma oncogene, Science, 220:1061.

Reddy, E.P., Lipman, D., Anderson, P.R., Tronick, S.R., and Aaronson, S.A., 1985, Nucleotide sequence analysis of the BALB-MSV transforming gene, J. Virol., in press.

Reddy, E.P., Reynolds, R.K., Santos, E., and Barbacid, M., 1982, A point mutation is responsible for the acquisition of transforming properties by the T24 human bladder carcinoma oncogene, Nature (London), 300:149.

Reddy, E.P., Smith, M.J., and Svinivasan, A., 1983, Nucleotide sequence of Abelson murine leukemia virus genome: Structural similarity of its transforming gene product to other onc gene products with tyrosine-specific kinase activity, Proc. Natl. Acad. Sci. USA, 80:3623.

Rosenberg, N., and Baltimore, D., 1980, Abelson virus, in: "Viral Oncology," pp. 187-203, G. Klein, ed., Raven, New York.

Rushlow, K.E., Lautenberger, J.A., Papas, T.S., Baluda, M.A., Perbal, B.J., Chirikjian, G., and Reddy, E.P., 1982, Nucleotide sequence of the transforming gene of avian myeloblastosis virus, Science, 216, 1421.

Santos, E., Marin-Zanca, D., Reddy, E.P., Pierotti, M.A., Porta, G.D., and Barbacid, M., 1984, Malignant activation of a k-ras oncogene in lung carcinoma but not in normal tissue of the same patient, Science, 223:661.

Santos, E., Tronick, S.R., Aaronson, S.A., Puciani, S., and Barbacid, M., 1982, T24 human bladder carcinoma oncogene is an activated form of the normal homologue of BALB and Harvey-MSV transforming genes, Nature (London), 298:343.

Shih, C., and Weinberg, R.A., 1982, Isolation of a transforming sequence from a human bladder carcinoma cell line, Cell, 29:161.

Shimizu, K., Birnbaum, D., Ruley, M.A., Fasano, O., Suard, Y., Edland, L., Taparowsky, E., Goldfarb, M., and Wigler, M., 1983, Structure of Ki-ras gene of the human lung carcinoma cell line Calu-1, Nature (London), 304:496.

Souza, L.M., Strommer, J.N., Hillyard, R.L., Komaromy, M.C., and Baluda, M.A., 1980, Cellular sequences are present in the presumptive avian myeloblastosis virus genome, Proc. Natl. Acad. Sci. USA, 77:5177.

Shimotohono, K., Muzutani, S., and Temin, H.M., 1980, Sequence of retrovirus provirus resembles that of bacterial transposable elements, Nature (London), 285:550.

Tabin, C.J., Brodley, S.M., Bargmann, C.I., Weinberg, R.A., Papageorge, A.G., Scolnick, E.M., Dhar, R., Long, D.R., and Chang, E.H., 1982, Mechanism of activation of a human oncogene, Nature (London), 300:143.

Taparowsky, E., Shimizu, K., Goldfarb, M., and Wiegler, M., 1983, Structure and activation of the human N-ras gene, Cell, 34:581.

Tsuchida, N., Ryder, T., and Ohtubo, E., 1982, Nucleotide sequence of the oncogene encoding the p21 transforming protein of Kirsten murine sarcoma virus, Science, 217:937.

Wang, J.Y.J., and Baltimore, D., 1983, Cellular RNA homologous to the Abelson murine leukemia virus transforming gene: Expression and relationship to the viral sequence, Mol. Cell. Biol., 3:773.

Wang, J.Y.J., Ledley, F., Goff, S., Lee, R., Grone, Y., and Baltimore, D., 1984, The mouse c-abl locus: Molecular cloning and characterization, Cell, 36:349.

Yuasa, Y., Srivastava, S.K., Dunn, C.Y., Rhim, J.S., Reddy, E.P., and Aaronson, S.A., 1983, Acquisition of transforming properties by alternative point mutations within c-bas/has human proto-oncogene, Nature (London), 303:775.

Yuasa, Y., Gol, R.A., Chang, A., Chiu, I., Reddy, E.P., Tronick, S.R., and Aaronson, S.A., 1984, Mechanism of activation of an N-ras oncogene of SW-1271 human lung carcinoma cells, Proc. Natl. Acad. Sci. USA, 81:3670.

DISCUSSION

Lee: Did I understand correctly that most of the single point mutations were transversions, from purine to pyrimidine?

Reddy: Yes.

Lee: Do you attach any significance to the observation of transversions as the predominant change in DNA for the point mutations?

Reddy: Genentech laboratory has claimed that they have brought about all possible changes in the 12th codon and every one of them, conversion of a glycine to any other amino acid, seem to activate the gene. So, in nature it may be a simple transversion, but not necessarily for activation of the gene.

Sutherland: Do you know, either from restriction polymorphism, or from other evidence of altered ras genes in tumors or tumor cell lines which do not transform NIH 3T3?

Reddy: We have screened all the cell lines that were found to be negative for restriction polymorphism. We have not actually screened the solid tumors, but none of the tumors that were negative showed the restriction polymorphism which indicates that there was a mutation. I strongly suspect that the 3T3s will be transformed by altered ras genes.

Sutherland: So one can conclude that there is a mutation at a different place, other than 12-61, but as far as you know there is no evidence that there is an activated unaltered ras in those nontransforming tumor cell lines.

Reddy: Yes, because if there is a mutated ras gene I think NIH 3T3 will pick it up easily.

Evans: Can I ask about the evidence you have got for the fact that there may be a deletion of the wild type Kirsten ras gene which helps the establishment of a cell in culture from a tumor?

Reddy: The very interesting observation first made with T24 cells was that there is not a nonmutated allele in this particular cell line. And with the exception of one recently established cell line that has been passed through not more than 10 generations, we have been unable to show the presence of nonmutated alleles in the well established cell lines. Whenever we look at solid tumors, we find the normal allele. One can pose an objection by saying that our tumor DNA's are contaminated with normal tissue, but the pathologists claim that 90% of the tumor mass they provided is tumorigenic tissue. Band intensities that we see in these solid tumors are same, between the normal and the mutated alleles. You can show the loss of normal allele, the wild-type allele, in most well established cell lines. These two types of observation suggest that this is a tissue culture phenomenon, and must be one of the requirements for establishment of the lines in culture.

Evans: It might be worth making the comment that you are looking and probing for one locus only, and that what you are seeing is a loss of one sequence, which might be part of the loss of a whole chromosome, or a whole arm.

Reddy: Yes. We think the whole chromosome is lost.

MUTATION IN SINGLE-CELL SYSTEMS INDUCED BY LOW-LEVEL MUTAGEN EXPOSURE

Howard L. Liber, Susan L. Danheiser and William G. Thilly

Genetic Toxicology Group
Massachusetts Institute of Technology
Cambridge, Massachusetts

INTRODUCTION

A principal question currently facing toxicologists and radiobiologists concerns how measurements can be made of mutations induced by low levels of chemicals or radiation. In this paper, we focus on measurements in single cell systems. We consider that cellular exposure to mutagens occurs via four modes.

(i) High dose rate x short time, yielding a high dose;
(ii) High dose rate x long time, yielding a very high dose;
(iii) Low dose rate x short time, yielding a low dose;
(iv) Low dose rate x long time, yielding a high dose.

Definitions of these terms are presented in Table I.

The first mode, high dose rate x short time, is the most commonly reported protocol for examining the mutational response of cells after exposure to radiation or chemical mutagens. Yet this mode, as well as high dose rate x long time exposures (Mode ii), would seem to reflect only rare circumstances in human experience. Thus, most of us would agree that such protocols are unrealistic in terms of mimicking human exposures. If our intention is to devise experiments which will be useful in predicting human health effects which result from living and working in the presence of low levels of environmental agents, then our interest must focus on low dose rates, in which exposure occurs intermittently or continuously (Mode iv).

In theory, low dose rate x short time effects can be assessed in two ways. First, direct measurements of induced responses can be made, utilizing standard microbiological approaches in mutation assays. Second, extrapolations from data obtained at high dose rate

Table I. Definition of Terminology

Dose Rate	Concentration of chemical or amount of radiation per unit time.
Dose	Dose rate x time of exposure; also called total dose, total accumulated dose, or integral dose.
Short Time	Short time of exposure; $\leq$ 1 cell generation, or sometimes used as relative to a longer time of exposure.
Long Time	Long time of exposure; >> 1 cell generation, or sometimes used as relative to a shorter time of exposure.
High Dose Rate	Dose rate in which significant toxic or mutagenic effects are observed in a short time in single cell systems. Variously used to mean a higher rate than common to human experience.
Low Dose Rate	Dose rate in which no significant effects are observed in a short time. Variously used to mean a rate within the range of human experience.
High Dose	A dose which yields a significant toxic or mutagenic response.
Low Dose	A dose which does not yield a significant toxic or mutagenic response.

x short time might conceivably yield approximations of the low dose rate response. In practice, there are drawbacks to both of these approaches, which we will discuss later in this paper.

A low dose rate x long time protocol is the most reasonable approach to study the effects of low levels of mutagens because such treatments permit the total accumulated dose to become high. This protocol makes feasible the detection of a significant increase in mutant fraction. However, we do not mean to imply that high doses accumulated in a low dose rate x long time protocol are equivalent to those accumulated in a high dose rate protocol. In fact, in the discussions which follow, we argue that any prediction about the relationship between effects caused by doses at high dose rate x short time versus low dose rate x long time is impossible.

We will discuss two variations of low dose rate x long time protocols which have been used to study mutation induced by low levels of environmental agents. In multiple low dose experiments, low dose rate x short time is repeated many times with a defined time interval separating the exposures. The continuous low dose experiments

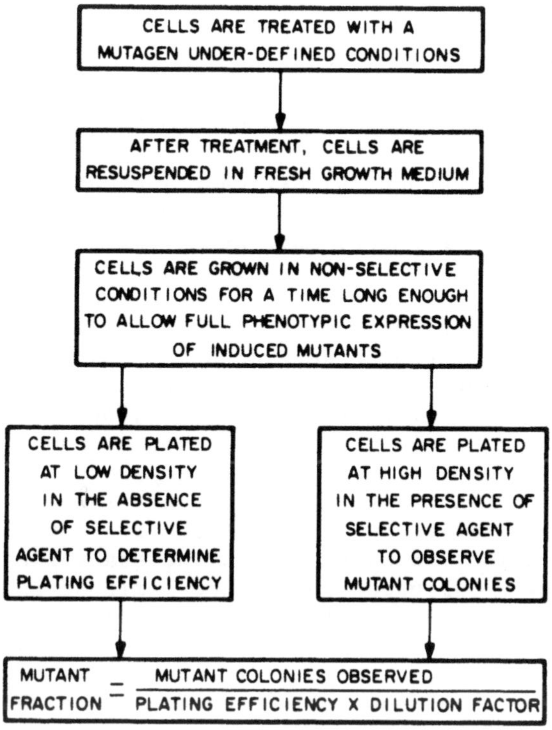

Fig. 1. General format of mutational assay in human lymphoblasts

utilize a low dose rate x long time exposure in which cells are grown in the presence of an approximately constant level of mutagen. In both protocols, the total dose is allowed to become high, and increases in mutant fraction are then measured.

Fig. 1 is a flow diagram outlining the principal steps involved in the performance of a mutational assay. Cells must be treated under defined conditions of dose rate and time of exposure so that the integral dose can be determined. The surviving fraction must be measured so that the investigator knows how many cells survived the treatment. Only in this way can one determine if a significant number of mutants survived the treatment. (Clearly, the statistical certainty for the mutant fraction depends on this number; unfortunately, measurement of the surviving fraction is not often performed, especially in experiments with bacteria.)

After treatment, cells must be grown in nonselective medium for a time long enough to allow full phenotypic expression of induced mutants. This parameter varies widely, depending on the locus, the cell system studied and even on the choice and dose of mutagen. Finally, cells can be plated in selective conditions to enumerate mutants and in nonselective conditions to determine plating efficiency. The mutant fraction (mutants per cell) is then calculated.

DIRECT MEASUREMENTS OF MUTATIONS FROM LOW DOSES OF CHEMICALS/RADIATION

Are there means to measure significant increases in mutant fraction after the application of a low dose of mutagen? The vast majority of mutation experiments have been performed at high dose because it is not easy to detect a mutational response after treating cells with a low dose. The background mutant fraction is never zero, and therefore the problem is to detect the signal above the noise. A series of independent determinations will yield a distribution of responses, both for the background and for treated populations. In our experience, with some 5,000 bacterial and several hundred human B cell mutation assays, we have found that the control values are distributed as if they were a multiple of Poisson distributions (that is, skewed to the right). Since this historical data base belies any expectation of a normally distributed control population, we have adopted the statistical expedient of using the observed 99% upper confidence limit of the untreated culture mutant fractions as a value that the mean mutant fraction of a treated culture must exceed in order to be considered significantly greater than the control. Additionally, we require that the treated and concurrent control mutant fractions demonstrate a statistically significant difference.

In bacterial forward mutation assays, our mean background mutant fractions for 8-azaguanine resistance are around 5×10^{-5} with a 99% upper confidence limit at 22×10^{-5}. For human mutation assays the mean background mutant fraction is about 3×10^{-6} with a 99% upper confidence limit of 6×10^{-6}. Thus, in our experience, a mutagen treatment of bacterial cultures would have to induce an average of $22 - 5 = 17 \times 10^{-5}$ mutants/survivor, and of human cells $6 - 3.2 = 2.8 \times 10^{-6}$ mutants/survivor in order to be detected as significantly mutagenic.

These, then, are the problems with using a single exposure to low doses of mutagens: (1) The background mutant fractions caused by relatively low rates of spontaneous mutations are significantly greater than those induced by exposures to mutagens at concentrations known to human experience; (2) Actual or traditional limitations in numbers of cells that can be conveniently handled lead to increases in the dispersion of estimates of the background mutant fractions.

Because of these problems of measuring a significant increase in mutant fraction at low dose, some investigators have attempted to increase sensitivity by using what has been called a "fluctuation test", but which, for reasons of accuracy, we feel shouldn't be termed a limiting dilution mutation assay. Based on the original fluctuation experiments of Luria and Delbruck (1943), these protocols involve mutagen treatment of a series of cultures at a time when each contains a very small number of cells so that the probability of a spontaneous mutant being present is small. These cultures are then grown to a large number of cells, and the mutant fractions of each is measured-- the magnitude of each mutant fraction depends on when the first mutant appeared during growth from low cell number. The pooled data from the series yields an estimate of the overall mutation rate. The reason that this protocol increases sensitivity is because it estimates mutation rate (mutation/cell/cell generation), rather than mutant fraction (mutation/cell). The latter value, obtained in the more "standard" mutation assay, includes in the background all mutations accumulated during previous generations, and it is always higher than the mutation rate. The lower noise value associated with limiting dilution assays thus allows the detection of mutation by lower doses.

In limiting dilution experiments performed in bacteria (Green et al., 1976; 1977a; 1977b) and yeast (Parry, 1977), a small number of amino acid auxotrophs were aliquoted into tubes containing a trace level of the required amino acid. A mutagen was added to one set of about fifty tubes, while another set served as control. Several days later, the tubes were screened for turbidity, which indicated the growth of prototrophs. In these assays, various mutagens have been reported to induce mutation at concentrations more than 100-fold less than in a typical plate assay. Care is required in interpreting results. In cases where the treatment is toxic, quantitative analysis of the data can become extremely difficult. For example, Green et al. (1976) found that 0.5 ug/ml potassium chromate caused an increase in the number of turbid tubes, while 2.5 ug/ml caused a significant decrease; undoubtedly this was due to toxicity to the mutants.

In mouse lymphoma cells, Cole et al. (1976) performed a limiting dilution experiment with ethylmethane sulfonate (EMS). For each experimental point, fifteen replicate samples of 500 cells were allowed to grow to about 10^6 cells and then treated for 30-48 hours. Cultures were plated to determine both the surviving fraction and the mutant fraction (measured as resistance to 1 mM ouabain). In terms of total accumulated dose, we find that the exposure which induced significant mutation in the limiting dilution assay was 1.2 mM x hr, as compared to 7.5 mM x hr in the standard assay in which cultures with large numbers of cells were treated. Thus, the increase in sensitivity from utilizing the limiting dilution protocol was 6.25-fold.

COMPARISONS OF MUTATIONS FROM HIGH AND LOW DOSE TREATMENTS

Almost all *in vitro* mutation experiments have been performed at high doses, where genetic changes can be more easily observed. However, this approach has made if necessary to extrapolate these data in order to predict low-level effects. Is this a reasonable exercise? We feel very strongly that the answer to this question is no. Our rationale for this belief is discussed below.

Mechanisms of Mutagenesis

In order to understand why exposure to low doses of mutagens might result in a different response than expected from extrapolation of high dose data, we shall first examine briefly a general outline of mechanisms leading to mutagenesis (Fig. 2). A mutagen interacts with DNA to produce a lesion, which, due to its potential to become a mutation, is termed a premutagenic lesion. The key point is that there are alternative pathways for the cellular metabolism of these DNA lesions.

1. Repair. A DNA repair system can recognize a DNA lesion and attempt to repair the damage. We can imagine that each repair system has an inherent probability of making an error during its repair function. Such an error might be lethal, but alternatively, it might be expressed as a gene locus mutation via a process we term misrepair.

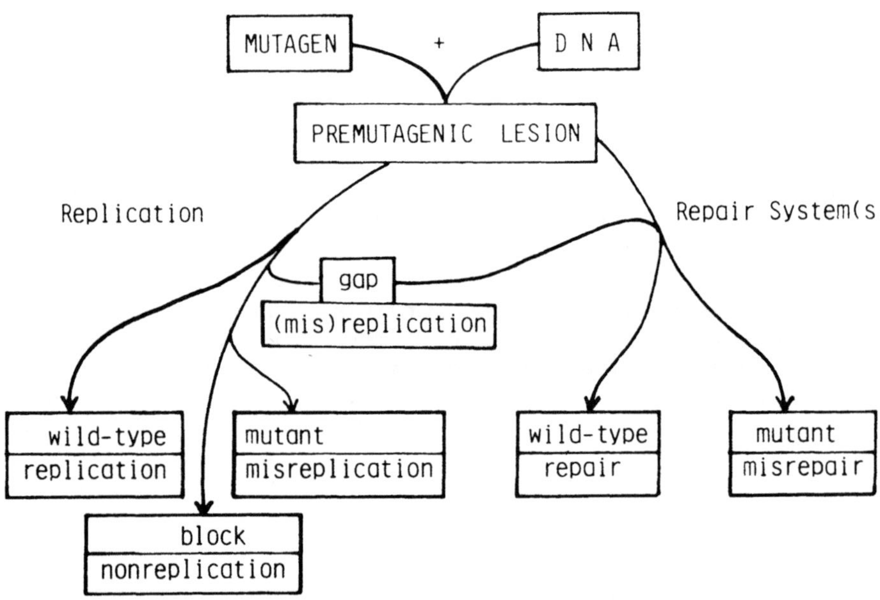

Fig. 2. Outline of mechanisms leading to mutagenesis

Consider that there is a set of different repair systems which may recognize the DNA lesion, termed X, and which may compete with one another to repair it. The different systems should have different affinities (Km) for X. Since the normal rules of enzyme kinetics would apply, the relative contribution of the various repair pathways to the total cellular metabolism of X will probably depend on the concentration of X (the number of lesions per cell). An additional complicating factor is that the expression of different repair systems in the cell may be either constitutive or inducible; furthermore, they may become saturated at very different concentrations of X.

2. Replication. If the DNA replication complex arrives at a lesion before repair has occurred, then there ought to be at least three possible outcomes. First, replication through the lesion may occur successfully, with maintenance of the correct DNA sequence. Second, the lesion may block replication which may result in cell death or perhaps merely cessation of growth awaiting the arrival of the repair system. Finally, replication may proceed with the DNA being copied incorrectly (misreplication). The polymerase may insert an incorrect base to produce a base-pair substitution mutation (reviewed in Loeb and Kunkel, 1982). If the region of DNA contains short repetitive sequences, slippage of the two strands, enhanced by the lesion, might result in a small deletion or insertion mutation (Streisinger et al., 1966). These two instances describe mutations which result directly in changes in the nucleotide sequence. Additionally, the replication complex may be able to continue through a lesion, leaving a gap in the nascent DNA strand. If the cell is to survive, the gap must later be repaired by DNA repair systems.

3. Mathematical Models for Mutational Dose-Response Curves. We cannot yet fully outline the possibilities for the metabolism of a single lesion; at the present time, we have only glimpses into the complexity of the cellular response to unwanted DNA lesions. The potential involvement of repair and replication systems in the generation of the shapes of mutational dose-response curves has been modelled algebraically for some simplified cases which are reviewed here (Thilly, 1983).

a. Assumptions

(1) DNA adducts are premutagenic lesions. There are two requirements for a DNA lesion to have the potential to cause mutation. First, it must occur in that part of the genome in which change would be detected by the mutation assay. This part of the genome may be considered the "target" for mutation and to have a dimension of β base-pairs. Secondly, even if a lesion forms in the target region, we should consider the fact that a mutagenic agent can form more than one kind of reaction product with DNA. That

fraction of all of its reaction products which are potentially able to cause mutation we denote f_m. The probability that these two requirements for mutation occur is simply equal to their product, $\beta \cdot f_m$.

(2) DNA lesions may be repaired, misrepaired or unrepaired. DNA repair is here defined as those processes which restore a section of DNA containing a lesion to the original DNA sequence. However, if the lesion is removed but a different sequence results, the process is termed misrepair. If no change is effected or if the process is abortive (for example, the lesion is removed, but no substrate for DNA replication results), then the process, or lack thereof, is called nonrepair. These definitions of repair, misrepair and nonrepair are useful because they are mutually exclusive in their end results. This logical construct aids in developing a mathematical description of what appears to be a complex mixture of molecular processes.

(3) Unrepaired DNA may be replicated, misreplicated or unreplicated. If DNA synthesis can proceed past a DNA lesion, then replication through the unrepaired site will have some finite probability of failing to duplicate the correct nucleotide sequence and thus give rise to a mutant sequence in one daughter strand.

What arises from this approach is a model in which a particular DNA lesion is either repaired (no mutation), misrepaired (probable mutation) or not repaired (probable lethality or mutation). Since the processes of repair and misrepair are assumed to be dependent on sets of enzymes acting on the area of a lesion as a substrate, we can imagine that the time available between the initial formation of the lesion and the initiation of attempted DNA replication represents the time available for repair or misrepair.

Lesions for which repair or misrepair systems exist will have increased probabilities of repair or misrepair as the time between lesion formation and DNA replication increases. A second prediction based on the model of the DNA lesion as substrate and the repair systems as enzymes is (i) that the probability of repair of a particular lesion will decrease as the number of altered sites increases, and (ii) that the probability of misrepair will also decrease with increased numbers of substrates for the repair/misrepair systems. A direct consequence of these expectations is that increased substrate (DNA lesions) is predicted to lead to a shift toward increased frequency of unrepaired lesions at the time of initiation of DNA synthesis.

b. Definitions and Explicit Formulation of Examples

In order to transform the model outlined above into something at least experimentally testable, the expectations must be expressed in quantitative terms relating cell mutation and survival to experimental variables.

A few definitions are required:

M mutant fraction, the fraction of those cells surviving treatment in which mutations have been induced.

S the fraction of cells surviving treatment.

D the average number of chemical adducts occurring in the cellular DNA (adducts/base pair).

β that portion of the cellular DNA which, when mutated, will be observed as a mutation (size of mutable target, base pairs).

f_m the fraction of adducts (DNA) which are potentially mutagenic by virtue of their chemical identity.

q the probability that an unrepaired lesion will give rise to a mutation during DNA replication.

$C_R(t)$ the capacity to repair a particular DNA lesion prior to DNA replication.

$C_M(t)$ the capacity to misrepair a particular DNA lesion prior to DNA replication.

$\emptyset$ the probability that an action of the misrepair system is a mutation.

Purists will already have noted that the definitions skip over many obvious possibilities. However, in order to keep things from becoming unnecessarily complicated, a few simplified situations are first proposed and their behavior considered.

The first simplification is to imagine a mutagen which reacts with DNA to give only one kind of lesion. The second is to consider that there is only one kind of system to remove this lesion. We can then combine some of them, such as multiple lesions and competition for substrates among recovery systems, and see if they make any sense in describing experimental observations.

Case 1: Nonrepair precedes mutation

For mutation by an imaginary chemical which creates only one kind of DNA lesion which could be repaired or not repaired, but not misrepaired, one should observe:

M = | Probability of Premutagenic Reaction in Target | $[\beta f_m D]$

x

| Probability of Nonrepair | $[e^{-C_R(t)/D}]$

x

| Probability of Misreplication | q

$$M = [\beta f_m D]\,[e^{-C_R(t)/D}] \cdot q$$

The probability that our genetic target has a potential mutation at all is $\beta f_m D$. The probability that this lesion goes unrepaired is $e^{-C_R(t)/D}$, where a capacity for $C_R(t)$ repair events is pictured as randomly distributed over D randomly distributed repairable lesions. The probability of nonrepair within the target is approximated by exponential [repair events/repairable events] via application of the Poisson distribution. Since the probability of the unrepaired site giving rise to mutation is q, and the three phenomena (DNA reaction, repair, DNA misreplication) are assumed to be independent events, the relationship between D and M, both directly measurable quantities, should be of the form derived. This model predicts a higher order dependence of mutation upon D than a simple linear response, as shown in Fig. 3.

This higher order curve has frequently been reported as a mutation dose response in which D is roughly approximated by knowledge of the initial concentration of chemical in the culture medium. We have observed examples of this kind of behavior for methylnitrosourea and ultraviolet light in human lymphoblast cultures.

Case 2: Misrepair precedes mutation

A second sample of this analytical process is the case in which an imaginary chemical creates only one kind of DNA lesion which can be misrepaired, but for which nonrepair does not lead to detectable mutation ($q = 0$). In such a case, we would observe the relation

M = | Probability of Premutagenic Reaction in Target | $[\beta f_m D]$

x

| Probability of Misrepair | $\emptyset\,[1-e^{-C_M/D}]$

$$M = \beta f_m D \times \emptyset \, [1-e^{-C_M(t)/D}]$$

where $\beta f_m D$ is again the probability of the genetic target sustaining a potentially mutagenic event. (However, the probability of misrepair $[1-e^{-C_M(t)/D}]$ is seen to be simply 1-[probability of not being misrepaired, $e^{-C_M(t)/D}$]. The probability of mutation by misrepair is proportional to the fraction of premutagenic lesions acted upon by the misrepair system $(1-e^{-C_M/D})$ multiplied by the frequency of error of the misrepair system, $\emptyset$. When $C_M(t)/D$ becomes a small number, as when D increases, the function $\emptyset\,(1-e^{-C_M(t)/D})$ approximates $\emptyset C_M(t)/D$ and leads to the prediction that at high values of D,

$$M = \beta f_m D \times C_M(t)/D$$

$$= \beta f_m \times C_M(t) = \text{a constant for any fixed value of } t.$$

This general form is a "saturation" type response as shown in Fig. 4. There are also many examples of this form of response in the literature of mutagenesis. ICR-191 has displayed this form of behavior in human lymphoblast experiments (DeLuca et al., 1977).

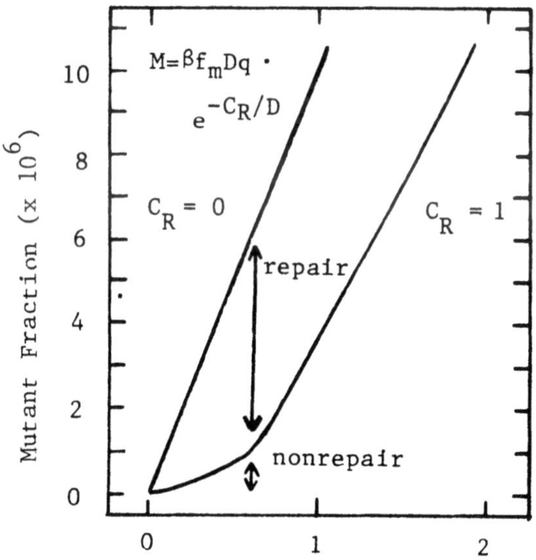

Fig. 3. Mutation as a function of initial DNA reaction products when mutation arises solely as a result of misreplication of unrepaired lesions. The product of the constants $\beta f_m q$ has been set at 10^{-5} for this example.

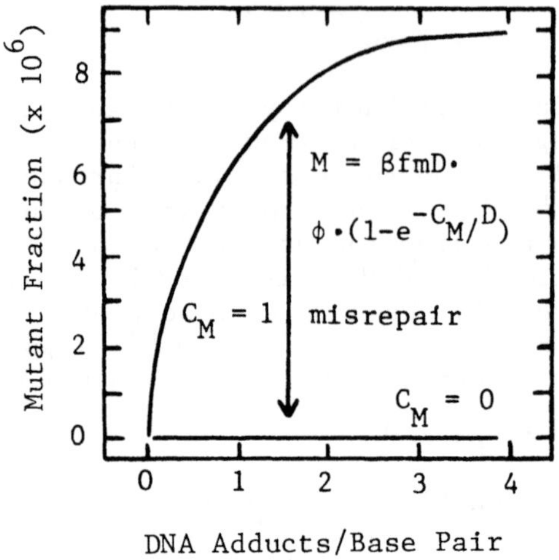

Fig. 4. Mutation as a function of initial DNA reaction products when mutation arises solely as a result of misrepair of lesions. The product of the constants βf_m has been set at 10^{-5} for this example. (Assumption for this example is that probability of misreplication of an unrepaired lesion, q, is zero.)

Case 3: Nonrepair and misrepair precede mutation

From our simple examples we can now make combinations of functions. For a case in which repair, misrepair and misreplication apply to the same single lesion formed by our single-adduct chemical, we can write

$$M = \beta f_m D \left([e^{-(C_M + C_R/D)} \cdot q] + \phi[1-e^{-C_M/D}] \right)$$

$$= \boxed{\text{Target Function}} \times \boxed{\text{Nonrepair \& Misreplication} + \text{Misrepair}}$$

In Fig. 5, the case of repair and misrepair systems of similar capacities is considered as an example. When q is relatively small (q = 0.01), then we can discern a more rapid rise in M at low D as misrepair makes its contribution. At higher D, both C_M and C_R are saturated, and mutation is seen to increase entirely as a result of nonrepair and misreplication.

Thus, our mathematical models lead to the obvious expectation that saturation of processing systems affect the shapes of mutational dose-response curves. The models suggest that observed nonlinearities can be accounted for by the interactions of (mis)repair and (mis)replication mechanisms, whose relative contributions to the overall metabolism of a particular lesion depend, among

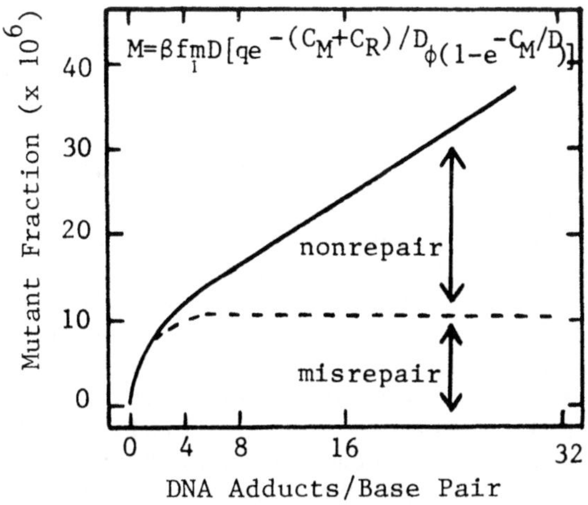

Figure 5. Mutation as a function of initial DNA reaction products when mutation arises as a combination of misrepair and misreplication of unrepaired lesions. The product of the constants βf_m have been set at 10^{-5} for this example with the probability of misreplication of unrepaired lesions, q, set at 0.10.

other things, on the abundance of the premutagenic lesion. Since this concentration varies with dose rate, we can expect that high dose-rate and low dose-rate exposures will yield considerably different results.

Factors Affecting the Pathways of Cellular Matabolism of DNA Damage

In this section, we consider some of the factors which determine the relative role of the different pathways of cellular metabolism of DNA damage.

1. Concentrations of DNA Lesions. A first set of factors are those that affect the chemical nature and distribution of the lesions. Treatment of cells with anything that ultimately reacts with DNA results in a spectrum of damage. If the agent interacted only with DNA, our first assumption might be that the spectrum would not change as a function of dose, unless the DNA became saturated with damage (a most unlikely possibility). However, damage might alter DNA conformation such that subsequent damage would result in a different spectrum. Also, of a chemical mutagen was first metabolized by cellular enzymes (such as transport elements of oxidative systems like P-450) before interacting with DNA, the damage spectrum would be expected to reflect any nonlinearity imposed by the kinetic constants of those enzymes. Therefore, the high dose and low dose situations could be quite different.

Another factor is the actual number of lesions in the DNA. This will especially affect the activity of repair systems which may be saturable at high doses, or even exhausted completely, a prime example of the latter case is $\underline{O}^6$-alkyl guanine transferase. A protein that removes the alkyl group from the $\underline{O}^6$-position stoichiometrically (Demple et al., 1982; Lindahl et al., 1982). Once the relatively small basal number of proteins per cell are used up, that mode of repair is no longer an option for the cell. $\underline{O}^6$-Alkylguanine is thought to be a premutagenic lesion because it can form base pairs with deoxythymidine as well as with deoxycytidine residues (Drake and Baltz, 1976; Dodson et al., 1982; Loeb and Kunkel, 1982). Therefore, it is clear that the elimination of the alkyltransferase protein can profoundly affect the subsequent mutagenicity of $\underline{O}^6$-alkylguanine lesions. The alkyltransferase mode of repair should be virtually 100% error-free because it seemingly abstracts the alkyl group without otherwise disturbing the DNA. The presence of this protein may explain why the mutational dose-response curves of some alkylating agents which could be mutagenic via an $\underline{O}^6$-alkylguanine misreplication mechanism are biphasic with little or no observed effect at low concentrations. Once the supply of alkyltransferase protein is gone, which would reproducibly occur at some particular dose, the mutational dose-response curve would rise abruptly.

In addition to the spectrum of lesions, and the number of lesions, a third factor which may affect the mutagenic potential of DNA damage is the relative abundances of the pathway components which modulate the response. The inducibility of repair systems is an important concept. Clearly, the preferential induction of one or more systems will alter the relative abundances of all the systems. Induction may be a function of lesion concentration, and its kinetics may be a function of time of continuous exposure. High dose versus low dose, or short time versus long time protocols may result in a completely different spectrum of repair enzymes, and consequently in an other than expected (as predicted by extrapolation from a different protocol) quantitative mutational response.

There is ample evidence in the literature to support the notion of inducible responses which modulate mutagenesis. In this brief summary, we will first consider systems found in *Escherichia coli*, and then review the evidence in mammalian cells.

2. Inducible Systems in Bacteria. The best studied instance of an inducible modulator is the SOS system in *E. coli* (reviewed by Witkin, 1976; Walker, 1984). Exposure to UV light and certain chemical agents induces a coordinated set of responses centered around the recA and lexA genes. Included among these responses are an error-prone pathway for repair of DNA damage, resulting in enhanced survival but with an increased yield of mutations. It appears as if the induction of the SOS response can also affect regions of the DNA which were not directly damaged (reviewed by Walker, 1984).

These mutations could reflect interactions on the SOS system with undamaged regions of DNA, or its effects on incidental damage such as spontaneous depurinations, or lesions caused by extraneous mutagens such as fluorescent light, oxygen radicals or endogenous alkylating agents.

Another inducible repair system in E. coli is termed the adaptive response (Samson and Cairns, 1977). When bacteria are exposed to a nontoxic and nonmutagenic dose of a methylating or ethylating agent, they become resistant to the subsequent toxic and mutagenic effects of a high concentration of the chemical. This response is independent of the SOS pathway and requires de novo protein synthesis in order to function. There are at least two proteins induced. One is the alkyltransferase, which has been described above. This protein, normally present at about 20-60 molecules per bacterium, increases to a level 100-200 times higher after induction (Lindahl et al., 1982; Mitra et al., 1982). The second protein is 3-methyladenine-DNA glycosylase II, which can release from DNA 3-methyladenine, 7-methyladenine, 3-methylguanine or 7-methylguanine. In his recent review, Walker (1984) has concluded that this enzyme removes potentially lethal lesions as well as premutagenic lesions which would otherwise undergo SOS processing.

A third set of inducible enzymes are the heat-shock proteins which were originally discovered to be induced in Drosophila (Ritossa, 1962; see Schlesinger et al., 1982, for review). Two proteins, which comigrate with two of the heat-shock proteins are induced in E. coli by UV light or nalidixic acid. The physiological significance of these proteins in chemical- or radiation-damaged cells is uncertain. Walker (1984) has speculated that they may play a role in the degradation of SOS enzymes, whose continued presence after their initial usefulness might prove detrimental to the cell.

Excision repair in E. coli is also at least partially inducible. The gene products of uvrA, uvrB and uvrC form an endonuclease which incises DNA near bulky lesions. These proteins appear to be coordinately expressed with the SOS system. The repair which the endonuclease initiates is thought to be highly accurate since variants lacking these loci are hypermutable (reviewed by Lindahl, 1982). Consequently, the induction of the SOS response seemingly results in an increase of both error-prone and error-free DNA repair systems. However, the error-prone system must predominate, since the end result is an increased mutation frequency (Witkin, 1976).

3. Inducible Systems in Mammalian Cells. The discovery of inducible repair systems which affect the levels of mutagenesis in bacteria has led investigators to search for similar systems in mammalian cells.

Studies describing a system which seems to be analogous to aspects of the *E. coli* SOS system have been reported (Das Gupta and Summers, 1978; Sarasin and Benoit, 1980; Sarasin et al., 1982). In the course of measuring mutation frequencies in viral genomes propagated through their mammalian hosts, it has been noted that UV irradiation of the host cells before infection with the UV-irradiated viral DNA caused a dramatic increase in the mutation frequency of the virus.

Certainly, the best-studied example of an inducible system in mammalian cells which may modulate mutagenesis is an adaptive response similar in some respects to the one described in *E. coli*. Like *E. coli*, mammalian cells contain an alkyltransferase activity which stoichiometrically removes alkyl groups, presumably in an essentially error-free manner. Initial reports indicated that a particular cell line would be either mex$^+$ or mex$^-$; it was supposed that this indicated that a cell would either remove *O*^{6}-alkylguanine lesions or it would not (e.g., Sklar and Strauss, 1981). However, in fact, Waldstein et al. (1982) have argued that all mammalian cells have about the same amounts of alkyltransferase (about 100,000 molecules per cell), but that mex$^+$ cells (1) have the ability to rapidly resynthesize the alkyltransferase acceptor protein (and thus remove more *O*^{6}-alkylguanine), and (2) that the activity is inducible after multiple low level exposures during which the concentration of alkyltransferase molecules per cell can increase up to three-fold above the basal level. In mex$^-$ cells, repetitive doses decrease the basal level of alkyltransferase, probably by simply exhausting the existing supply faster than it can be replaced.

Samson and Schwartz (1980) first reported evidence for such an adaptive DNA repair pathway in Chinese hamster ovary (CHO) and SV40-transformed human fibroblast cell lines. Chronic pretreatment with nontoxic concentrations of methylnitronitrosoguanidine (MNNG) rendered both cell lines resistant to the induction of sister chromatid exchanges by further doses of MNNG or methylnitrosourea (MNU). The pretreated CHO cells were also more resistant to killing by MNNG. However, later work by these same investigators revealed that the adaptive response did not reduce gene mutation in CHO cells (Schwartz and Samson, 1983).

At least two laboratories have sought evidence for the adaptation of V79 Chinese hamster cells to bring about altered responses to mutagens after previous exposure to low doses of methylating agents. Durrant et al. (1981) exposed cells to a pretreatment dose of 0.3 mM MNU for an unspecified time. They stated, but did not show, that this treatment allowed 100% survival. Five days later, they treated the cells with 1-5 mM MNU for 30 minutes and observed that pretreated cells were relatively resistant to killing, but equally sensitive to mutation (measured by 6TG-resistance) relative to the controls that had not been pretreated with MNU.

On the other hand, Kaina (1982; 1983) pretreated cells with MNNG (0.5-4.0 nM) or MNU (0.5-40 uM) for 6 hours and then added a challenge dose of 0.3 uM MNNG or 60 uM MNU for 60 minutes. His data showed a trend toward the pretreatment resulting in resistance to mutation. For MNNG, the intermediate pretreatment concentrations (1-10 nM) seemed to have the most pronounced effects, while at the higher levels (10-40 nM), pretreatment was apparently ineffective.

The discrepancy between these experiments may be accounted for by the fact that the treatment conditions varied significantly and different clonal derivatives of V79 were used. The mex phenotype is unpredictable (Sklar and Strauss, 1981) so it is not valid to assume that the repair capacities for $\underline{O}^6$-methylguanine were the same.

How can the presence of the alkyltransferase be expected to affect mutation experiments? If indeed the $\underline{O}^6$-alkylguanine is a premutagenic lesion, then its persistence in DNA would be a major factor in determining the efficiency of a particular chemical treatment. Mex$^+$ cells should have an advantage over mex$^-$ cells because they can resynthesize the alkyltransferase much more quickly. Chronic exposures of mex$^+$ cells to alkylating agents causes an increase of alkyltransferase levels over basal value, contrasted to a decrease for mex$^-$. Again assuming alkyltransferase mediated repair to be nonmutagenic under most treatment conditions, mex$^-$ cells should be hypermutable compared to mex$^+$. This hypothesis has not yet been tested.

Medcalfe and Lawley (1981) showed that repair of $\underline{O}^6$-alkylguanine lesions in mex$^+$ cells was very fast for the first 1-2 hours, and then it slowed to 1/5 the rate. At low doses of MNNG (0.1 mM), 90% of the $\underline{O}^6$-methylguanine lesions were removed during the initial time period; at 0.5 mM, only 50% of the lesions were excised. This is consistent with the idea of a limited number of nonrecyclable proteins performing the repair function. The shape of a mutational dose-response curve after various concentrations of an alkylating agent (as single doses) would be expected to be nonlinear. The damage from low doses would be largely repaired so the initial slope of the curve would be small. After exhaustion of the alkyltransferase, mutation would be expected to rise abruptly as a function of increasing dose. This effect should be more noticeable for mex$^+$ cells since they have the capacity to remove more $\underline{O}^6$-alkylguanine lesions before their supply of alkyltransferase is exhausted.

Dose-Fractionation Studies

One way to examine whether inducible or saturable systems play a role in mutagenesis is to perform dose-fractionation, or split-dose, experiments. Here, two or more pulse doses are separated by a time interval in order to determine if cellular factors can modulate the effects of subsequent doses. Generally, each pulse dose is high enough to induce a significant response. An early paper by Elkind and Sutton (1959) showed that administering a dose of ionizing radiation in two fractions resulted in increased survival. Such a reduction in a biological response due to dose fractionation is generally considered as evidence of a cellular repair process (reviewed in Elkind and Whitmore, 1967). If the split dose protocol leads to an enhanced effect, then one might conclude that the first dose induced some process which potentiates the effects of the second dose. Finally, when two treatments separated by time interval yield the same effect as when the total dose is given as one fraction, then the conclusion is that the effects are additive. Only in this latter case could extrapolation from high-pulse doses to low-protracted doses even be considered. A few selected examples of dose fractionation studies with radiation are discussed below.

Chang et al. (1978) studied ultraviolet light split-dose mutagenesis in V79 Chinese hamster cells. They reported that splitting the dose into fractions separated by time intervals of 4-24 hours resulted in the same or lower mutant fractions as the unfractionated control. DeLuca et al. (1983) studied UV split-dose mutagenesis in human lymphoblasts. He fractionated nontoxic (100% survival), moderately toxic (about 60% survival) and toxic (about 30% survival) fluences, varying the time intervals between doses from 0-24 hours. With the nontoxic fluences, the fractionated mutational response was reduced compared to the unfractionated control. There was little or no effect at the moderately toxic fluence, but at the toxic fluence, a time-dependent increase in mutant fraction was observed when the time interval was greater than 24 hours.

While interpreting the UV split-dose experiments, it is tempting to postulate the existence of inducible systems which are expressed after the first dose and can then modulate the second dose such that the summed response of the two treatments is not equal to the response of the same total dose given as a single exposure, Chang et al. (1978) suggested that an inducible error-free postreplication repair system exists in Chinese hamster cells. DeLuca et al. (1983) reached the same conclusion for human lymphoblasts, at least for the low-fluence response. These interpretations clearly argue that, if one cannot predict the mutant fraction expected by splitting a high dose into only two high fractionated doses, then one could never hope to predict what would occur if the high dose were split into low-dose fractions.

Jostes and Painter (1981) treated CHO cells with two 450-rad doses of x-rays separated by one or two hours. The observed mutant fractions at the 6TG-resistance locus were intermediate between the responses seen with either 450 or 900 rad. Thus, as in the case of

UV-treated cells, fractionated dose studies with x-rays indicate that repair systems may be inducible and that they may affect the quantitative response of the cell to a subsequent dose.

From the above discussion, it should be clear that a mutational response may vary greatly, depending on the circumstances of the particular exposure. We cannot expect to extrapolate high-dose data in order to predict a low-dose response.

Split-dose protocols are useful because they can indicate the presence of repair systems that can modulate a mutagenic response. Still, they utilize high doses and thus do not necessarily indicate the situation which will exist at low dose. Furthermore, one must be wary when interpreting these experiments because the first dose can cause cell cycle perturbations which may affect the mutability of the second dose, due to the sensitivity of certain stage(s) of the cell cycle.

PROTRACTED EXPOSURES TO LOW LEVELS OF MUTAGENS

We feel that the most reasonable approach to the study of effects of low levels of mutagens is to perform low dose x long time experiments with the goal of observing a statistically significant response. Humans are exposed to low doses of mutagens either in an intermittent or in a continuous fashion. Each of these situations can be mimicked for single cells in culture. An intermittent exposure consists of treating cells with multiple low doses, each separated by a defined time period, Δt. In essence, it is a dose-fractionation experiment; the major difference is that low doses are used. As Δt goes to zero, this protocol becomes a continuous exposure mode.

In studies of these kinds, the exposure to mutagen can be expressed as dose per unit time, or dose rate. When comparing the low-dose protocol with high-dose experiments, dose rate multiplied by time of exposure yields a measure of the total dose. Once again, we can ask the question: is a short term, high dose-rate exposure equivalent to a long term, low dose-rate exposure?

Dose-Response Studies with Multiple Low Doses

In this section, we review the literature, including our own work on long term, low dose-rate mutation experiments. Wherever possible, we have compared the data to higher dose-rate experiments.

There are only a few reports of treatments of mammalian cells with multiple low doses of mutagens.

In the course of looking for a mammalian cell adaptive response, Schwartz and Samson (1983) treated CHO cells with MNNG as 13 doses of 0.01 ug/ml each, spaced at 6-hour intervals. They noted an increase in the $6TG^R$ mutant fraction which was approximately equal to that observed after a single exposure to 0.1 ug/ml MNNG.

Nakamura and Okada (1983) treated L5178Y mouse lymphoma cells with multiple low doses of EMS. They utilized a treatment protocol in which they exposed cells to 1 mM EMS for 20 hours (survival > 80%), followed by a 6-day recovery period; that cycle was repeated up to 6 times. They found, at the 6TG-resistance locus, that the multiple treatments were strictly additive.

Our group has performed experiments to explore the mutagenicity of multiple exposures to low concentrations of chemical mutagens. We will review those here in some detail. The two questions we posed initially were: (1) Do prior exposures to low concentrations of mutagens affect subsequent treatments or is each an independent event? and (2) How do the dose-response curves obtained at low levels of mutagen exposure compare to those obtained at high levels? In our first experiments, we treated TK6 human lymphoblast cells with either MNU, EMS or 4-nitroquinoline-N-oxide (4NQO) (Penman et al., 1983). Two protocols were used. First, single exposure, high dose experiments were performed, utilizing at least five different concentrations of mutagens for only one day. Second, the cells were exposed once each day for 1, 5, 10, 15 or 20 days to each of the mutagens at concentrations which did not induce statistically significant mutation or toxicity in a single exposure. Mutation was measured at two distinct genetic loci. Resistance to the purine analog 6TG measured mutation at the X-linked hgprt locus, while resistance to the pyrimidine analog trifluorothymidine (F_3TdR) measured mutation at the heterozygous, autosomal thymidine kinase locus (Thilly et al., 1980; Furth et al., 1981; Liber and Thilly, 1982).

The results from these experiments are plotted in Figs.6-10. The three mutagens were chosen for study because the respective dose-response curves in the high exposure condition have three distinct shapes (Fig. 6). MNU is linear, 4NQO plateaus and EMS has a concave upward shape. The concentrations used daily in the multiple exposure protocol all resulted in surviving fractions greater than 95% for a 1-day treatment. The plots of mutant fraction as a function of time of exposure for different concentrations are shown for each mutagen in Figs. 7-9. In each case, after a least squares power curve fit revealed that there was no tendency to deviate from linearity, straight lines were fitted through the data.

Finally, the data were transformed so that a comparison between the multiple low-dose protocol and the single exposure, high-dose protocol could be made. The daily increase in the mutant fraction (average of 6TG-resistance and F_3TdR resistance); that is, the slopes of Figs. 7-9 were plotted versus the daily mutagen concentration (Fig. 10). The extrapolation from the high-dose experiment is shown as a dashed

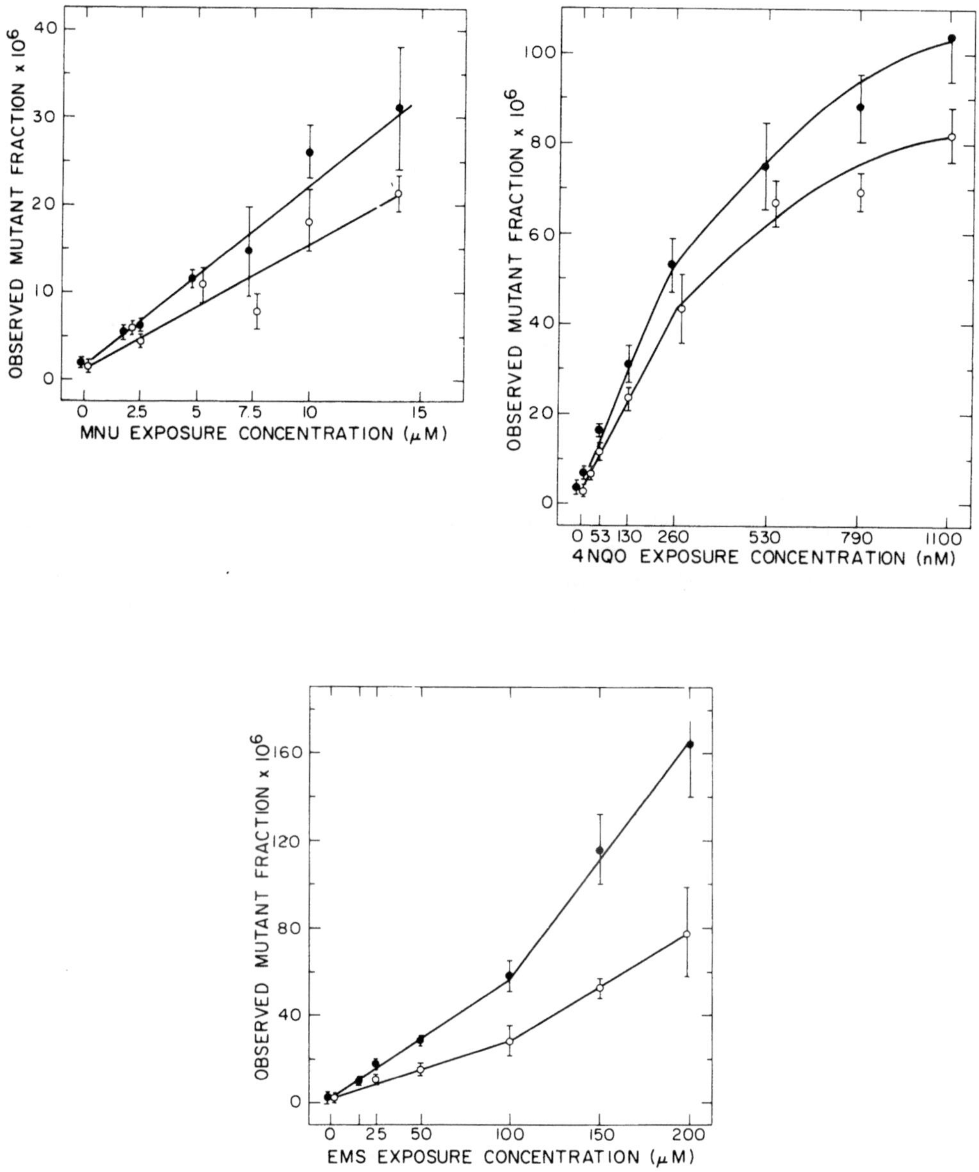

Fig. 6. Mutation of TK6 by methylnitrosourea (MNU), 4-nitroquinoline-N-oxide (4NQO) and ethylmethane sulfonate (EMS). Cultures were exposed to the indicated concentrations and resuspended in fresh medium after 24 hr. The trifluorothymidine-resistant (F_3TdR^R) (●) and 6-thioguanine-resistant (o) mutant fractions were determined 8 days later. Error bars are standard error of the mean (from Penman et al., 1983).

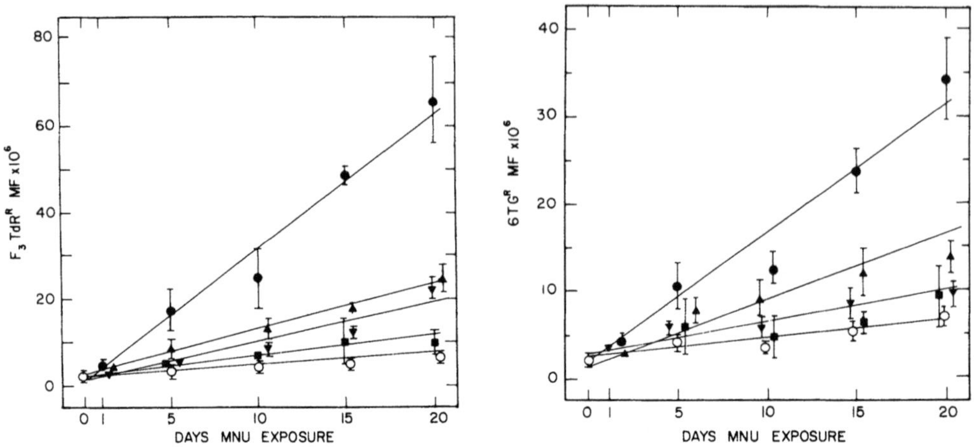

Fig. 7. Mutation of TK6 to F_3TdR^R or $6TG^R$ as a function of days of exposure to MNU. Cultures were exposed to 1.0 (●), 0.5 (▲), 0.25 (▼), 0.125 (■) or 0 (o) uM MNU for the indicated number of days. Cultures were treated daily. Error bars are the standard error of the mean (from Penman et al., 1983).

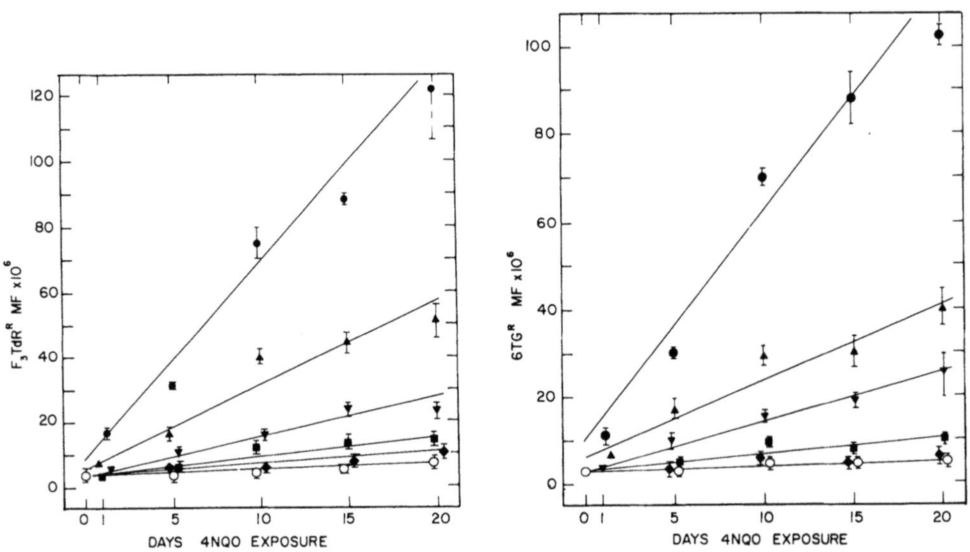

Fig. 8. Mutation of TK6 to F_3TdR^R or $6TG^R$ as a function of days of exposure to 4NQO. Cultures were exposed to 53 (●), 26 (▲), 13 (▼), 6.6 (■), 3.3 (◆) or 0(o) nM 4NQO for the number of days indicated. Cultures were treated every day. Error bars are the standard error of the mean (from Penman et al., 1983).

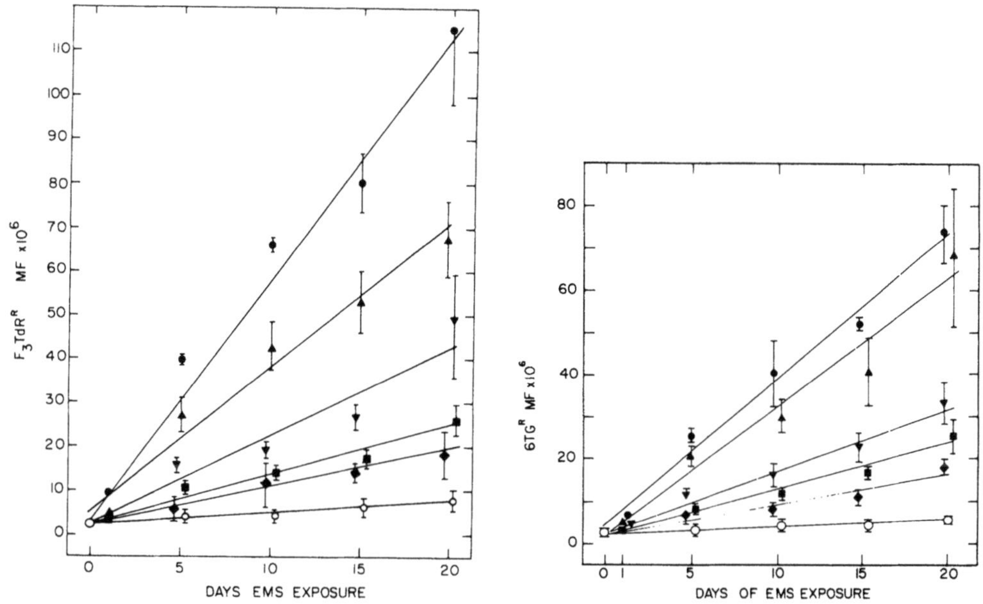

Fig. 9. Mutation of TK6 to F_3TdR^R or $6TG^R$ as a function of days of exposure to EMS. Cultures were exposed to 16 (●), 8 (▲), 4 (▼), 2 (■), 1 (◆) or 9 (o) uM EMS for the number of days indicated. Error bars are the standard error of the mean (from Penman et al., 1983).

line. For EMS and MNU, the multiple low dose protocol clearly yielded the same response as did the high dose procedure; that is, each treatment was independent and the effects were additive. For 4NQO, the multiple low dose protocol was somewhat less efficient at producing mutations than was predicted by the extrapolation of the high dose curve. Our major conclusion from this study was that, in the human lymphoblast cells used, there was no cellular response induced by low levels of exposure to the chemicals that could significantly influence the response to subsequent treatments 24 hours later.

The TK6 cell line has been shown to be mex^- (Sklar and Strauss, 1981), and so we cannot predict how the dose-response curves for a mex^+ cell line would have looked with agents such as EMS and MNU, which induce significant amounts of $\underline{O}^6$-alkylguanine.

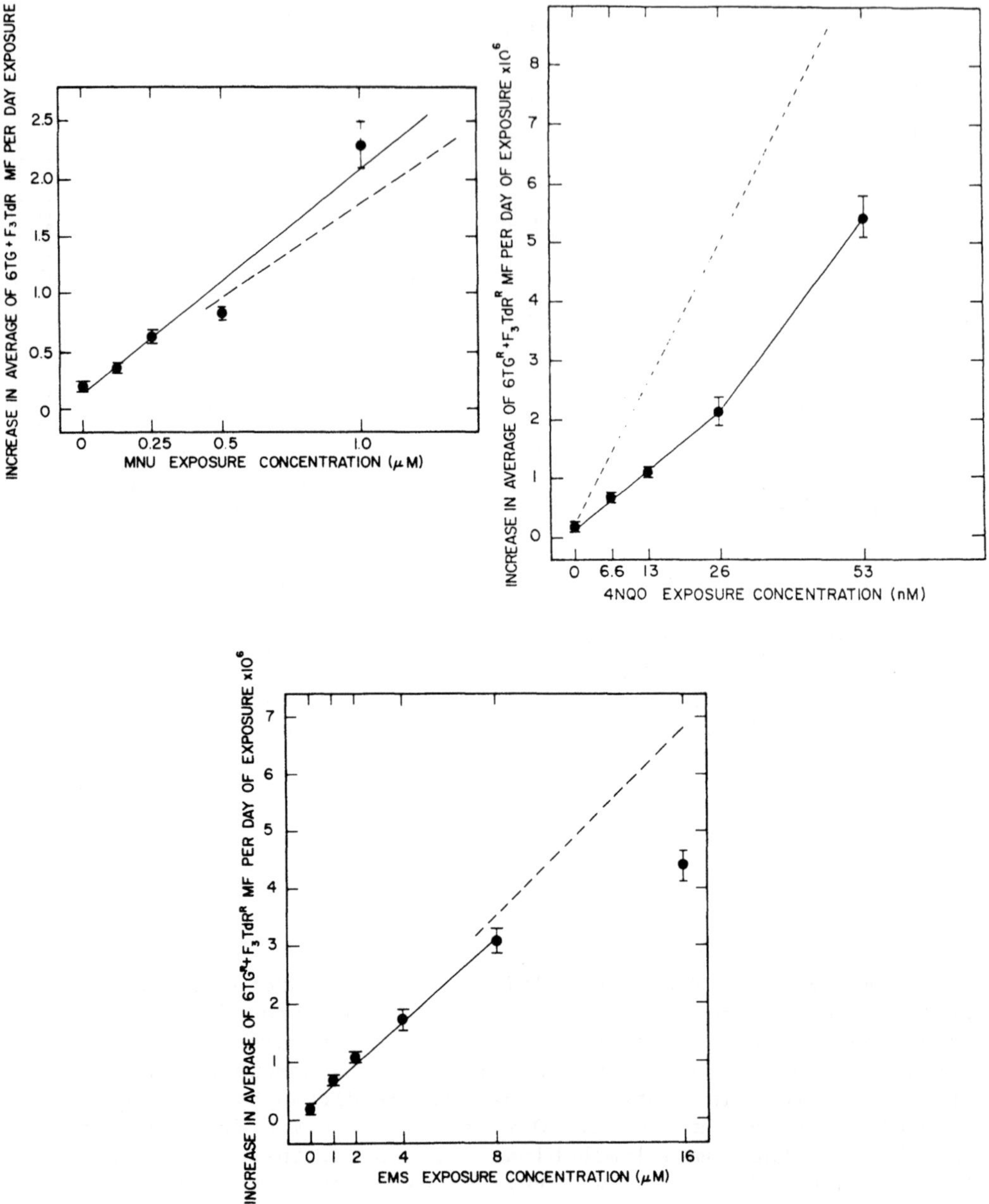

Fig. 10. Increase in the average of the TK6 F_3TdR^R and $6TG^R$ mutant fractions per day of mutagen exposure as a function of exposure concentration. Values plotted are the slopes of regression lines ± S.D. through the average of F_3TdR^R and $6TG^R$ mutant fractions from Figs. 7-9. The dashed line is regression line through average of F_3TdR^R and $6TG^R$ data from single exposure experiments of Fig. 6. The solid curve was fitted by eye.

Dose-Response Studies with Continuous Doses

Perhaps a disadvantage to studying mutagenicity using discrete multiple low-level treatments is that the time interval between doses may become an important contributor to the observed response. For example, if a relatively short-lived modulating system were induced, its potential effects on subsequent doses would not be seen if the interval were too long. Therefore, protocols which involve continuous exposures are valuable experimental tools. We review below those available reports which utilize such procedures for measuring mutation caused by different dose rates of radiation or chemicals.

1. Radiation. The effect of x-ray dose rate on survival and mutation in Neurospora crassa was studied by deSerres et al. (1968). Forward mutation to adenine auxotrophy yielded pigmented colonies, and the mutants consisted of gene-locus mutations and multilocus deletions which could be distinguished by genetic analysis. Dose rates of 10 rad/min or 1000 rad/min were equally effective in reducing survival and in inducing multilocus deletions.

Nakamura and Okada (1981) studied mutational dose-rate effects of gamma-rays in L5178Y mouse lymphoma cells and expressed their data as mutant fraction versus total dose (dose rate x time). For the 6TG-resistance locus, they fit their data to a linear quadratic curve at the high dose rate of 50 rad/min, but to a linear curve with a lower slope at the low dose rate of 0.8 rad/min. For the methotrexate-resistance locus, both curves were fitted to a linear plot; the slope of the high dose rate curve was about 2.5 times larger than the slope of the low dose rate curve. These results argue that different mutational mechanisms may be occuring at high versus low dose rates as well as between different loci. Ueno et al. (1982) also performed mutation experiments at the 6TG-resistance locus in L5178Y cells, utilizing low dose rates (0.2-0.8 rad/min) of gamma radiation. Their data are nearly identical to that of Nakamura and Okada (1981).

Thacker and Stretch (1983) performed similar experiments in V79 Chinese hamster cells, utilizing γ-rays at 169 rad/min as the high dose rate and at 0.34 rad/min as the low dose rate. Their endpoint also was mutation to 6TG-resistance. The low dose rate was considerably less efficient at producing mutations than the high dose rate, on a total exposure basis. Unlike the results with mouse lymphoma cells, both dose rates led to curves that fit a linear quadratic equation.

Our laboratory has modelled continuous exposure to ionizing radiation by growing human lymphoblast cells in the presence of tritiated water (HTO) (H. Liber, unpublished observations). HTO concentrations of 33-400 uCi/ml were used, corresponding to dose rates of 0.0054-0.064 rad/min of β-radiation. The exposure times were 2, 4, 6 and 8 days for each concentration. Preliminary results have indicated the following: (1) The mutant fraction (6TG-resistance and TFT-resistance) as a function of concentration for each time of exposure was apparently linear; (2) The mutant fraction as a function of time of exposure showed a tendency toward saturation at longer times. If these observations are confirmed, it could represent evidence of cellular adaptability to the mutagenic effects of ionizing radiation.

2. Chemicals. Probably the earliest studies utilizing continuous exposures to low levels of mutagens were the experiments of Kubitschek and his colleagues (Kubitschek, 1960; Kubitschek and Bendigketi, 1961, 1964a, b; Kubitschek, 1970). They measured mutation of *E. coli* to resistance to the toxic effects of bacteriophage T5, using an assay system originally described by Novick and Szilard (1950). These experiments were done in a chemostat, using glucose-limited growth to control the generation time of the bacterial cultures. The data were plotted as mutants/day versus cell generation/day. In this way, the shapes of the curves were potentially indicative of the mechanism of mutation.

> "(1) If mutagenesis occurs during synthesis of the gene, the mutation rate will be proportional to the frequency of replication, and therefore to the generation rate.
>
> (2) If mutagenesis occurs after synthesis of the gene, the mutation rate will be independent of the replication rate, and therefore mutants will accumulate at a constant rate."*

Studies with 2-aminopurine nitrate (25 ug/ml) or caffeine (450 ug/ml) yielded curves that fit category (1), while experiments using UV light (0.5 erg/mm^2/sec) or acridine orange plus white fluorescent light yielded curves that fit category (2).

The effect of dose rate on gene locus mutations by diepoxybutane (DEB) has been carefully studied in *Neurospora crassa* (Kolmark and Kilbey, 1968; Auerbach and Ramsey, 1970; Kilbey, 1973, 1974). Since DEB has a long half-life relative to the exposure times used in those experiments, the dose rate is equivalent to the initial concentration per unit time. At an initial DEB concentration of 0.01 M, mutation

*Quoted from Kubitschek (1970).

as a function of exposure time was linear, but at higher initial concentrations of > 0.03 M, the time-response curve was curvilinear with positive curvature. Initial concentrations of 0.03, 0.10 or 0.31 M were equally efficient at producing mutations in that they induced the same mutant fraction at equivalent total doses (concentration x time) (Kolmark and Kilbey, 1968).

Auerbach and Ramsey (1970) went on to investigate this effect further. Pretreatment with 0.01 M DEB was found to increase the mutant fraction induced by subsequent exposure to 0.1 M DEB. Also, a 60-minute exposure to 0.037 M DEB induced more than twice as many mutants per survivor as a 30-minute exposure to 0.037 M. If the 60-minute exposure was interrupted briefly, an intermediate number of mutants per survivor was obtained. If two 30-minute exposures were separated by 2 hours, the effect on the mutant fraction was additive.

The authors' interpretation of these results was that DEB treatment inactivated or depleted a cellular repair system that could reduce DEB-induced mutation and that, after exposure ended, this repair system could be regenerated. Subsequent experiments supported this hypothesis. Inhibition of protein synthesis resulted in a condition in which the cells were more easily mutated by subsequent DEB treatments, indicating that protein synthesis was required for the regeneration of the cellular repair system (Kilbey, 1973). When cultures were treated at a low DEB concentration of 0.005 M, the dose-response curve for mutation was linear as a function of total dose. However, if protein synthesis was inhibited during treatment at this concentration, the dose-response curve once again bent upward (Kilbey, 1974). This finding suggested that at low concentrations of DEB, the repair system was being inactivated but that this inactivation was balanced by resynthesis.

We have performed some preliminary experiments in which human lymphoblast cells competent for the metabolic activation of certain polycyclic aromatic hydrocarbons have been grown in the continuous presence of low levels of benzo(a)pyrene (BP). The cells were maintained daily with fresh BP-containing medium. This compound was used by the cells and degraded by the medium very slowly, so that >90% (est) remained after each 24-hour period. The exposure was thus continuous. The results of these experiments are presented in Figs. 11-13. When mutant fraction is plotted as a function of treatment time, the curve rose approximately linearly for the first five days, but there was little or no change upon further incubation (Fig. 11). We also noted an obvious saturation of response between the low dose of 0.1 uM and the higher doses of 0.5 or 1 uM. When the data were replotted as mutant fraction versus concentration, we again visulaize this saturation phenomenon (Fig. 12). There is no apparent difference between 5- and 10-day treatments, nor between 0.5 and 1.0 uM concentrations. Earlier experiments had been done with these cells at higher concentrations of BP for shorter

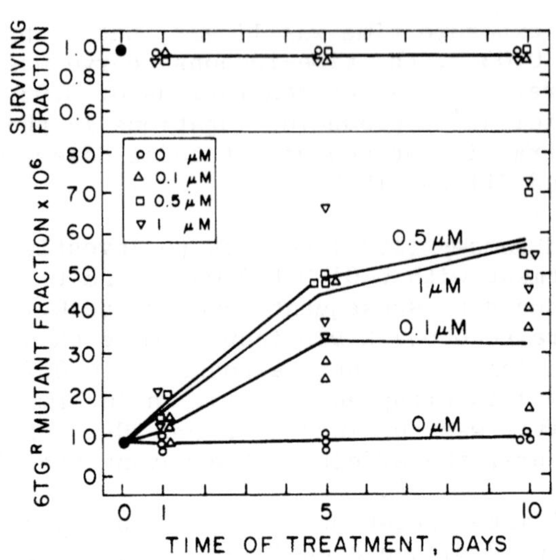

Fig. 11. Mutagenicity of benzo(a)pyrene (BP) to AHH-1 human lymphoblasts as a function of time. Cells were incubated for the indicated times with various concentrations of BP. The 6TG^R mutant fractions were determined 8 days after the end of treatment.

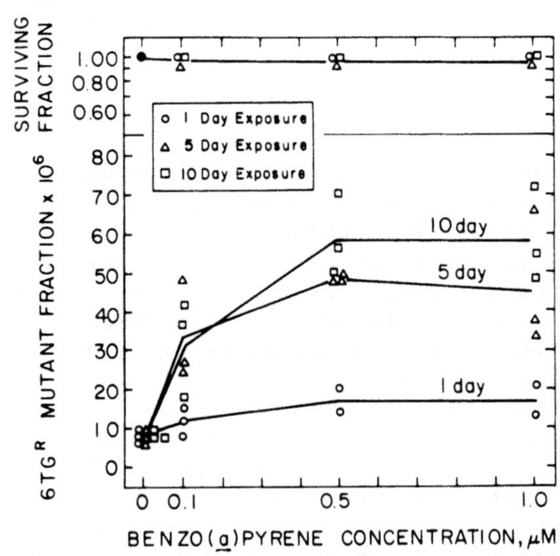

Fig. 12. Mutagenicity of BP to AHH-1 cells as a function of concentration. The data from Fig. 11 are replotted here.

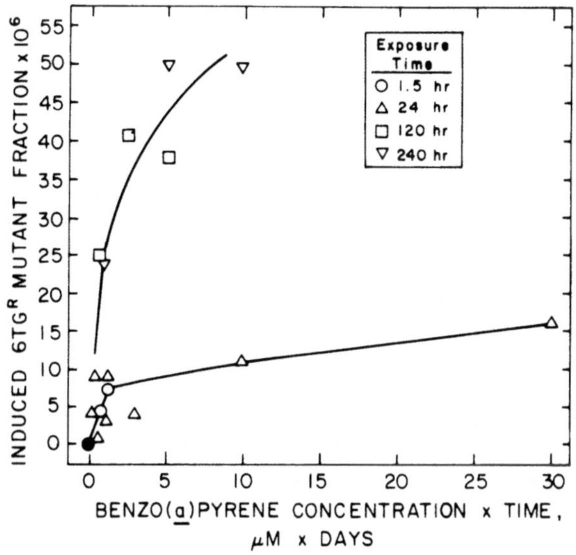

Fig. 13. The mutagenicity of BP as a function of total dose (concentration x time of exposure).

exposure times (1.5 or 24 hours) (Crespi, 1982). Thus, we were able to compare high and low dose-rate data by plotting mutant fraction versus the total exposure to BP, which is equal to concentration x time (Fig. 13). The data very clearly divide into two regions. The high dose-rate data appear to follow a biphasic curve, while the low dose-rate data are much more efficient at producing mutation on the basis of this total dose plot.

Interpretation of these experiments is complicated by the fact that BP induces the oxidative metabolizing enzymes in these cells, and we have not yet examined those levels nor the overall binding levels of BP to DNA beyond an initial 24-hour exposure. However, whether it be BP metabolizing systems or DNA repair systems, it is clear that <u>something</u> is modulating the mutagenic response over time.

CONCLUSIONS

The importance of understanding the mechanisms of low-dose mutagenesis lies mainly in the fact that human exposure to most environmental chemicals and radiation usually occurs chronically in a continuous or intermittent mode.

Extrapolation of data obtained under high-dose conditions cannot be assumed to represent what would be observed under low-dose conditions because the overall yield of mutants following a particular treatment is dependent on the interactions of a number of cellular systems, including mutagen metabolism, DNA replication and DNA repair. These systems may be constitutive or inducible, saturable (or exhaustable) or present in essentially unlimited quantity. The summation of the effects of these processes on the mutagen and on the lesions it produces determines the mutagenic response. We expect that the summations will be different with different treatment regimens. This expectation is realized by noting that low-dose protocols are apparently either more, less, or equally efficient (on the basis of total dose) at inducing mutations than high-dose protocols!

Consequently, low-dose studies of mutagens must employ low dose. We have reviewed and discussed the use of multiple and continuous low-dose protocols, and we would certainly urge that research in genetic toxicology and radiobiology give major emphasis to these more difficult but considerably more realistic modes of experimentation.

REFERENCES

Auerbach, C., and Ramsey, D., 1970, Analysis of a case of mutation specificity in Neurospora crassa. III. Fractionated treatment with diepoxybutane, Mol. Gen. Genet., 190:285.

Chang, C.-C., D'Ambrosio, S.M., Schultz, R., Trosko, J.E., and Setlow, R.B., 1978, Modification of UV-induced mutation frequencies in Chinese hamster cells by dose fractionation, cycloheximide and caffeine treatments, Mutat. Res., 52:231.

Cole, J., Arlett, C.F., and Green, M.H.L., 1976, The fluctuation test as a more sensitive system for determining induced mutation in L5178Y mouse lymphoma cells, Mutat. Res., 41:377.

Crespi, C.L., 1982, Xenobiotic metabolism and mutation in diploid human lymphoblasts, Ph.D. Thesis, Massachusetts Institute of Technology, Cambridge.

Das Gupta, U.B., and Summers, W.C., 1978, Ultraviolet reactivation of herpes simplex virus is mutagenic and inducible in mammalian cells, Proc. Natl. Acad. Sci.,USA, 75:2378.

DeLuca, J.G., Kaden, D.A., Krolewski, J., Skopek, T.R., and Thilly, W.G., 1977, Comparative mutagenicity of ICR-191 to S. typhimurium and diploid human lymphoblasts, Mutat. Res., 46:11.

DeLuca, J.G., Weinstein, L., and Thilly, W.G., 1983, Ultraviolet light induced mutation of diploid human lymphoblasts, Mutat. Res., 107:347.

Demple, B., Jacobsson, A., Olsson, M., Robins, P., and Lindahl, T., 1982, Repair of alkylated DNA in E. coli, J. Biol. Chem., 257: 13776.

deSerres, F.J., Malling, H.V., and Weber, B.B., 1968, Dose-rate effects on inactivation and mutation induction in Neurospora

crassa, Brookhaven Symp. Biol., 20:56.
Dodson, L., Foote, R.S., Mitra, S., and Masker, W.E., 1982, Mutagenesis of bacteriophage T7 in vitro by incorporation of O^6-methylguanine during DNA synthesis, Proc. Natl., Acad., Sci., USA, 79:7440.
Drake, J.W., and Baltz, R.H., 1976, The biochemistry of mutagenesis, Ann. Rev. Biochem., 45:11.
Durrant, L.G., Margison, G.P., and Boyle, J.M., 1981, Pretreatment of Chinese hamster V79 cells with MNU increases survival without affecting DNA repair or mutagenicity, Carcinogenesis, 2:55.
Elkind, M.M., and Sutton, H., 1959, X-ray damage and recovery in mammalian cells in culture, Nature, (London), 184:1293.
Elkind, M.M., and Whitmore, G.F., 1967, "The Radiobiology of Cultured Mammalian Cells," Gordon and Breach, N.Y.
Furth, E.E., Thilly, W.G., Penman, B.W., Liber, H.L., and Rand, W.M., 1981, Quantitative assay for mutation in diploid human lymphoblasts using microtiter plates, Anal. Biochem., 110:1.
Green, M.H.L., Bridges, B.A., Rogers, A.M., Horspool, G., Muriel, W. J., Bridges, J.W., and Fry, J.R., 1977b, Mutagen screening by a simplified bacterial fluctuation test: Use of microsomal preparations and whole liver cells for metabolic activation, Mutat. Res., 48:287.
Green, M.H.L., Muriel, W.J., and Bridges, B.A., 1976, Use of a simplified fluctuation test to detect low levels of mutagens, Mutat. Res., 38:33.
Green, M.H.L., Rogers, A.M., Muriel, W.J., Ward, A.C., and McCalla, D.R., 1977a, Use of a simplified fluctuation test to detect and characterize mutagenesis by nitrofurans, Mutat. Res., 44:139.
Jostes, R.F., and Painter, R.B., 1980, Evidence for repair of premutational damage from split X-ray doses, Radiat. Res., 85:586.
Kaina, B., 1982, Enhanced survival and reduced mutation and aberration frequencies induced in V79 Chinese hamster cells pre-exposed to low levels of methylating agents, Mutat. Res., 93:195.
Kaina, B., 1983, Studies on adaptation of V79 Chinese hamster cells to low doses of methylating agents, Carcinogenesis, 4:1437.
Kilbey, B.J., 1973, The stabilization of a transient mutagen-sensitive state in Neurospora by the protein synthesis inhibitor actidione, Mol. Gen. Genet., 123:67.
Kilbey, B.J., 1974, The analysis of a dose-rate effect found with a mutagenic chemical, Mutat. Res., 26:249.
Kolmark, H.G., and Kilbey, B.J., 1968, Kinetic studies of mutation induction by epoxides in Neurospora crassa, Mol. Gen. Genet., 101:89.
Kubitschek, H.E., 1960, The error hypothesis of mutation, Science, 131:730.
Kubitschek, H.E., 1970, "Introduction to Research with Continuous Cultures," Prentice Hall, Inc., Englewood Cliffs, N.J.
Kubitschek, H.E., and Bendigketi, H.E., 1961, Latent mutants in chemostats, Genetics, 46:105.

Kubitschek, H.E., and Bendigketi, H.E., 1964a, Mutation in continuous cultures. I. Dependence of mutational response upon growth limiting factors, Mutat. Res., 1:113.

Kubitschek, H.E., and Bendigketi, H.E., 1964b, Mutation in continuous cultures. II. Mutations induced with ultraviolet light and 2-amino-purine, Mutat. Res., 1:209.

Liber, H.L., and Thilly, W.G., 1982, Mutation assay at the thymidine kinase locus in diploid human lymphoblasts, Mutat. Res., 94:467.

Lindahl, T., 1982, DNA repair enzymes, Ann. Rev. Biochem., 51:61.

Lindahl, T., Demple, B., and Robins, P., 1982, Suicide inactivation of the E. coli O^6-methylguanine-DNA methyltransferase, Eur. Mol. Biol. Org. J., 1:1359.

Loeb, L.A., and Kunkel, T.A., 1982, Fidelity of DNA synthesis, Ann. Rev. Biochem., 51:429.

Luria, S., and Delbruck, M., 1943, Mutations of bacteria from virus sensitivity to virus resistance, Genetics, 28:491.

Medcalfe, A.S.C., and Lawley, P.D., 1981, Time course of O^6-methylguanine removal from DNA of N-methyl-N-nitrosourea-treated human fibroblasts, Nature (London), 289:796.

Mitra, S., Pal, B.C., and Foote, R.S., 1982, O^6-methylguanine-DNA methyltransferase in wild-type and ada mutants of E. coli, J. Bacteriol., 152:534.

Nakamura, N., and Okada, S., 1981, Dose-rate effects of gamma-ray-induced mutations in cultured mammalian cells, Mutat. Res., 83:127.

Nakamura, N. and Okada, S., 1983, Mutations resistant to bromodeoxyuridine in mouse lymphoma cells selected by repeated exposure to EMS. Characteristics of phenotypic instability and reversion to HAT resistance by 5-azacytidine, Mutat. Res., 111:353.

Novick, A., and Szilard, L., 1950, Experiments with the chemostat on spontaneous mutations of bacteria, Proc. Natl. Acad. Sci., USA, 36:708.

Parry, J.M., 1977, The use of yeast cultures for the detection of environmental mutagens using a fluctuation test, Mutat. Res., 46:165.

Penman, B.W., Crespi, C.L., Komives, E.A., Liber, H.L., and Thilly, W.G., 1983, Mutation of human lymphoblasts exposed to low concentrations of chemical mutagens for long periods of time, Mutat. Res., 108:417.

Ritossa, F., 1962, A new puffing pattern induced by temperature shock and DNP in Drosophila, Experentia, 18:571.

Samson, L., and Cairns, J., 1977, A new pathway for DNA repair in E. coli, Nature (London), 267:281.

Samson, L., and Schwartz, J.L., 1980, Evidence for an adaptive DNA repair pathway in CHO and human skin fibroblast cell lines, Nature (London), 287:861.

Sarasin, A., and Benoit, A., 1980, Induction of an error-prone mode of DNA repair in UV-irradiated monkey kidney cells, Mutat. Res., 70:71.

Sarasin, A., Bourre, F., and Benoit, A., 1982, Error-prone replication of ultraviolet-irradiated Simian virus 40 in carcinogen-treated monkey kidney cells, Biochimie, 64:815.

Schlesinger, M.J., Ashburner, M., and Tissieres, A., 1982, "Heat Shock. From Bacteria to Man." Cold Spring Harbor Laboratory, Cold Spring Harbor, N.Y.

Schwartz, J.L., and Samson, L., 1983, Mutation induction in Chinese Hamster ovary cells after chronic pretreatment with MNNG, Mutat. Res., 119:393.

Sklar, R., and Strauss, B., 1981, Removal of O^6-methylguanine from DNA of normal and xeroderma pigmentosum-derived lymphoblastoid lines, Nature (London), 289:417.

Streisinger, G., Okada, Y., Emrich, J., Newton, J., Tsugita, A., Terzaghi, E., and Inouye, M., 1966, Frameshift mutations and the genetic code, Cold Spring Harbor Symp. Quant. Biol., 31:77.

Thacker, J., and Stretch, A., 1983, Recovery from lethal and mutagenic damage during postirradiation holding and low dose-rate irradiations of cultured hamster cells, Radiat. Res., 96:380.

Thilly, W.G., 1983, Analysis of chemically induced mutation in single cell populations, in "Induced Mutagenesis, Molecular Mechanisms and Their Implications for Environmental Protection," p. 337, C.W. Lawrence, ed., Plenum Press, New York and London.

Thilly, W.G., DeLuca, J.G., Furth, E.E., Hoppe, H., IV, Kaden, D.A., Krolewski, J.J., Liber, H.L., Skopek, T.R., Slapikoff, S., Tizard, R.J., and Penman, B.W., 1980, Gene locus mutation assays in diploid human lymphoblast lines, in: "Chemical Mutagens," Vol. 6, p. 331, F.J. deSerres and A. Hollaender, eds., Plenum Press. N.Y.

Ueno, A.M., Furuno-Fukushi, I., and Matsudaira, H., 1982, Induction of cell killing, micronuclei and mutation to 6-thioguanine resistance after exposure to low dose-rate γ-rays and tritiated water in cultured mammalian cells (L5178Y), Radiat. Res., 91:447.

Waldstein, E.A., Cao, E.-H., and Setlow, R.B., 1982, Adaptive re-synthesis of O^6-methylguanine-accepting protein can explain the differences between mammalian cells proficient and deficient in methyl excision repair, Proc. Natl. Acad. Sci., USA, 79:5117.

Walker, G.C., 1984, Mutagenesis and inducible responses to deoxyribonucleic acid damage in Escherichia coli, Microbiol. Rev., 48:60.

Witkin, E.M., 1976, Ultraviolet mutagenesis and inducible DNA repair in Escherichia coli, Bacteriol. Rev., 40:869.

DISCUSSION

Painter: Have you other evidence for the induction of a repair enzyme by high levels (> 1,000 molecules/cell) of aflatoxin?

Thilly: Not yet.

Painter: This point is rather important for the concept of risk assessment. If you are saying that as long as exposure is at a low level, you are not going to induce the fast repair system, it follows that the effects low-level exposure over a long time may be worse than a single high exposure.

Thilly: It seems to me that the effects of a long-term low dose may be far more dangerous to our genetic health than a single high dose for some chemicals such as aflatoxin.

Painter: The burden is on you to prove your hypothesis. Nobody has yet seen an induced repair system in human cells.

Thilly: At the low doses of aflatoxin we see the slow removal of DNA-bound aflatoxin. Fast removal is seen only in cells which receive higher amounts of DNA adduction. We will try to design experiments to verify our hypothesis (Kaden, 1983).

Trosko: Have you used ouabain-resistance as a mutation marker in your low-dose/long exposure studies?

Thilly: No.

Trosko: As you are aware, the $6TG^R$ and TK^- markers are phenotypes which are inferred to indicate mutations which inactivate the respective enzymes responsible for the drug-resistant phenotypes. Might it not be possible that at high doses of the chemicals, physiological changes could occur in the cell which turn off the gene, thereby giving rise to the $6TG^R$ and TK^- phenotypes? In other words, might these resistant clones reflect some epigenetic changes, not mutational changes?

Thilly: The Capecchi's have shown the continued synthesis of HGPRT cross-reacting material in $6TG^R$ mammalian cells (Capecchi _et al._, 1974; Hughes _et al._, 1975; Wahl _et al._, 1975). However, it is possible to imagine stable changes in phenotype which do not involve primary DNA sequence changes.

Borg: Dr. Painter, I believe you said that you know of no evidence of repair induction in human cells. Is that right?

Painter: Yes.

Borg: What then is your response to the recently published work of Olivieri, Bodycotte and Wolff (1984)? The last sentence of their abstract states: "This response is analogous to the adaptive response to alkylating agents whereby prior treatment with small doses for a long period reduces the damage occurring from large doses of similar agents given for a short time." Furthermore, an earlier sentence suggests that there really can be net protection: "Often fewer aberrations were found after exposure to radiation from both sources (the presumptive inducing irradiation and the challenging irradiation) than were found after exposure to x-rays alone".

Painter: Those experiments were done with human lymphocyte cultures, incubated with tritiated thymidine for 24-48 hrs, and then treated with x-rays. The investigators compared the incidence of aberrations in those cells, with cells that were not incubated with tritiated thymidine.

Lymphocytes are notorious for their variable radiosensitivity. In our own experiments, we found that exposure of HeLa cells for 48 h to 0.1 microcurie of thymidine killed almost all the cells. What I think happened is that the tritiated thymidine killed all of the radiosensitive cells, so that all that was left to look at were the radioresistant cells. Consequently, the aberration frequencies are lower. I do not think the experiment shows any induction of a repair system.

Borg: With regard to your belief that constitutively resistant T-cells may have been selected, I cannot offer a directly relevant rejoinder. Indirectly relevant, however, is the demonstration in E. coli (there's the rub!) of inducible repair to both x-ray and H_2O_2; that is, the same repair induction (true induction, actually active versus added predamaged DNA) protecting against both agents in terms of survival. At least induced x-ray repair is known in nature (Demple and Halbrook, 1983).

Painter: As you point out, that's E. coli for which most of the inducible repair data exists, and has nothing to say about mammalian cells. Moreover, that experiment was done only in one direction; H_2O_2 to induce and x-rays to challenge. What is more, those data were for survival only, and actual enhanced removal of DNA damage induced by x-rays has not yet been demonstrated.

Tice: Why would you not expect cell stage specificity in your experiments with aflatoxin at high doses if a fast rate of repair is operational?

Thilly: Dr. Kaden synchronized the human lymphoblasts and found that for cell killing and mutation there was no cell cycle specificity. The non-linearities in the dose-responses cannot, therefore, be due to differences in the stage of the cell cycle.

Tice: Depending upon how fast the fast rate of repair is I might expect that you are going to end up with much more removal of damage if the cells are held in liquid medium before DNA synthesis, and therefore the mutation frequency should be higher.

Thilly: Ours were not liquid holding experiments, and we do not have any data regarding that possibility.

REFERENCES

Capecchi, M.R., Capecchi, N.E., Hughes, S.H., and Wahl, G.M., 1974, Selective degradation of abnormal proteins in mammalian tissue culture cells, Proc. Natl. Acad. Sci., USA, 71:4732.

Demple, B., and Halbrook, J., 1983, Inducible repair of oxidative DNA damage in Escherichia coli, Nature (London), 304:466).

Hughes, S.H., Wahl, G.M., and Capecchi, M.R., 1975, Purification and characterization of mouse hypoxanthine-guanine phosphoribosyl transferase, J. Biol. Chem., 250:120.

Kaden, D.A., 1983, Mutation of human lymphoblast cells by aflatoxin B_1: Comparison of normal and hypermutable cells, Ph.D. Thesis, Massachusetts Institute of Technology, Cambridge.

Olivieri, G., Bodycote, J., and Wolff, S., 1984, Adaptive response of human lymphocytes to low concentrations of radioactive thymidine, Science, 223:594.

Wahl, G.M., Hughes, S.H., and Capecchi, M.R., 1975, Immunological characterization of hypoxanthine-guanine phosphoribosyl transferase mutants of mouse L cells: evidence for mutations at different loci in the HGPRT gene, J. Cell. Physiol., 85:307.

QUANTITATIVE NEOPLASTIC TRANSFORMATION IN C3H/10T1/2 CELLS[1]

John S. Bertram[2] and John E. Martner[3]

Grace Cancer Drug Center
Roswell Park Memorial Institute
666 Elm Street
Buffalo, NY 14263

INTRODUCTION

The development of mammalian cell culture systems in which neoplastic transformation can be induced by exposure to chemical and physical carcinogens provided a major stimulus to the study of carcinogenic mechanisms. Furthermore, the potential ability to quantitate the induction of neoplasia on a per cell basis has allowed the study of factors modulating carcinogenesis with a quantitative precision hitherto impossible in animal models. However, problems arise in quantitation of the induced response when the number of cells at risk is varied or when treated cells are replated at various densities. Since in most cases the original rationale behind these cell-number manipulations was to increase the number of cells at risk and thus increase the probability of detecting low level exposure to radiation and chemicals, these problems and the various hypotheses generated to explain it, will be the major focus of this paper.

[1] Supported by USPHS Grant CA 21359
[2] Current address: Cancer Research Center of Hawaii, 1236 Lauhala St., Honolulu, HI 96813
[3] Current address: McArdle Laboratory for Cancer Research, Madison, WI 53706

CHARACTERISTICS OF THE C3H/10T1/2 CELL LINE

The line was developed in 1972 in the laboratory of Dr. Heidelberger from C3H mouse embryo fibroblasts (Reznikoff et al., 1973a). Primary cells that had been dissociated from embryos, spontaneously became nonsenescent with repeated passaging. From these survivors were cloned lines possessing a high degree of post-confluence inhibition of cell division. One of these clones (clone 8) was found to form a highly uniform, stable, low-density monolayer at confluence, and also to undergo morphological transformation after exposure to a variety of polycyclic aromatic hydrocarbon carcinogens (Reznikoff et al., 1973b). This clone C3H/10T1/2 Cl 8 (or 10T1/2) was expanded, cryopreserved at low passage number and now appears to be the single most widely used *in vitro* model for the study of physical and chemical carcinogenesis.

Morphological transformation resulting from carcinogen exposure only occurs after a latent period of 4-5 weeks postcarcinogen when cultures exist as a contact-inhibited monolayer. Transformation is characterized by the ability of cells to re-initiate DNA synthesis and form foci of densely packed cells contrasting markedly with the low-density monolayer formed by nontransformed cells. Two types of foci have been described as being neoplastically transformed (i.e., tumorigenic in animals). Type II foci are comprised of cells with rounded morphology which form extremely dense foci with well-defined edges; Type III foci contain polar, highly basophilic cells which exhibit much criss-crossing and which form foci with jagged invasive edges. Some 60% of the former and 90% of the latter foci are tumorigenic in immunosuppressed syngeneic mice (Reznikoff et al., 1973b). Because foci only appear many weeks after seeding of cells, when cultures have multiplied from between 2-5 x 10^2 to 5-8 x 10^5 cells/60 mm dish, problems have arisen in interpreting the results on a quantitative basis. Specifically, should results be calculated as the transformation frequency (i.e., transformed foci/surviving cell) or is it more proper to express results as foci/dish regardless of initial number of cells at risk? The scientific debate over these issues has yet to be resolved and has sparked further experimentation in a number of laboratories. We believe that problems in interpretation of the type described below are not unique to the 10T1/2 line, but will be encountered in other systems as their behavior is placed under comparable scrutiny.

QUANTITATIVE NEOPLASTIC TRANSFORMATION IN 10T1/2 CELLS

a. Dose-response Relationships

In our original publication (Reznikoff et al., 1973b), we

presented the results as percent transformation frequency (TF), calculated as the ratio of number of Type II and III foci/number of surviving cells determined separately by colony counting. Two major factors were found to influence TF, dose of carcinogen and cell density at the time of treatment. For a fixed seeding density of about 250 viable cells, increasing the dose of methylcholanthrene (MCA) or dimethylbenzanthracene (DMBA) up to limits determined by solubility and toxicity, respectively, gave a straight-line,dose-response relationship when plotted on log/log coordinates. Similar results have been obtained with benzo-(a)pyrene (Haber et al., 1977). For x-rays (Terzaghi and Little, 1976; Han and Elkind, 1979) and neutrons (Han and Elkind, 1979, 1982), increasing the dosage gives a log/linear increase in TF until a plateau is reached at radiation doses giving a surviving fraction of 0.1 or less. Proposed explanations for this plateau are that at this dose all surviving cells are initiated (Kennedy et al., 1980). We would, however, propose that all competent cells are initiated at this dose, competency being perhaps determined by a cell-cycle parameter (Bertram and Heidelberger, 1974), and that competent cells comprise only a small fraction (1%) of the survivors.

b. Enhancement by 12,13-Tetradecanoyl-phorbolacetate (TPA)

In animal model systems of skin carcinogenesis, TPA and related compounds cause tumors to develop when applied repeatedly after a low dose of a skin carcinogen which is subcarcinogenic when applied alone (Boutwell, 1974). In a similar manner repeated applications of TPA to 10T1/2 cells previously exposed to MCA (Mondal et al., 1976), x-rays (Kennedy and Little, 1980) or neutrons, enhances the number of foci, especially at low dosage. While TPA itself appears to cause the expression of a low background of transformation, its effects are greater than additive when combined with an initiating carcinogen (Han and Elkind, 1982). Use of TPA has also been shown to allow the detection of direct, rapidly acting carcinogens, such as *N*-methyl-*N*'-nitro-*N*-nitrosoguanidine, which are not transforming when tested in the conventional assay system (Boreiko et al., 1982).

The mechanism of action of TPA is obscure; however, it should be noted that TPA has recently been shown to activate a Ca^{++} dependent protein kinase (Kikkawa et al., 1983), and our group has recently reported that increases in phosphorylation of specific proteins occur concomitantly with morphological transformation of carcinogen-initiated cells (Martner and Bertram, 1984). In these same cells TPA accelerates the development of morphological transformation (Mordan et al., 1982). A second, perhaps unrelated, potential mechanism is that TPA, by disrupting cellular communications (Yancey et al., 1982), inhibits growth-inhibitory signals emanating from adjacent nontransformed cells which act to suppress the expression of transformation (Bertram et al., 1982). Regardless of mechanism it is clear that TPA is a useful adjunct in the detection of low-level carcinogenic insults.

c. Influence of Seeding Density on Transformation

One approach to the detection of low-frequency events is to increase the number of detectors, in this case, the number of cells at risk. When this is done with 10T1/2 cells, paradoxical results are obtained in that, within limits, the TF varies inversely with the seeding density (Reznikoff et al., 1973b; Haber et al., 1977; Fernandez et al., 1980, Han and Elkind, 1979). Thus, as the cell density is increased in carcinogen-exposed dishes, the ratio of surviving cells to induced transformed foci progressively decreases. This has been observed over a 400-fold range in seeding densities and a 200-fold range of TF. While the shape of the cell density/TF curve follows a log/log straight line relationship at intermediate cell densities, there is dispute as to its shape at low densities. While Kennedy and Little (1980) maintain that TF continues to rise to yield a theoretical maximum of 100% in dishes seeded at the practical minimum of 1 cell/dish, Han and Elkind (1979) have clearly shown a plateau at about 0.3% TF to exist for seeding densities below about 300 cells/100 mm dish. Since both laboratories utilized virtually identical x-ray exposures, these different results are perplexing. As will be discussed later, the existence or non-existence of this plateau is of crucial significance to the understanding of quantitative aspects of transformation in this line.

While the ratio of transformed foci/total cells at risk decreases with increasing exposure or seeding densities for both chemical and physical carcinogens, differences are observed between these two types of carcinogens when the absolute number of transformed foci/dish are calculated. In an extensive series of experiments utilizing x-rays, Kennedy and Little (1980) and Kennedy et al. (1980) have presented evidence that this parameter is constant over a range of initial cell densities from 10 to 10^5 cells/dish. In contrast, when MCA was used as carcinogen in the studies of Fernandez et al. (1980), the number of foci/dish increased from about 0.1 to 1.0 when the seeding cell density was increased from about 100 to 5 x 10^4 cells/dish. Clearly, however, this 10-fold increase in foci/dish, did not match the 500-fold increase in cell number, leading to the aforementioned decrease in TF with seeding density. A possible explanation for this discrepancy may be found in the recent results of Huband et al. (1984), in which MCA has been detected to remain on or in the plastic Petri dishes after removal of carcinogen-containing medium and washing of cells. The amount of residual MCA has been shown to be sufficient to induce transformation over the prolonged time period during which this chemical can be released to proliferating cultures. Complex kinetics of release of MCA from residues on dishes can be expected as initial seeding density is altered, and as the cell density increases with proliferation. Thus interpretation of data demonstrating seeding density effects on transformation induced by lipophilic stable compounds like MCA must be viewed with caution.

INTERPRETATION OF QUANTITATIVE ASPECTS OF 10T1/2 CELL TRANSFORMATION

a. Probabilistic Models

One of the problems immediately encountered in the analysis of the behavior of 10T1/2 cells in response to carcinogen is that the proportion of cells initiated cannot be measured directly, because transformed foci only arise from a confluent monolayer of cells 4-5 weeks after exposure to carcinogen. Since, after usual seeding densities of 200-300 cells/dish, the progeny of one initiated cell can be calculated to have formed a colony of 2000-4000 cells at confluence, from how many such cells does a focus arise? The problem is further complicated by the observation that if, prior to transformation, confluent monolayers of carcinogen-exposed cultures are trypsinized and reseeded at various densities, the yield of transformants does not increase, as might be expected by the dispersal of a colony of 2000-4000 initiated cells to separate locations over the culture dish (Haber et al., 1977; Kennedy et al., 1980).

Prompted by these findings two groups published alternative hypotheses to explain in mathematical terms the behavior of 10T1/2 cells with respect to seeding density. These hypotheses have much in common. Fernandez et al. (1980) propose that three probabilities govern the yield of foci/dish: P_1 is the probability of initiation, P_2 the probability per generation that an initiated cell will transform, and P_3 the probability per generation that an initiated cell will become noninitiated. In this model P_1 is assumed to be between 0.02 and 1, P_2 about 10^{-6} and P_3 0.24. Thus, the frequency of initiation is presumed to be high while that of transformation is extremely low. To explain their data on x-ray transformation, Kennedy et al. 1980 have proposed that two events occur, the first being induced in all surviving cells in the dose range used (600 rads), and transmitted to their progeny; the indirect consequence of the primary event(s) and results in a cell with the ability to produce a transformed focus.

Both models have in common the requirement for a high-frequency initiating event, and a low-frequency probability that the progeny of that initiated cell will transform. Also implicit in these models is the assumption that once produced, every transformed cell will produce a focus. We were uncomfortable with many of the assumptions made. For example, we have direct evidence that expression of transformation in this system can be inhibited by contact with nontransformed 10T1/2 cells (Bertram et al., 1982). Additionally we had seen no evidence for a frequency of initiation of greater than about 2% in our isolation of initiated cells from MCA-treated cultures (Mordan et al., 1982). Furthermore, such high-frequency induction is difficult to reconcile with mutational mechanisms of carcinogen action and calls for inheritable epigenetic perturbations of cellular behavior (Kennedy et al., 1984).

b. Cell Interaction Model

As long as the number of initiated cells originally present in the carcinogen-treated culture could only be deduced from the development of transformed foci many weeks and many cell divisions later, resolution of these conflicting models was not possible. In 1982, as part of our research program on retinoids, we isolated what we believe to be a clone of cells which had been initiated by methylcholanthrene and in which the progression of these initiated cells to neoplasia had been stabilized by a completely nontoxic concentration of retinyl acetate. In the absence of retinoid these cells spontaneously transform after a latent period of about 4 weeks; again, as in the aforementioned assay system for carcinogens, transformed foci appear from a confluent monolayer of growth-inhibited cells. In reconstruction experiments about 70% of clonogenic cells will produce transformed foci (Mordan et al., 1982). With this new line, designated INIT/10T1/2, at our disposal we could now add known numbers of initiated cells to parental 10T1/2 cells and investigate the dependency of transformation frequency on variables such as generation number, colony size and initiated cells/culture postulated to influence the yield of transformants (Fernandez et al., 1980; Kennedy et al., 1980).

i. Effect of Seeding Density on Transformation. To examine the effects of seeding density on transformation frequency, a constant number of initiated cells (100) were co-cultured with increasing numbers (10^3-10^5) of 10T1/2 cells. Since INIT/10T1/2 and 10T1/2 cells have identical rates and saturation densities (Mordan et al., 1982), and since the total growth surface area is constant regardless of seeding density, the progressively increasing seeding density of the co-cultured 10T1/2 cells will lead to progressively smaller colonies of initiated cells at confluence. It was found that the percentage of INIT/10T1/2 cells forming transformed loci was directly dependent on the size of the initiated cell colony at confluence (Fig. 1). As the colony size at confluence increased from approximately 40 cells (highest seeding density) to about 3300 cells (lowest seeding density), the percentage of

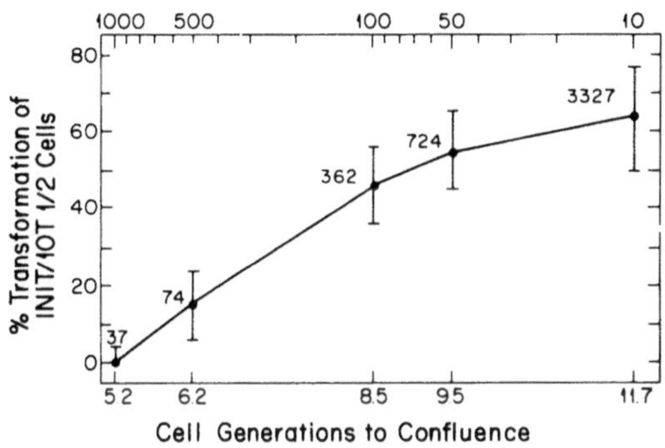

Fig. 1 Effect of colony size on transformation of initiated 10T1/2 cells. INIT/10T1/2 cells (100/dish) were cultured with increasing numbers of 10T1/2 cells (10^3-10^5/dish), thus controlling the number of cell generations required to attain a confluent monolayer (600,000 cells/dish). Each point represents the mean $\pm$ standard deviation of the percentage of colonogenic INIT/10T1/2 cells forming transformed foci, in 10-12 cultures, 5 weeks after seeding. The number next to each point is the mean number of cells per colony at confuence.

INIT/10T1/2 cell colonies forming types II and III foci increased from 0.8% to 65%. This is the maximum percentage of INIT/10T1/2 cells that were able to form types II and III foci (Mordan et al., 1982). It is readily apparent that INIT/10T1/2 cells must form a colony at confluence in excess of approximately 40 cells in order to attain the competence to transform. Thereafter, there is a direct relationship between colony size at confluence and the probability of transformation. Maximum expression of the transformed phenotype was reached with colonies of approximately 740-3300 cells (Mordan et al., 1983).

ii. Effect of Accumulated Generations. Because an increase in colony size at confluence cannot be produced without an increase in the number of cell generations accumulated, it was not clear which of these variables influenced the expression of transformation. An experiment was therefore performed in which cell generations were allowed to accumulate by weekly passaging. At each passage, reconstruction experiments were set up at increasing seeding densities so that we had colonies of different size, total accumulated generations, and generations to confluence. The results of this experiment (Fig. 2) again showed clearly that the percentage of INIT/10T1/2 cells forming transformed foci was consistently dependent on the seeding density, which determines colony size as we have described above, and was not affected by the total number of cell generations accumulated over the five-week period of the experiment. The maximum percentage of INIT/10T1/2 cells able to form transformed foci at each weekly interval was consistently 60-65% throughout the experiment, in spite of the accumulation of 40 generations. Thus, no decrease in the pool of initiated cells capable of neoplastic transformation was detected in these studies as would be predicted by the model of Fernandez et al. (1980). Transformation of INIT/10T1/2 cells was completely inhibited at the highest seeding density as would be predicted from the results presented in Fig. 1, for colonies of from 20-50 cells.

iii. Effects of Serial Dilution of a Mixed Cell Suspension. The results so far obtained indicated clearly the requirement for large colonies for efficient expression of transformation. However, the results might simply reflect an increased probability of transformation as a consequence of the increased numbers of initiated cells contained in those large colonies. To control this variable, we plated serial dilutions of a mixture of INIT/10T1/2 and 10T1/2 cells. Because the final saturation density is constant, regardless of seeding density, and since both cell types have the same growth rate (Mordan et al., 1982), the relative proportions of initiated and normal cells at confluence will be the same as at the time of seeding. All cultures at confluence will contain the same number of INIT/10T1/2 cells; they will, however, differ in their distribution. Cultures seeded at high density (undiluted) will contain many small colonies, those seeded at lower densities (diluted) will contain fewer but larger colonies, with the mathematical product, colony number x cells/colony = constant. These concentrations also apply to the cocultured 10T1/2 cells.

Serial dilution of the mixed cell suspension produced no

significant variation in the number of transformed foci/dish in cultures receiving 1:4 to 1:64 dilutions (Table 1), in spite of a 15-fold variation in the numbers of initiated cells seeded. These results are similar to those reported by Kennedy et al. (1980) using de novo x-ray-induced transformation. However, for cultures seeded undiluted at the highest cell density of 2.5 x 10^5 cells/dish, in which initiated cells were only capable of producing colonies of about 11 cells, no transformed foci developed. This effect was consistent with previous results (Fig. 1). However, in this experiment, since the total number of initiated cells was constant within the various groups, as was the total number of generations accumulated by all initiated cells as they proliferated to form a confluent monolayer, the only variables were the colony size attained by the initiated cells and the generations accumulated by individual cells. We have demonstrated that the latter variable does not influence transformation frequency (Fig. 2); we conclude therefore that it is the colony size that is influencing the transformation frequency.

When we plotted transformation frequency against colony size of the initiated cells, we again found a direct relationship between colony size and the expression of transformation. Thus, the decrease in number of initiated cells seeded with increasing dilution of the cell suspension appears to be offset by a closely corresponding increase in the frequency of initiated cells forming transformed foci. This result is a consequence of the relationship between transformation frequency, seeding density, and colony size (Mordan et al., 1983).

DISCUSSION

Others have also shown that nontransformed cells can inhibit the expression of transformation of initiated cells or of established transformed cells. Stoker and co-workers (1966) demonstrated that confluent 3T3 cells would inhibit growth of polyoma-transformed 3T3 cells, and similar effects were reported by Sivak and Van Duuran (1970) in chemically transformed Balb/3T3 cells. In the 10T1/2 system we have shown that if we increase the confluent density of 10T1/2 cells by increasing the serum concentration in the growth medium, expression of de novo-initiated or of established transformed cells can be reversibly inhibited (Bertram, 1977). Alternatively, if we elevate cAMP levels in confluent monolayers by inhibiting cAMP phosphodiesterase, expression of transformation in these two situations can again be reversibly inhibited (Bertram, 1979). In both cases growth inhibition of neoplastic cells was absolutely dependent upon the simultaneous presence of and, as far as we can judge, contact with growth-inhibited 10T1/2 cells. Our working hypothesis, currently being tested, is that the mediator of growth inhibition is transferred across cell membranes,

TABLE 1. Effects of Serial Dilution of a Mixed Suspension of Initiated and Nontransformed 10T1/2 Cells on the Production of Transformed Foci.

Total Cells Seeded[1]	Clonogenic Initiated Cells[2]	Cells/Colony at Confluence	Generations to Confluence	Initiated Cells at Confluence[3]	Transformed Foci/Dish[4]	% Transformation
2.5×10^5	1205	11	3.4	13260	0	0
6.0×10^4	301	42	5.4	12500	10.5 ± 1.7	3.5
1.5×10^4	75	169	7.4	12700	8.3 ± 1.2	11.0
4.0×10^3	19	676	9.4	12800	9.1 ± 1.8	48.0
1.0×10^3	5	2702	11.4	13500	3.0 ± 0	60.0

[1] By serial dilution of a mixture of initiated and 10T1/2 cells.

[2] Determined from separate measurements of plating efficiency.

[3] Product of colony number (column 2) and cells/colony (column 3).

[4] Mean of 10–12 dishes/group.

possibly by gap junctions. This conclusion is based on our failure to detect the release of a growth inhibitory factor into the culture medium (Bertram,1977, 1979) and our failure to induce growth inhibition in normal or malignant cells exposed to isolated membranes from confluent 10T1/2 cultures, nonpermissive for the growth of malignant cells (Faletto and Bertram, unpublished results).

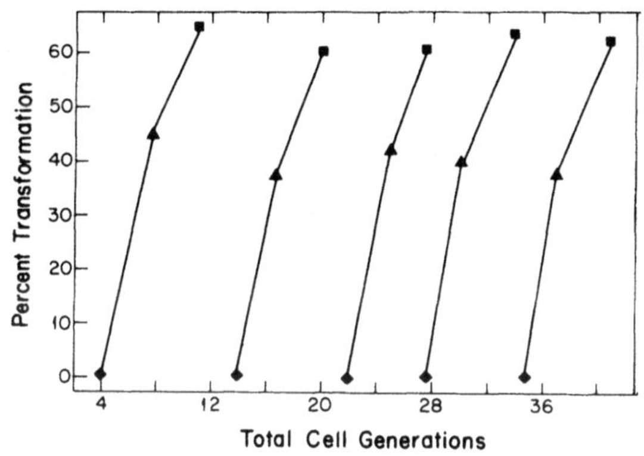

Fig. 2 Effect of accumulated cell generations on the transformation of initiated 10T1/2 cells. INIT/10T1/2 cells were cultured in exponential growth phase for approximately 40 generations by weekly passaging at low density. At each passage, 100 INIT/10T1/2 cells were co-cultured with 10^3 (■), 10^4 (▲), or 10^5 (◆) 10T1/2 cells thus reproducing low, medium, and high seeding densities, respectively. By determining the plating efficiency and the number of cells in representative cultures, the number of cell generations accumulated between passage cultures could be calculated. Each point represents the mean transformation frequency of clonogenic INIT/101/2 cells in 8-12 cultures.

In conclusion, when the 10T1/2 assay system is used to detect weak carcinogens or carcinogens present in low concentration, it is not a simple matter to increase the sensitivity of the system by increasing the number of cells at risk, since unless the area available for growth is proportionately expanded, the unavoidable decrease in colony size will suppress the expression of transformation in any initiated cell present. While increasing the area available for growth is possible, this soon becomes prohibitive in terms of incubator space and technician time. The use of a tumor promoter such as TPA has the potential to limit junctional communication and decrease the colony size dependency (Yancy et al., 1982). However, in our hands, TPA has rather minor effects on cellular interactions of the type described. Other techniques such as prolonged or delayed exposure as described by Nesnow et al. (1982) may increase sensitivity, but compromise the quantitation of carcinogen exposure.

Finally, while our model satisfactorily explains most aspects of quantitation in the 10T1/2 system and stresses that the colony size attained by survivors of carcinogen-treatment must be controlled for, it does not satisfactorily explain the extremely high frequencies of transformation recently reported by Kennedy and Little (1980). However, it should be noted that others using similar protocols and x-ray exposures did not report similar high frequencies (Han and Elkind, 1979) nor were they found in earlier work from Little's group (Terzaghi and Little, 1976).

REFERENCES

Bertram, J.S., 1977, Effects of serum concentration on the expression of carcinogen-induced transformation in the C3H/10T1/2 CL8 cell line, Cancer Res., 37:514.

Bertram, J.S., 1979, Modulation of cellular interactions between C3H/10T1/2 cells and their transformed counterparts by phosphodiesterase inhibitors, Cancer Res., 39:3502.

Bertram, J.S., and Heidelberger, C., 1974, Cell cycle dependency of oncogenic transformation induced by N-methyl-N'-nitro-N-nitrosoguanidine in culture, Cancer Res., 34:526.

Bertram, J.S., Bertram, B.B., and Janik, P., 1982, Inhibition of neoplastic cell growth by quiescent cells is mediated by serum concentration and cAMP phosphodiesterase inhibitors, J. Cell Biochem., 18:515.

Boreiko, C.J., Ragan, D.L., Abernethy, D.J., and Frazelle, J.H., 1982, Initiation of C3H/10T1/2 cell transformation by N-methyl-N'-nitro-N-nitrosoguanidine and aflatoxin B, Carcinogenesis, 3:391.

Boutwell, R.K., 1974, The function and mechanism of promoters of carcinogenesis, CRC Crit. Rev. Toxicol., 2:419.

Fernandez, A., Mondal, S., and Heidelberger, C., 1980, Probabilistic view of the transformation of cultured C3H/10T1/2 mouse embryo fibroblasts by 3-methylcholanthrene, Proc. Natl. Acad. Sci. USA, 77:7272.

Haber, D.A., Fox, D.A., Dynan, W.S., and Thilly, W.G., 1977, Cell density dependence of focus formation in the C3H/101/2 transformation assay, Cancer Res., 37:1644.

Han, A., and Elkind, M.M., 1979, Transformation of mouse C3H/10T1/2 cells by single and fractionated doses of x-rays or fission-spectrum neutrons, Cancer Res., 39:123.

Han, A., and Elkind, M.M., 1982, Enhanced transformation of mouse 10T1/2 cells by 12-O-tetradecanoylphorbol-13-acetate following exposure to x-rays or fission-spectrum neutrons, Cancer Res., 42:477.

Huband, J.C., Abernethy, D.J., and Boreiko, C.J., 1984, High frequency of C3H/10T1/2 cell transformation by 3-methylcholanthrene: a treatment artifact, Proc. Am. Assoc. Cancer Res., 25:136.

Kennedy, A.R., Cairns, J., and Little, J.B., 1984, Timing of the steps in transformation of C3H/10T1/2 cells by x-irradiation, Nature (London), 307:85.

Kennedy, A.R., and Little, J.B., 1980, Investigation of the mechanism for enhancement of radiation transformation in vitro by 12-O-tetradecanoylphorbol-13-acetate, Carcinogenesis, 1:1039.

Kennedy, A.R., Fox, M., Murphy, G., and Little, J.B., 1980, Relationship between x-ray exposure and malignant transformation in C3H/10T1/2 cells, Proc. Natl. Acad. Sci. USA, 77:7262.

Kikkawa, U., Takai, Y., Tanaka, Y., Miyake, R., and Nishizuka, Y., 1983, Protein kinase C as a possible receptor protein of tumor-promoting phorbol esters, J. Biol. Chem., 258:11442.

Martner, J.E. and Bertram, J.S., 1984, Appearance of tyrosine phosphorylation correlates with time of transformation in 10T1/2 cells: inhibition by retinyl acetate, Proc. Am. Assoc. Cancer Res., 25:146.

Mondal, S., Brankow, D.W., and Heidelberger, C., 1976, Two-stage chemical oncogenesis in cultures of C3H/10T1/2 cells, Cancer Res., 36:2254.

Mordan, L.J., Bergin, L.M., Budnick, J.L., Meegan, R.R., and Bertram, J.S., 1982, Isolation of methylcholanthrene-"initiated" C3H/10T1/2 cells by inhibiting neoplastic progression with retinyl acetate, Carcinogenesis, 3:279.

Mordan, L.J., Martner, J.E., and Bertram, J.S., 1983, Quantitative neoplastic transformation of C3H/10T1/2 fibroblasts: dependence upon the size of the initiated cell colony at confluence, Cancer Res., 43:4062.

Nesnow, S., Garland, H., and Curtis, G., 1982, Improved transformation of C3H/10T1/2 Cl 8 cells by direct- and indirect-acting carcinogens, Carcinogenesis, 3:377.

Reznikoff, C.A., Brankow, D.W., and Heidelberger, C., 1973a, Establishment and characterization of a cloned line of C3H mouse embryo cells sensitive to postconfluence inhibition of cell division, Cancer Res., 33:3231.

Reznikoff, C.A., Bertram, J.S., Brankow, D.W., and Heidelberger, C., 1973b, Quantitative and qualitative studies on chemical transformation of cloned C3H mouse embryo cells sensitive to postconfluence inhibition of cell division, Cancer Res., 33:3239.

Sivak, A., and Van Duuran, B.L., 1970, A cell culture system for the assessment of tumor-promoting activity, J. Natl. Cancer Inst., 44:1091.

Stoker, M.G.P., Shearer, M., and O'Neill, C., 1966, Growth inhibition of polyoma-transformed cells by contact with static normal fibroblasts, J. Cell Sci., 1:297.

Terzaghi, M., and Little, J.B., 1976, X-radiation-induced transformation in a C3H mouse embryo-derived cell line, Cancer Res., 36:1367.

Yancy, S.B., Edens, J.E., Trosko, J.E., Chang, C.C., and Revel, J.P., 1982, Decreased incidence of gap-junctions between Chinese hamster V-79 cells upon exposure to the tumor promoter 12-O-tetradecanoylphorbol-13-acetate, Exp. Cell Res., 139:329.

DISCUSSION

Thilly: Would you comment, relative to your gap junction model, on the finding of Reznikoff et al. (1973) that transformation of the CH3 10T1/2 cells is inhibited by fungizone.

Bertram: I know that fungizone increases the leakiness of cell membranes.

Thilly: I believe that it causes holes in membranes.

Bertram: It may thus increase the degree of transfer of this putative growth inhibitor. What we did not show in our original paper on fungizone was that its effects were reversible. Thus, in order for this to be a possible mechanism for the action of fungizone, the effects have to be reversible, since we have shown the growth inhibition created by normal cells is reversible.

Hall: From the published literature, our findings tend to agree with yours, rather than those of the Harvard group (Kennedy et al., 1980). However, in the absence of anyone from Dr. Kennedy and Dr. Little's group, I will play the part of the devil's advocate. You admitted that it was very difficult to sort out between the two hypotheses, whether the initiated colony had to be a certain size to be expressed, or whether a certain number of generations were necessary, with a certain finite probability of the second step of transformation per cell cycle (the Kennedy and Little model). The principle experiment that you advance in favor of your model is the one in which the initiated cells go through 40 generations before they are used, yet that does ot alter your yield. From the point of view of the Kennedy and Little model, this experiment would not answer the question, because you have suppressed cells all this time with Vitamin A.

Bertram: They were not cultured with Vitamin A at this time; they were passaged in the absence of Vitamin A.

Hall: Even if you carry parental 10T1/2 cells through 40 generations, is not the spontaneous rate greatly up?

Bertram: No. We find that it is either an all-or-none effect. Fortunately, in this situation, it was none. That experiment you discuss does address, but cannot answer, the problem of generation number which is required to distinguish the Kennedy model from ours. Simply because, upon each serial passage, the Kennedy model would predict there would be a probability of a transformant arising. But when one calculates that probability on the basis of how many cells are present at that passage, the probability was always less than one. At each passage, we would only be keeping about 10% of the total accumulated cells, so even had transformation occurred, we would not have seen it.

The experiment that did answer the crucial question of generations was the final one in which we had the same number of cells and a constant number of cell generations. In my view, I think that Kennedy and J. Cairns have a problem explaining the lack of transformation of low colony size, and fitting it into their model.

Sonneblick: In your culture did you see any differential cell death? Were there fragments of nuclear material in the cell culture?

Bertram: There is always cell debris in the culture but we do not find differential cell death. We believe that the death rates of initiated cells are virtually identical to the generation rates of control 10T1/2 cells, because they have similar proliferation rates.

Painter: Have your experiments separated the variables, cells per colony from the number of generations that initiated cells experience in the presence of suppressor cells? Could the latter factor be important?

Bertram: I do not think we can completely exclude that, although it would be a rather bizarre finding.

Martner: If you plate the initiated cells by themselves in the dish and allow them to achieve confluence and remain there, the dish will entirely transform, at least well above the 60% level. The transformants will appear to be type I, II, and III foci, but there is no contact whatsoever with normal cells, so it is not a requirement that there is contact.

Mendelsohn: Your hexagonal model suggests that a seven cell colony would produce a central cell out of contact with its environment. Obviously this is very simplified. But what is your explanation for colony sizes up to 1000 cells being important in this phenomena?

Bertram: If you look at the dye injection experiments, in which lucifer yellow is injected into transformed foci, you will see that if you inject one cell the dye will spread throughout the entire focus. So it is not just an adjacent cell which will pick up a message, but the message is transferred from cell, to cell, to cell. What one then must know is the rate at which such a message might be destroyed by malignant cells. Bell, at Los Alamos, fitted a very nice diffusional model of this type (Bell, 1976).

Thilly: It would be useful to extend Lowenstein's basic hypothesis with regard to the relationship between potentially transformed cells and non-transformed cells as a diffusion-based system. Basically, as the potentially transformed cells surround themselves, they create a diffusional barrier to receive some signal from normal cells. But we should also point out that Part A of Lowenstein's hypothesis is a stochastic model. He felt there was a growth signal which was generated by normal cells randomly in time, and that if the chemical remained at high concentration within that cell, that cell would undergo cell division. However, if the cell were permitted to let that self-generated signal diffuse away by gap junctions or by holes in the membrane, the signal would reach a low concentration, and potentially would not lead to cell growth or continued growth, i.e., transformation.

Bertram: One can look at is as a positive or as a negative signal. I would agree, it is going to be very hard to distinguish what is going on.

Evans: Would you expect a big difference between epithelial cells and fibroblasts on this basis, in terms of their connections?

Bertram: Epithelial cells are known to be very good producers of gap junction connections, so I would not expect a major difference on that basis. The question has never been examined in cultures of epithelial cells. In the old mutation experiments, similar cell density restrictions on yield of revertents was discovered, and that was again due to metabolic cooperation--the kiss of death or kiss of life.

Weiner: Can you comment on the molecular mechanism of Vitamin A action on initiated cells.

Martner: My investigations of the initiated 10T1/2 cells and fully transformed 10T1/2 cells centered on phosphorylation events taking place within them during the process of transformation of an entire culture. In the absence of retinoids in cultures of initiated cells, you see simultaneously the appearance of a newly phosphorylated 35,000 molecular weight protein and regions of high labelling index. Both of these events are totally inhibited by retinoids, as well as the appearance of morphologically transformed cells, which comes about several days later in these same cultures. The 35,000 molecular weight newly phosphorylated protein was also detected in transformed cells by cloning them, and establishing pure cultures of transformant, and it seems to be a marker of the establishment and maintenance of transformation in the cell. There is another event in fully transformed cells, which is inhibitable by retinoids after the transforming event has taken place: you cannot make the cells revert to normal by treatment with retinoids. You can, however, effect tyrosine phosphorylation of a second 35,000 molecular weight protein. Tryosine phosphorylation was not seen on that protein in the normal cell, nor in the initiated cell, but it is seen in the transformant, and that phosphorylation can be inhibited by retinoids. I cannot yet comment on the mechanism involved.

Evans: Can you comment further on the 60% level of transformation being the maximum possible level. One interpretation would be that one normal cell is required for every one transforming cell. What thoughts do you have about this limitation?

Bertram: I do not know how common the finding is. We have been looking at only one initiated clone. As John Martner mentioned, we do not get just one type of focus produced, but have Type II and Type III foci, so the initiated cells are not programmed down one straight road. There seems to be some "choice" that initiated cells can make upon transformation. I do not know whether one of those choices is to become normal.

REFERENCES

Bell, G. I., 1976, Models of carcinogenesis as an escape from mitotic inhibitors, Science, 192:569.

Kennedy, A. R., Fox, M., Murphy, G., and Little, J. B., 1980, Relationship between x-ray exposure and malignant transformation in C3H 10T1/2 cells, Proc. Natl. Acad. Sci. USA, 77:7262.

Reznikoff, C. A., Brarkow, D. W., and Heidelberger, C., 1973, Establishment and characterization of a cloned line of C3H mouse embryo cells sensitive to post-confluence inhibition of cell division, Cancer Res., 33:3231.

RISK ESTIMATE FOR GENETIC EFFECTS

Seymour Abrahamson

Department of Zoology
University of Wisconsin
Madison, WI 53706

It is not my intention to reinvent the wheel in this presentation, but perhaps I can straighten a few spokes, here and there. I have been a member of both the National Academy's BEIR and CCEM committees concerning risk from ionizing radiation (National Research Council, 1972, 1980) and environmental chemical mutagens (National Research Council, 1983) and will provide a general summary of approaches used in their reports. In the case of radiation, Bender, Denniston, Schull and myself have recently prepared a detailed risk report for the Nuclear Regulatory Commission's updated reactor safety study and so I also will incorporate major elements of that unpublished report into this discussion.

Over the past fifty years we have developed a slow but steady understanding of basic biological issues associated with the interaction of a few comparatively direct-acting physical agents on genetic material. The problem is horrendous when we deal with a minimum of 50,000 diverse-acting chemicals. The basic strategy proposed by the National Academy's Committee on Chemical Environmental Mutagens (National Research Council, 1983) suggested five steps in the process of identification of mutagens and estimation of risk. The first stage recommends a tier-testing system for the identification of potential mutagens and suggests the types of quick and relatively sensitive tests that are already available for this purpose and those that are in need of development; namely, the aneuploid detection tests. The second stage involves semiquantitative ranking of the mutagenic potency of the chemicals in each testing system used. Stage three is an estimate of the damage to germ cells induced per unit of exposure (dose, if known) in mammalian test systems and of the impact of such damage over future generations. This stage would be employed only if a decision to limit or exclude the chemical could not be made on the basis of the information developed in the first stages. This information would be augmented by independently developed carcinogenicity data. Stage 3 would also be invoked if the chemical is either so important for society or is in such widespread use that it is imperative that additional information be obtained. Stage four is the risk assessment relative to the expected human exposure, and stage five would attempt a risk-benefit analysis for the chemicals that enter the last stages of analysis.

Two points emerged from the Committees' recommendations that are worth noting. Protection against carcinogens is likely to mean protection against mutagens, and data from animal experiments and from humans will have a greater cogency than a negative mouse germ-cell test. Finally, the largest impact of genetic damage will be expressed in the first half-dozen generations, and so our concern should focus on dominant and chromosomal diseases. Recessive diseases will be so diluted by the passage of perhaps hundreds of generations before their contribution is manifested that other elements, environmental or technological, probably will have interceded to alter their impact.

PRESUMED NATURE OF MUTATIONS INDUCED BY IONIZATION

In order to draw biologically sound conclusions from the genetic effects of irradiation and chemical mutagens requires that we have an understanding of the nature of the recovered mutational event. With respect to ionizing radiation, it is becoming increasingly clear that the "induced gene mutational" endpoint is not of the simple base substitutional type but predominantly involves a more complex type of aberration as reviewed in UNSCEAR Report (United Nations, 1982). Ionizing radiation cannot induce ouabain-resistance mutations in cultured mammalian cells although such mutations occur both spontaneously and are induced by a variety of chemical agents. Why is this? To develop ouabain-resistance, the gene must be altered in a very restricted nucleotide site; deletions of the gene or other types of gross aberration destroy the sodium-potassium ATPase membrane transport system and thus the viability of the cell. Technically speaking, such mutations are not recovered, but it clearly would be inappropriate to conclude that irradiation is nonmutagenic. The endpoint is an improper one to detect the mutagenic potential of ionizing radiation. Zero survival is not equivalent to either complete repair or zero mutation-induction rate. Lack of ouabain-resistance is not the only line of evidence that points to the conclusion that nucleotide base substitutions are not a major component, if a component at all, of ionizing radiation-induced mutations at least in higher eucaryotes. Mutations at the HGPRT locus, which show measurable enzyme activity or even cross reactivity with anti-HGPRT sera, are not increased by increasing doses of radiation, but null mutants, those with no protein product activity, are.

There are as yet no mutations recovered from mouse spermatogonial irradiation experiments that screen for electrophoretic variants at 18 enzyme loci, although such mutations do occur spontaneously or when induced by chemical mutagens, for example, ethylnitrosourea (ENU). Just as with ouabain-resistance, enzyme electrophoretic mutants result from the most subtle kinds of DNA changes leading to amino acid substitutions.

Racine et al. (1980) screened for gamma-ray-induced enzyme mutants in Drosophila, and they found that, of the five intralocus mutational events, three were CRM^+ and could not be differentiated from small deletions, frame shifts or base substitutions. Kelley et al. (1983) analyzed some 31 X-ray-induced alcohol dehydrogenase gene mutations, 23 of which were deletions by classical mapping studies, 2 were large intralocus rearrangements, and the remaining appear to be events changes involving less than fifty nucleotide units. Moreover, none of these mutants give proteins recognizable by 2-D gel electrophoresis, which implies that they are not single base-substitution events.

It seems to me that these are not exceptional cases but point the way to an inescapable conclusion that may be applicable to other instances of supposedly nonmutagenic responses of certain genetic endpoints analyzed in the mouse, as, for example, the inability to induce mutations at tissue antigen histocompatibility loci. The conclusion simply stated is that mutations induced by ionizing radiation involve a variety of classical chromosome aberration types. The corollary to this conclusion is that the dose-response kinetics of these mutational events should be identical to that observed for recognized chromosome aberrations.

DOSE-RESPONSE KINETICS UNDER A VARIETY OF EXPOSURE REGIMES

Let us briefly reexamine chromosome aberration dose-response kinetics in higher organisms (see National Council for Radiation Protection, 1980 for detailed discussion of such data). Firstly, for acutely delivered low LET radiations (x and gamma rays), the dose response is not linear over the dose-response range (unless serious selective elimination of cells occurs) but is usually well fitted by the linear quadratic model over a wide dose range before saturation or falling off of the response occurs. It also appears that for different species, the transition from predominantly linear to predominantly dose squared (D^2) kinetics occurs at very different doses. Within the linear region (D) of the curve, neither dose-rate changes nor dose-fractionation regimens lead to reduction of yield below the single acute-dose value. When the events scored are reciprocal two-break translocations, inversions, deletions or dicentric chromosomes, the conclusion reached is that the two-break events (substitute lesions, if you prefer) are the result of a single ionization track, that is, they do not require the interaction of two independent tracks. It also follows that the two events initiated by the single ionization must be temporally concurrent and have some spatial probability of joining with each other. The joining process is mediated by repair enzymes.

In the D^2 region of the curve (βD^2), the aberrations are predominantly the result of the interaction of "breaks" produced by the traversal of two independent tracks. Beta is the probability of two independent track events interacting in the cell system under study with respect to its specific physical and physiological conditions. Thus again, there are temporal and spatial restrictions, but they are no different than before provided the dose is acutely delivered. In most metabolically active eucaryote cells, the "repair process" does not last more than a few hours and is probably completed in under an hour in many cases. This length of time has been determined in many species by comparing the results of fractionated exposures to a single acute exposure. When, however, the dose is delivered in a protracted manner over many hours or days, the interaction of breaks resulting from independent ionization tracks is prevented because such breaks are no longer contemporaneous. Thus the yield of events in this region decreases as the time of delivery is protracted until the yield is derived exclusively from single-track events. In other words, the β coefficient decreases to zero.

In the higher dose regions for single acute doses where saturation of effect occurs, the interaction of events leads more often to inviable cell products, and the yield reaches a plateau or may even decrease with increasing dose. It is in this region that fractionated doses can produce greater yields than single acute doses. More cells are spared because of the dose fractionation, and the yield is the sum of ($\alpha D + \beta D^2$) fractions (as it was in the D^2 region), but now the yield will exceed that obtained from single acute doses.

I believe that it is highly probable that the single-track, two-break events (α) result from a different component of the ionizing track than do the two-track, two-break events (β) (Abrahamson, 1976). In the latter case, the probability of event one ($\beta^{1/2}$) interacting with event two ($\beta^{1/2}$) leads to the final rearrangement. It invariably turns out that the probability of

$\beta^{1/2}$ occurring is between one and two orders of magnitude greater than the probability of α happening. This ratio suggests that single-track, one-break events are much more common than single-track, two-break events and thus that the former are probably the result of a single ionization from the track while the latter events (α) are produced by the clusters of ionizations in the more densely ionizing (and thus more efficient) tails of the track. This component of the ionizing track is analogous to the densely ionizing clusters of most high LET radiations which produce all of the breaks necessary for rearrangements within the single tracks and thus show only a linear dose-response relationship. If this were not the case, then the efficiency of rearrangement production should be greater with low LET radiations. Since in the above discussion we have laid out the observations recorded for chromosome rearrangement, dose-response kinetics, and the interpretations that have been widely accepted by cytogeneticists, I need only add that exactly the same dose-response kinetics apply to those events characterized as gene mutations in higher organisms. Within each region of the dose-response curve, the same types of observations with respect to dose, dose rate, dose fractionation apply for both low and high LET-induced mutations as apply to chromosome rearrangements, which is as expected if they are derived from the same initial cause, chromosome breakage.

It has been argued that the data are amenable to analysis by models other than linear-quadratic, namely, quadratic. Of course, a given set of acute data is not often robust enough to allow a satisfactory discrimination between two or more models; however, low dose-rate studies on the same endpoint frequently yield a linear response with the same slope regardless of the dose rates and doses employed, which serves to eliminate the quadratic (D^2) response curve from consideration.

HUMAN RISK ESTIMATES

There appears as yet to be few human data that permit development of solid risk estimates for the genetic endpoints of interest. None of the different analyses carried out on the offspring of the Hiroshima and Nagasaki A-bomb survivors have shown a significant increase in any effect (Schull et al., 1981a and b). This outcome is not surprising, because the total population that was exposed to radiation is relatively small, particularly that component which received a substantial exposure. An independently derived set of risk estimates (discussed in detail below) suggests that there would not have been more than about 50 or so induced genetic cases in over 16,000 children examined for genetic disorders. Assuming the correctness of the analysis, there would simply be too few cases to show a significant increase. These data do, however, help in determining the upper bounds of the risk and, as sharper genetic endpoints are found, they may ultimately serve to refine or replace the risk estimates derived from nonhuman or less extensive data.

The one element of human data that serves as a basis for a human risk estimate derives from cytogenetic analysis of spermatocyte cells of prisoners whose spermatogonial cells received x-ray doses. Beyond this, risk assessment for gene mutational endpoints are derived from mouse experiments.

The extensive radiation studies on mouse spermatogonial cells have shown that the specific locus mutation rates are responsive to dose, dose rate and dose fractionation (as discussed earlier, they follow the same rules as chromosome aberrations) and have served as a basis for extrapolation to human risk. However, the data derived from the immature mouse oocyte, which shows no induced mutation response to either low LET or high LET radiations, has been the subject of considerable skepticism in regard to its employment for risk estimation purposes; indeed, the UNSCEAR reports, United Nations (1977, 1982), have refrained from producing risk estimates for female germ cells. This lack of response from the irradiation of the mouse oocyte may not be as real as it appears according to recent studies by Dobson et al. (1983). These researchers have provided evidence that the lack of mutagenic response results exclusively from the fact that radiation-induced cell membrane damage causes cell lethality; indeed, the LD^{50} for the immature mouse oocyte is about 9 rad, marking them as perhaps the most sensitive cell type in the animal kingdom. Since the immature oocytes of primates, including humans, respond differently, at least to cell killing, it would be prudent not to use these mouse data in deriving central or upper estimates of human risk. In effect then, we have two choices, not to attempt risk estimates for one-half the human population, or to assume until better evidence is developed that the induced male rate approximates the induced female rate. Here is a place where newer data from the Japanese A-bomb study could ultimately play an important and perhaps unequivocal role in demonstrating transmission of newly induced mutations or aberrations from exposed mothers under the appropriate conditions.

Estimates of Human Mutation Rates

Dominant disorders. Two types of mouse mutation studies have been applied to the problem of estimating human risk. The studies of Selby and Selby (1977) on dominant skeletal abnormalities and the dominant cataract disorders studied by Ehling et al. (1982), both induced in spermatogonia at high acute doses of x-rays, provide the initial data for extrapolation to man. The observed frequencies must be corrected for dose, dose rate, severity, and for the relative contribution made by these disorders to the total population of induced dominant disorders. After all these manipulations, the central risk estimate for an induced dominant mutation rate per rad is 15×10^{-6} for males and females (Table 1). We also present the range of uncertainties for each sex in Table 1 as the "Lower" and "Upper" values. The value of 15×10^{-6} represents the α contribution for low dose or dose-rate exposure. If the exposure were acutely delivered and large enough to have a D^2 contribution, we would use a β value of 15×10^{-8} because the average α/β ratio for most mammalian genetic endpoints is about 100 (Abrahamson, 1976).

X-linked disorders. For single gene, X-linked recessive disorders we employ the average specific locus induced rate and the number of loci on the X-chromosome that produce genetic disorders (McKusick, 1983) (about 25% of these are associated with mental retardation phenotypes, so there can be little doubt that there is a substantive genetic detriment associated with these cases). The induced rate is 7.2×10^{-8} and is derived from low dose-rate studies (Russell and Kelly, 1982), and the number of genes is 250. Thus the central risk estimate for this class of genetic disease is 18×10^{-6}/rad/gamete for low dose-rate exposure (Table 1) and the β coefficient is 18×10^{-8} when high dose rates are employed.

Aneuploidy. Because human epidemiological data are equivocal and mouse studies on the induction of aneuploidy have been negative in the male, an International Commission on Radiological Protection task group again used an indirect doubling-dose approach to derive an upper estimate of aneuploidy induction (Oftedal and Searle, 1980); assuming a doubling dose of 100 rad, their estimate was 30 cases/10^6/rad which is in close agreement to an independently derived estimate by Schull et al. (1981a). Zero is not excluded at least for male induction in spermatogonia because such aneuploid cells may be inviable and thus unable to proliferate. We again use an approximate geometric mean estimate (central estimate) of 5 cases/10^6/rad for each sex (Table 1).

Unbalanced chromosome rearrangements. This class of genetic disorder results from the induction of chromosome rearrangements, usually but not exclusively, a reciprocal translocation in which nonhomologous chromosomes (or, in the case of oocytes, nonhomologous chromatids) exchange parts after breakage. Following this exchange, meiosis may result in gametes which contain deficient and/or duplicated chromosome segments as well as balanced products.

Cytogenetic studies on spermatocyte cells (prior to undergoing meiosis) of humans and marmosets whose spermatogonia received measured doses of x-rays serve as the basis for establishing induced translocation rates for both reciprocal and

Table 1. Genetic Mutation Rate Assumptions for Low Dose-rate Exposure

Disease Class	Central* Estimate/Rad α value x 10^{-6}	Lower	Upper
Dominants			
male	15	5	45
female	15	0	45
X-linked			
male	18	7.2	72
female	18	0	72
Irregularly inherited, equilibrium			
male	70	45	450
female	70	0	450
Aneuploid			
male	5.0	0	15
female	5.0	0	15
Unbalanced translocations			
male	7.4	0.82	18.5
female	5.6	0	14

*The central estimate is the geometric mean ($\sqrt{\text{lower range x upper range values}}$) and is chosen as the preferred estimate.

---

unbalanced events. Again it is necessary to correct for dose rate and gamma-ray RBE to establish a per rad estimate of induced translocations in germ cells and then apply the additional correction factors for transmission of a balanced or unbalanced event after meiosis to a gamete. For unbalanced gametes, an additional correction factor is required, namely, an estimate of the proportion of chromosomally unbalanced zygotes which can survive gestation. Recent studies suggest that this value represents about 10% of the cases. (Trunka, unpublished). Most liveborn individuals with recognized unbalanced translocations are severely affected with rather limited survival capability.

While the induction rates of translocations in the two sexes may be taken as the same, it should be pointed out that the transmission rates would be different in the two sexes with fewer balanced or unbalanced translocations being recoverable from females. See Box below to determine method of calculation.

For testicular gamma irradiation, we have developed a central estimate for transmitted balanced translocations of 3.7×10^{-5}/rad, and for transmitted, viable but unbalanced translocations the central estimate is 7.4×10^{-6}/rad (Table 1). From ovarian gamma-ray exposure, the estimates for balanced and unbalanced translocation, respectively, are 9.3×10^{-6}/rad and 5.6×10^{-6}/rad (Table 1).

Example of Calculation

Uncorrected Rate (A) = 7.4×10^{-4}/Rad

CORRECTIONS

Dose Rate (B) = 1/2
RBE (C) = 1/2.5

Transmission

Males Balanced Translocations (D) = 1/4
Females Balanced Translocations (E) = 1/16
Males Unbalanced Translocations (F) = 1/2
Females Unbalanced Translocations (G) = 6/16
Survival-Unbalanced Translocations (H) = 1/10

Frequency

Male Balanced Translocations = (A)•(B)•(C)•(D)
Female Balanced Translocations = (A)•(B)•(C)•(E)
Male Unbalanced Translocations = (A)•(B)•(C)•(F)•(H)
Female Unbalanced Translocations = (A)•(B)•(C)•(G)•(H)

We provide both balanced and unbalanced rates because it is becoming increasingly clear that not all balanced translocations are benign, some have been associated with sterility and others are implicated in some heritable forms of cancer (Murphree and Benedict, 1984). We should also note that for x-rays relative to gamma rays there may be an RBE of approximately 2.5 by which our risk estimate for chromosome rearrangements should be multiplied.

Irregularly inherited diseases. The only other risk estimate derived from mouse studies involves the class of irregularly and complexly inherited diseases which contribute the largest component of the current genetic disease incidence. This class is least well understood and our risk estimates here are subject to greatest uncertainty. We have a poor understanding of the equilibrium time for these diseases, that is, their persistence over many generations. Equilibrium time, in turn, is determined by how many mutant genes must interact to give a phenotype and how that phenotype is modifiable by various environmental influences. The persistence time has arbitrarily been taken as ten generations which is about twice that for simple single gene dominant diseases.

That portion of the mutation incidence that is directly proportional to the mutation rate has been defined as the mutational component of irregularly inherited diseases. Unlike single gene diseases and most chromosome diseases which have a mutation component of one, irregularly inherited diseases are believed to have mutation components well below one; the range is assumed to lie between .05-0.5. We employ 0.16, the geometric mean of the range.

The estimate of risk is an indirect one, involving the assumption of the relative mutation risk, which is the inverse of the doubling dose, and is calculated to be 1/100 from mouse data. The incidence of irregularly inherited disease has been estimated to be 90,000 cases per million liveborn. Thus the central estimate of induced cases at equilibrium is approximately 140 cases per rad of exposure (90,000 x 1/100 x .16). The male and female rates per rad are approximately 70 cases/10^6/rad, respectively, at equilibrium (Table 1). Since ten generations are taken as the average equilibrium time, the number of cases expected would average about 14 each generation. Considerably greater theoretical and experimental studies are required to establish this risk on a firmer basis.

Ranges for the Human Risk Estimates

All of the above risk estimates are summarized in Table 1. We also provide upper and lower risk estimates for each sex. In each case the lower estimate for the female is taken as zero, on the grounds that the human immature oocyte might respond like the mouse oocyte which appears to be unable to survive the traversal of an ionization track that damages the plasma membrane (Dobson et al., 1983). For dominant and irregularly inherited diseases, the lower and upper estimates employ the ranges suggested in the BEIR III report (National Research Council, 1980). For the X-linked disorders, the lower risk assumes that only 100 loci on the X chromosome may be responsible for sex-linked disorders while the upper risk assumes that 1000 loci (about 5% of the total genome) may exist on the X chromosome. For aneuploidy, zero is not excluded as a lower estimate and the upper range is that developed by Oftedal and Searle (1980). The unbalanced translocation lower estimate employs the UNSCEAR (United Nations, 1982) dose-rate reduction factor of 9 for gamma radiation and the upper range assumes no RBE differences between gamma and x-rays.

FIRST GENERATION GENETIC EFFECTS OF PARENTAL IRRADIATION

In Table 2 we present the expected first generation effects on one million liveborn after a conjoint parental 1-rad exposure using our approach and that of the BEIR III committee (National Research Council, 1980). These estimates are contrasted with the current incidence of genetic disease of approximately 106,000 cases. It is clear that while the two central estimates of risk differ by about a factor of two, both indicate that the normal disease incidence has a trivial contribution from radiation.

Multi-generational Effects of a Single Parental Exposure

For our recent committee report (unpublished), Denniston developed computer programs designed to simulate the consequences of a given radiation exposure to future generations. The programs allow us to vary dose, dose rate, manner of delivery, mutation rate by sex, viability and maternity functions and can be applied to present U.S. demographic conditions or any other desired demographic composition.

With the model, we have examined several different viability conditions for dominant and sex-linked disorders from such extremes as all mutants die prior to reproduction or all mutants survive to reproduce to the more likely situation in which, on average, there is about a 20% selective disadvantage for the newly induced gene mutation, sex linked or dominant. It is the latter situation which we have used to develop the cumulative multi-generation consequences for a single rad exposure. It should be recognized that the numbers generated by these procedures provide only an approximate rough estimate of what might occur. Our crystal ball does not include knowledge of future reproductive performance, nor does it project major advances in medical technology. Changes in either of these situations can have rather profound influence on such predictions. There is value in developing risk estimates based on conditions as they presently exist and assuming reasonable population size stability. Moreover, the majority of the genetic disorders would, in fact, be observed in approximately the first three generations or 90 years postparental exposure as shown in Fig. 1. Table 3 contrasts these predictions with those of the BEIR report (National Research Council, 1980). The differences result from the different starting assumptions of risk that I discussed earlier.

Table 2. First Generation Genetic Diseases per Million Liveborn per Million Parent Rad

		Induced Incidence (Estimated)	
Disease Class	Current Incidence	Reactor Safety Study	BEIR 1980
Dominants and X-linked	10,000	30	18
		9	a
Aneuploid and Unbalanced Translocations	6,000	10	0
		13	<10
Irregularly Inherited	90,000	a	a
	Σ 106,000	62	<28

[a]Assumed to be included in dominants.

medical technology. Changes in either of these situations can have rather profound influence on such predictions. There is value in developing risk estimates based on conditions as they presently exist and assuming reasonable population size stability. Moreover, the majority of the genetic disorders would, in fact, be observed in approximately the first three generations or 90 years postparental exposure as shown in Fig. 1. Table 3 contrasts these predictions with those of the BEIR report

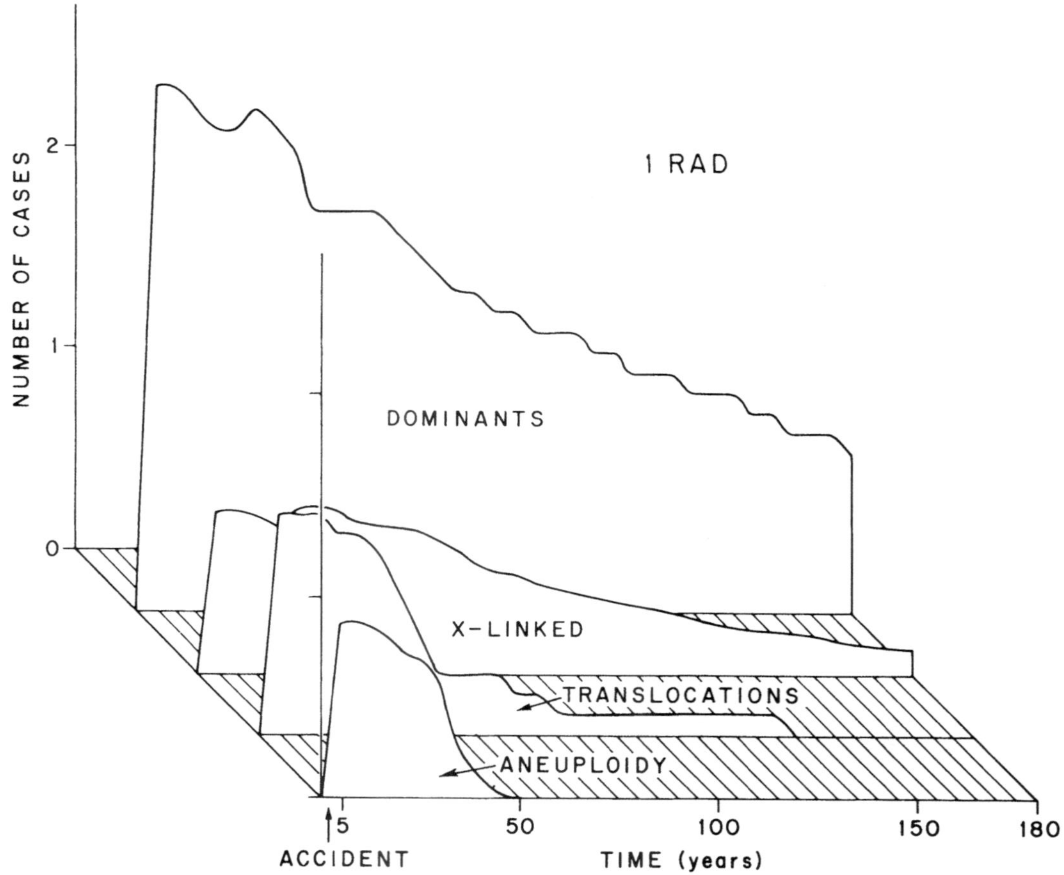

Fig. 1. Projected distribution and number of genetic disorders over five generations from a conjoint 1 rad parental exposure, assuming 480,000 births per generation from a population of one million people (1978 vital statistics). Thus for one million births the number of cases expected would be approximately twice the number shown.

(National Research Council, 1980). The differences result from the different starting assumptions of risk that I discussed earlier.

Table 3. Cumulative Genetic Effect of 1 Parental Rad Over All Time

	Central Estimates	
Disease Class	Reactor Safety Study	BEIR 1980
Dominants	125	90
X-linked	90	(a)
Aneuploid	10	0
Unbalanced Translocations	21	<10
Irregularly Inherited	140	140
Σ	385[b]	240

a)Included in dominants.

b)Totals rounded off to avoid implication of precision, and assumes 20% selective disadvantage for dominant and sex-linked mutants.

THE RELATIVE IMPACT OF INDUCED GENETIC DISORDERS

The final points in the summary chapter of the BEIR III Report was on the extreme difficulty of attempting "to compare the societal impact of a cancer with that of a serious genetic disorder" (National Research Council, 1980).

Some progress has been made along these lines since the publication of the UNSCEAR 1982 Committee Report and particularly through the efforts of Cedric Carter who has developed estimates of the years of life lost, years of life impaired, and the degree of impairment for the major categories of genetic disease that we have been considering (United Nations, 1982). This approach is a real start in developing a form of comparative societal costs.

Table 4 presents the weighted values for the five major classes of genetic disease in terms of the effective years of life lost, a composite of the sum of years lost and years impaired times the degree of impairment; in Table 5 I have applied these values to the predicted first and cumulative multi-generation effects of the one rad exposure. What emerges from this approach, not unexpectedly, is the observation that the total genetic impact on the first generation is about equal to the somatic impact in years of life lost from cancer on the parent generation. In other words, though fewer genetic than cancer cases are induced per rad, their impact is greater. This result quite interestingly agrees with an independent cost estimate prepared ten years ago for the EPA by Arthur D. Little Inc. in a report entitled "Development of Common Indices for Radiation Health Effects" which was based on the 1972 BEIR Report (Little, 1974). Using a cost-benefit economic approach to the health effects, they concluded that society spent about four times as much in health-care dollars for genetic versus cancer disorders. It has been suggested that some scaling consideration be given to reduce the weight given to infant and childhood deaths in terms of years of life lost relative to deaths from maturity onwards.

Table 4. Effective Years of Life Lost (After United Nations, 1982)

	Effective Years Life Lost	= Years Lost + Years Impaired x Severity
Dominants	21	13 + (25 x 33%)
X-linked	44	28 + (40 x 40%)
Aneuploid	46	
Unbalanced Translocations	70	
Irregularly Inherited*	35	30 + (20 x 25-30%)

Assumes normal 70-year lifespan.

*The severity estimate for this class was produced by the Nuclear Regulatory Commission's Reactor Safety Committee.

Table 5. Effective Years of Life Lost per Million Liveborn per Million Parent Rad

Disease Class	Current Incidence (x 10^4)	Incidence 1st Generation (x 10^2)	All Generations (x10^2)
Dominants and X-linked	23	6.3	26
		4.0	40
Aneuploid	23	4.6	5
Unbalanced Translocations	7	9.1	15
Irregularly Inherited	329	a	49
Σ	3,820,000	2,400	13,500

[a]Included within dominants.

THE APPLICATION OF RISK ESTIMATES TO THE A-BOMB FIRST GENERATION DATA

We have attempted to see how well our central risk estimates might serve as a prediction of what could be anticipated in the ongoing study of the children of the A-bomb survivors at Hiroshima and Nagasaki. Schull et al. (1981a and b) have presented their analysis of the genetic effects observed in their study of the children born to the exposed and to the nonexposed groups. Presently, the dosimetry is in a state of reanalysis, and our predictions may seem as no more than a futile attempt until that problem is resolved. Nevertheless, it may be illuminating to present our approach in the hope that it might encourage a sharpening-up of the risk estimate entities.

Our general model for risk involves the linear-quadratic equation discussed earlier when dealing with low LET radiations acutely delivered for doses above about 50 rad. Based on the observations that for many of the mammalian genetic endpoints studied the ratio of the α coefficient to β coefficients of the linear-quadratic equation is in the range of 100 rad (Abrahamson, 1976, Fig. 1), the β coefficients for dominant, x-linked and chromosome rearrangement disorders were chosen as 1/100 of the α coefficients presented in Table 1.

In their paper Schull et al. (1981b) presented the estimated rad doses encountered by the parents and the genetic outcomes. Table 7 of that paper which examines mortality up to age 17 for 16,713 children of exposed parents serves as the base line for our analysis. The exposed parents belonged to four dose groups, 1-9, 10-49, 50-99 and 100+ rad groups. We have assumed the average gonad dose for these groups to be 5, 29.5, 74.5 and 200+. The "true" average doses were unavailable to us, but I am led to believe that they actually were lower. In Table 6 we present the expected number of cases generated by our model for four of the

Table 6. Predicted First Generation Cases of Genetic Disease in 16,713 A-Bomb Offspring

Disease Class	Central Estimates (Linear-Quadratic)				
	Average Parental Dose (Rad)				
	5	29.5	74.5	200	Σ
Dominants	0.7	3.3	4.9	19.0	27.9
X-linked	0.2	0.5	0.9	0.8	2.4
Aneuploid	0.2	0.8	0.9	2.1	4.0
Translocations Unbalanced	0.3	1.4	2.1	8.5	12.3
Central Estimate					Σ = 47
Range					(3-136)
Translocations Balanced	0.8	3.6	4.9	15.9	25.0
Range					(3-63)

classes of genetic disease referred to above under Estimates of Mutation Rates and in Tables 1-5. Our central estimate predicts somewhat less than 50 total cases of disease, about 60% of these occurring in the 200-rad dose group. The dominant disorders are the major component and unbalanced rearrangements the next largest component; all of the latter but only some of the former diseases would be part of the early mortality group. The range on our central estimate varies from a low of three cases to a high of 136 cases. Because the ongoing studies include a cytogenetic analysis of the children and their parents, I have also included an estimate of the total number of induced balanced translocation cases, namely 25, in Table 6. As stated earlier, the carriers of this condition usually should not suffer early selective disadvantage and the true induced frequency, when established, along with accurate dosimetry, could provide an effective check on this risk estimate. Moreover, should any of the truly induced cases also be of maternal origin, this observation could be critical evidence in support of the view that the human oocyte responds differently from the mouse oocyte to irradiation.

As of now, our crude predictions support the conclusions of Schull et al. (1981a and b) that there is no significant nor detectable increase in genetic disease in the offspring of A-bomb victims. I feel reasonably confident that our conservative approach has not seriously overestimated the outcomes and is unlikely to have underestimated them.

Up to this point we have discussed the nature of radiation-induced mutational events, the shape of the dose-response curve, and the two kinds of approaches employed to derive a set of risks for the various endpoints that comprise the human genetic burden after exposure to ionizing radiation. For the two approaches, the "direct" measure of dominants, or the indirect doubling-dose approach, a set of assumptions and subsequent extrapolations from mouse to man are required. I feel on safer ground with radiation as it is a physical mutagen and is less subject to the physiological and biochemical interspecies vagaries in extrapolation that exist for a chemical mutagen. Nevertheless, it would be reasonable to expect incorporation of the radiation-risk methodologies for chemicals. At least one international agency, the International Commission for Protection against Environmental Mutagens and Carcinogens, will be recommending extrapolation with caution from mouse to man by employing the doubling-dose approach or, more accurately, its inverse, the relative mutation risk per unit does as the basis for risk estimation. This extrapolation is to be done independently with each endpoint for which mutagenicity data exists for the specific chemical mutagen. The report by the Committee on Chemical Environmental Mutagens (National Research Council, 1983) is in agreement to the extent that extrapolation to man should be done from mammalian germ cell studies, and not from lower systems if quantitative risk estimates are required. However, the report takes great pains to enumerate the serious problems that will be encountered when such an extrapolation is performed, particularly when a well-established mutagen elicits a negative response in the mouse tests or when only a single datum point is available.

REFERENCES

Abrahamson, S., 1976, Mutation process at low or high radiation doses, in: "Biological and Environmental Effects of Low-Level Radiation, pp. 3-7, Vol. I. International Atomic Energy Agency, Vienna.

Denniston, C., 1982, Low level radiation and genetic risk estimation in man. Annu. Rev. Genet., 16:329.

Dobson, R.L., Straume, J., Felton, J.S., and Kwan, T.C., 1983, Mechanism of radiation and chemical oocyte killing in mice and possible implications for genetic risk estimation, Environ. Mutagen., 5:498.

Ehling, U.H., Favor, J., Kratochvilova, J., Neuhauser-Klaus, A., 1982, Dominant cataract mutations and specific-locus mutations in mice induced by radiation or ethylnitrosourea, Mutat. Res., 92:181.

Kelley, M.R., Farnet, C.M., Mims, I.E., and Lee, W.R., 1983, Molecular analysis of Adh null mutants induced by X-rays in Drosophila melanogaster, Environ. Mutagen., 5:456.

Little, Arthur D., Inc., 1974, "Development of Common Indices for Radiation Health Effects," A final report to the E.P.A. (Contract No. 68-01-0496, Task Order No. 68-01-1126 Case 76101).

McKusick, V.A., 1983, "Mendelian Inheritance in Man: Catalogs of autosomal dominant, autosomal recessive, and X-linked phenotypes," 6th Ed., The Johns Hopkins University Press, Baltimore.

Murphree, A.L., and Benedict, W.F., 1984, Retinoblastoma: Clues to human oncogenesis, Science, 223:1028.

National Council on Radiation Protection, 1980, "Influence of dose and its distribution in time on dose-response relationships for low LET radiations," NCRP report No. 64., National Council on Radiation Protection, Washington, D.C.

National Research Council, Committee on the Biological Effects of Ionizing Radiations, 1972, "The Effects on Populations of Exposure to Low Levels of Ionizing Radiation," National Academy of Sciences, Washington, D.C.

National Research Council, Committee on the Biological Effects of Ionizing Radiations, 1980, "The Effects on Populations of Exposure to Low Levels of Ionizing Radiation," National Academy of Sciences, Washington, D.C.

National Research Council, Committee on Chemical Environmental Mutagens, 1983, "Identifying and Estimating the Genetic Impact of Chemical Mutagens," National Academy Press, Washington, D.C.

Oftedal, P., and Searle, A.G., 1980, An overall genetic risk assessment for radiobiological protection purposes, J. Med. Genet., 17:15.

Racine, R.R., Langley, C.H., and Voelker, R.A., 1980, Enzyme mutants induced by low dose rate gamma irradiation in Drosophila: frequency and characterization, Environ. Mutagen., 2:167.

Russell, W.L., and Kelly, E.M., 1982, Mutation frequencies in male mice and the estimation of genetic hazards of radiation in men. Proc. Natl. Acad. Sci. USA, 79:542.

Schull, W.J., Otake, M., and Neel, J.V., 1981a, Genetic effects of the atomic bomb: A reappraisal. Science, 213:1220.

Schull, W.J., Otake, M., and Neel, J.V., 1981b, Hiroshima and Nagasaki: A reassessment of the mutagenic effect of exposure to ionizing radiation, in: "Population and Biological Aspects of Human Mutation," pp. 277-303, E.B. Hook and I.H. Porter, eds., Academic Press, Inc., New York.

Selby, P.B., and Selby, P.R., 1977, Gamma-ray-induced dominant mutations that cause skeletal abnormalities in mice. Mutat. Res., 43:357.

United Nations Scientific Committee on the Effects of Atomic Radiation, 1977, "Ionizing Radiation: Sources and Biological Effects," United Nations Publication A/32/40, United Nations, New York.

United Nations Scientific Committee on the Effects of Atomic Radiation, 1982, "Ionizing Radiation: Sources and Biological Effects," United Nations Publication A/36/49, United Nations, New York.

DISCUSSION

Neel: Your formulation seems to leave no place in radiation genetics for a one-track, one-hit event. Is this so?

Abrahamson: Yes, that is what I was describing as the square root of β, and I do not think that the square root of β gives rise to a mutational endpoint.

Neel: Is that an article of faith right now?

Abrahamson: Yes. The exponent demonstrates that the square root of β is always one to two orders of magnitude larger than α, and would give us a much higher mutation rate. I am suggesting that one-track, one-break events never give rise to mutations, unless they interact with other systems.

Haynes: It seems to me that you have not just two terms--αD and βD^2--but really an infinite series, because there will be three-track events, four-track events, five-track events, and so on, even though they become rarer and rarer as the number of "tracks" or "hits" increases. You have shown a finite polynomial, an approximation to what is, in effect, an infinite series. When you approximate an infinite series by a finite polynomial, the values of the coefficients will depend upon the number of terms used in the approximation. Furthermore, the ratio between α and β will change dramatically as the number of terms increases, so that one cannot attach much significance to the numerical values of α and β, or even their ratios.

Abrahamson: That may be the case. All of the mammalian systems that have been looked at, regardless of endpoint, have α over β values that are in the range of 100. For _Drosophila_, the α over β range is 3,000, and, for _Neurospora_, it is closer to 40,000. These findings are describing somethings about targets to us.

Haynes: I am not disputing the facts, but rather their interpretation and significance in terms of targets, simply because the numerical values of α and β will change very dramatically, depending on how many terms there are in your series.

Abrahamson: I do not disagree with that statement, but we rarely see third-order or fourth-order terms in biology. If there is evidence for it, we ought to introduce these factors into our calculations.

Schneiderman: There is an analogy (but not an implication for the identical mechanism) with the multi-stage model of chemical carcinogenesis. There one has an equation in the form:

$$P\ (ca.) = a_o + a_1\ d + a_2 d^2 + \ldots$$

Operationally, one fits the data and retains the a_1's which differ signficantly from zero. Richard Peto has remarked on the need for the a_o, a background effect. Considering formaldehyde, the maximum likelihood fitting of the multi-stage model led to an equation with only the a_3 coefficient as significantly different from zero:

$$P\ (ca.) = a_3 d^3$$

This is consistent with the material presented by Starr. Coincidentally, if there is truly no "background", i.e. $a_o = 0$, then the appearance of two tumor bearing animals each (in the mice and rats) at equivalent "real" doses, is statistically signficantly greater than zero - i.e., there is an effect, at the specific "low" dose.

Abrahamson: The recommendation of the National Academy of Sciences to the EPA at the time when it prepared the BEIR 1972 risk etimates was to use the lowest dose for which there were valid significant data, and extrapolate linearly for radiation effects.

Thilly: You have stuck your neck out to demonstrate that the model of the dose-squared phenomena is true at the molecular level, so I will stick my neck out and say that I think that your assumption that point mutations are not occurring in mammalian cells under ionizing radiation conditions is, in fact, ill-founded and is based on a misinterpretation of data, discussed yesterday, concerning oubain resistance. I think that new kinds of experiments are possible to examine the contention that ionizing irradiation causes only chromosome breakage which masquerades, in some cases, for gene mutation. I understand that, for example, dihydrothymine is being incorporated directly into plasmids for the purpose of discovering whether or not the events which can be measured in DNA immediately after ionizing radiation are mutagenic. So probably our differences may be resolved.

Abrahamson: I am delighted to stick my neck out if it will encourage some of these experiments, but I would argue for now that, if there are point mutations as you and I are defining them, they are base substitution events.

Thilly: No, I am not defining point mutations. I use the term point mutations to refer to a loss within a very small locus within the DNA, as opposed to a loss of a large amount of genetic material. A frameshift mutation to me is another type of point mutation, involving deletion of a small number of base pairs not evenly divisible by three.

Abrahamson: My definition is a spectrum of deletions and small rearrangements either within or outside the locus - multi-locus and intralocus.

Thilly: I was first introduced to this use of polynomials to describe dose-response curves from a paper in which Drosophila was responding to aflatoxin; the dose-response curve went up and very quickly reached a plateau. The authors continued the series of $\alpha D + \beta D^2$ and on to αD^3, and, sure enough, when αD^3 was assigned a negative value the curve flattened out. This is basic curve fitting at its worst. Behavior of systems which, within a narrow range of values, appears quadratic does not necessarily demonstrate two-hit phenomena. Better to recall that the two-hit model is a hypothesis to account for one form of nonlinear response. So also are the systems which are induced by a reasonable amount of damage in the DNA, as in the SOS system in bacteria. So also would be the saturation of systems involved in the repair or recovery of the cells. They would predict transitions apparently from one slope to another, just as the model you have suggested here. I think we will require a new level of sophistication in studying genetic changes in order to be able to resolve these interpretations.

Abrahamson: The basic point I want to make is that any linear quadratic curve that you draw for the induced mutation events in Drosophila or mammalian germ cell systems is essentially the same curve you see when you score for chromosome aberrations. There are the same infection values, and the same dose-rate phenomena.

Thilly: That is an excellent point, and a few years ago Dr. T. Skopek pointed out to me a calculation that he had made with data from cytogenetic experiments and gene mutation experiments, where he found that within an order of magnitude one way or the other, the frequency of chromosomal breaks per base pair (something which you do not often see calculated) was about equal to the amount of gene mutations per base pair. Therefore, there may be saturable steps in common in clastogenesis and gene mutation which could give this remarkable similarity.

Liber: With respect to Dr. Thilly's statements yesterday about the possibility of ionizing radiation inducing base pair substitution mutations, I would like to add that I have looked at the ability of gamma rays to induce mutation to DRB^R, PPT^R, OUA^R, $6TG^R$ and TFT^R. It induced no OUA^R; however, it did induce mutation at the other loci to about the same degree. Both the DRB^R and PPT^R resistance loci behave in a fashion similar to OUA - they bind and inactivate essential cellular components of the transcription system. Therefore, some form of base substitution mutations may well be occurring in some genomic positions after x- or gamma-ray treatment of human cells. This work was done with Phaik-Mooi Leong, a graduate student in our laboratory at MIT.

Abrahamson: Are you characterizing those as base substitution events?

Liber: I think it is probably a little early to characterize them, but they do behave in the same way as the oubain resistance locus. they apparently bind to functional proteins, so it is hard to imagine that a deletion will result in such a change.

Abrahamson: And they showed a dose-response relationship?

Liber: Yes. The dose-response curve had about the same efficiency for production of mutations at these loci as at the 6-thioguanine resistance or trifluorothymidine resistance loci, which is just as interesting.

Starr: Did you say that the linear term arises from densely ionized tracks?

Abrahamson: From the dense cluster of a single ionization track, usually called a delta component.

Starr: Does the quadratic term arise from a one-track, one-hit event?

Abrahamson: Two-track, two separate tracks.

Starr: A pair of them? So in fact, you are discussing two qualitatively different processes.

Abrahamson: The point I am making is that the linear term and the quadratic term are describing two qualitatively different interactions between radiation and matter. This is different from the way chemical carcinogenesis is conceptualized, where the linear term describes a one-hit phenomenon, and the quadratic term describes a two-hit phenomenon. If there is an analog, it is at best an analog, and it is important to keep that distinction clear.

Borg: You have referred to the densely ionizing parts of the ionization paths of low-LET radiations as the probable sites for the one-track, two-hit events which you propose as the physical counterpart of the α coefficient in the linear term of linear-quadratic dose-response curves. I have no real argument with that, but I point out that you do not have to seek recourse to the high-density parts of low-LET radiation. As John Ward pointed out at the NCI conference, every primary ionization in the aqueous "matrix" gives rise to an average of four to five closely spaced hydroxyl radicals. A single hydroxyl radical reaction with a deoxysugar can give rise to chain breakage (as in an indirect way can some of the hydroxyl radical-base adducts). Hence the possibility of cooperative action in a small volume (i.e., two "hits" close in space and coincident in time) always exist with radiation. This is not an intrinsic property of hydroxyl radicals independent of their spatial disposition, because hydroxyl radicals produced homogeneously by chemical ractions (the Fenton reaction) behave quite differently.

Neel: After all this discussion, do you not feel a real need for more data on the ability of radiation to produce electromorphic mutations?

Abrahamson: The experiments have been done. Researchers at the NIEHS (Valcovis and Malling, personal communication) could not obtain x-ray induced electrophoretic mutants from mouse spermatogonia; they did obtain null mutants, however.

Ehling: We can get very large electrophoretic mutants with x-rays.

Abrahamson: Did you get a very significant increase over control values? The NIEHS group did not find any increase with x-rays while they got lots of mutants after treatment with ethylnitrosourea (Lewis and Johnson, 1982).

Ehling: That the mutation rate is low there is no question; enzyme activity counteracts the mutation rate.

Abrahamson: That was background radiation, estimated U.S. millirem (mr) dose just to put you in perspective of what the genetic dose would be from terrestrial and man-made radiation. We are talking about 100 mr per year as background. I am going to make risk estimates; about 100,000 cases per million liveborn is the current incidence of genetic disease, and is probably a minimum estimate at this time.

Sonnenblick: Earlier you gave various genetic bases for congenital diseases which had different degrees of morbidity, slight to severe. In teratology, how do you distinguish the impact of genetic effects and non-genetic influences? And where did you get the data?

Abrahamson: The information was from the British Columbia Data by Trimble and Doughty, published in 1975, that looked at 750,000 children born and studied up to the age of 9 (Trimble and Doughty, 1974).

The only way that you can distinguish between genetic and nongenetic effects is to say that there is a genetic basis to most of this kind of disorder: that is the way that the BEIR committee argued. Secondly, the committee said that the mutational component, that is the induced component, which increases with increasing dose, is not well known but lies somewhere between five and fifty percent. So the slope for the dose-response curve is different essentially from the slope of the other events. For these induced events I computed the geometric mean for the two ranges chosen by BEIR and the same range chosen by UNSCEAR, i.e., five to fifty, while UNSCEAR chose to use 5% as the mutation component for most of congenital anomalies. The paper in Science by Crow and Dennison (1981) shows that, for a certain class of congenital anomalies and irregularly inherited diseases, 50% does represent a mutational component for a part of irregularly inherited disease.

Neel: If one can reconsider your data on first generation defects in Hiroshima and Nagasaki, in due time, when we have the revised dosage estimates from Japan, we will be in a position to compare our actual findings with your projection. Second point, in Awa's study, the numbers of balanced reciprocal translocations are thus far the same in the children of distally and proximally exposed (Awa, 1975). The family studies are not yet completed.

Abrahamson: You can see that clearly the most important families to look at are those whose parents (either one or both) were exposed in the 200 rad range, because that is where all the data are going to be. If you cannot find evidence in that set of parents, I do not think you are going to find it in others.

Neel: Well, that is where Dr. Awa is looking, examining preferentially children in the higher dose families.

Evans: What did the data show concerning the ratio of unbalanced to balanced translocations? Was there a relative increase in the numbers of unbalanced forms?

Neel: Blood samples are not drawn for Awa's study until the children are aged 13, so the data re of little value with respect to unbalanced translocations.

Abrahamson: There have not been too many cytogenetic analyses. Seven thousand children that have been looked at in the exposed group totally, that is, in all doses, and 7,000 in the unexposed group.

Evans: Balanced translocations ascertained through a phenotypically normal carrier, and therefore part of the normal background of inherited translocations, are not usually associated with any phenotypical abnormalities. However, a significant proportion of balanced translocations arising de novo, that is as fresh mutations, are associated with an abnormal phenotype. Thus, in a population living in a "normal environment" there will be a proportion of individuals carrying balanced translocations which have been inherited through their families and which in the vast majority of cases are not in any way associated with disease manifestation, and a proportion with an associated clinical abnormality; this latter class will contain the majority of all new mutations. We would, therefore, expect that in a population exposed to a mutagen, the proportion of balanced translocations that are associated with a disease state should be increased if there is a significant increase in the background mutation rate. I understand that there was a total of 25 balanced translocations, 12 unbalanced translocations and 4 aneuploids detected in the population studied, and my question is whether one could discern any increase in the proportion of balanced translocations associated with an abnormal phenotype within those exposed populations?

Abrahamson: The survey measures actual mortality. This group would not be seen other than as mortality, because when the cytogenetic studies started most of the children were twelve or thirteen years of age. This class is basically gone except in the mortality records; they only see one aneuploid in all of 15,000 children.

Neel: In the early days of our study we did look for Down's Syndrome (Schull and Neel, 1962). This was in the "pre-cytogenetic" days, and in fact, we saw fewer children with Down's in the children of the proximally exposed than in the distally exposed. Now Awa is not picking them up in his cytogenetic studies either because of a very high death rate in these children in Japan, or else because of a real reticence on the part of the parents to subject children to the studies. There are no good data on the sex chromosomes aneuploids (Awa, unpublished).

Starr: I would have preferred not to have seen a summary total life loss aggregated across different diseases. Apples are not oranges: benign and malignant neoplasms are not equal.

Abrahamson: We add them together because we are looking at the total genetic impact. If our calculations are right, even by factors of two or three, what we are really saying is that very low doses of radiation have trivial impact on society. It is one of society's less hazardous environmental pollutants.

REFERENCES

Awa, A.A., 1975, Cytogenetic study, J. Radiat. Res., 16 (Suppl.):75.

Crow, J.F., and Denniston, C., 1981, The mutation component of genetic damage, Science, 212:888.

Lewis, S.E., and Johnson, F.M., 1982, The nature of electrophoretically expressed mutations induced by ethylnitrosourea in the mouse, Environ. Mut., 4:338.

Schull, W.J., and Neel, J.V., 1962, Maternal radiation and mongolism, Lancet, 1962 (ii):537.

Trimble, B.K., and Doughty, J.H., 1974, The amount of hereditary disease in human populations, Ann. Hum. Genet., 38:199.

SATURATION OF REPAIR

R.B. Setlow

Biology Department
Brookhaven National Laboratory
Upton, NY 11973-5000

INTRODUCTION

The deleterious effects of the exposure of cells, tissues or animals to exogenous radiations and chemicals are ameliorated by DNA repair processes. Many repair pathways are constitutive, and some, such as the SOS system in bacteria, are inducible while others, such as the UvrABC excision system in *E. coli*, are both constitutive and inducible (Friedberg, 1985). The ability of bacteria to adapt to low chronic doses of *N*-methyl-*N*-nitro-*N*-nitrosoguanine (MNNG) so as to become resistant to the cytotoxic or mutagenic effects of a large challenge dose is an important example illustrating that low chronic exposures not only may protect against subsequent high acute ones but also that extrapolation from high acute doses to low chronic ones may grossly overestimate the biological effects of the low chronic doses (Robins and Cairns, 1979). Another example is where low doses of nitrosamines protect rats against the carcinogenic effects of a later high dose (Montesano et al., 1980; Margison, 1982). In bacteria, chronic exposure to MNNG involves adaptation along separate pathways against mutagenic and cytotoxic agents. The former pathway entails the repair of O^6-methyldeoxyguanosine (m^6dGuo) by means of an alkyl transfer, whereas the latter involves the induction of glycosylases (Karran et al., 1982; Evensen and Seeberg, 1982).

Whatever mechanisms play a role in the processes of DNA repair, it is important to recognize that the key events are the rate of repair relative to the rates of replication or transcription of DNA. If a lesion is repaired before a replication fork or an RNA polymerase passes through it, then there will be no deleterious effect. A delay in replication or transcription enhances the effect of DNA repair. It is also important to realize that in dividing repair-proficient cells, many lesions are not repaired before they become fixed by replication or transcription. Thus, for example, *E. coli* cells proficient in excision repair of pyrimidine dimers show a marked enhancement of survival as a result of photoreactivation--the removal of pyrimidine dimers by an enzyme system utilizing near-ultraviolet and visible radiation. The damage that can be repaired by photoreactivation is often very high (60 to 80 percent) indicating that excision lags far behind the fixation of mutations or gene expression. Of course, excision-deficient cells are much more sensitive than excision-proficient ones and, therefore, show an even higher degree of photoreactivation (Harm, 1980).

Damages to DNA can produce biological effects as a result of molecular mechanisms that take place at different times in the cell cycle, and DNA repair may be more effective for one mechanism compared to another. For example, both the mutagenic and cytotoxic effects of UV radiation are decreased by excision repair (Konze-Thomas et al., 1982), but the mutagenic effect is determined by the time when DNA is replicated compared to repair whereas the cytotoxic effect is probably more concerned with when RNA is synthesized compared to DNA repair (Kantor and Hull, 1979). Thus, nondividing human cells in culture are killed by UV radiation (they no longer remain attached to the solid surface on which they grow), and killing seems to arise from the fact that dimers in DNA interfere with RNA synthesis. As a result, there is improper protein synthesis, and the cells die. On the other hand, doses of X-rays that would be lethal to rapidly proliferating cell populations do not affect RNA synthesis in nondividing cultures and the cells are not killed (Kantor and Hull, 1978; and personal communication). Here is a case in which damage to DNA is not lethal unless the cells are dividing.

ENDOGENOUS VERSUS EXOGENOUS DAMAGE

Cells growing at 37° C sustain large amounts of damage resulting from normal chemical and metabolic reactions associated with necessary endogenous methylation reactions. The risk from exogenous agents that cause the same kinds of damage as endogenous reactions must be measured relative to the endogenous background rate of damage and repair. Table 1 shows the rates of introduction of some endogenously formed chemical damage into DNA as well as injury inflicted by two environmental agents--sunlight, in terms of its ability to make pyrimidine dimers in skin cells, and X-rays, in terms of their ability to make single-strand breaks. Two points emerge from these data. The first is that the number of alterations per cell resulting from endogenous reactions is appreciable. The number of damages per cell per lifetime is approximately 2 x 10^9, and, since there are only 10^{10} nucleotides per cell, endogenous damage, if not repaired, would make a mockery out of the utility of DNA as the mainstay of genetic information. The second point is that the environmental agent that does the most damage is sunlight. The amount of injury per exposed cell far outweighs anything else humans are liable to encounter in the environment. Hence, it is not surprising that sunlight-induced skin cancer among whites in the United States amounts to about 500,000 new cases per year (Scotto et al., 1983), approximately equal to all other cancers combined.

The data in Table 1 indicate that there must be effective DNA repair mechanisms if we are to survive the wear and tear of living at 37° C. Table 2 gives estimated repair rates for several types of damage. It is apparent that the repair rate for endogenous damage far exceeds the rate of production of such damages. Nevertheless, over a lifetime, a replicational or transcriptional growing-point is bound to pass over some of these alterations an appreciable number of times. For example, even if repair were 10^4 times more rapid than the introduction of damage, there would still be a small amount remaining at any instant of time, and, over a

Table 1. Appearance of Damages in the DNA of a Mammalian Cell at 37°C

Damage	Events/hr	Reference[a]
Depurination	580	1
Depyrimidination	29	1
Deamination of cytosine	8	1
Single-strand breaks (ssb)	2300	1
Ssb after depurination	580	1
_O_^{6}-methylguanine	130	2
Pyrimidine dimers in skin (noon Texas sun)	5×10^4	3
Ssb from ionizing radiation background	10^{-4}	4

[a] The values given are estimates from data on (1) DNA in solution (Shapiro, 1981); (2) alkylation of DNA by _S_-adenosylmethionine (Barrows and Magee, 1982); (3) UV dosimetry from the inactivation of bacteria corrected for the transmission of skin (Harm, 1969; Setlow, 1982); and (4) ssb at high dose (Setlow and Setlow, 1972) extrapolated to 1 mGy/yr (1 ssb/yr)

Table 2. Typical Repair Rates in Human Cells

Damage	Number/hr	Reference[a]
Single-strand breaks	2×10^5	1
Pyrimidine dimers		
normal cells	5×10^4	1
xeroderma pigmentosum group C cells	5×10^3	1
$\underline{O}^6$-methylguanine	$10^4 - 10^5$	2

[a]Data from (1) Setlow (1982) and (2) Waldstein et al. (1982a).

lifetime, such damage could account for an alteration of perhaps 1 in 10^5 nucleotides that might be reflected in an error in replication or transcription. It is also evident that the rate of repair of UV-induced pyrimidine dimers in human skin cells is of the same order of magnitude as the rate of their introduction by sunlight (Table 2). On the other hand, in the case of repair-deficient human cells (xeroderma pigmentosum, Group C cells), the rate of repair is appreciably less than the rate of introduction.

REPAIR-DEFICIENT HUMAN DISEASES

Much of our understanding of the relationship between damage to DNA and carcinogenesis comes from investigations of human cancer-prone diseases. The best understood of these is xeroderma pigmentosum (XP) (Friedberg, 1985). It is a complicated disease at the molecular level having seven complementation groups, none of which is absolutely deficient in excision repair. All XP individuals are very sensitive to sunlight-induced skin cancer, and, on the average, show a skin cancer prevalence approximately 10^4-fold greater than the general population, despite the fact that their repair defects, while not 100 percent, amount to ~80 percent of normal (Setlow, 1982). The reason for the big amplification of cancer prevalence relative to the small deficit in damage removal is due to the fact that ultraviolet makes tremendous numbers of damages per cell, and that skin cancer incidence among normal individuals, as shown by the available epidemiologic data, increases exponentially with the average yearly ultraviolet fluence at the surface of the earth. Because of this exponential relation, it has been estimated that the dose-modification factor of DNA repair is only of the order of tenfold--a clear instance in which small changes in exposure may produce large changes in biological effect.

Numerous measurements have been made of the rate of repair versus the amount of initial damage. The data indicate that the rate of excision repair saturates at a dose of 20 J/m^2 of 254 nm radiation, a dose that makes approximately 2×10^6 pyrimidine dimers per cell (Setlow, 1982). This amount of injury far exceeds that produced by sunlight, where the repair system for normal cells is not saturated. An increase in sunlight exposure results in an increase in the number of dimers removed per unit time. Of course, this is not the case for most XP cells (Tables 1 and 2). Under constant UV illumination, the number of dimers in XP cells would be expected to increase dramatically as a function of time, but, in normal cells, it would be approximately constant with the constant being proportional to the incident exposure rate.

It is important to understand that the repair mechanisms working on pyrimidine dimers in human cells also act upon other types of bulky damages and that XP individuals seem to have a significantly enhanced level of internal cancers--cancers not ascribable to UV exposure (Kraemer et al., 1984). The approximately tenfold excess of internal cancers is much less than those for sunlight-induced ones, presumably because of the relatively high dose rate for UV compared to the agents making bulky damages internally.

Interestingly, there appears to be a synergistic interaction between X-ray and ultraviolet radiation. Patients treated for psoriasis with X-rays and PUVA therapy show a much enhanced incidence of skin cancer compared to the normal population or compared to those given only PUVA with no X-rays (Stern et al., 1982). Individuals given X-irradiation therapy to the scalp for ringworm experience an increase in skin cancer incidence in the sunlight-exposed areas of the scalp and face (Shore et al., 1984). This marked increase has not been observed among black patients.

A second relevant repair-deficient disease is ataxia telangiectasia (AT). The affected individuals and their cells in culture are sensitive to the cytotoxic effects of X-rays. The nature of the presumptive repair defect is not known; some AT cell strains are deficient in unscheduled DNA synthesis after irradiation, but others are not. The cytotoxic effects may be associated with the fact that X-irradiation does not inhibit the initiation of replicons as it does in normal cells (Painter, 1981). Hence, one could claim that "repair" in normal cells inhibits the initiation of replicons. There is no evidence for the saturation of any such repair system at doses less than 5 Gy.

A number of XP complementation groups consist of individuals with neurological abnormalities. Ataxia telangiectasia also is a disease with many neurological symptoms. Cells from individuals with other neurological disorders, such as Huntington's disease, Parkinson's disease, and Alzheimer's disease, show small increases in sensitivity to X-rays or to methylating agents (Robbins, 1983; Kidson et al., 1983). The nature, if any, of the repair defects in such conditions is not known. However, if we recall that relatively small decreases in DNA repair can have large biological consequences, it is reasonable to look for some causal relation between the disease and the defect. As a matter of fact, there are relatively wide variations in DNA repair capabilities among apparently normal populations (Setlow, 1983). Such variations have been found for UV repair in fibroblasts, epithelial cells and lymphocytes; for presumptive X-ray repair as measured by cytotoxicity; and for the level of alkyl transferase repairing m^6dGuo in lymphocytes and many other human tissues (see below). Thus, relatively small changes in DNA repair in the normal population may be reflected in large changes in sensitivity to endogenous and exogenous agents that damage DNA.

ALKYLATION DAMAGE

The story of potential alkylating agents provides an excellent example of the complications encountered in estimating exposures at the molecular level and the effects that result from such exposures. Most environmental alkylating agents arise from innocuous compounds in the environment that may be activated chemically to form nitrosamines. The latter, in turn, are metabolically activated to produce intermediates that are capable of reacting with DNA. Thus, the pharmokokinetics are complicated. The methylation of DNA results in many products, only a small subset of which is important for mutagenesis and carcinogenesis. For example, after treatment with MNNG, m^6dGuo amounts to approximately 10 percent of m^7dGuo, but seems to be responsible for most of the mutagenic and carcinogenic activities (Pegg, 1984a). On the other hand, m^7dGuo is relatively innocuous, but its repair by glycosolase activity or chemical depurination probably accounts for the great majority of the observed unscheduled DNA synthesis and for the single-strand breaks that are seen (Lindahl, 1982) (the situation is more complicated for ethylating agents, however, since the mutagenic effects seem to be associated with several O-ethylation products [Singer, 1984]). The repair of m^6dGuo in bacteria and in mammalian cells is accomplished by the transfer of the methyl group to an acceptor protein; no attack on the polynucleotide backbone of DNA is involved, and there is no unscheduled DNA synthesis (Friedberg, 1985). The acceptor group can only be used once; it does not turn over. Table 3 gives some characteristics of the transferase reaction.

Since the transfer reaction is stoichiometric, it is a relatively simple matter to determine the number of acceptor sites per cell. In the normal human cell, these numbers vary from 10^4 to 10^5 (Waldstein et al., 1982a; Pegg, 1984b). The repair of m^6dGuo is much more rapid than the repair of pyrimidine dimers, and a comparison of the number of acceptors per cell and the number of m^6dGuo molecules made endogenously (Table 1) indicates that this repair system would not be saturated at the chronic dose rates encountered in the environment.

Table 3. Characteristics of m^6dGuo Repair in Human Cells

Rapid	15-30 min
Stoichiometric	Suicide reaction
S-methylcysteine	Only reaction product

Most repair measurements have been made on cells in culture, and because of the correlation between *in vivo* and *in vitro* measurements for XP and AT, it has been assumed that *in vitro* measurements are good estimates of the capability of human tissues to repair DNA. Such is not necessarily the case, however, because although many cell strains grown from tumor tissue are defective in the repair of m^6dGuo (Yarosh et al., 1983), the cells in the tumor tissue itself are not (Wiestler et al., 1984). The difference could be the result of the kind of cell types that are able to continue to grow in culture or that the relevant gene may often not be expressed in culture. The direct measurement of repair activity in various human tissues such as lymphocytes, brain and liver indicate that there are wide variations among the population; variations of several fold. These points are summarized in Table 4.

CONCLUSIONS

Only a relatively small number of DNA damages have been identified as producing biological effects. Pyrimidine dimers and m^6dGuo are the best documented ones. We still do not understand well the molecular mechanisms of most DNA repair systems. Nevertheless, it is clear that DNA repair does not saturate at environmental doses, nor is repair 100 percent efficient in that all the damage is not removed before a replicational or a transcriptional event takes place. Proficient repair decreases the cytotoxic, mutagenic and carcinogenic effects of radiations and chemicals, and it is possible that variations in DNA repair among individuals in the normal population could account for large differences in cancer susceptibility.

Table 4. m^6dGuo Methyl Transferase in Extracts of Adult Human Tissues (Several Series of Experiments)

Tissue	Activity (fmol/mg protein) Average	Range	Average	Range
Liver[1,2]	1070 (9)	7.8-fold	870 (5)	4.4-fold
Colon[1,2]	140 (10)	9.6	260 (10)	3.1
Small intestine[1]	210 (12)	42		
Stomach[1]	200 (5)	1.7		
Lung[2]			120 (13)	4.7
Brain[2]			76 (5)	3.3
Lymphocytes[3]	1000 (38)	7.9	1280 (10)	2.9
Stimulated lymphocytes[3]	1750 (36)	3.4		

Data from (1) Myrnes et al. (1983); (2) Grafstrom et al. (1984); and (3) Waldstein et al. (1982a,b).

Numbers in parentheses are the numbers of samples.

REFERENCES

Barrows, L.R. and Magee, P.N., 1982, Nonenzymatic methylation of DNA by S-adenosylmethionine in vitro, Carcinogenesis, 3:349.

Evensen, G., and Seeberg, E., 1982, Adaptation to alkylation resistance involves induction of a DNA glycosylase, Nature (London), 296:773.

Friedberg, E.C., 1985, "DNA Repair," W.H. Freeman, New York.

Grafstrom, R. C., Pegg, A.E., Trump, B.F., and Harris, C.C., 1984, O^6-alkylguanine-DNA alkyltransferase activity in normal human tissues and cells, Cancer Res., 44:2855.

Harm, W., 1969, Biological determination of the germicidal activity of sunlight, Radiat. Res., 40:63.

Harm, W., 1980, "Biological Effects of Ultraviolet Radiation," Cambridge Univ. Press, Cambridge.

Kantor, G.J., and Hull, D.R., 1978, A comparison of the responses of arrested HDF populations to UV and X-rays, J. Supramol. Struct. Suppl. 2, 8:81.

Kantor, G.J., and Hull, D.R., 1979, An effect of ultraviolet light on RNA and protein synthesis in nondividing human diploid fibroblasts, Biophys. J., 27:359.
Karran, P., Hjelmgren, T., and Lindahl, T., 1982, Induction of a DNA glycosylase for N-methylated purines is a part of the adaptive response to alkylating agents, Nature (London), 296:770.
Kidson, C., Chen, P., Imray, F.P., and Gipps, E., 1983, Nervous system disease associated with dominant cellular radiosensitivity, in: "Cellular Responses to DNA Damage," pp. 721-729, E.C. Friedberg and B.A. Bridges, eds., A.R. Liss, New York.
Konze-Thomas, B., Hazard, R.M., Maher, V.M., and McCormick, J.J., 1982, Extent of excision repair before DNA synthesis determines the mutagenic but not the lethal effect of UV radiation, Mutat. Res., 94:421.
Kraemer, K.H., Lee, M.Y., and Scotto, J., 1984, DNA repair protects against cutaneous and internal neoplasia: evidence from xeroderma pigmentosum, Carcinogenesis, 5:511.
Lindahl, T., 1982, DNA repair enzymes, Annu. Rev. Biochem., 51:61.
Margison, G.P., 1982, Chronic or acute administration of various dialkylnitrosamines enhances the removal of O^6-methylguanine from rat liver DNA in vivo, Chem.-Biol. Interact., 38:189.
Montesano, R., Bresil, H., Planche-Martel, G., Margison, G.P., and Pegg, A.E., 1980, Effect of chronic treatment of rats with dimethylnitrosamine on the removal of O^6-methylguanine from DNA, Cancer Res., 40:452.
Myrnes, B., Giercksky, K.E., and Krokan, H., 1983, Interindividual variation in the activity of O^6-methylguanine-DNA methyltransferase and uracil-DNA glycosylase in human organs, Carcinogenesis, 4:1565.
Painter, R.B., 1981, Radioresistant DNA synthesis: an intrinsic feature of ataxia telangiectasia, Mutat. Res., 84:183.
Pegg, A.E., 1984a, Methylation of the O^6 position of guanine in DNA is the most likely event in carcinogenesis by methylating agents, Cancer Invest., 2:223.
Pegg, A.E., 1984b, Repair of O^6-methylguanine in DNA by mammalian tissues, in: "Biochemical Basis of Chemical Carcinogenesis," pp. 265-274, H. Greim, R. Jung, M. Kramer, M. Marquardt, and F. Oesch, eds., Raven Press, New York.
Robbins, J.H., 1983, Hypersensitivity to DNA-damaging agents in primary degenerations of excitable tissue, in: "Cellular Responses to DNA Damage," pp. 671-700, E.C. Friedberg and B.A. Bridges, eds., A.R. Liss, New York.
Robins, P., and Cairns, J., 1979, Quantitation of the adaptive response to alkylating agents, Nature (London), 280:74.
Scotto, J., Fears, T.R., and Fraumeni, J.F., Jr., 1983, "Incidence of Nonmelanoma Skin Cancer in the United States," DHHS Pub. No. (NIH) 76-1029, Bethesda, MD.
Setlow, R.B., 1982, DNA repair, aging, and cancer, Natl. Cancer

Inst. Monogr., 60:249.

Setlow, R.B., 1983, Variations in DNA repair among humans, in: "Human Carcinogenesis," pp. 231-254, C.C. Harris and H.N. Autrup, eds., Academic Press, New York.

Setlow, R.B., and Setlow, J.K., 1972, Effects of radiation on polynucleotides, Annu. Rev. Biophys. Bioengineer., 1:293.

Shapiro, R., 1981, DNA damage caused by hydrolysis, in: "Chromosome Damage and Repair," pp. 3-18, E. Seeberg and K. Kleppe, eds., Plenum Academic, New York.

Shore, R.E., Albert, R.E., Reed, M., Harley, N., and Pasternack, B.S., 1984, Skin cancer incidence among children irradiated for ringworm of the scalp, Radiat. Res., 100:192.

Singer, B., 1984, Alkylation of the O^6 of guanine is only one of many chemical events that initiate carcinogenesis, Cancer Invest., 2:233.

Stern, R.S., Zierler, S., and Parrish, J.A., 1982, Psoriasis and the risk of cancer, J. Invest. Dermatol., 78:147.

Waldstein, E.A., Cao, E.-H., Bender, M.A., and Setlow, R.B, 1982a, Abilities of extracts of human lymphocytes to remove O^6-methylguanine from DNA, Mutat. Res., 95:405.

Waldstein, E.A., Cao, E.-H., Miller, M.E., Cronkite, E.P., and Setlow, R.B., 1982b, Extracts of chronic lymphocytic leukemia lymphocytes have a high level of DNA repair activity for O^6-methylguanine, Proc. Natl. Acad. Sci. USA, 79:4786.

Wiestler, O., Kleihues, P., and Pegg, A.E., 1984, O^6-alkylguanine-DNA alkyltransferase activity in human brain and brain tumors, Carcinogenesis, 5:121.

Yarosh, D.B., Foote, R.S., Mitra, S., and Day, R.S., III., 1983, Repair of O^6-methylguanine in DNA by demethylation is lacking in Mer^- human tumor cell strains, Carcinogenesis, 4:199.

ROLE OF TUMOR PROMOTION IN AFFECTING THE MULTI-HIT NATURE OF CARCINOGENESIS

James E. Trosko and Chia-cheng Chang

Department of Pediatrics and Human Development
Michigan State University
East Lansing, Michigan 48824

INTRODUCTION: MULTI-STAGE NATURE OF CARCINOGENESIS

A variety of radiations and chemicals can induce a wide spectrum of acute and chronic disease states, including cancer, when animals or human beings are acutely exposed to moderate or high levels of these agents. However, with the exception of accidents, exaggerated lifestyles and genetic predispositions, the common real-life situation is one of chronic low-level exposure.

Exposure of cells to radiation and chemicals can be expected to cause a diversity of molecular damage. The response to that damage could be restoration to normal function, mutations (viable and lethal; recessive and dominant), cell death, and modulation of gene expression (Trosko and Chang, 1981). The biological or physiological consequences of mutations, cell death or gene modulation depend on many factors (Trosko et al., 1982c). For example, a single dominant or recessive mutation in one postmitotic cell in which this locus is repressed will not influence the organism. A single recessive mutation in an expressed locus of a stem cell also would have little impact, unless the opposite allele is also mutated or repressed. Even if this single cell had both alleles mutated, its physiological impact would be inconsequential unless this cell multiplied to a large critical mass (Trosko and Chang, 1980; Trosko and Chang, 1984a).

Low-level chronic exposure to mutagens, cytotoxic agents and gene modulators might be expected to have different consequences than would acute moderate or high-level exposure. The complex interplay of molecular and cellular repair or restoration mechanisms, and differences in cellular and physiological consequences of molecular lesions induced at low or high levels might dramatically influence the biological damage of radiation or chemicals (Trosko and Chang, 1981; Trosko and Chang, 1983; Jones et al., 1983). For example, induction of few DNA lesions in nondividing cells might be repaired and have no biological affects. On the other hand, multiple DNA lesions in dividing cells might not all be repaired, leading to mutated cells, cell death, and cells with altered gene expression. Exposure to non-mutagenic, cytotoxic agents would have different consequences depending on when the organism was exposed, how many cells of a tissue died, and what effects tissue regeneration had upon the type of surviving stem cells.

Using carcinogenesis as one model of a chronic disease process, and keeping within the multi-stage theory of carcinogenesis, we will try to examine how the various factors influencing carcinogenesis might contribute to cancer formation at chronic low-level exposures. Carcinogenesis has been conceptualized as a complex multi-step process from the results of experimental studies on animals (Boutwell, 1974; Pitot and Sirica, 1980; Slaga, 1983), from epidemiological studies on human cancer (Armitage and Doll, 1954; Moolgavkar et al., 1980), from the study of the pathogenesis of human tumorigenesis (Foulds, 1975), from studies on oncogenes (Land et al., 1983; Newbold and Overell, 1983), from *in vitro* transformation studies (Barrett and Ts'o, 1978), and from genetic theory (Trosko and Chang, 1978; Potter, 1983). The concepts of initiation, promotion and progression have evolved to explain the observation that tumors could be induced by application of a subthreshold dose of a carcinogen (the initiation phase) followed by repetitive treatment with a noncarcinogen (the promotion phase). In this description, we had to use terms (subthreshold, carcinogen, and noncarcinogen) which are clearly operationally and mechanistically vague (see review Trosko et al., 1982c). Empirically, the initiation phase appears to involve an irreversible change in the genome, while the promotion phase is reversible at least initially. In experimental carcinogenesis, this initiation/promotion concept seems to be applicable to skin, liver, lung, colon, mammary gland, bladder, Harderian gland, and thyroid gland (Slaga, 1983).

Although some controversy exists on the theories of the mechanisms of initiation and promotion, in general, mutagens appear to be effective initiators or complete carcinogens (see Trosko et al., 1982c for the distinction between initiators and complete carcinogens) while substances providing sustained mitogenic stimuli appear to be effective promoters (Argyris, 1982). For the sake of stimulating discussion, we will that the initiation process involves the irreversible conversion of a normal cell to a premalignant cell, and that promotion is the process causing the clonal amplification of this cell to a critical mass during which time conversion to the malignant phenotype occurs (Trosko et al., 1983). Progression refers to the rapid variation in phenotypes of the malignant cells, possibly due to an unstable maintenance of the karyotypes (Pitot et al., 1981).

The stable or irreversible nature of the initiation process has been noted (Loehrke et al., 1983). The observation that animals treated with promoters can also produce some tumors suggests either that there are no pure promoters (that is, promoters can be weak initiators or complete carcinogens) or that promoters can activate, can promote spontaneously initiated cells in the absence of application of known initiators (Schulte-Hermann et al., 1983). Theoretically and experimentally, the initiation/promotion model of carcinogenesis suggests that there are of modifiers of these processes (that is, anti-initiators, anti-promoters, co-initiators, co-carcinogens) (Slaga et al., 1982).

Recently, molecular studies on the role of oncogenes in the *in vitro* transformation of rodent cells indicate that several different phenotypic alterations must occur in overcoming senescence and then in the transformation of the immortalized cell to a malignant cell (Land et al., 1983; Newbold and Overell, 1983). Therefore, the concepts of oncogenes must now be integrated into the multi-stage phenomenon of carcinogenesis and the concepts of initiation and promotion (Trosko et al., 1984b). The biological function of the various oncogenes in the role of initiation and promotion are undetermined; however, it is tempting to speculate that during initiation a stable alteration affecting the ability of a stem cell to differentiate terminally would block its ability to senesce and the abnormal production or function of some growth factor might act to stimulate the expansion of this abnormal stem cell (that is, act as a promoter).

ROLE OF PROMOTION IN CARCINOGENESIS

The concepts of initiation and promotion were derived from empirical observations of experimental tumorigenesis. The terms are operational, in that they refer to a situation in which the administration of an ineffective dose of a known carcinogen, followed by repetitive treatment with another agent elicits the appearance of many tumors. Application of this second agent alone causes only a few tumors. The appearance of papillomas in skin and enyzyme-altered foci and hyperplastic nodules in the liver precede the ultimate appearance of carcinomas.

A general hypothesis to explain the function of promoters is that it involves the clonal expansion of the initiated cell (Trosko and Chang, 1980; Potter, 1980). Several assumptions are made within this hypothesis (Trosko et al., 1983). In brief, the carcinogenic process is assumed to involve multiple hits or so-called permanent genetic alterations. Initiation is assumed to induce a mutation in a stem cell which then is unable to terminally differentiate. It should also be noted that not all cells exposed to initiators are so affected.

If some initiators are mutagens and if all the genes of a cell are potentially mutable, only specific genes may set the cell down the path to malignancy (Trosko et al., 1982c). Promotion somehow gives the initiated cell a selective advantage over the non-initiated: this might be due to either a growth advantage or to the inability of the initiated cell to differentiate terminally, thereby proliferating into a mass of dysfunctional cells, or both (Trosko and Chang, 1980; Potter, 1983; Trosko et al., 1983). This explanation appears to have been experimentally confirmed (Yuspa et al., 1982).

A major function of the promotion process is to enhance the proliferation of cells with one specific type of mutation, thereby increasing the probability of obtaining a cell with two or more relevant genetic changes (Trosko and Chang, 1980; Potter, 1983; Trosko et al., 1983). Clearly, this would explain why initiation must precede promotion.

Recently, Potter (1981) postulated a new protocol to test this hypothesis. It is specifically based on the assumptions that (a) at least two genetic changes must take place in a single cell before it is converted from a normal to a premalignant, to a malignant phenotype; and (b) promotion enhances that probability.

Potter predicted that malignant conversion could be enhanced by exposing initiated cells which had already been promoted to more mutagens. Experimental confirmation of this prediction has been made (Hennings et al., 1983; Reddy and Fialkow, 1983). In addition, several other experimental studies seem to be consistent with this hypothesis (see Fry et al., 1982b).

NATURE AND POTENTIAL MECHANISMS OF TUMOR PROMOTION

If the role of promotion is to enhance additional genetic changes during the clonal expansion of the initiated cell, then one obligatory cellular function of promoters is that of mitogenesis. Cell proliferation would then facilitate the occurrence of other phenomena during tumor promotion, namely (a) increasing the target size of initiated cells containing one mutational change; (b) ensuring the fixation of other mutations by DNA replication (Trosko and Chang, 1981; Trosko and Chang, 1984b); and (c) increasing the mass of dysfunctional cells in an organ. Additional cellular changes occur during tumor promotion, including the modulation of gene expression and altered differentiation (see Trosko et al., 1984c). These effects probably ought not to be perceived as independent functions of tumor promoters, since the responses of stem cells could involve altered gene modulation in order to stimulate cell proliferation in both normal and initiated cells. After promoter-induced proliferation, the uninitiated proliferated stem cell might be able to differentiate, whereas the initiated cell might only have altered gene expression.

Several different modes of tumor promotion have been implicated, such as cell removal, surgery, cell death, growth factors or hormones, physical irritation and exogenous tumor promoters (Frei, 1976; Trosko et al., 1982b; Trosko et al., 1983; Jones et al., 1983), and it is significant that the one thing these various modes of tumor promotion have in common is the fact that they all can induce sustained hyperplasia.

Assuming cell proliferation to be obligatory for tumor promotion, the next major question is "What is the mechanism inducing cell proliferation in quiescent stem cells?" In the context of tumor promotion, we can ask "What is the mechanism of tumor promotion?" At this stage, definitive answers to either question cannot be given. However, there are several hypotheses regarding the mechanisms of triggering cell division (Berridge, 1975;

Whitfield et al., 1979; Rasmussen and Waisman, 1983) and tumor promotion (Trosko et al., 1983). Converging lines of evidence in support of both phenomena suggest their possible unification (Trosko et al., 1984b).

Several theories of tumor promotion have been postulated, for instance, gene amplification (Varshavasky, 1981); free-radical production (Troll et al., 1982); clastogenic action of promoters (Emerit and Cerutti, 1982); inhibition of intercellular communication (Yotti et al., 1979; Murray and Fitzgerald, 1979); altered differentiation (reviewed by Yamasaki, 1984). Many of these theories are not mutually exclusive. However, any theory of tumor promotion must explain several fundamental observations concerning the complex carcinogenic process: (a) cancer appears to be a product of dysfunctional homeostasis (Iversen, 1965), a disease of differentiation (Pierce, 1974), or a manifestation of "oncogeny as blocked or partially blocked ontogeny" (Potter, 1978); (b) tumor promoters affect proliferation and differentiation (Diamond et al, 1980; Yamasaki, 1984), and the primary cellular target appears to be cell membranes (Weinstein et al., 1981), not DNA (see review by Trosko and Chang, 1984a); (c) carcinogenesis involves the conversion of a contact-inhibited normal cell to a premalignant cell and then to a noncontact-inhibited malignant cell; (d) many, if not all, tumor promoters inhibit gap-junctional intercellular communication (Trosko et al., 1983).

It is this latter hypothesis which appears to integrate many of the known observations related to carcinogenesis. Intercellular communication within and between cells of various tissues is a fundamental biological process required to orchestrate complex homeostatic mechanisms regulating cell proliferation and differentiation (Loewenstein, 1979). Gap junctions are membrane structures which facilitate the transfer of ions and small molecules from cell to cell (Revel et al., 1980). Inhibition of intercellular communication caused by cell death, cell removal, or endogenous or exogenous chemicals can trigger either adaptive or nonadaptive responses of cells, depending on circumstances (Trosko and Chang, 1984c). Chemical and viral inhibition of intercellular communication may play a role in the mechanisms of teratogenesis, tumor promotion, atherosclerosis and reproductive dysfunction (Trosko and Chang, 1980; Trosko et al., 1982a; Trosko et al., 1984b). Decrease of gap junction function has been observed during tumor promotion (Enomoto et al., 1981; Yancey et al., 1982), viral transformation (Atkinson et al., 1981), and in malignant cells (Fentiman et al., 1979); Larsen (1983) has discussed the exceptions and has given possible explanations for them. Although the biochemical basis for the modulation of gap-junctional intercellular communication is not known, it has been speculated that it might be related to the phosphorylation of various proteins including the gap junction protein by either c-AMP-dependent protein kinase or the Ca^{++}-dependent, phospholipid-sensitive protein kinase protein kinase c (Johnson and Johnson, 1982; Wiener and Loewenstein, 1983; Trosko et al., 1984b, 1984c).

Mitotic arrest for cells in solid tissue is associated with contact inhibition (Levine et al., 1965). Gap-junctional communication occurs during contact inhibition. Mitogenesis seems to be associated with the elimination of contact inhibition and gap-junctional communication (Borek and Sachs, 1966; Corsaro and Migeon, 1977), although some apparent exceptions have been reported (Finbow and Yancey, 1981). The role of Ca^{++}, protein kinase C, and gap-junctional communication in both mitogenesis and tumor promotion has been noted (for review, see Trosko et al., 1984c). Consequently, it appears that chemicals and conditions which stimulate <u>mitogenesis</u> can amplify the potential cellular damage caused by <u>mutagenesis</u> in a single cell. On the other hand, chemicals which inhibit mitogenesis in certain cells and stimulate c-AMP-dependent protein kinase, for example, theophylline, caffeine, and steroid hormones, tend to act as anti-tumor agents (Trosko and Chang, 1978; Chang et al., 1978).

POSSIBLE BIOLOGICAL ROLE OF PROMOTION IN AMPLIFYING THE EFFECTS OF LOW-LEVEL CARCINOGEN EXPOSURE

There a number of experimental and clinical examples which suggest that tumor promotion may be a significant factor in both modulating the number and time of appearance of tumors. Put another way, in the absence of known exposure to tumor promoters, the number of initiation events or initiated cells induced by low-dose levels of radiations or by low concentrations of mutagens-carcinogens exceeds the number of malignant tumors that grow to a detectable size within the average lifetime of the individual.

Classical initiation/promotion studies in mouse skin, rat liver, mammary glands, bladder, thyroid glands and lung bear out the observation that a single or short-term exposure to mutagens carcinogens at levels which are not usually necrobiotic produce few tumors. However, if these initiated organs are repetitively exposed to nonmutagenic chemicals or physical conditions which can induce hyperplasia of those initiated cells in those target organs, there is both an earlier appearance and an enhanced frequency of benign and malignant tumors (reviewed by Boutwell, 1974; Pitot and Sirica. 1980; Slaga, 1983).

Fry and his co-workers (1980,1981,1982a) have shown that Harderian gland tumors in mice, initiated by fission neutrons, were promoted by hormonal stimulation from post-irradiation pituitary isografts. In addition, initiation of mouse skin by the mutagenic treatment of 8-methoxypsoralen and long wavelength ultraviolet radiation (UVR) (320-400 nm), followed by promotion with 12-O-tetradecanoyl-13-O-phorbol acetate (TPA), not only enhanced the time of appearance and frequency of skin carcinomas, but lowered the apparent no-effect level (Fig. 1).

Xeroderma pigmentosum (XP), a genetic syndrome predisposing patients to sunlight-induced skin cancer (Kraemer, 1980a), might also provide evidence consistent with the importance of promotion in enhancing low-dose damage. Cells from XP patients have been characterized as hypersensitive to the killing and mutagenic effects of ultraviolet light and many chemical mutagens/carcinogens (Maher and McCormick, 1976; Glover et al., 1979; Myhr et al., 1979). Cairns (1981) has argued that although XP patients have multiple skin cancers, mutations play little or no role in carcinogenesis since the patients do not seem to suffer from any excess of internal cancers. While this claim is not entirely valid (Kraemer, 1980b), there might be a reason for the general observation that most of the cancers are on the skin (Trosko, 1981). If we assume that the initiation/promotion mechanism of carcinogenesis is valid for human tumors, then we must seriously consider the role of cell death as a promoter in the case of XP skin cancers (Trosko, 1981; Trosko and Chang, 1983).

In the context of the hypothesis that inhibited intercellular communication is involved in tumor promotion, cell killing could lead to compensatory hyperplasia. Therefore, the observation that more cancers appear in the skin of XP individuals than in other tissues is not surprising. DNA lesions which are not repaired can lead not only to mutations, but to cell death. Although all the cells of the classic form of xeroderma pigmentosum patients are deficient in their ability to remove certain kinds of DNA lesions, only the cells of the skin are exposed to constant cytotoxic fluxes of environmental mutagens like the ultraviolet component of sunlight. In all likelihood, internal cells are never exposed to equivalent levels of chemicals, mutagens carcinogens. Therefore, although internal cells of XP patients will have chemically induced DNA

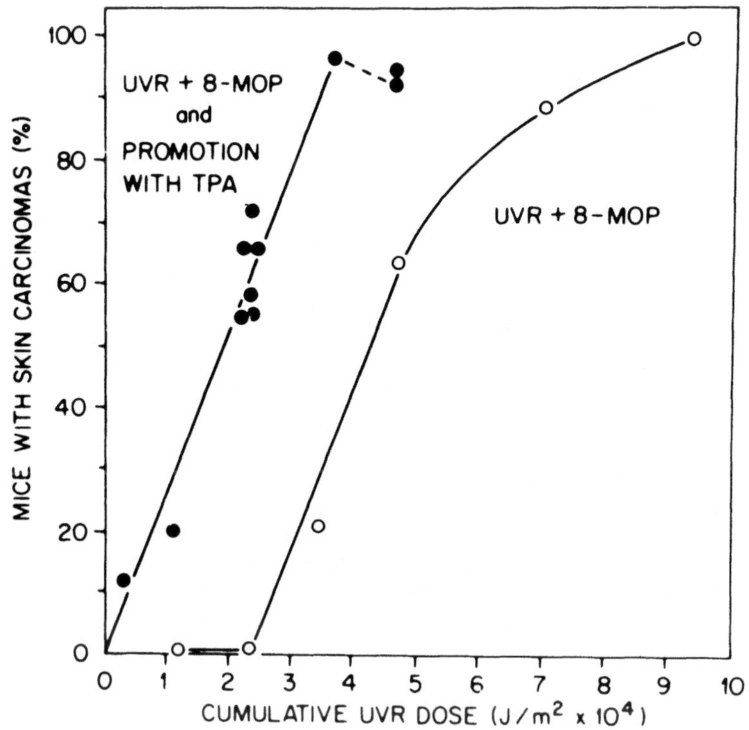

Fig. 1 The percent of mice with squamous cell carcinomas as a function of total dose of 320-400 nm UVR given in various fractions plus 8-methoxypsoralen (o——o); and similar exposures but followed at the end of the fractionation regimen by treatment with 5 g of 12-O-tetradecanoyl-13-O-phorbol acetate three times per week (●——●). (Reproduced from Fry et al., 1980, with permission from Elsevier/North Holland Biomedical Press.)

lesions which are not repaired as efficiently as in normal cells, the amount of cell killing such as to act as an indirect promoter to force the proliferation of surviving initiated stem cells would not be as great as in the skin. In other words, UV light would be a complete carcinogen in the skin, initiating by its mutagenic capacity and promoting by its cytotoxic effects. The large clones of initiated cells of the skin are then further exposed, increasing the chances of incurring additional genetic changes for the malignant phenotype. The relative lack of internal tumors in patients with xeroderma pigmentosum might be explicable within the initiation/promotion model. Possibly, DNA damage in internal cells caused by chemical mutagens is far less than that caused in skin cells by sunlight. In that case, any internal initiated cell would have to be promoted either by exogenous promoters or by cell killing.

This same explanation might also be applicable to the observation that patients given PUVA treatment for psoriasis are at a significant risk for skin cancer (Bridges and Strauss, 1980). PUVA treatment is known to be both mutagenic and cytotoxic (Burger and Simons, 1979). The induction of mutations in surviving stem cells would repopulate and replace cells killed by the PUVA treatment. In addition, the multiple exposures to the PUVA radiation could induce more genetic changes in these initiated clones of cells.

In effect, the experiments reported by Fry (1982b) and Fry et al. (1981), seem to bear out the prediction of Potter (1981) and Trosko and Chang (1980) that by initiating cells, followed by promotion to increase the number of these target cells, and then treatment with more mutagens, hastened the conversion of premalignant lesions to malignant lesions. When initiated and promoted mouse skin was followed by irradiation of various wavelengths, only the shorter wavelengths (those that were capable of inducing DNA damage) enhanced the tumor incidence.

To illustrate how natural, nonmutagenic chemicals can influence the production of tumors initiated by mutagens/carcinogens, Aylsworth et al. (1984) have shown that high levels of dietary fat can influence the time of appearance and frequency of dimethylbenz(a)anthracene-induced mammary tumors in rats. Hormones also have been implicated in promoting tumors of the breast, liver, prostate, vagina, and thyroid glands (Takizawa and Hirose, 1978; Yoshida et al., 1980; Sheehan et al., 1982). They do not act as genotoxins (Yager and Yager, 1980; Yager and Fifield, 1982), nor as initiators (Schuppler et al., 1983).

To summarize: promoters, while not being mutagenic at the DNA level in and of themselves, can facilitate the chances of mutagenesis by virtue of their mitogenic effects and enhance the chances of multiple genetic changes in the initiated cells by the clonal expansion of the target cells (Trosko et al., 1977, 1983; Trosko and Chang, 1984b).

Finally, can these observations shed light on the thorny issue of the possible existence of thresholds in carcinogenesis?

Within the context of the initiation/promotion model of carcinogenesis, the question of thresholds must be refined by answering the following: "Are there thresholds for either, both, or neither of the initiation and promotion phases?". Clearly, in the absence of rigorous empirical data, one can only speculate (Trosko and Chang, 1983; Trosko et al., 1983). If one assumes that mutagenesis can contribute to initiation, then we cannot state that there are no-effect levels by which cells can incur some DNA damage without being initiated. Because of multimodes of mutagenesis in eukaryotic cells and of the many uncontrollable factors which could influence the amount or kind of DNA lesion induced by mutagens, the repair of these lesions, and the fidelity of replication (Trosko et al., 1984a), it would seem highly unlikely that threshold levels exist for initiators .

Every time a cell divides there is a finite probability that a spontaneous error of replication might occur, as well as errors due to the incomplete repair of lesions in the template DNA. Assuming each gene, including oncogenes, are not immune to mutagens, no gene is safe from mutagenesis.

In a recent extensive review, the International Commission for Protection Against Environmental Mutagens and Carcinogens, after critically assessing almost 100 dose-response curves of mammalian mutagenesis from the literature, concluded that "one cannot at this state be predictive about the genotoxic thresholds of chemical mutagens (Ehling et al., 1983)." Scherer and Emmelot (1975) and Emmelot and Scherer (1980) showed that, at doses from 0.3 to 30 mg diethylnitrosamine, the number of induced enzyme altered foci in rat liver (a measure of initiated cells) was directly proportional, suggesting a one-hit mechanism of induction. Similar observations have been made with mouse skin tumors induced by benzo(a)pyrene, 7,12-dimethylbenz(a)anthracene (DMBA), nitroquinoline oxide and beta-propriolactone (Burns et al., 1983), as well as with UVR-induced mouse skin tumors (Fry et al., 1982b). In practical terms, these experimental observations suggest that mutagenic/initiating events might be induced as a linear, non-threshold function of dose at noncytotoxic levels of exposure.

Since our understanding of tumor promotion is equally limited, one cannot infer that there are thresholds for promoters. Based on studies of TPA promotion of DMBA-initiated mouse skin and of phenobarbital promotion of diethylnitrosamine-initiated rat livers, Verma and Boutwell (1980) and Goldsworthy et al., (1984) have demonstrated no apparent effect or threshold levels for these two promoting agents [Fig. 2].

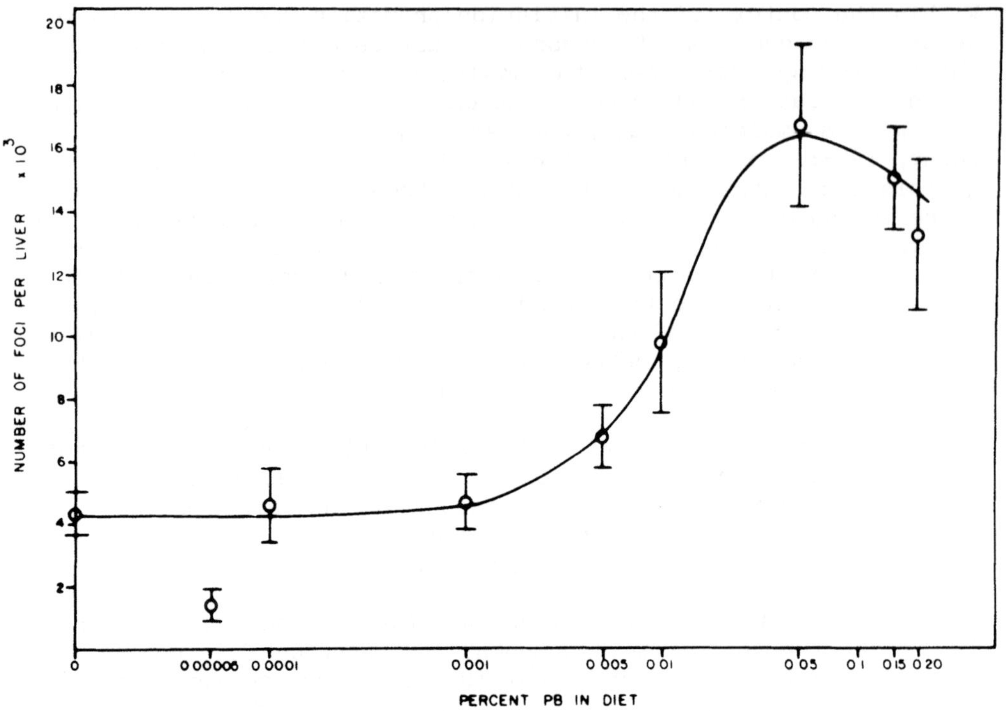

Fig.2 The promotion of enzyme-altered foci/liver as a function of the level of phenobarbital (PB) in the diet. Female Sprague-Dawley rats (4-12 at each data point) weighing 200 g were initiated by partial hepatectomy/diethylnitrosamine and 2-6 weeks later started on laboratory chow diets containing the levels of PB indicated above. The promoting agent was fed in the diet for 7-8 months. At the end of that time the animals were killed and the livers processed. Data points are plotted logarithmically along the abscissa. Reproduced from Goldsworthy et al. (1983) with permission from IRL Press.

Considering initiation as a linear-dependent process, contingent on promotion before carcinogenesis, and promotion as a threshold-dependent process, one might reason that, above the promoter threshold for the initiated tissue, carcinogenesis might appear as a linear function of dose. Fry et al. (1982) have made observations that seem to be consistent with this idea, in that initiated mouse skin promoted by effective levels of TPA seems to lack a threshold for carcinogenesis induced by 8-methoxypsoralen and long-wavelength UVR (320 - 400 nm). If these observations can be extended to other tumors and promoters, then promotion might indeed be most important at low doses of carcinogenic initiators. At high doses of initiators, cytotoxicity induced by the initiator could be the indirect promoter of any surviving initiated cell (Trosko et al., 1983).

Therefore, low-level exposure to mutagen/initiators would have dramatically different consequences in terms of multi-step, clonally derived diseases, such as cancer or atherosclerosis if the initiated tissue is exposed to sustained and effective levels of promoters. Supposing that intercellular communication does play a role in tumor promotion, there may be some theoretical basis for postulating threshold levels for single promoters, since the modulation of gap junction function is probably the result of exceeding a critical mass of regulatory molecules or ions (Trosko et al., 1983). To alter a cell, which is communicating with other cells in a 3-dimensional environment via many gap junctions so that it cannot communicate would probably necessitate elimination of more than one of its gap junctions. However, even if there were theoretical thresholds or practical no-effect levels for individual promoters, there would still be the possibility of potentiation, synergism or interaction of several promoters, each of which might be at subthreshold levels.

SUMMARY

The assessment of risk from low-level exposure to radiation and chemicals is hindered by the basic lack of scientific understanding of the complex nature of the multiple levels of protective or synergistic interactions. Radiation and chemicals have the potential of inducing mutations, cell death and altered gene expression. The biological consequences of each of these effects is again complex, since many factors can enhance or mask the effect of a single mutation, cytotoxic or gene modulatory change.

Carcinogenesis, representing but one chronic disease state, is a multi-step process, involving the clonal expansion of a single altered cell (that is, the initiation event). Radiation and chemicals appear to contribute to the initiation process by their ability to induce viable mutations. Insufficient theoretical and empirical knowledge precludes a determination of whether mutagenesis and, hence, initiation exhibit a threshold phenomenon. Because of a variety of redundancy mechanisms on the genetic and cellular levels, the physiological impact of a single dysfunctional cell is felt only after it is amplified to a large number (the promotion phase of carcinogenesis). Radiation and chemical-induced cytotoxicity, as well as noncytotoxic chemical induction of mitogenesis, can induce surviving single dysfunctional stem cells to multiply. If during this multiplication of dysfunctional cells, the initiated cells are further exposed chronically to low levels of mutagens, there is an enhanced probability of additional genetic changes. The accumulation of foci of dysfunctional cells can occur in any stem cell population of any tissue. As with mutagens, the existence of threshold levels for promoting conditions and chemicals is still not yet scientifically validated. However, specific examples of the actions of a few promoters does seem to be consistent with that idea. The concepts of initiation and promotion imply the existence of anti-initiation and antipromotion conditions. Together with genetic factors, the complex and unpredictable interactions of radiation and chemicals as initiators, anti-initiators, promoters and antipromoters, defy prospects of an easy means to predict the consequences of exposure to chronic low levels of radiation and chemicals.

ACKNOWLEDGEMENTS

Research on which this manuscript was based was supported by grants from the EPA [R808587] and [CA21104].

Although the information described in this article has been funded wholly or in part by the United States Environmental Protection Agency under assistance agreement [R808587] to JET, it has not been subjected to the Agency's required peer and administrative review and therefore does not necessarily reflect the views of the Agency and no official endorsement should be inferred.

REFERENCES

Armitage, P. and Doll, R., 1954, The age distribution of cancer and a multi-stage theory of carcinogenesis, Brit. J. Cancer, 8:1

Atkinson, M.M., Mendo, A.S., Johnson, R.G., Sheppard, J.R., and Sheridan, J.D., 1981, Rapid and reversible reduction of junctional permeability in cells infected with a temperature-sensitive mutant of avian sarcoma virus, J. Cell Biol., 91:573.

Aylsworth, C.F., Jone, C., Trosko, J.E., Meites, J., and Welsch, C.W., 1984, Promotion of dimethylbenz(a)anthracene-induced mammary tumorigenesis by high dietary fat in the rat: Possible role of intercellular communication, J. Natl. Cancer Inst., 72:637.

Argyris, T.S., 1982, Tumor promotion by regenerative epidermal hyperplasia in mouse skin, J. Cutan., Pathol., 9:1.

Barrett, J.C., and Ts'o, P.O.P., 1978, Evidence for the progressive nature of neoplastic transformation in vitro, Proc. Natl. Acad. Sci. USA, 75:3761.

Berridge, M.J., 1975, Control of cell division: A unifying hypothesis, J. Cyclic Nucleotide Res., 1:305.

Borek, C. and Sachs, L., 1966, The difference in contact inhibition of cell replication between normal cells and cells transformed by different carcinogens, Proc. Natl. Acad. Sci. USA, 56:1705.

Boutwell, R.K., 1974, The function and mechanism of promoters of carcinogens, Crit. Rev. Toxicol., 2:419.

Bridges, B.A. and Strauss, G.H., 1980, Possible hazards of photochemotherapy for psoriasis, Nature (London), 283:523.

Burger, P.M. and Simons, J.W.I.M., 1979, Mutagenicity of 8-meth oxypsoralen and long-wave ultraviolet irradiation in diploid human skin fibroblasts, Mutat. Res., 63:371.

Burns, F., Albert, R., Ahshuler, B., and Morris, E., 1983, Approach to risk assessment for genotoxic carcinogens based on data from the mouse skin initiation-promotion model. Environ. Health Perspect., 50:309.

Cairns, J., 1981, The origin of human cancer, Nature (London), 289:353.

Chang, C.C., Trosko, J.E., and Warren, S.T., 1978, In vitro assay for tumor promoters and antipromoters, J. Environ. Pathol., 2:43.

Corsaro, C.M., and Migeon, B.R., 1977, Comparison of contact-mediated communication in normal and transformed human cells in culture, Proc. Natl. Acad. Sci. USA, 74:4476.

Diamond, L., O'Brien, T.G., and Baird, W.M., 1980, Tumor promoters and the mechanism of tumor promotion, Adv. Cancer Res., 32:1.

Ehling, U.G., Averbeck, D., Cerutti, P.A., Friedman, J., Greim, H., Kolbye, A.C., and Mendelsohn, M.L., 1983, Review of the evidence for the presence or absence of thresholds in the induction of genetic effects by genotoxic chemicals. Mutat. Res., 123:281.

Emmelot, P., and Scherer, E., 1980, First relevant cell stage in rat liver carcinogenesis, Biochim. Biophys. Acta, 605:247. 247.

Emerit, I., and Cerutti, P.A., 1982, Tumor promoters phorbol-12-myristate-13-acetate induces a clastogenic factor in human lymphocytes, Proc. Natl. Acad. Sci. USA, 79:7509.

Enomoto, T., Sasaki, Y., Kanno, Y., and Yamasaki, H., 1981, Tumor promoters cause a rapid and reversible inhibition of the formation and maintenance of electrical cell coupling in culture, Proc. Natl. Acad. Sci. USA, 78:5628.

Fentiman, I.S., Hurst, J., Ceriani, R.L., and Taylor-Papadimitriou, J., 1979, Junctional intercellular communication pattern of cultured human breast cancer cells. Cancer Res., 39:4739.

Finbow, M.E., and Yancey, S.B., 1981, The roles of intercellular junctions, in: "Biochemistry of Cellular Regulation," Vol. IV, p. 215, M.J. Clemens and P. Knox, eds., CRC Press, Boca Raton.

Foulds, L., 1975, "Neoplastic Development," Vol. 2, Academic Press, London.

Frei, J.V., 1976, Some mechanisms operative in carcinogenesis: A review, Chem.Biol. Interact., 12:1.

Fry, R.J.M., 1981, Experimental radiation carcinogenesis: What have we learned?, Radiat. Res., 87:224.

Fry, R.J.M., Ley, R.D., and Grube, D., 1982a, Ultraviolet radiation skin carcinogenesis, Proc. Am. Assoc. Cancer Res., 23:95.

Fry, R.J.M., Ley, R.D., Grube, D., and Staffeldt, E., 1982b, Studies on the multistage nature of radiation, in: "Carcinogenesis," Vol. 7, p.155, E. Hecker, N.E. Fusenig, W. Kunz, F. Marks, H.W. Thielmann, eds., Raven Press, New York.

Fry, R.J.M., Storer, J.B., and Ullrich, R.L., 1980, Radiation toxicology: Carcinogenesis, in: "The Scientific Basis of Toxicity Assessment," p. 291, H.B. Witschi, ed., Elsevier/North Holland Biomedical Press, New York.

Glover, T.W., Chang, C.C., Trosko, J.E., and Li, S.L., 1979, Ultraviolet light induction of diphtheria toxin-resistant mutants in normal and xeroderma pigmentosum human fibroblasts, Proc. Natl. Acad. Sci. USA, 76:3982.

Goldsworthy, T., Campbell, H.A., and Pitot, H.C., 1984, The natural history and dose-response characteristics of enzyme-altered foci in rat liver following phenobarbital and diethylnitrosamine administration, Carcinogenesis, 5:67.

Hennings, H., Shores, R., Wenk, M.L., Spangler, E.F., Tarone, R. and S.H. Yuspa, 1983, Malignant conversion of mouse skin tumors is increased by tumor initiators and unaffected by tumor promoters, Nature (London), 304:67.

Iversen, D.H., 1965, Cybernetic aspects of the cancer problem, in: "Progress in Biocybernetics," Vol. 2, p. 76, N. Wiener and J.P. Schade, eds., Elsevier Publ. Co., Amsterdam.

Johnson, K.R., and Johnson, R., 1982, Bovine lens MP26 is phorsphorylated in vitro by an endogenous c-AMP-dependent protein kinase, Fed. Proc., 41:755.

Jones, T.D., Griffin G.D., and Walsh, P.J., 1983, Unifying concept for carcinogenic risk assessments, J. Theor. Biol., 105:35.

Kraemer, K.H., 1980a, Xeroderma pigmentosum, in: "Clinical Dermatology," Vol. 4, p. 1, D.J. Dennis, R.L. Dobson, and J. McGuire, eds., Harper and Row, Hagerstown.

Kraemer, K.H., 1980b, Oculo-cutaneous and internal neoplasms in xeroderma pigmentosum: Implications for theories of carcinogenesis, in: "Carcinogenesis: Fundamental Mechanisms and Environmental Effects," p. 503, B. Pullman, P.O.P. Ts'o, and H. Gelboin, eds., D. Reidel Publishing, Amsterdam.

Land, H., Parada, L.F., and Weinberg, R.A., 1983, Tumorigenic conversion of primary embryo fibroblasts requires at least two cooperating oncogenes, Nature (London), 304:596.

Larsen, W.J., 1983, Biological implications of gap junction structure, distribution and composition: A review, Tissue Cell, 15:645.

Levine, E.M., Becker, Y., Boone, C.W., and Eagle, H., 1965, Contact inhibition, macromolecular synthesis and polyribosomes in cultured human diploid fibroblasts, Proc. Natl. Acad. Sci. USA, 53:350.

Loehrke, H., Schweizer, J., Dederer, E., Hesse, B., Rosenkranz, G., and Goerttler, K., 1983, On the persistence of tumor initiation on two-stage carcinogenesis on mouse skin, Carcinogenesis, 4:771.

Loewenstein, W.R., 1979, Junctional intercellular communication and the control of growth, Biochim. Biophys. Acta, 560:1.

Maher, V.M., and McCormick, J.J., 1976, Effect of DNA repair on cytotoxicity and mutagenicity of UV irradiation and of chemical carcinogens in normal and xeroderma pigmentosum cells, in: "Biology of Radiation Carcinogenesis," p. 129, J.M. Yuhas, R.W. Tennant, and J.D. Regan, eds., Raven Press, New York.

Moolgavkar, S.H., Day, N.E., and Stevens, R.G., 1980, Two-stage model for carcinogenesis: Epidemiology of breast cancer in females, J. Natl. Cancer Inst., 65:559.

Murray, A.W., and Fitzgerald, D.J., 1979, Tumor promoters inhibit metabolic cooperation in co-cultures of epidermal and 3T3 cells, Biochem. Biophys. Res. Comm., 91:395.
Myhr, B.C., Turnbull, D., and DiPaolo, J.A., 1979, Ultraviolet mutagenesis of normal and xeroderma pigmentosum variant human fibroblast, Mutat. Res., 62:341.
Newbold, R.F., and Overell, R.W., 1983, Fibroblast immortality is a prerequisite for transformation by EJ c-Ha-ras oncogene, Nature (London), 304:648.
Pierce, G.G., 1974, Neoplasms, differentiation, and mutations, Am. J. Pathol., 77:103.
Pitot, H.C., Goldsworth, T., and Moran, S., 1981, The natural history of carcinogenesis: Implications of experimental carcinogenesis in the genesis of human cancer, J. Supramol. Struct. Cell. Biochem., 17:133.
Pitot, H.C., and Sirica, A.E., 1980, The stages of initiation and promotion in hepatocarcinogenesis, Biochim. Biophys. Acta, 605:191.
Potter, V.R., 1978, Phenotypic diversity in experimental hepatomas: The concept of partially blocked ontogeny, Br. J. Cancer, 38:1.
Potter, V.R., 1980, Initiation and promotion in cancer formation: The importance of studies on intercellular communication, Yale J. Biol. Med., 53:367.
Potter, V.R., 1981, A new protocol and its rationale for the study of initiation and promotion of carcinogenesis in rat liver, Carcinogenesis, 2:1375.
Potter, V.R., 1983, Alternative hypotheses for the role of promotion in chemical carcinogenesis, Environ. Health. Perspect., 50:139.
Rasmussen, H., and Waisman, D.M., 1983, Modulation of cell function in the calcium messenger system, Rev. Physiol. Biochem. Pharmacol.., 95:111.
Reddy, A.L., and Fialkow, P.J., 1983, Papillomas induced by initiation-promotion differ from those induced by carcinogen alone, Nature (London), 304:69.
Revel, J.-P., Yancey, S.B., Meyer, D.J., and Nicholson, B., 1980, Cell junctions and intercellular communications, In Vitro, 16:1010.
Scherer, E., and Emmelot, P., 1975, Kinetics of induction and growth of precancerous liver cell foci and liver tumor formation by diethylnitrosamine in the rat, Eur. J. Cancer, 11:689.
Schuppler, J., Damme, J., and Schulte-Hermann, R., 1983, Assay of some endogenous and synthetic sex steroids for tumor initiating activity in rat liver using the Solt-Farber system, Carcinogenesis, 4:239.

Schulte-Hermann, R., Timmermann-Trosiener, I., and Schuppler, J., 1983, Promotion of spontaneous preneoplastic cells in rat liver as a possible explanation of tumor production by non mutagenic compounds, Cancer Res., 43:839.

Sheehan, D.M., Frederick, C.B., Branhan, W.S., and Heath, J.E., 1982, Evidence of estradiol promotion of neoplastic lesions in the rat vagina after initiation with N-methyl-N-nitrosourea, Carcinogenesis, 3:957.

Slaga, T.J., 1983, "Mechanisms of Tumor Promotion," Vol. 1, CRC Press, Boca Raton.

Slaga, T.J., Fischer, S.M., Weeks, C.E., Nelson, K., Mamrack, M., and Klein-Szanto, A.J.P., 1982, Specificity and mechanisms of promoter inhibitors in multistage promotion, in: "Carcinogenesis," Vol. 7, p. 19, E. Hecker, N.E. Fusenig, W. Kunz, F. Marks, and H.W. Thielmann, eds., Raven Press, New York.

Takizawa, S., and Hirose, F., 1978, Role of testosterone in the development of radiation-induced prostate carcinoma in rats, Gann, 69:723.

Troll, W., Witz, G., Goldstein, B., Stone, D., and Sugiwara, T., 1982, The role of free oxygen radicals in tumor promotion and carcinogenesis, in: "Carcinogenesis", Vol. 7, p. 593, E. Hecker, N.E. Fusenig, W. Kunz, F. Marks and H.W. Thielmann, eds., Raven Press, New York.

Trosko, J.E., 1981, Cancer causation, Nature (London), 290:356.

Trosko, J.E., and Chang, C.C., 1978, Environmental and carcinogenesis: An integrative model, Q. Rev. Biol., 53:115.

Trosko, J.E., and Chang, C.C., 1980, An integrative hypothesis linking cancer, diabetes and atherosclerosis: The role of mutations and epigenetic changes, Med. Hypotheses, 6:455.

Trosko, J.E., and Chang, C.C., 1981, The role of radiation and chemicals in the induction of mutations and epigenetic changes during carcinogenesis, in: "Advances in Radiation Biology," p. 1, J.T. Lett and H.Adler, eds., Academic Press, New York.

Trosko, J.E., and Chang, C.C., 1983, Potential role of intercellular communication in the rate-limiting step in carcinogenesis, J. Am. Coll. Toxicol., 2:5.

Trosko, J.E., and Chang, C.C., 1984a, Implications of genotoxic and non-genotoxic mechanisms in carcinogenesis to risk assessment, in: "Quantitative Estimation of Risks to Human Health from Chemicals," N. Nelson, ed., John Wiley and Sons, Inc., New York, in press.

Trosko, J.E., and Chang, C.C., 1984b, Error-prone DNA repair and replication in relation to malignant transformation, Transplant. Proc., in press.

Trosko, J.E., and Chang, C.C., 1984c, Adaptive and non-adaptive consequences of chemical inhibition of intercellular communication, Pharmacol. Rev., in press.

Trosko, J.E., Chang, C.C., and Medcalf, A., 1983, Mechanisms of tumor promotion: Potential role of intercellular communication, Cancer Invest., 1:511.

Trosko, J.E., Chang, C.C., and Netzloff, M., 1982a, The role of inhibited cell-cell communication in teratogenesis, Teratogenesis Carcinog. Mutagen., 2:31.

Trosko, J.E., Chang, C.C., and Wade, M.H., 1984a, Mammalian mutagenesis, Genetics, in press.

Trosko, J.E., Chang, C.C., Yotti, L.P., and Chu, E.H.Y., 1977, Effect of phorbol myristate acetate on the recovery of spontaneous and ultraviolet light-induced 6-thioguanine- and ouabain-resistant Chinese hamster cells, Cancer Res., 37:188.

Trosko, J.E., Jone, C., Aylsworth, C., and Tsushimoto, G., 1982b, Elimination of metabolic cooperation is associated with the tumor promoters, oleic acid and anthralin, Carcinogenesis, 3:1101.

Trosko, J.E., Jone, C., and Chang, C.C., 1982c, The role of tumor promoters on phenotypic alterations affecting intercellular communication and tumorigenesis, Ann. New York Acad. Sci., 407:316.

Trosko, J.E., Jone, C., and Chang, C.C., 1984b, Oncogenes, inhibited intercellular communication and tumor promotion, in: "Princess Takamatsu Symposium," Vol. 14, H. Fujiki, ed., Japan Scientific Press, Tokyo, in press.

Trosko, J.E., Jone, C., Rintel, R.A., and Chang, C.C., 1984c, Potential role of calmodulin in tumor promotion: Modulator of gap junctional intercellular communication?, in: "Calmodulin Antagonists and Cellular Physiology," H. Hidaka and D.J. Hartshorne, eds., Academic Press, New York, in press.

Varshavsky, A., 1981, Phorbol ester dramatically increases incidence of methotrexate-resistant mouse cells: Possible mechanisms and relevance to tumor promotion, Cell, 25:561.

Verma, A.K., and Boutwell, R.K., 1980, Effects of dose and duration of treatment with the tumor-promoting agent, 12-O-tetradecanoylphorbol-13-acetate on mouse skin carcinogenesis, Carcinogenesis, 1:271.

Weinstein, I.B., 1981, Current concepts and controversies in chemical carcinogenesis, J. Supramol. Struct. Cell. Biochem., 17:99.

Whitfield, J.F., Boynton, A.L., MacManus, J.P., Sikorska, M., and Tsang, B.K., 1979, The regulation of cell proliferation by calcium and cyclic AMP, Mol. Cell. Biochem., 27:155.

Wiener, R.C., and Loewenstein, W.R., 1983, Correction of cell-cell communication defect by introduction of a protein kinase into mutant cells, Nature (London), 305:433

Yager, J.D., and Fifield, D.S., Jr., 1982, Lack of hepatogenotoxicity of oral contraceptive steroids, Carcinogenesis, 3:625.

Yager, J.D., and Yager, R., 1980, Oral contraceptive steroids as promoters of hepatocarcinogenesis in female Sprague-Dawley rats, Cancer Res., 40:3680.

Yamasaki, H., 1984, Modulation of cell differentiation by tumor promoters, in: "Mechanisms of Tumor Promotion," Vol. 3, T.J. Slaga, ed., CRC Press, Boca Raton, in press.

Yancey, S.B., Edens, J.E., Trosko, J.E., Chang, C.C., and Revel, J.-P., 1982, Decreased incidence of gap junctions between Chinese hamster V79 cells upon exposure to the tumor promoter 12-O-tetradecanoyl-phorbol-13-acetate, Exp. Cell Res., 139:329.

Yoshida, H., Fukunishi, R., Kato, Y., and Matsumoto, K., 1980, Progesterone-stimulated growth of mammary carcinomas induced by 7,12-dimethylbenz(a)anthracene in neonatally androgenized rats, J. Natl. Cancer Inst., 65:823.

Yotti, L.P., Chang, C.C., and Trosko, J.E., 1979, Elimination of metabolic cooperation in Chinese hamster cells by a tumor promoter, Science, 206:1089.

Yuspa, S.H., Ben, T., Hennings, H., and Lichti, U., 1982, Divergent responses in epidermal basal cells exposed to the tumor promoter 12-O-tetradecanoylphorbol-13-acetate, Cancer Res., 42:2344.

DISCUSSION

Ehling: I have a comment regarding the suggestion that the multistage progression of a tumor indicates that there have been multihits. I do not think that is accurate. Each stage can be by itself a single hit, but the fact that you have a multistage does not necessarily indicate several hits. Can I ask you about this idea of cell expansion. You may see papillomas that have already undergone an enormous amount of cell expansion, if you stop treatment with phorbolester, 99% of the papillomas will regress. If it is just a question of cell expansion, why do the papillomas regress?

Trosko: As long as the cells have the genetic ability to communicate, but are in the presence of chemicals that block that ability, they will grow. If you remove the promoter, the ability to communicate is still there, and if the tumor has not reached Bell's critical size, factors can diffuse between cells, and the tissue will return to its normal condition. It is only when a few of the cells in the mass constitutively lose the ability to communicate that no amount of growth factor suppressors in the surrounding environment will stop it from proliferating or differentiating. Promoters have the interesting property in many of the systems that have been tested of altering the differentiation pattern. One of the functions of cell--cell communication is that it allows cells to differentiate normally. When you block cell communication, either genetically or chemically, you alter their ability to differentiate.

Portier: I have three questions. Could you discuss how your theory relates to radiation. Secondly, has there been any research on the differences in gap junctions for different tissues? Why, in fact, is liver affected by this promoter, but the pancreas is not, and vice versa? Thirdly, how does your theory deal with tumor regression?

Trosko: To the first point, radiations are good mutagens, but at higher doses they also are good cytotoxic agents. From that standpoint, enough radiation, both UV or x-rays, can mutate cells and, by killing cells, act as indirect promoters facing the surviving initiated stem cells to go into compulsory hyperplasia.

Secondly, chemical carcinogens are both species and organ specific, and I would have thought _a priori_ that there should be no problem in accepting the same kind of specificity for promoters either due to the distribution of the chemical, or to the receptors for given chemicals. There are at least two classes of promoters; those that are dependent on receptors, like hormones, growth factors, and TPA, and those which are not dependent on receptors, such as DDT and phenobarbital. The pharmaco-dynamics that are involved in humans versus animals may in some way explain the organ specificity.

Portier: Have you done any research to see if there were differences in gap junctions in the different tissues?

Trosko: Gap junctions are found, virtually, in all cells, save red blood cells, striated smooth muscle cells, and few others. Although they have not been studied in detail, there does not seem to be that much difference between a gap junction in liver, bovine lens, or Chines hamster lung. We have antibodies to chick lens which cross-reacts to Chinese hamster gap junctions.

I cannot satisfactorily explain tumor size regression in terms of our theory. I know that if you promote mouse skin that has been initiated and the papillomas appear, you can stop before a critical threshold level is reached and the papillomas regress to a point where, if you looked at the mouse a month later or so, you would never know that they had been there. But, if you wait another year, and reapply TPA, the papillomas reappear. The cells have not died, but they somehow regressed and their phenotype reverted back to a quasi-normal state.

Hei: I just want to echo the point you made with regard to tumor regression in the mouse skin model. We know that once you apply a hepatocarcinogen such as aflotoxin or diethylnitrosamine to the animals, focal areas of altered enzyme activity develop. Not all of these foci, however, are capable of developing into carcinomas; in fact Farber et al have shown that once the carcinogen or promoter is removed, over 70% of the foci will regress. The outcome depends on the conditions of the experiment, the dosage and timing of the treatments. My second remark is a question actually: do we know anything with regard to the biochemical and functional effects of Phenobarbital and TPA types of promoter on plasma membranes?

Phenobarbital has been postulated to alter membrane permeability by altering its cholesterol content thereby increasing the fluidity of the actyl moiety of Phospholipids. Is there any evidence that suggests similar effects due to phorbol esters?

Trosko: I have a whole new lecture just waiting for that talk, but let me summarize. The phorbol ester receptor appears to be a calcium-dependent phospholipid-dependent protein kinase (protein kinase C) and once activated by phorbol esters, it phosphorylates many things. Recently we have speculated that one of the things that happens when the protein kinase is activated by TPA, the gap junction protein is phosphorylated. Phenobarbital apparently does not bind to the protein kinase C. However, phenobarbital affects the membrane, and alters the calcium flux. In turn, calcium also activates the protein kinase C, which is a phospholipid sensitive, calcium dependent protein kinase. So it may be that the action of the receptor type of promoter and the nonreceptor type are converging. The receptor type promotor somehow affects the phospholipid component activating the protein kinase C, while the nonreceptor type, such as PBB and saccharin, affect calcium levels, that then activate the protein kinase C. This is our present hypothesis.

Hughes: I would like to add a word to your last statement regarding synergisms and additivities among promoters and carcinogens by noting that we have a paper in press reporting the effect of NTA combined with three different nitrosamines, NBBN in one case, DPN in another, and MNNG in a third. These nitrosamines are organ specific, and in a study conducted by NCI at the Stamford Research Institute we have looked at different organs. The results clearly show that there is a dose/response relationship between inhibition of the normal nitrosamine tumor development in the presence of NTA.

Trosko: There are several other examples such as that of phenobarbital and PBB. If PBB is given to a rat prior to administration of a carcinogen, for example, AAF or DMBA, it will actually protect the animal from the initiating potential of these particular compounds. The same compound, given in the exact same way, but after initiation, acts as a promoter. So here we have a real dilemma in that it is going to be impossible to put red flag or green flag on a molecule just by virtue of its structure. We have to make our assessment in the context of the biological behavior of the compound. Why is PBB, phenobarbital, retinoic acid or any of these kinds of chemicals acting as anti-initiators under one set of conditions, and as anti-promoters in one organ system, but as promoters in another organ system of the same species? All I can say is that I think they act as promoters by virtue of the ability to block cell to cell communication. The ability to induce drug metabolizing enzymes might explain why a substance acts as an anti-initiator.

Patel: Do you consider that asbestos acts as a promoter by blocking intercellular communication?

Trosko: I think so, but the few experiments that have been made have not demonstrated this effect. These results could be taken as evidence that asbestos has no effect, but the assay conditions may not have been quite right; at present I would like to believe the latter.

GENETIC AND EPIGENETIC ASPECTS OF TUMOR PROGRESSION AND TUMOR HETEROGENEITY

R.G. Liteplo[1], P. Frost[2] and R.S. Kerbel[1]

[1]Cancer Res. Labs., Dept. Pathology, Queen's Univ., Kingston, Ont., Can., K7L 3N6
[2]Dept. Medicine, VAH Hospital, Long Beach California

INTRODUCTION

It is obvious that in order to reduce the incidence of cancer we need to identify those factors in our environment which may be responsible for the formation of neoplasms. What is not so obvious is that the threat of exposure to mutagens or carcinogens may persist even after a tumor has formed for the reason that these agents can affect the progression of the disease. This is especially important to bear in mind as far as cancer therapy is concerned, since exposure of tumors to radiation and chemicals in the form of radiotherapy and chemotherapy increases dramatically after diagnosis. It is therefore important to ask if exposure to these agents can influence the biological behavior of residual tumor cells. This question is not trivial, since the failure of most therapeutic regimens to "cure" cancer may be related to two important factors which, at least in theory, can be affected by exposure to mutagenic agents.

The first factor is that any given tumor is usually heterogeneous with regard to many different cellular phenotypes (Fidler and Hart, 1982). One of the best examples of the cellular heterogeneity within tumors is their sensitivity to chemotherapeutic agents (Trsuro and Fidler, 1981). The second factor is that most deaths from cancer are not due to the growth of the primary tumor, but rather to the growth of distant metastases. Since individual metastases may also exhibit intra- or inter-lesional phenotypic heterogeneity (Poste et al., 1982), the problem of designing a therapeutic program which will eliminate all tumor cells becomes much more difficult. Moreover, only a very small number of tumor cells populating a primary tumor have all of the characteristics that are supposedly required for completion of all of the steps in the "metastatic cascade" (Fidler and Kripke, 1977; Poste and Fidler, 1980; Kozlowski et al., 1984). As such, the appearance of highly malignant (metastatic) cells represents perhaps the most dramatic and certainly the most lethal aspect of tumor heterogeneity.

However, most tumors are unicellular (clonal) in origin (Fialkow, 1979) and in their earliest (benign) stages of development are probably phenotypically and genotypically homogeneous. With time, however, and in the context of host selection pressures, they tend to become much more phenotypically heterogeneous and increasingly malignant (that is, invasive and metastatic). They gain increased autonomy from the host in relation to their ability to proliferate and spread, eventually killing the host. This evolution or progression of tumor cell populations from a benign, phenotypically homogeneous state to a more autonomous form has been discussed extensively by Foulds (1975) and Nowell (1976, 1983). If effective treatment modalities are to be developed and utilized in the best way, it is extremely important to be able to understand the mechanism(s) responsible for the diversification of tumor cell populations and their progression from a relatively benign to a highly malignant state. Indeed, if current modes of cancer therapy promote the biological mechanisms that are responsible for the generation of tumor cell heterogeneity and tumor progression, there may be a risk that these therapies may actually increase tumor cell diversification, and promote the evolution of highly malignant tumor cell subpopulations with increased metastatic ability (Kerbel and Davies, 1982; Poupon et al., 1984).

The mechanisms responsible for the generation of tumor cell heterogeneity and tumor progression are not clearly understood. However, there has been a general trend in thinking towards "genetic" or "mutational" changes as being primarily, if not solely, responsible (Nowell, 1983). These changes include small-scale point mutations and large-scale mutations involving gene rearrangements and gene amplification (Nowell, 1983; Hill et al., 1984). Recently we proposed that an "epigenetic" mechanism, namely DNA methylation, may also play a role in this process (Frost and Kerbel, 1983). In this chapter we will review the results obtained in our own laboratories, as well as the work of others, which gave rise to this hypothesis. Initially, however, we will briefly describe the genetic mechanisms that are thought be play a role in tumor progression and present a short summary of the role of DNA methylation in regulating gene expression. Then we will discuss why it is reasonable to think that such epigenetic changes may be important in many aspects of tumor progression, heterogeneity and metastasis.

GENETIC MECHANISMS OF TUMOR HETEROGENEITY AND TUMOR PROGRESSION

One of the central themes to emerge from the concept of tumor progression was its association with genetic (mutational) events taking place within the tumor-cell population, which was itself subject to host-selection pressures. The effects of this host-selection pressure meant that those tumor cells which had a relative growth advantage, (that is, increased autonomy from the host) would be able to survive and proliferate at the expense of their "less fit" counterparts (Nowell, 1976, 1983). These variant cells were presumed to arise as a result of somatic mutations in the cellular genome, and tumors progressed, as a consequence of an increase in their genetic instability. Thus, highly malignant tumor cell subpopulations might be more genetically unstable than their less malignant counterparts. There is evidence supporting this argument. It has been known for some time that highly malignant tumors and tumor cell subpopulations exhibit a greater degree of chromosomal aberrations in terms of ploidy levels, chromosome breaks, and sister-chromatid exchanges, than do less malignant tumors and tumor cell subpopulations (reviewed in Nowell, 1983). Moreover, Cifone and Fidler (1981) have reported that highly metastatic clones of various murine tumors have higher mutation rates than do poorly metastatic clones from the same tumors, supporting Nowell's hypothesis that increased genetic instability is associated with a more malignant phenotype. Unfortunately the molecular mechanisms associated with increased genetic instability have not been clearly demonstrated although Nowell (1983) has listed a series of inherited (such as, chromosome-breakage syndromes) and acquired (such as, alterations in DNA repair or replication) defects that may be responsible for this genetic instability. Nowell (1983) and Kerbel and Davies (1982) have suggested that therapeutic agents used in cancer therapy may actually contribute to the process of tumor progression and tumor heterogeneity as a consequence of their mutagenic effects.

There are, however, a number of observations that support the argument that epigenetic mechanisms also contribute to the generation of tumor cell heterogeneity and tumor progression. By "epigenetic", we mean a heritable phenotypic alteration brought about by a change in gene expression that is not due to a mutation in a structural gene. These observations include the phenotypic instability or "drift" observed in many tumor cell populations, as well as the generation of new phenotypes at rates which are much higher than can be accounted for merely on the basis of conventional somatic mutational mechanisms (reviewed in Nicolson, 1984). We have proposed that changes in DNA methylation may be one such epigenetic mechanism involved in tumor heterogeneity and tumor progression. We will first review, however, some of the characteristics of DNA methylation in relation to the control of gene expression.

DNA METHYLATION AND GENE REGULATION

The subject of DNA (cytosine) methylation and gene regulation has been reviewed recently (Razin and Riggs, 1980; Ehrlich and Wang, 1981; Cooper, 1983; Doerfler, 1983; and Riggs and Jones, 1983), and therefore only a brief summary will be presented here.

While eukaryotic DNA has only a small proportion of cytosine residues methylated, mainly in the sequence 5'-CpG-3' in the 5-position of the pyrimidine ring, recent results have provided strong evidence suggesting that there is an inverse relationship between the transcriptional activity of a gene and the extent to which the cytosine residues in or around that gene are methylated. In general, the less methylated the DNA is in the area of a given gene, the more likely that the gene will be transcriptionally active. Predictably, there are many exceptions to this finding (see Cooper, 1983 for a review), and results obtained from studies at a variety of genetic loci, indicate that it is the methylation of the DNA in the 5' promoter region of a gene, which may be critical in controlling its transcriptional activity. DNA is methylated following replication by the transfer of a methyl group from S-adenosylmethionine to the appropriate cytosine residue in the newly replicated strand. This transfer is mediated through the action of the S-adenosylmethionine DNA-cytosine methyltransferase enzyme. Since this enzyme uses the pattern of cytosine methylation on the parental strand of DNA as a template for the methylation of the newly synthesized daughter strand, the pattern of DNA methylation in a given cell type is a somatically heritable trait. However, Wigler et al. (1981) and Schmookler-Reis and Goldstein (1982) have reported that although the pattern is heritable, it is not passed on with 100% fidelity and some instability in the pattern may arise. The pattern in which DNA is methylated may be altered through the action of specific demethylases (Gjerset and Martin, 1982), by de novo methylation (Gasson et al, 1983), or during excision-repair of DNA (Kastan et al, 1982).

Many studies examining the role of DNA methylation in the regulation of gene expression have utilized the cytosine analogue 5-azacytidine (5-Aza-CR). Treatment of cells with 5-Aza-CR produces a dramatic decrease in the level of DNA (cytosine) methylation (Jones and Taylor 1980; Creusot et al., 1982, Taylor and Jones, 1982). The analogue must be incorporated into the newly synthesized daughter DNA strand in order to inhibit the methylation of the appropriate cytosine residues in that strand (Creusot et al, 1982; Taylor and Jones, 1982). It is believed that 5-Aza-CR inactivates the DNA-methyl-transferase enzyme (Creusot et al, 1982; Taylor and Jones, 1982; Santi et al, 1983). Finally, treatment of a variety of cell populations with 5-Aza-CR has been shown to result in the increased expression of a variety of genes (reviewed in Riggs and Jones, 1983).

DNA METHYLATION IN TUMOR PROGRESSION AND IN THE GENERATION OF TUMOR CELL HETEROGENEITY

Our own studies on the role of DNA methylation in tumor progression and tumor cell heterogeneity (Frost et al., 1984; Kerbel et al., 1984) were based on experiments evaluating the effects of in vitro mutagen treatment of tumor cells on the growth properties of the these cells in vivo (Frost et al., 1983). We reported that exposure of highly tumorigenic (tum^+) mouse tumor cell lines to either ethyl methanesulphonate (EMS) or N-methyl-N'-nitro-N-nitrosoguanidine (MNNG), both of which are powerful mutagens, followed by cloning of the tumor cell populations, resulted in the generation of nontumorigenic (tum^-) clones. Clones were designated tum^- when injection of large doses (many orders of magnitude more than normally required to give a tumor take of the untreated cell population) of these clones fails to give rise to a continuously growing tumor in the normal syngeneic host. This inability of the tum^- clones to produce progressively growing lethal tumors in the syngeneic host is likely due to the increased immunogenicity of the variants, since these clones do grow in highly immunosuppressed nude mice, and are usually able to generate strong cytotoxic T-cell immune responses *in vitro*. The ability of mutagens to increase the immunogenicity of nonimmunogenic or poorly immunogenic tumors had previously been reported by Boon and his colleagues (Boon, 1983).

Certain aspects of our results and those of Boon's were particularly interesting. The first was that the proportion of tum^- clones generated following mutagen treatment was extraordinarily high depending upon the tumor cell line used (Table 1). These frequencies were much greater than could be accounted for by any known mutational mechanism. Indeed, in the case of the murine mammary adenocarcinoma TA3, more than 90% of the clones isolated following mutagen treatment had acquired the tum^- phenotype.

The other aspect found by us, was the instability of the tum^- phenotype (Table 1). With increasing time in culture, there was a gradual reduction in the proportion of clones exhibiting the tum^- phenotype; that is, eventually the tum^- clones re-acquired their tum^+ phenotype. This phenotypic instability in the mutagen-induced tum^- clones struck us as surprising, since phenotypic alterations induced by mutagens causing point mutations are usually very stable. That is, the altered phenotype (such as, drug resistance) may still be exhibited when the cells are maintained in tissue culture for years, even in the absence of any selection pressure (Siminovitch, 1976).

The unusually high frequency at which the tum^- clones were generated after mutagen treatment as well as their phenotypic instability led us to wonder whether these results could be due to the mutagens having an epigenetic effect on these tumor cells. We therefore tested the ability of 5-Aza-CR to induce the formation of tum^- clones in the same tumor cell lines. 5-Azacytidine had been reported to be weakly (if at all) mutagenic in a variety of other

TABLE 1. Proportion of Tum$^-$ Variants Obtained after Treatment with Mutagen or 5-Aza-CR and Their Phenotypic Instability

Tumor	Treatment	Percent of tum$^-$ clones after weeks in culture (1-2)	(4-6)	(12-16)	(26)
TA3	EMS	95	76	10	0
(Adenocarcinoma	MNNG	96	n.d.	n.d.	n.d.
	5-Aza-CR	87	75	62	13
P815	EMS	43	20	0	n.d.
(Mastocytoma)	MNNG	0	n.d.	n.d.	n.d.
	5-Aza-CR	17	17	0	n.d.
MDAY-D2	EMS	0	n.d.	n.d.	n.d.
(Undifferentiated	MNNG	13	n.d.	n.d.	n.d.
Tumor)	5-Aza-CR	0	n.d.	n.d.	n.d.

Data adapted from Frost et al., (1984) and Kerbel et al., (1984). Cells were treated once with EMS, MNNG, or 5-Aza-CR. One week later the cells were cloned by limiting dilution, and individual clones injected subcutaneously into the appropriate syngeneic host. Tumor growth was assessed at weekly intervals following inoculation. Injections were performed after various times in culture. n.d., not done.

cultured mammalian cell lines (Landolph and Jones, 1982; Marquardt and Marquardt, 1977), but presumably due to its DNA hypomethylating action, it was able to induce alterations in gene expression at a variety of genetic loci at very high frequencies. The genetic loci included those relevant to the expression of thymidine kinase (Harris, 1982; Liteplo et al., 1984), dexamethasone sensitivity (Gasson et al., 1983), and emetine resistance (Worton et al., 1984). Moreover, these alterations were sometimes found to be highly unstable, in that the treated cells reverted to their original phenotypes within a few months in culture (Gasson et al., 1983).

As shown in Table 1, treatment of our tumour cells lines with 5-Aza-CR resulted in the generation of tum$^-$ clones at a frequency remarkably similar to that which we obtained using known mutagens. Where mutagen treatment produced a high frequency of tum clones, such as in TA3, 5-Aza-CR produced a similar result. Likewise, single EMS or 5-Aza-CR treatment of the MDAY-D2 tumor cell line did not give rise to any tum$^-$ variants. Moreover, the 5-Aza-CR induced tum$^-$ variants of TA3 were phenotypically unstable, like those of the mutagen-induced tum$^-$ variants.

Because of the similarity of the results obtained after mutagen and 5-Aza-CR treatment, we sought to determine whether 5-Aza-CR was having any obvious "mutagenic effect" on our tumor cell populations. Following treatment with either EMS or 5-Aza-CR, the frequency of ouabain, 6-thioguanine and 2-deoxygalactose-resistant mutants was examined. We found that 5-Aza-CR was insignificantly mutagenic in our tumor cell lines, in agreement with the results of Landolph and Jones (1982) and Marquardt and Marquardt (1977) (Table 2). Similar results showing 5-Aza-CR to have a negligible mutagenic effect have also been reported by Bouck et al. (1984) and Delers et al. (1984). We should point out that the loci used to determine mutagenic activity could only detect the formation of point mutations. It is therefore possible that 5-Aza-CR could induce other effects (such as, gene deletions, rearrangements, or amplifications) that would not be picked up in these assays. Indeed, agents which inhibit DNA replication can induce gene amplification in the form of so-called "double minute" chromosomes (Mariani and Schimke, 1984). 5-Aza-CR has also been shown to produce a small increase in the frequency of sister-chromatid exchange, chromosome decondensation, and endoreduplication (Banerjee and Benedict, 1979; Hori, 1983), but it is not clear if these effects are due to the 5-Aza-CR itself or rather to the DNA hypomethylation which it produces (Chambers and Taylor, 1982).

TABLE 2. Frequency of Drug-Resistant Mutants Obtained after Treatment of MDAY-D2 or TA3 Tumor Cells with 5-Aza-CR or EMS

		Frequency of Drug-Resistant Mutants		
Cell Line	Treatment	Ouabain	6-thioguanine	2-deoxygalactose
MDAY-D2	None	n.f.	$8x10^{-6}$	n.f.
MDAY-D2	5-Aza-CR	$2x10^{-7}$	$8x10^{-6}$	$3x10^{-7}$
MDAY-D2	EMS	$2.9x10^{-5}$	$7x10^{-5}$	$1x10^{-5}$
TA3	None	n.f.	n.f.	n.d.
TA3	5-Aza-CR	$1x10^{-6}$	n.f.	n.d.
TA3	EMS	$1.5x10^{-5}$	$1.3x10^{-6}$	n.d.

Data adapted from Frost et al. (1984). n.f. indicates that no resistant colonies were found in 9.6×10^6 cells plated. n.d., not done.

Our contention that the effects of the mutagens on the generation of the tum^- variants was due to an epigenetic mechanism, most likely DNA hypomethylation, is supported by the observations of Wilson and Jones (1983) and Boehm and Drahovsky (1981a, 1981b). These authors have reported that known mutagens, including EMS and MNNG, are able to inhibit DNA methylation. Thus, particularly in the case of the TA3 tumor cell line, the ability of these mutagens to induce the formation of tum^- variants, that are phenotypically unstable during prolonged cell culture may have been due to their ability to alter DNA methylation, rather than to their mutagenic activities. Definite proof of our hypothesis, however, is still required. The putative new tumor antigens which are likely to be present on the surface of the tum^- variants have not yet been identified nor have the genes coding for them. Only when this has been done will we be able to conclusively determine if DNA hypomethylation is the mechanism responsible for the altered tumorigenic phenotype of these variants. It is also worth noting that Ivarie and Morris (1982) have reported that EMS can induce tremendously high frequencies (20-30%) of prolactin-deficient variants in a rat pituitary tumor. They speculated that the effects were due to EMS-induced hypermethylation (not hypomethylation) of the prolactin gene on the basis of subsequent results with 5-Aza-CR.

5-Azacytidine treatment not only decreased the malignant capacity of the tumor cell lines (that is, $tum^+ \rightarrow tum^-$) but it was also able to increase the metastatic capacity ($met^- \rightarrow met^+$) of these cells as well. A small number, 16 of 124 of the 5-Aza-CR-treated TA3 tumor clones retained the tum^+ phenotype, and three of the 16 metastasized to the liver, spleen or kidneys of animals following subcutaneous inoculation. This finding was rather startling since the parental TA3 tumor line had never been observed to metastasize spontaneously in A strain mice, even though we had injected hundreds of animals. Again, however, this tum^+met^+ phenotype proved to be phenotypically unstable, since after 4 months in culture all of these tum^+ met^+ clones had reverted to their original tum^+ met^- phenotype. Interestingly, a similar result has been recently reported by Olsson and Forchhammer (1984), who indicated that 5-Aza-CR was able to induce a metastatic phenotype in tum^+ met^- subpopulations of Lewis lung carcinoma cells.

Taken together, the results from our laboratory as well as those of Olsson and Forchhammer (1984) provide indirect evidence that changes in DNA methylation may indeed play a role in the generation of tumor cell heterogeneity and tumor progression.* Other investigators, using entirely different approaches, have come to a similar conclusion. Diala et al. (1983) have found that the DNA isolated from human tumor cell lines was less methylated than that derived from normal cells. Gama-Sosa et al. (1983) studying the DNA from human tumor tissue samples found that, in general, the DNA isolated from metastases was less methylated than that isolated from benign tumors or normal tissue. Similarly Feinberg and Vogelstein (1983a, 1983b) again using human tissue samples, have indicated that certain genes (including the ras-oncogene) are less methylated in tumor tissue than in normal tissue. More importantly, the level of methylation of the DNA isolated from a metastasis in one patient, was less than that observed for the primary tumor. However, Flatau et al. (1983) have reported no consistent decreases in the DNA-methylcytosine content in some human tumor lines nor in fresh pediatric tumor explants.

Parenthetically, we should point out that DNA hypomethylation has also been suggested to play a role in carcinogenesis itself (Holliday, 1979; Boehm and Drahovsky, 1983; Nyce et al., 1983). Indeed, 5-Aza-CR is itself able to induce neoplastic transformation in baby hamster kidney cells (Bouck et al., 1984), Chinese hamster embryo fibroblasts (Harrison et al., 1983) and mouse C3H/10T1/2 clone 8 cells (Benedict et al., 1977). Ethionine, a potent liver carcinogen (Farber, 1963) produces hypomethylated DNA (Boehm and Drahovsky, 1981c). Moreover, Wilson and Jones (1983) have shown that many carcinogens are able to inhibit the methylation of DNA. Thus it appears that DNA-methylation may play an important role in the formation as well as the progression of tumors.

---

*see DISCUSSION of this chapter for an alternative, genetic, explanation (unstable gene amplifications) put forward recently by Hill et al. (1984) to account for the emergence and instability of the metastatic phenotype.

CANCER THERAPY, DNA-METHYLATION AND TUMOR PROGRESSION/ TUMOR HETEROGENEITY

In any assessment of the risk from exposure to radiation and chemicals, two new issues may need to be examined. First, can exposure to any of these agents affect the overall biological behavior of already established tumors in such a way so as to promote, accelerate, or amplify their genotypic and phenotypic diversification and/or enhance their progression from a relatively benign to a more malignant state? Secondly, in identifying the mechanisms associated with these risks, it is important to examine both genetic and epigenetic processes. Since most commonly employed anticancer therapeutic agents are mutagenic and mutagens can induce alterations in DNA methylation, these therapies may enhance tumor cell diversification not only by the induction of mutations, but by increasing or inducing the transcriptional activity of certain genes or, by a combination of these. These genetic and epigenetic mechanisms need not be mutually exclusive. Indeed, it is conceivable that alterations in DNA methylation at a particular site in the genome might render it more likely to undergo genetic rearrangement or amplification.

It should be pointed out that DNA hypomethylating agents have already been used therapeutically in clinical settings. 5-Azacytidine has been utilized in the treatment of β+ thalassemia (Ley et al., 1982), and 5-aza-2-deoxy-cytidine (the deoxy-analogue of 5-aza-CR) has undergone clinical trials for the treatment of childhood acute leukemia (Rivard et al, 1981). Since these agents may promote tumor cell diversification and tumor progression through changes in DNA methylation, the risks of increasing the malignant capacity of an established tumor, or actually inducing a new tumor must be weighed against their potential therapeutic value.

REFERENCES

Banerjee, A., and Benedict, W.F., 1979, Production of sister-chromatid exchanges by various cancer chemotherapeutic agents, Cancer Res., 39:797.

Benedict, W.F., Banerjee, A., Gardner, A., and Jones, P.A., 1977, Induction of morphological transformation in mouse C3H/10T 1/2 CL8 cells and chromosomal damage in hamster $A(T_1)Cl-3$ cells by cancer chemotherapeutic agents, Cancer Res., 37:2202.

Boehm, T.L., and Drahovsky, D., 1981a, Hypomethylation of DNA in Raji cells after treatment with N-methyl-N-nitrosourea, Carcinogenesis, 2:39.

Boehm, T.L.,and Drahovsky, D., 1981b, Inhibition of enzymatic DNA methylation by N-methyl-N-nitro-N-nitrosoguanidine in human Raji lymphoblast-like cells, Int .J. Biochem, 13:1225.

Boehm, T.L., and Drahovsky, D., 1981c, Elevated transcriptional complexity and decrease in enzymatic DNA methylation in cells treated with L-ethionine, Cancer Res., 41:4101.

Boehm, T.L., and Drahovsky, D., 1983, Alteration of enzymatic methylation of DNA cytosines by chemical carcinogens: A mechanism involved in the initiation of carcinogenesis, J. Natl. Can. Inst., 71:429.

Boon, T., 1983, Antigenic tumor variants obtained by mutagens, Adv. Cancer Res., 39:121.

Bouck, N., Kokkinakis, D., and Ostrowsky, J., 1984, Induction of a step in carcinogenesis normally associated with mutagenesis by non-mutagenic concentrations of 5-azacytidine, Mol. Cell Biol., in press.

Chambers, J.C., and Taylor, J.H., 1982, Induction of sister-chromatid exchanges by 5-fluorodeoxycytidine: correlation with DNA methylation, Chromosoma, 85:603.

Cifone, M., and Fidler, I.J. 1981, Increasing metastatic potential is associated with increasing genetic instability of clones isolated from murine neoplasms, Proc. Natl. Acad. Sci. U.S.A., 78: 6949.

Cooper, D.N., 1983, Eukaryotic DNA methylation, Human Genetics, 64:315.

Creusot, F., Acs, G., and Christman, J.K., 1982, Inhibitior of DNA methyltransferase and induction of Friend erythroleukemia cell differentiation by 5-azacytidine and 5-aza-2-deoxycytidine, J. Biol. Chem., 257:2041.

Delers, A., Szpirer, J., Szpirer, C., and Saggorio, O., 1984, Sontaneous and 5-azacytidine-induced reexpression of orinthine carbamoyl transferase in hepatoma cells, Mol. Cell Biol., 4:809.

Diala, E.S., Cheah, M.S.C., Rowitch, D., and Hoffman, R.M., 1983, Extent of DNA methylation in human tumor cells, J. Natl. Cancer Inst., 71:755.

Doerfler, W., 1983, DNA methylation and gene activity, Ann. Rev. Biochem., 52:93.

Ehrlich, M., and Wang, R. Y.-H., 1981, 5-methylcytosine in eukaryotic DNA, Science, 212:1350.

Farber, E., Ethionine carcinogenesis, 1963, Adv. Cancer Res., 7:383.

Feinberg, A.P., and Vogelstein, B., 1983a, Hypomethylation distinguishes genes of some human cancers from their normal counterparts, Nature (London), 301:89.
Feinberg, A.P., and Vogelstein, B., 1983b, Hypomethylation of ras-oncogenes in primary human cancers, Biochem. Biophys. Res. Commun., 111:47.
Fialkow, P.J., 1979, Clonal origin of human tumors, Annu. Rev. Med., 30:135.
Fidler, I.J., and Kripke, M.L., 1977, Metastasis results from pre-existing variant cells within a malignant tumor, Science, 197:893.
Fidler, I.J., and Hart, I.R., 1982, Biologic diversity in metastatic neoplasms: origins and implications, Science, 217:990.
Flatau, E., Bogenmann, E., and Jones, P.A., 1983, Variable 5-methylcytosine levels in human tumor cell lines and fresh pediatric tumor explants, Cancer Res., 43: 4901.
Foulds, L., 1975, "Neoplastic Development", Academic Press, New York.
Frost, P., and Kerbel, R.S., 1983, On a possible epigenetic mechanism(s) of tumor cell heterogeneity, Cancer Metastasis Rev., 2:375.
Frost, P., Kerbel, R.S., Bauer, F., Tartamella-Biondo, R., and Celfalu, W., 1983, Mutagen treatment as a means of selecting immunogenic variants from otherwise poorly immunogenic malignant tumors, Cancer Res., 43:125.
Frost, P., Liteplo, R.G., Donaghue, T.P., and Kerbel, R.S., 1984. The selection of strongly immunogenic "tum$^-$" variants from tumors at high frequency using 5-azacytidine, J. Exp. Med., 159:1491.
Gama-Sosa, M.A., Slagel, V.A., Trewyn, R.W., Oxenhandler, R., Kuo, K.C., Gehrke, C.W., and Ehrlich, M., 1983, The 5-methylcytosine content of DNA from human tumors, Nucleic Acids Res., 11:6883.
Gasson, J.C., Ryden, T., and Bourgeois, S., 1983, Role of de novo DNA methylation in the glucocorticoid resistance of a T-lymphoid cell line, Nature (London), 302:621.
Gjerset, R.A., and Martin, D.W., 1982, Presence of DNA demethylating activity in the nucleus of murine erthyroleukemia cells, J. Biol. Chem, 257:8581.
Harris, M., 1982, Induction of thymidine kinase in enzyme-deficient Chinese hamster cells, Cell, 29:483.

Harrison, J.J., Anisowicz, A., Gadi, I.K., Raffeld, M., and Sager, R., 1983, Azacytidine-induced tumorigenesis of CHEF/18 cells: correlated DNA methylation and chromosome changes, Proc. Natl. Acad. Sci.U.S.A., 80:6606.

Hill, R.P., Chambers, A.F., and Ling, V., 1984, Dynamic heterogeneity: rapid generation of metastatic variants in mouse B16 melanoma cells, Science, 224:998.

Holliday, R., 1979, A new theory of carcinogenesis, Br. J. Cancer, 40:513.

Hori, T.-A., 1983, Induction of chromosome decondensation, sister-chromatid exchanges and endoreduplications by 5-azacytidine, an inhibitor of DNA methylation, Mutat. Res., 121:47.

Ivarie, R.D., and Morris, J.A., 1982, Induction of prolactin-deficient variants of GH_3 rat pituitary tumor cells by ethyl methanesulfonate: reversion by 5-azacytidine, a DNA methylation inhibitor, Proc. Natl. Acad. Sci. U.S.A., 79:2967.

Jones, P.A., and Taylor, S.M., 1980, Cellular differentiation, cytidine analogs and DNA methylation, Cell, 20:85.

Kastan, M.B., Gowans, B.J., and Lieberman, M.W., 1982, Methylation of deoxycytidine incorporated by excision repair synthesis of DNA, Cell 30:509.

Kerbel, R.S., and Davies, A.J.S., 1982, Facilitation of tumor progression by cancer therapy, Lancet (ii):977.

Kerbel, R.S., Frost, P., Liteplo, R., Carlow, D.A., and Elliott, B.E., 1984, Possible epigenetic mechanisms of tumor progression: Induction of high-frequency heritable but phenotypically unstable changes in the tumorigenic and metastatic properties of tumor cell populations by 5-azacytidine treatment, J. Cell. Phys in press.

Kozlowski, J.M., Hart, I.R., Fidler, I.J., and Hanna, N., 1984, A human melanoma line heterogeneous with respect to metastatic capacity in athymic nude mice, J. Natl. Cancer Inst. 72:913.

Landolph, J.R., and Jones, P.A., 1982, Mutagenicity of 5-azacytidine and related nucleosides in C3H/10T 1/2 clone 8 and V79 cells, Cancer Res., 42:817.

Ley, T.J., De Simone, J., Anagnov, N.P., Keller, G.H., Humphries, R.K., Turner, P.M., Young, N.S., Keller, P. and Nienkuis, S.W., 1982, 5-azacytidine selectively increases γ-globin synthesis in a patient with β+ thalassemia, N. Engl.J. Med., 307:1469.

Liteplo, R.G., Frost, P., and Kerbel, R.S., 1984, 5-azacytidine induction of thymidine kinase in a spontaneously enzyme-deficient murine tumor line, Exp.Cell Res., 150:499.

Mariani, B.D., and Schimke, R.T., 1984, Gene amplification in a single cell cycle in Chinese hamster ovary cells. J. Biol. Chem., 259:1901.

Marquardt, H., and Marquardt, H., 1977, Induction of malignant transformation and mutagenesis in cell cultures by cancer chemotherapeutic agents, Cancer, 40:1930.

Nicolson, G.L., 1984, Generation of phenotypic diversity and progression in metastatic tumor cells, Cancer Metastasis Rev., 3:25.

Nowell, P.C., 1976, The clonal evolution of tumor cell populations, Science, 194:23.

Nowell, P.C., 1983, Tumor progression and clonal evolution: the role of genetic instability, in: "Chromosome Mutation and Neoplasia", pages 413-432, J. German, ed., A.R. Liss Inc., New York.

Nyce, J., Weinhouse, S., and Magee, P.N., 1983, 5-Methylcytosine depletion during tumor development: An extension of the miscoding concept, Br. J. Cancer, 48:463.

Olsson, L., and Forchhammer, J., 1984, Induction of the metastatic phenotype in a mouse tumor model by 5-azacytidine and characterization of an antigen associated with metastatic activity, Proc. Natl. Acad.Sci. U.S.A., 81:3389.

Poste, G., and Fidler, I.J., 1980, The pathogenesis of cancer metastasis, Nature (London), 283:139.

Poste, G., Doll, J., Brown, A.E., Tzeng, J., and Zeidman, I., 1982, Comparison of metastatic properties of B16 melanoma clones isolated from cultured cell lines, subcutaneous tumors, and individual metastases, Cancer Res., 42:2770.

Poupon, M.-F., Pauwels, C., Jasmin, C., Antoine, E., Lascaux, V., and Rosa, B., 1984, Amplified pulmonary metastases of a rat rhabdomyosarcoma in response to nitrosourea treatment, Cancer Treatment Reports, 68:749.

Razin,A., and Riggs, A.D., 1980, DNA methylation and gene function, Science, 210:604.

Riggs, A.D., and Jones, P.A., 1983, 5-Methylcytosine, gene regulation, and cancer, Adv. Cancer Res., 40:1.

Rivard, G.E., Momparler, R.L., Demers, J., Benoit, P., Raymond, R., Lin, K.-T., and Momparler, L.F., 1981, Phase I study on 5-aza-2-deoxycytidine in children with acute leukemia, Leukemia Res., 5:453.

Santi, D.V., Garrett, C.E., and Barr, P.J., 1983, On the mechanism of inhibition of DNA-cytosine methyltransferase by cytosine analogues, Cell, 33:9.

Schmookler-Reis, R.J., and Goldstein, S., 1982, Variability of DNA methylation patterns during serial passage of human diploid fibroblasts, Proc. Natl. Acad. Sci. U.S.A., 79:3949.

Siminovitch, L., 1976, On the nature of heritable variation in cultured somatic cells, Cell, 7:1.

Taylor, S.M., and Jones, P.A., 1982, Mechanism of action of eukaryotic DNA methyltransferase, J. Mol. Biol., 162:679.

Trsuro, T., and Fidler, I.J., 1981, Differences in drug sensitivities among tumor cells from parental tumors, selected variants, and spontaneous metastases, Cancer Res., 41:3058.

Weiss, L., 1980, Metastasis: differences between cancer cells in primary and secondary tumors, Pathobiol. Ann., 10:51.

Wigler, M., Ley, D., and Perucho, M., 1981, The somatic replication of DNA methylation, Cell, 24:33.

Wilson, V.L., and Jones, P.A., 1983, Inhibition of DNA methylation by chemical carcinogens *in vitro*, Cell, 32:239.

Worton, R.G., Grant, S.G., and Duff, C., 1984, Gene inactivation and reactivation at the EMT locus in Chinese hamster cells, *in*: "Gene Transfer and Cancer", H.L. Steinberg, and M.L. Pearson, eds., Raven Press, New York.

DISCUSSION

Woodhead: Could you comment further on the differences between your theory and Dr. Fidler's concept that one metastatic cell from a tumor proliferates to give rise to all the rest. Do you now give less weight to the latter idea?

Kerbel: No, not necessarily. I think it's fair to say there are essentially two schools of thought. One school holds the view that metastatic cells are distinct entities which pre-exist while the opposing school maintains that the metastatic phenotype is essentially ephemeral (Weiss, 1980; Hill et al., 1984). On the face of it our DNA methylation hypothesis supports the latter idea and also provides a plausible explanation for the transient nature of the metastatic phenotype (Kerbel et al., 1984a). Hill et al. (1984) prefer a genetic explanation, namely the idea that unstable forms of gene amplification could accomplish the same thing. Either way you have a situation where a gene that is required for a tumor cell to complete successfully a particular step in the multi-step process of metastasis is expressed (or amplified) at a moment in time when it is needed, and then is 'silenced' when it is no longer required, that is, after the step has been completed and the cell can go on to the next step. In this situation, recovery of cells that form a metastasis may yield a cell population which behaves no differently from the parent (primary) tumor cell population.

However, I also think it's reasonable to suppose that if true metastatic mutants pre-exist which are phenotypically stable, their frequency is probably quite low (10^{-6} to 10^{-7}). In order to isolate such authentic mutants you would need to apply a very stringent selection pressure. In my view, recovery of lung colonies from mice that had been injected intravenously with a large dose of tumor cells will not in many cases necessarily provide the appropriate selection pressure. On the other hand, if you inject an inherently poorly metastatic tumor cell population subcutaneously and recover some infrequent spontaneous metastases some time later, they could well represent true metastatic mutants. We've adopted this approach by injection of human tumors into nude mice (Kerbel et al., 1984b). It is well known that metastases in this situation are rarely observed. When they do arise, they may be few in number and take many months to manifest themselves. Thus, in this situation, one is probably dealing with a "rare event." And sure enough, we have found cytogenetic evidence that the metastases in this situation do not appear to arise randomly, but represent selected subpopulations of cells (Kerbel et al., 1984b).

Thus, I think metastases can arise through both processes. But it is not at all clear which of the two is the more common, particularly in the context of human neoplasias in their natural hosts.

Shellabarger: What are the effects of x-rays?

Kerbel: So far as I know, no one has looked at x-rays in terms of the tum^- behavioral phenomenon that I've outlined. Experiments have been done with UV light and all sorts of chemotherapeutic drugs and mutagens (see Boon, 1983), and very similar results have been reported.

Bertram: Do you have any comment to make on the surprising observations made by George Poste et al. (1981), that if one subclones from a population of tumor cells with median metastatic potential one can obtain clones of low metastatic potential and clones of a high metastatic potential; but then, if these are maintained separately, they all re-establish a parental-type metastatic phenotype, as if they're capable of monitoring the behavior of the cell population as a whole.

Kerbel: In retrospect the results may not be surprising, they certainly are interesting. More and more laboratories are reporting that if you take cell populations, clone them, and then look for a certain phenotype characteristic (such as relative drug resistance, relative hormone sensitivity, ability to metastasize or invade), you find recently derived clones, far from being homogeneous, show heterogeneity within any one clone (Welch and Nicolson, 1983). The very process of cloning itself seems to act as a stimulus for the rapid generation of phenotypic diversity, almost as if heterogeneity is the norm, or an "equilibrium" state for the cell population as a whole. What is not yet clear is how general this clonal interaction phenomenon is, and what accounts for its origin,

Sonnenblick: Suppose you have a human metastatic carcinoma, which has invaded normal tissues. What do you expect the pathologist to see in the metastatic tissue? There is evidence that the dividing cells of that metastasis have a wide array of cellular phenotypes. Some conclude that an appreciable number of varying malignancies are contained in a specific metastatic tissue.

Kerbel: What pathologists and others have noted can be extremely variable. If one looks within a metastasis, you can sometimes find evidence for striking phenotype diversity, for example, in cellular morphology, in chromosome levels and markers, and so forth. Also, to make the situation more complicated, if you look within the same patient at different metastases located either in the same organ, or in different organs, you can also find evidence of what is called interlesional heterogeneity. Obviously we are trying to sort out some of the ways which can account for the evolution of this process and we are especially interested in high frequency/phenotypically unstable phenomena I referred to in my lecture. Are we seeing an epigenetic type of event and if so, what is it due to? How might cancer therapy affect this process? I think that it is important to know the answers to these questions.

Sonnenblick: Metastasis need not be an inevitable condition of neoplasia, illustrated, for example, by the carcinoma-*in-situ* situation. Your work is providing conceptual, perhaps ultimately clinical, approaches to the heterogeneity problem. Increasingly there are findings of several malignancies and different primary tumors in single patients. One possible origin of these are from segregated cells from the original heterogenous mass.

Kerbel: That fact is clearly the bane of oncologists. As long as this kind of extreme cellular heterogeneity exists, it will essentially be impossible to successfully treat the disease. In effect what you have to do is to find a strategy that will circumvent the existence of this cellular heterogeneity; for example, a drug, or combination of drugs with which you would be able to kill off every cell, or at least every cancer stem cell. It seems that it would be almost impossible to do this *in vivo*. And another disturbing point is that any agent that you use may itself bring about either genetic or epigenetic changes which could actually promote the evolution of tumor cell heterogeneity.

Borg: I have some both hopeful and some discouraging information to support your statements about the heterogeneity of metastases, particularly of melanomas. One of my collaborators has a series of monoclonal antibodies to various antigenic determinants. On the one hand, the heterogeneity you speak of is shown all of the time, whether you are looking at the surgical biopsies grown in nude mice, or looking directly, with a panel of monoclonal antibodies. On the other hand, we have been able to collect and combine enough monoclonal antibodies to different determinants, so that we have not found any melanomas (including amelanotic ones) to which we cannot get some response. If one can find the right vehicles to use as therapeutic agents, the so-called "therapeutic cocktails", we may well be successful.

Kerbel: Using "cocktails" is obviously the wave of the future; and a cocktail may not just consist of a panel of different monoclonal antibodies, but in fact, a combination of very different therapeutic modalities. The present trends are to employ a series of drugs in a particular sequence, or combination; or you may combine chemotherapy with another form of therapy, such as radiotherapy or immunotherapy. But a lot of this work is still at the level of trial-and-error, and I fear it is going to be a very slow, very frustrating process to achieve the results we all desire so much for the cancer patient.

REFERENCES

Boon, T., 1983, Antigenic tumor variants obtained by mutagens, Adv. Cancer Res., 39:121.

Hill, R.P., Chambers, A.F., and Ling, V., 1984, Dynamic heterogeneity: rapid generation of metastatic variants in mouse B16 melanoma cells, Science, 224:998.

Kerbel, R.S., Frost, P., Liteplo, R., Carlow, D.A., and Elliott, B.E., 1984a, Possible epigenetic mechanisms of tumor progression: Induction of high-frequency heritable but phenotypically unstable changes in the tumorigenic and metastatic properties of tumor cell populations by 5-azacytidine, J. Cell. Physiol., in press.

Kerbel, R.S., Man, M.S., and Dexter, D., 1984b, A model of human cancer metastasis: extensive spontaneous and derived variant sublines in nude mice, J. Natl. Cancer Inst., 72:93.

Poste, G., Doll, J., and Fidler, I.J., 1981, Interaction between clonal subpopulations affect the stability of the metastatic phenotype in polyclonal populations of B16 melanoma cells, Proc. Natl. Acad. Sci. U.S.A., 78:6226.

Weiss, L., 1980, Metastasis: differences between cancer cells in primary and secondary tumors, Pathobiol. Ann., 10:51.

Welch, D.R. and Nicolson, G.L., 1983, Phenotypic drift and heterogeneity in response of metastatic mammary adenocarcinoma cell clones to Adriamycin, 5-fluoro-2'-deoxyuridine and methoprexate treatment in vitro, Clin. Exp. Metastasis, 10:317.

INTERACTION OF IONIZING RADIATION AND 8-METHOXYPSORALEN PHOTOSENSITIZATION: SOME IMPLICATIONS FOR RISK ASSESSMENT

B. A. Bridges

MRC Cell Mutation Unit, University of Sussex
Falmer, Brighton BN1 9RR,
England

> "I have never encountered any problem however complicated which, when looked at in the proper way, did not become still more complicated".
>
> Poul Andersson

Much effort, including the greater part of the present Workshop, is devoted to consideration of the likely consequences of human exposure to low concentrations or fluences of single genotoxic agents. Yet humans do not live in a one-toxin environment. A multitude of mutagens and promoters are ingested, inhaled, or generated endogenously as an obligatory component of everyday life, and we know they cannot be ignored. There are those who would draw attention to the existence of detoxifying pathways and repair systems which might be induced by this abundance of everyday mutagens, so that any dose of a single additional agent would encounter these systems at high levels of activity. Low doses of the additional agents might then be rapidly removed or the damage repaired, resulting in an effective threshold or quasi-threshold. Others would have us focus on the error-prone pathways and metabolic activation systems that might be induced, or detoxification pathways and repair pathways that might be saturated by ambient exposure to mutagens. A small dose of an additional mutagen might then cause more damage than expected. How then should we proceed, given that interactions between agents are likely to be crucial in any real-life situation? I fear that the burden of this essay is essentially pessimistic. I shall describe how consideration of the interaction of two apparently simple genotoxic agents revealed mechanistic possibilities of unsuspected complexity. Lack of mechanistic understanding precludes confident estimation of effects at low doses. If such complexity emerges with just two agents, what can be expected from the interactions of a multitude of ambient mutagens?

TWO SIMPLE GENOTOXIC AGENTS

The two agents to be considered are among the best studied mutagens and carcinogens. Ionizing radiation has been found to be a potent mammalian carcinogen and a mutagen in all organisms which have received adequate study. In DNA it causes mainly single- and double-strand breaks and a smaller yield of various types of base damage. Human exposure occurs from natural sources, from environmental contamination by the nuclear industries, and, to a considerable extent, from the medical use of X-rays and other ionizing radiations. The other agent, the combination of 8-methoxypsoralen plus near ultraviolet light (PUVA), is a probable human skin carcinogen, a proven mouse skin carcinogen, and a mutagen in all systems that have been studied. Under the action of near ultraviolet light, 8-methoxypsoralen forms adducts with pyrimidine bases in DNA: a proportion of these adducts react further to form interstrand crosslinks. Human exposure occurs in the treatment of skin diseases such as psoriasis, vitiligo and alopecia; often quite large doses are given. The related compound 5-methoxypsoralen has similar photobiological properties and in some countries is widely used in preparations of bergamot oil to promote sun-tanning.

Metabolic processes do not enter significantly into the action of either ionizing radiation or PUVA, and one might have expected that interaction of these two direct-acting mutagens and initiators of carcinogenesis might have been rather simple to elucidate, perhaps involving only the various pathways of DNA repair. This expectation has not, however, been realized.

THE PSEUDOPROMOTER HYPOTHESIS

The first indication of mechanistic complexity came from a consideration of the report of Reed (1976) and Reed et al. (1977) who treated xeroderma pigmentosum patients with PUVA and observed the very early appearance of skin tumors, in one case within a month of commencement of treatment. This onset was too soon to be attributed to an increased rate of tumor initiation and we postulated that PUVA might be acting as a promoter (Bridges and Strauss, 1980). This idea was also consistent with the report of Stern et al. (1979) that a higher incidence of squamous cell carcinoma of the skin occurred in patients with psoriasis given PUVA who had previously been exposed to ionizing radiation.

Table 1. Inhibition by PUVA of Delayed Cellular Hypersensitivity (DCH) Response to Dinitrochlorobenzene Sensitization (from Strauss et al., 1980).

Subjects	Number	Subnormal	Normal
PUVA-treated psoriatics	102	55	47
Non-PUVA-treated psoriatics	9	0	9
Untreated normal subjects	12	0	12

In looking for a mechanism for this effect, we were impressed by evidence that human squamous cell carcinomas were subject to immune surveillance (Walder et al., 1971; Kinlen et al., 1979) and by several lines of evidence that PUVA could cause damage to the cell-mediated immune system in animals (for example, Spellman, 1979; Morison et al., 1979). It seemed to us quite plausible that promotion of squamous cell carcinomas by PUVA could result from a breakdown in tumor control by the immune system in the skin.

Subsequently, for a serendipitous reason described elsewhere (Bridges, 1983), we were able to provide evidence in humans for suppression by PUVA of cell-mediated immunity . We measured the delayed cellular hypersensitivity response in more than one hundred patients with psoriasis undergoing PUVA therapy and showed that the response was impaired in half of them (Table 1). The impairment was greatest in patients with skin types I and II (fair skin) and in those receiving greater exposures (Strauss et al., 1980). A substantial body of evidence has now accumulated that exposure of the skin to PUVA and UV alone (both DNA-damaging agents) may have profound effects on the immune system (Lynch et al., 1981; Morison, 1981; Kripke et al., 1981; Kripke, 1982). Because this type of action of PUVA would be different from that classically considered as promotion, we have tended latterly to use the term pseudopromotion.

INTERACTION IN MOUSE SKIN

Following upon these ideas, a collaborative study was undertaken with Dr. Antony Young of the Institute of Dermatology in London to establish whether or not PUVA could act as a (pseudo)promoter for radiation-induced carcinogenesis in mouse skin. The results are not yet definitive but are considered to be sufficiently well founded to warrant some discussion here.

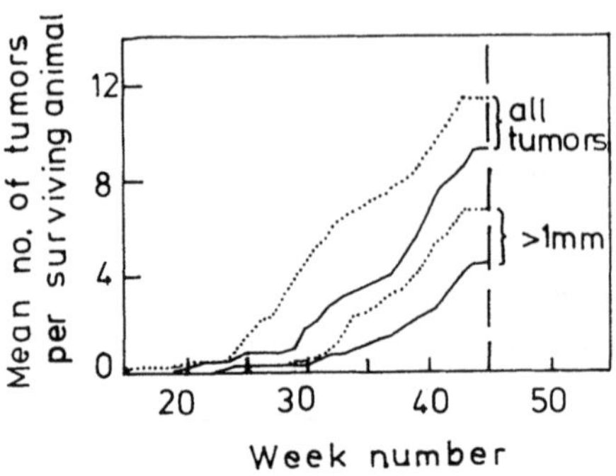

Fig. 1. Induction of skin tumors in mice continually exposed to PUVA following beta irradiation (nominal dose 10 Gy) at time zero (dotted line) or without prior irradiation (solid line).

Ionizing radiation exhibits an unusual dose response for the induction of skin tumors in mice, with an apparent or real threshold of around 15 Gy (Hulse et al., 1983). We selected a dose of 10 Gy beta radiation, below the threshold. After eight weeks, animals (Skh-1 mice) were given a course of PUVA involving thrice-weekly topical administration of 50 μl of a 0.01% ethanolic solution of 8-methoxypsoralen to each flank followed by 1 J/m^2 UVA filtered through 4 mm glass for forty-five weeks, at which time they were killed.

No tumors were induced by beta particles followed by ethanol and UVA, or the beta particles followed by 8-methoxypsoralen without UVA. Animals not exposed to beta particles but given PUVA showed a clear induction of tumors. PUVA-treated animals previously given beta radiation showed earlier appearance of tumors and there were approximately twice as many as on unirradiated animals at a given stage of PUVA treatment (Fig. 1). The effect was quite clear visually, although Peto's Heterogeneity test, based on the appearance of the first tumor per mouse, did not find the difference significant. The pathology of the tumors has yet to be done. We are now trying to obtain more definitive data using more mice and a somewhat higher dose of beta particles.

THE SPANNERS IN THE WORKS

Given that these data are preliminary they are nevertheless indicative of a synergistic interaction and are consistent with the hypothesis that PUVA may, in addition to its own activity as a complete carcinogen, possess a (pseudo)promoting activity for tumors initiated by ionizing radiaton. Unfortunately, while this experiment, when complete, may serve to establish a synergism between these agents, the effect may well be explained in at least two other ways.

The first alternative explanation stems from an exceedingly interesting observation by Frank and Williams (1982) with V79 Chinese hamster cells. They observed that exposure of these cells to 9 Gy X-rays resulted in the induction of a long-lasting, perhaps permanent, state of hypermutability towards a subsequent PUVA treatment. The hypersensitivity persisted for up to 108 days and was most evident with low doses of PUVA that were effectively nonmutagenic in unirradiated cells (Fig. 2). Subsequent work (Williams and Frank, 1982) showed that the hypermutable state was induced by X-rays with a frequency 100 to 1,000 times greater than mutation to 6-thioguanine resistance. Hypermutable cells were also mutable by 12-*O*-tetradecanoyl-phorbol-13-acetate (TPA), a nonmutagenic agent in normal cells. Extrapolating this phenomenon to the mouse and human skin cancer studies, one might hypothesize that exposure to ionizing radiation initiated a hypermutable state in the cells of the epidermis such that subsequent exposure to PUVA resulted in the initiation of a larger number of tumorigenic cells.

In the context of risk assessment generally, the concept of the induction of an indefinite hypermutable state by ionizing radiation (and presumably by at least some other agents) must have profound consequences which would exacerbate the already intractable problems with which we are faced.

If it should be possible to show that the synergism of ionizing radiation and PUVA in skin carcinogenesis is due to an effect of PUVA on tumor development, it would still not establish the existence of pseudopromotion by interference with immune processes. Recent findings point to a further credible hypothesis that PUVA might act as a classical promoter (in addition to its undoubted initiation ability). One of the major hypotheses concerning tumor promotion currently competing for attention is that the process is mediated by active oxygen

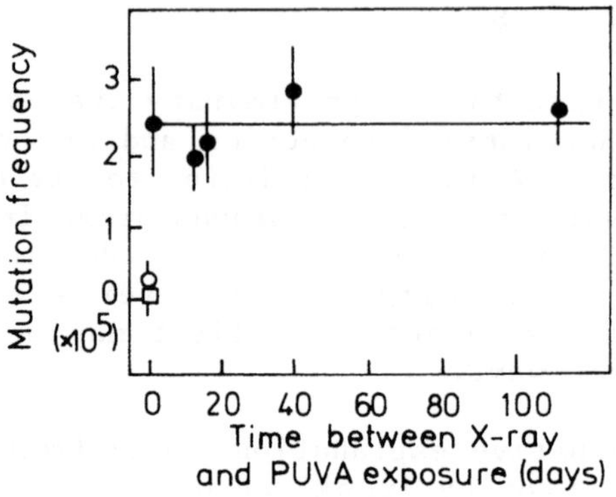

Fig. 2. Mutation frequency induced by PUVA (10^{-5} M 8-methoxypsoralen plus 216 J m^{-2} UVA) (●) as a function of time after 9 Gy X-radiation; PUVA alone (O); spontaneous mutation frequency in unexposed cells (□). Redrawn from Frank and Williams (1982).

species (Kensler et al., 1983; Marx, 1983). This hitherto neglected aspect of the photobiology of 8-methoxypsoralen is now becoming apparent. De Mol et al. (1981) first provided evidence in E.coli for the involvement of singlet oxygen in a component of PUVA mutagenesis; subsequently Wunder (University of Heidelberg, personal communication) implicated active oxygen species in the killing of lymphocytes by PUVA by showing a pronounced inhibitory effect of the radical scavenger cysteine. More recently Joshi and Pathak (1983) have shown that PUVA can generate both singlet oxygen and superoxide radicals.

EPILOGUE

To make defensible regulatory decisions in the area of risk assessment at low doses, one needs to have more than a mathematical model that fits the observable (higher) dose range for a single agent. One also needs to have a mechanistically credible understanding of the interaction of the agent with other agents to which humans may be exposed. The problems encountered in the interaction of ionizing radiation and PUVA should serve to caution us against relying on facile risk assessments that ignore these precepts. It is likely that empirical decisions based on pragmatic considerations may be the best that our science has to offer the regulator for some time to come.

ACKNOWLEDGEMENTS

I thank Antony Young for allowing me to discuss the preliminary data of our mouse skin work, and Gary Strauss who was involved in much of the work from my laboratory.

REFERENCES

Bridges, B.A., 1983, Psoralens and serendipity: aspects of the genetic toxicology of 8-methoxypsoralen, Environ. Mutagenesis, 5:329.

Bridges, B.A., and Strauss, G.H., 1980, Possible hazards of photochemotherapy for psoriasis, Nature, (London); 283:523.

De Mol, N.J., Beijersbergen van Henegouwen, G.M.J., Mohn, G.R., Glickman, B.W., and van Kleef, P.M., 1981, On the involvement of singlet oxygen in mutation induction by 8-methoxypsoralen and UVA irradiation in Escherichia coli K-12, Mutation Res., 82:23.

Frank, J.P., and Williams, J.R., 1982, X-ray induction of persistent hypersensitivity to mutation, Science., 216:307.

Hulse, E.V., Lewkowicz, S.J., Batchelor, A.L., and Papworth, D.G., 1983, Incidence of radiation-induced skin tumours in mice and variations with dose rate, Int. J. Radiat. Biol., 44:197.

Joshi, P.C., and Pathak, M.A., 1983, Production of singlet oxygen and superoxide radials by psoralens and their biological significance, Biochem. Biophys. Res. Commun., 112:638.

Kensler, T.W., Bush, D.M., and Kozumbo, W.J., 1983, Inhibition of tumor promotion by a biomimetic superoxide dismutase, Science, 221:75.

Kinlen, L.J., Sheil, A.G.R, Peto, J., and Doll, R., 1979, Collaborative United Kingdom/Australasian study of cancer in patients treated with immunosuppressive drugs, Brit. Med. J., 2:1461.

Kripke, M.L., 1982, Immunologic mechanisms in UV radiation carcinogenesis, Adv. Cancer Res., 34:69.

Kripke, M.L., Morison, W.L., and Parrish, J.A., 1981, Differences in the immunologic reactivity of mice treated with UVB or methoxypsoralen plus UVA radiation, J. Invest. Dermatol, 76:445.

Lynch, D.H., Gurish, M.F., and Daynes, R.A., 1981, Relationship between epidermal Langerhans cell density, ATPase activity and the induction of contact hypersensitivity, J. Immunol., 126:1892.

Marx, J.L., 1983, Do tumor promoters affect DNA after all? Science, 219:158.

Morison, W.L., Woehler, M.E., and Parrish, J.A., 1979, PUVA and systemic immunosuppression in guinea pigs, J. Invest. Dermatol., 72:273.

Morison, W.L., 1981, Photoimmunology, J. Invest. Dermatol., 77:71.

Reed, W.B., 1976, Treatment of psoriasis with oral psoralens and long-wave ultraviolet light, Acta Derm. Venereol. (Stockh), 56:315.

Reed, W.B., Sugarman, G.I., and Mathis, R.A., 1977, De Sanctis-Cacchione Syndrome. A case report with autopsy findings, Arch. Dermatol., 113:1561.

Spellman, C.W., 1979, Skin cancer after PUVA treatment for psoriasis, N. Engl. J. Med., 301:554.

Stern, R.S. Thibodeau, L.A., Kleinarman, R.A., Parrish, J.A., and Fitzpatrick, T.B., (1979) Risk of cutaneous carcinoma in patients treated with oral methoxypsoralen photochemotherapy for psoriasis, N. Engl. J. Med., 399:809.

Strauss. G.H., Bridges, B.A., Greaves, N., Hall-Smith, P., and Vella-Briffa, D., 1980, Inhibition of delayed cellular hypersensitivity reaction in skin (DNCB test) by 8-methoxypsoralen photochemotherapy, Lancet, 2:556.

Walder, B.K., Robertson, M.R., and Jeremy, D., 1971 (ii): Skin cancer and immunosuppression, Lancet, 1282.

Williams, J. R., and Frank, J.P., 1982, Induced hypermutability: a newly observed cellular effect of X-rays which may be important in carcinogenesis, Radiat. Res., 83:368.

DISCUSSION

Holtzman: I agree with your statement of the need for mechanistic understanding when dealing with interactions. There is an analogous situation to support the concept of a promotion mechanism in the interaction between two mutagens or carcinogens. Extensive work performed here at Brookhaven National Laboratory by Claire Shellabarger, Pat Stone, and myself on rat mammary carcinogenesis resulting from synergistic interactions of both physical and chemical agents with estrogens strongly indicates that to a large extent the estrogens act as indirect promoters.

Abrahamson: If all the data, beta rays (1000 rads) + PUVA are at this high dose, what do you think is happening in terms of synergism at low doses? How would you project the yield down to doses in the range of 100 mrad to 10 mrads?

Bridges: In most systems you cannot do experiments at those low doses with single agents. We studied synergism because it is a real problem; there are patients who may have had doses both of radiation and of PUVA in this range, so this is a "real-life" problem, just as "real-life" as the 1 rad area that one concentrates on in populations. Sufficient patients have had radiation therapy for psoriasis in the past to affect the epidemiology.

Sonneblick: There have been a number of successes in curing cancer using combinational therapy without very much basic knowledge. Would you agree that we do not have to wait until the ultimate mechanism is described?

Bridges: Yes. Cancer can be cured by "cocktails" and we do not know why they work. I just do not know how to begin to try and predict how agents are going to act synergistically at 1 rad and such low dose levels, except that I am sure that synergisms exist. The work that I have presented was attacking the problem of synergism at what are essentially massive dose levels and massive amounts of DNA damage. The problems are immense when you approach synergism at tiny doses.

Trosko: You referred to the PUVA phenomenon as indicating a "pseudo-promotion" effect. I bring to your attention the fact that several effective tumor promoters are known to modulate the immune response, such as TPA, phenobarbital and DDT.

Also, I believe an alternative explanation as to how PUVA might be acting as a promoter could be its cytotoxic effects. In other words, at low fluxes of PUVA, it acts as an initiator by its DNA-damaging/mutating activity. At high doses, some of the unrepaired DNA lesions, as well as other cellular damage, contribute to cell death. This scenario would allow surviving initiated stem cells to go into compensatory hyperplasia, expanding the "target size" of the initiated cell population. Further PUVA exposure on this expanded clone of initiated cell would enhance the probability of additional genetic hits needed for neoplastic conversion.

Bridges: In reply to your first point, yes, I have been told that. We used the word promoter in our original paper in Nature (Bridges and Strauss, 1980), and then Peter Brooke questioned this (personal communication) saying that we could not use the word promoter for something which is basically interfering with the immune system. He referred me to a paper by Isaac Berenblum where some of these terms were defined (Berenblum, 1979), and it seemed to me that a better word to use was pseudo-promoter. I am aware that TPA affects almost everything if you look hard enough, including the immune system.

With regrd to your second point, PUVA is a very damaging treatment. Long after the tan is gone you can recognize someone who has had PUVA treatment, for their skin has a thin, transparent appearance. There has been a lot of cell death, and certainly, Langerhan cells are known to have been killed off in PUVA therapy, and Langerhan cells are important in antigen presentation.

Fry: The skin is very efficient in "suppressing" the expression of initiated cells in exposure regimens involving the many exposures of PUVA or UVR that are required to produce tumors. It is possible that the effect of many of the later exposures are on the expression of tumors, rather than induction.

The expression of initiated cells in skin can probably be influenced in a number of different ways, just as Dr. Bridges has indicated. We have examined the UVR-wavelength dependence for "promoting" the cells initiated with PUVA. Exposure to 365 nm did not "promote" the cells, whereas exposure to 280-400 nm did. Since exposure to 320-400 nm had little effect, the effective wavelengths in the broad spectrum exposure must be the 280-320 nm which, of course, is also the tumorigenic spectrum of UVR. By using monochromatic light, it should be possible to dissect further the mechanisms involved. We have also studied the interaction of x-rays and PUVA. It appears that there is a x-ray dose-dependent interaction and, just as Dr. Bridges showed, the doses of x-ray required even to initiate, never mind produce skin tumors, is high.

The fact that high doses of ionizing radiation are required to induce skin tumors, even with subsequent exposure to "promoter", suggests that the number of people at risk will be small. In the interesting study of the tinea capitis patients by Shore et al. (1984), it is increasingly clear that UVR from sunlight is an important factor in the excess of skin cancers in patients exposed to x-rays over 20 years ago. Support for the idea that skin cancer involves an interaction between x-rays and UVR is that black patients treated with x-rays have not shown an excess of skin cancer.

Hei: Is there any information available on the effects of PUVA on the second component of the immune surveillance system, namely the antibody-producing B cell population? Secondly, was the animal you showed on the slide an athymic nude mouse or simply a normal hairless, or an animal with hair shaved off to show the skin lesions?

Bridges: We use a hairless type of mouse, SKH1, which is one of the standard mice. I think Dr. Fry knows more about that than I do. I cannot answer your question about B lymphocytes. I do not think that there is a major effect on antibody production by PUVA. Morrison et al. (1981) have found that in humans given PUVA, there are definite changes in the proportions of subsets of the lymphocyte population. We found too much individual-to-individual variability to distinguish such an effect. I think the techniques which now are becoming available will make this experiment much easier in the future.

Tice: Are you aware of any relationship between treatment with x-rays and subsequent PUVA treatment for psoriasis and tumor incidence?

Fry: May I reply to that? One of the reasons why it was thought that the PUVA was playing a part was that all of the tumors in the psoriasis patients were squamous cell carcinomas. This finding is rather unusual because you would expect both squamous cell carcinomas and basal cell carcinomas. It is not known why the PUVA treatment should cause a specific type of skin tumor.

Borg: It seems to me that the considerable capacity of skin to maintain the initiated state in an occult form has really been known to us for a long time. The original Berenblum experiments separated initiation from promotion by periods of up to weeks or months.

Because there is a discussion of "promotion" at this time and Dr. Trosko has already commented on the possible role of cell death and proliferation in the present context, I want to ask him a question. At meetings where the subject of carcinogenic promotion is a major topic, it is usually noted that although most promoters cause hyperplasia, and hyperplasia may be a necessary component of promotion, it seems clear that hyperplasia is an insufficient condition for promotion: not all agents causing hyperplasia are promoters. How do you reconcile this with the picture of promotion you described yesterday?

Trosko: Very briefly, I do not agree with that, because the few examples quoted have been challenged by pathologists. The critical point in promotion is sustained hyperplasia. For example, mezerein will induce hyperplasia, but is a poor tumor promoter of the skin. However, if you look at the skin after chronic exposure to mezerine, with time there is less and less hyperplasia.

Fry: I think that hyperplasia is a red herring. We cannot do quantitative experiments with UV or PUVA if we use doses high enough to get hyperplasia. You can get promotion without the hyperplasia. In fact, it is interesting that you only go to a hyperplastic state when the skin cannot accomodate the damage by shortening the cell cycle to balance the production and loss. If you keep the doses at a level which shorten the cell cycle, but not elicit hyperplasia, you will get promotion.

REFERENCES

Bridges, B. A., and Strauss, G. H., 1980, Possible hazards of photochemotherapy for psoriasis, Nature (London), 283:523.

Berenblum, I., 1979, Theoretical and practical aspects of the two-stage mechanism of carcinogenesis, in: "Carcinogenesis: A Comprehensive Study", Vol. 2, "Mechanisms of Tumor Promotion and Carcinogenesis", pp. 1-10, T. J. Slaga, A. Sivak, and R. K. Boutwell, eds., Raven Press, New York.

Morison, W. L., Parrish, J. A., Moscicki, R., and Bloch, K. J., 1981, Abnormal lymphocyte function following long-term PUVA therapy for psoriasis, J. Invest. Dermatol., 76:303.

Shore, R. E., Albert, R. E., Reed, M., Harley, N., and Pastenack, B. S., 1984, Skin cancer incidence among children irradiated for ringworm of the scalp, Radiat. Res., 100 (in press).

LOW-LEVEL EXPOSURES TO CHEMICAL CARCINOGENS:

MUTATIONAL END POINTS

William R. Lee

Department of Zoology
Louisiana State University
Baton Rouge, LA 70803

The choice of a mutational end point has an important bearing on both: (1) extrapolation from relatively high doses in laboratory animals to the much lower doses that are typical in environmental situations and (2) extrapolation from laboratory test systems to humans. Extrapolation from relatively high doses to the low doses typical in environmental situations is dependent on the shape of the dose-response curve. *In vivo* mutagenesis in eukaryotes usually consists of describing the relation of exposure of the whole organism to the mutation frequency. This conventional approach has the effect of placing the entire organism in a "black box" in which there are no intermediate determinants between exposure of the entire organism and observed mutation frequency in succeeding generations. To avoid combining physiological differences with mutagenic differences, I will distinguish between the dose to the target of interest - DNA - and exposure of a whole organism. In previous publications (Lee, 1976, 1978 and 1983) we have discussed the partitioning of the exposure to genetic response curve into an exposure-dose response curve and a dose-genetic response curve. The importance of this partitioning is to enable one to study physiology separately from the mutation process. The importance of separating physiological studies from genetic studies can be easily seen when experimental material is chosen for experimental models with inhalation, transport and physiology similar to man. It is readily observed that experimental materials suitable for determining the relation of exposure to dose are not suitable for genetic end points, whereas materials selected for experimental studies in genetics - cells, insects, and small mammals - do not have the appropriate physiological characteristics for extrapolation of the exposure-dose relation to man. Hence, the partitioning of exposure-dose studies from dose-genetic effect studies is necessary to the experimentalist.

The shape of the dose-response curve measuring dose to the target of interest and response to mutational end point has been shown to vary according to the choice of the mutational end point. Very early in the study of mutagenesis it was shown with X-rays that reciprocal translocations required two independent breaks in nonhomologous chromosomes and yielded a nonlinear dose-response curve, whereas sex-linked recessive lethals yielded a linear dose-response curve in repair-incompetent Drosophila sperm cells (Muller, 1940). For the chemical mutagen ethyl methane sulfonate (EMS) a linear relation was shown between ethyl adducts on DNA - dose - and sex-linked recessive lethals (Aaron and Lee, 1978). Reciprocal translocations and sex-linked recessive lethals were detected from the same group of EMS-treated Drosophila males (Vogel and Natarajan, 1979). Using the sex-linked recessive lethal test as the common denominator, one can compute from the Aaron and Lee (1978) data the dose as ethylations per nucleotide used in the reciprocal translocation experiments and plot the dose-response curve for both reciprocal translocations and sex-linked recessive lethals (Fig. 1). A linear model is fit to the data of Aaron and Lee (1978) and accounts for 99% of the variance. The best fit equation for the reciprocal translocation data is $m = D^2$ and accounts for 96% of the variance. Reciprocal translocations have also been determined in the mouse (Generoso et al., 1974) using exposures (mg/kg, ip injections) that are similar to those used by Sega et al. (1974) for molecular dosimetry determination. Therefore, a dose-response curve for reciprocal translocations in the mouse was prepared and found to be nonlinear and is best fit by a curve $m = D^2$ (Lee, 1978). The genetic test for determining the end point of reciprocal translocations between the mouse and Drosophila is very different. In the case of the Drosophila data, only reciprocal translocations between the second and third chromosomes were detected, but this detection is very objective depending upon the detection of missing classes of segregates. In the case of the mouse the initial data observed are a reduction in fertility due to aneuploidy which are then verified as reciprocal translocations from cytological observations. The mouse data potentially detect translocations between any two nonhomologous chromosomes in a genome of twenty pairs in contrast to the Drosophila data where only translocations between the second and third are recorded. Therefore, an absolute comparison between species is not feasible. The relative shape of the dose-response curve for reciprocal translocations in both species is similar and nonlinear.

In the Drosophila data an interval of 4 - 13 days between the end of treatment and fertilization of the egg was used (storage experiment) and is necessary in order to detect translocations (Vogel and Natarajan, 1979), whereas in the case of the mouse this interval of time was always a part of the experiment due to the longer life cycle.

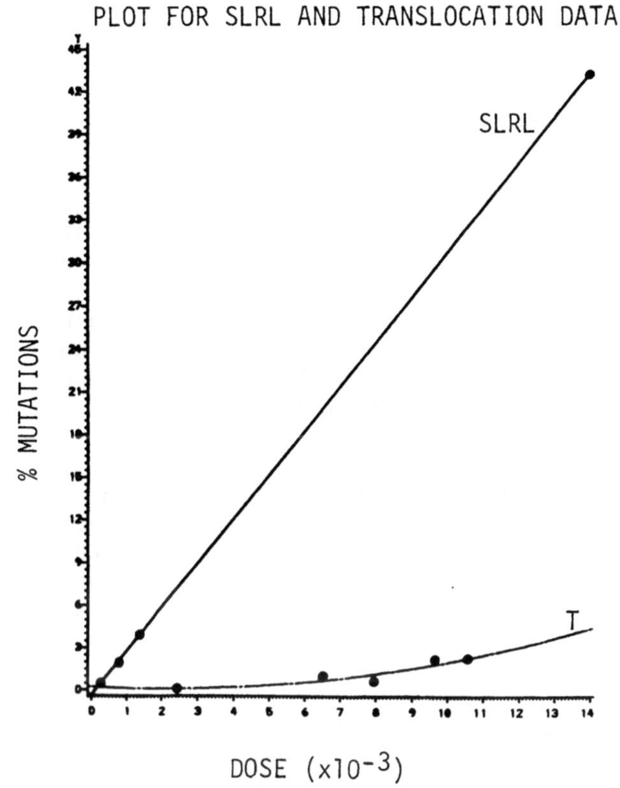

Fig. 1. Sex-linked recessive lethals, SLRL (Aaron and Lee, 1978) and reciprocal translocations, T (Vogel and Natarajan, 1979) plotted against ethylations per nucleotide.

It would be expected on the basis of theory that ethyl groups on one chromosome would not be able to interact with ethyl groups on nonhomologous chromosomes; therefore, the coincidence of breaks in two nonhomologous chromosomes should be a nonlinear relation of dose to reciprocal translocation.

The consequence of an exponential term of 2 or more in the dose-response relation of the ethylating agent EMS to reciprocal translocations is that at high doses the difference between the mutation frequency of the treated (2×10^{-2}) and spontaneous mutation frequency (10^{-4}) is easily detected, making the reciprocal translocation tests appear as an economical assay for the ethylating agent EMS. However, at low doses a significant difference between treated and spontaneous populations is difficult to establish. This leads to an anomaly in the data if one is using translocations as an assay for mutagenicity. For mutagenicity can readily and economically be detected for compounds that produce a high dose, whereas compounds that are either slow in their reaction with DNA or highly toxic cannot be used to produce a sufficiently high dose to be detected in a system that has such a high exponential dose-response curve. Therefore, it is not surprising that in the same test organism, Drosophila melanogaster, the Gene-Tox committees found that while all compounds that were positive in the reciprocal translocation tests were also positive in the recessive sex-linked tests, the converse was not true. There were a number of compounds positive in the sex-linked recessive lethal tests but not detected as mutagens in the reciprocal translocation test (Lee et al., 1983 and Valencia et al., in press). Any genetic mutational end point that yields a steep exponential dose-response curve should not be used in screening for mutagenicity nor interspecific comparisons. The assumption of proportionality required for risk assessment (Crow et al., 1982) would not be appropriate for data using translocations as a genetic end point. Furthermore, reciprocal translocations would not be expected to be important in the low dose levels that are typical of environmental situations. However, reciprocal translocations might be a significant monitor of acute accidental exposures, especially if acute high doses are achieved in humans as a result of an accident. The use of reciprocal translocations for monitoring has an advantage in that the spontaneous plus the background frequency of long-term chronic exposures should be very low in relation to the response to a high dose acute exposure.

The sex-linked recessive lethal test has been shown to be consistent with a linear dose-response curve for EMS (Aaron and Lee, 1978) and 1,2-dibromoethane (Lee, 1983) within the dose range in which it is feasible to conduct tests. However, the toxicity of compounds may severely restrict the upper limit of dose. The most useful information for extrapolation to low levels is at the low end of the dose-response curve where, unfortunately, the experimentalist is limited by the sensitivity of the genetic test system. The sex-linked recessive lethal test in Drosophila is an objective end point that permits scoring low frequencies of events with accuracy, and for this reason efforts have been made to measure the sex-linked recessive lethal frequency at only a fraction of the spontaneous mutation frequency (Abrahamson and Meyer, 1976). Unfortunately, at these very low levels of genetic response the change in spontaneous mutation frequency that occurs over time in the same laboratory limits the resolution of the system (Abrahamson and Meyer, 1976). When one considers the cost and the limitation of stability of the spontaneous frequency, estimates of mutation frequency at less than the doubling dose are not practical.

Consequently the experimentalist is left, in the case of toxic substances, with a relatively narrow range in which valid data for a dose-response curve can be obtained. As a result of this limitation in the range in which one can study the dose-response curve, it is not feasible to attempt to prove a mechanism of mutation by means of the dose-response curve. In this paper I suggest that the dose-response curve be used to test a limited number of models suggested from studies of the molecular mechanisms of mutagenesis. If, for example, one can show from a study of the molecular mechanism of mutagenesis that a large fraction of the mutations are the result of a combination of events (too far apart to be accounted for by a single adduct), then it is reasonable to propose that at low levels the response would be linear and that at higher levels there would be a departure in the direction of an exponent greater than one. If for a compound the range of exposures which are feasible to use for determining a dose-response curve yields a line whose slope extrapolates to the origin (after correction for the spontaneous frequency), then at least consistency with a linear model has been demonstrated. The rationale for accepting such an extrapolation of the induced frequency to the origin will have to be based on studies at the molecular level of mutation, and the dose-response curve is only used to show that the data range tested is not so high as to cause a significant contribution of the multi-hit mutational events.

The frequency of adducts per nucleotide of about one per thousand (Fig. 1.) is a frequency too low to give competition at the nucleotide level. Therefore, one expects a pseudo, first-order reaction on the sites of alkylation on DNA (Aaron, 1976) and consequently a constant distribution of adducts for each alkylating agent. Furthermore, the frequency of adducts per one thousand nucleotides would be beyond the range in which individual adducts could directly interact at the molecular level. However, one could anticipate interaction at the genetic level because, for example, a break initiated at one adduct and rejoined to another one several hundred or more nucleotides away could produce intragenic deletions with multi-hit kinetics. These intragenic deletions could not be distinguished from point mutations by the classical genetic test of complementation. New methods of analysis of mutant genes using the recombinant DNA technology now permit the analysis of mutations at the nucleotide level. We have organized these new methods of molecular analysis of mutant genes into a system that can be economically used with a locus previously sequenced. The significance of this analysis is that we can distinguish between changes affecting one or a few nucleotides from deletions of fifty or more nucleotides that should give alterations in restriction mapping. If one finds that a mutagen induces a large proportion of mutants that produce protein with near normal molecular weight, as in the case of EMS where 11 out of 13 mutants induced at the Adh locus produced a detectable protein (Hollocher and Place, personal communication), then the extrapolation of the line from the available data points to the origin seems justifiable (Fig. 1.). In contrast, we found (Kelley, et al., in press) that among the mutants induced at the Adh locus by Aaron (1979) using x-rays, only 2 of 31 produce a detectable protein. Of the 31 x-ray-induced mutants, at least 20 were deletions that included the Adh locus and one or more neighboring loci (Ashburner et al., 1982). We found that in 7 of Aaron's x-ray-induced mutants that complemented adjacent loci all showed normal restriction patterns for the promotor and Adh structural gene region. However, 4 of the 7 mutants showed alterations in restriction patterns outside the promotor and Adh structural gene region when probed with a 4.8kb genomic probe that has over a thousand base pairs on each side of the Adh locus (Kelley et al., in press). All 7 mutants produced a normal-sized, hybridizable mRNA when probed with the genomic clones, suggesting the mutant phenotype was not due to transcriptional inhibition or mRNA splicing. At high x-ray exposure it is expected that departures from linearity will be observed. Therefore, by properly characterizing mutations induced by an agent at the molecular level, a model may be proposed that can test adequately within the limitations of the dose range usable for a dose-response curve.

Extrapolation from laboratory data to human health effects requires, first of all, a look at the health effects data in humans. At this time, there is no clear set of data in which an individual mutagen can be identified as inducing genetic health effects in humans. The difficulty in looking for induced mutagenic health effects in humans is that only effects due to dominance are expressed in the F_1. Therefore, only a fraction of the induced mutations would be observed in the F_1 generation. Consequently, the risk estimates made in the BEIR III Report (National Research Council, 1980) for ionizing radiation and those that we would compute for chemical mutagens must be based on the ratio of induced mutations to spontaneous mutants with the health effects per induced mutation are equivalent to health effects per spontaneous mutant.

The question of how well induced mutations mimic spontaneous mutants at the molecular level can now be asked using current techniques of gene cloning. It has been dogma in textbooks that spontaneous mutants were the results of transitions. However, recent work with transposable elements suggests that either insertions or deletions may also play a significant role in spontaneous mutations. The data base is too limited at present to reach a conclusion except to say that clearly spontaneous mutants consist of a mixed group of mutagenic mechanisms. If, for example, the spectrum of spontaneous mutants is shown to mimic induced mutants by 50%, one is left to describe the fit as half-correct or half-wrong. The important point in risk estimation is that a 50% error is well within the limits that would be useful. A fit of only 50% in molecular mechanisms may actually be either better or worse in its estimation of health effects when one considers that the important attribute of a mutant gene on health effects is the dominance of the gene. Dominance is the most important characteristic of the mutant gene from the consideration of potential health effects to future generations because with a low frequency of mutant genes in the population only a small degree of dominance (Crow, et al., 1982) will cause the mutant to be eliminated as a heterozygote rather than a homozygote. The greater the degree of dominance, the shorter the time until equilibrium is reached in the frequency of the mutant genes in a population. If a mutagen is released at high concentrations to a limited population, it is the dominance of the mutant genes that will be important for health effects among the children and grandchildren of the exposed population, whereas a completely recessive gene will be diluted into the world population.

Methods are now available to characterize the mechanisms of mutation and it is imperative to use these methods in conjunction with classical, statistical methods to extrapolate from high to low-dose and from laboratory experiments to man.

REFERENCES

Aaron, C. S., 1976, Molecular dosimetry of chemical mutagens. Selection of appropriate target molecules for determining molecular dose to the germ line, Mutat. Res., 38: 303.

Aaron, C. S., 1979, X-ray induced mutations affects the level of enzyme alcohol dehydrogenase in Drosophia melanogaster: Frequency and genetic analysis of the null enzyme mutants, Mutat. Res., 63: 127.

Aaron, C. S., and Lee, W. R., 1978, Molecular dosimetry of the mutagen ethyl methanesulfonate in Drosophila melanogaster spermatozoa: Linear relation of DNA alkylation per sperm cell (dose) to sex-linked recessive lethals, Mutat. Res., 49: 27.

Abrahamson, S., and Meyer, H. U., 1976, Quadratic analysis for the induction of recessive lethal mutations in Drosophila oogonia by x-irradiation, in: "Biological and Environmental Effects of Low Level Radiation," pp. 9-17, IAEA-SM-202/100, Vienna.

Ashburner, M., Aaron, C. S., and Tsubota, S., 1982, The genetics of a small autosomal region of Drosophila melanogaster, including the structural gene for alcohol dehydrogenase. V. Characterization of x-ray-induced Adh null mutations, Genetics, 102: 421.

Crow, J. F., 1982, Chairman, Committee on Chemical Environmental Mutagens, Identifying and Estimating the Genetic Impact of Chemical Environmental Mutagens, National Academy Press, Washington, D.C.

Generoso, W. M., Russell, W. L., Huff, S. W., Stout, S. K., and Gosselee, D. G., 1974, Effects of dose on the induction of dominant-lethal mutations and heritable translocations with ethyl methanesulfonate in male mice, Genetics, 77: 741.

Kelley, M. R., Mims, I. P., Farnet, C. M., Dicharry, S. A., and Lee, W. R., Molecular analysis of x-ray induced alcohol dehydrogenase (Adh) null mutations in Drosophila melanogaster, Genetics (in press).

Lee, W. R., 1976, Molecular dosimetry of chemical mutagens: Determination of molecular dose to the germ line, Mutat. Res., 38: 311.

Lee, W. R., 1978, Dosimetry of chemical mutagens in eukaryote germ cells, in: "Chemical Mutagens," pp. 177-202, A. Hollaender and F. J. de Serres, eds., Plenum, New York.

Lee, W. R., 1983, Macromolecular dosimetry, in: "Developments in the Science and Practice of Toxicology," pp. 281-288, A. W. Hayes, R. C. Schnell, and T. S. Miya, eds., Elsevier, Amsterdam.

Lee, W. R. (Leader), Abrahamson, S., Valencia, R., von Halle, E. S., Wurgler, F. E., and Zimmering, S., 1983, The sex-linked recessive lethal test for mutagenesis in Drosophila melanogaster: A report of the U.S. Environmental Protection Agency Gene-Tox Program, Mutat. Res., 123: 183.

Muller, H. J., 1940, An analysis of the process of structural change in chromosomes of Drosophila, J. Genetics, 40: 1.

National Research Council, Committee on the Biological Effects of Ionizing Radiation, 1980, "The Effects on Populations of Exposure to Low Levels of Ionizing Radiation," National Academy of Sciences, Washington, D.C.

Sega, G. A., Cumming, R. B., and Walton, M. F., 1974, Dosimetry studies on the ethylation of mouse sperm DNA after in vivo exposure to (^{3}H)-ethyl methanesulfonate, Mutat. Res., 24: 317.

Valencia, R. (Leader), Abrahamson, S., Lee, W. R., von Halle, E. S., Woodruff, R. C., Wurgler, F. E., and Zimmering, S., Chromosome mutation tests for mutagenesis in Drosophila melanogaster: A report of the U.S. Environmental Protection Agency Gene-Tox Program. Submitted to Mutat. Res.

Vogel, E., and Natarajan, A. T., 1979, The relation between reaction kinetics and mutagenic action of mono-functional alkylating agents in higher eukaryotic systems. I. Recessive lethal mutations and translocations in Drosophila, Mutat. Res., 62: 51.

DISCUSSION

Abrahamson: Is it possible that the mutant that you just described with a lesion within Adh and one within the next 4000 bases may be part of an inversion incorporating both lesions that would be detectable by standard techniques, and, in fact, would not alter the RNA transcript?

Lee: If the particular breakpoints of an inversion consist of one in the Adh locus and one outside the locus, the mRNA should be altered and detectable by the Northern Blot technique. There could be no detectable difference in the restriction map if the distances between restriction site and breakpoints differed by less than 50 bases (unless a new restriction site was formed); breakpoints not of equal distance from the restriction site should produce differences in mobility of the bands associated with the Adh locus including both promoter and structural regions.

Abrahamson: If I understand correctly, you are saying that it could be an inversion with one breakpoint in the Adh locus?

Lee: Yes, if the distance on both sides of the restriction site was equal within the limits of resolution of about 50 bases. Otherwise an inversion with one breakpoint within the locus and one without should show a band migrating at a different rate. An inversion consisting of two breaks outside the locus would not alter the restriction map of bands associated with the locus.

Neel: One specific question. In 7 out of 23 of the induced mutations, you picked up messenger RNA and the remainder are presumably what you might call classic "nulls".

Lee: That is an interesting definition of a "null". I record a null when there is a lack of activity of the ADH enzyme.

Neel: The classical definition is no cross-reacting protein. I would certainly like to see a similar analysis of spontaneous mutations. You probably recall that more than half of Russell's spontaneous mutations in the mouse are homozygous lethals. We need to define further what the spontaneous picture is in order to get a better idea of how radiation-induced mutations depart from the spontaneous.

Lee: That is a point that I should have made. In order to make comparisons of the extrapolation from an experimental system to humans, we need to look at the spontaneous mutants, both in Drosophila and in humans. Often, in any extrapolation to human health effects, we are dealing with the spontaneous mutational health effects since these are the only extensive data available other than the Hiroshima/Nagasaki data.

Ehling: There is a discrepancy between the results of Oak Ridge and Neuherberg. In the control group in Neuherberg, only 9% of specific locus mutations are lethal in homozygous conditions. In experiments with mitomycin, procarbazine and ENU between 17-25% are lethal in the homozygous condition; in contrast, radiation-induced mutations were 75% lethal in homozygous conditions.

Lee: Let me ask a question here. When are we going to get clones of these mutants, and will we have their actual sequence, so that we can find out whether they are really deletions or not?

Ehling: We will do that with the cataracts, where it is much easier.

RADIOGENIC NEOPLASIA IN THE THYROID AND MAMMARY CLONOGENS: PROGRESS, PROBLEMS AND POSSIBILITIES*

Kelly H. Clifton, Ken Kamiya,
R. Timothy Mulcahy and Michael N. Gould

Department of Human Oncology
Wisconsin Clinical Cancer Center
University of Wisconsin Medical School
Madison, Wisconsin 53792

INTRODUCTION

It appears unlikely that the human epidemiological data available, or apt to become available, will allow distinction among the various mathematical models of radiation dose-neoplasia risk. Hence, risk estimates in the critical low dose and low dose-rate regions based on the human experience alone remain open to question.

An attractive approach to this problem is to develop experimental systems which allow study of the complex process of carcinogenesis from radiogenic neoplastic initiation through progression to overt disease at the cellular level *in vivo*. Ideally, in order to relate radiogenic initiation to biophysical theory, the size and survival fractions of the specific irradiated cell populations of interest should be knowable. And, importantly, every initiated cell should give rise to an overt neoplasm.

We have developed two experimental models, the rat mammary and thyroid epithelial clonogen transplant systems designed to approximate these ideals, and Jirtle et al. (1981) have developed a comparable rat hepatocyte system. All allow morphologic and

*This work supported by contract DE-AC02-84ER60195 from the U.S. Department of Energy, and by grant R01-CA13881 from the National Cancer Institute, DHHS.

physiologic investigation of normal clonogen biology (c.f. Clifton and Gould, 1984) and study of radiation dose-clonogen survival after exposure *in vivo*. All three clonogen types are capable of a tissue-environment dependent intracellular repair process (*in situ* repair, ISR) when left in their usual tissue surroundings *in vivo* for several hours after exposure.

The following discussion deals with early results of studies on the risk of neoplasia per irradiated clonogenic cell, and with problems which these results highlight. The major findings are: a) the high frequency of radiogenic initiation under conditions designed to maximize its expression in the form of overt neoplasia, and b) evidence of cell-cell interactions which alter the expression of initiation.

METHODOLOGY

The thyroid and mammary epithelial clonogen transplant assays have been described in detail (Clifton and Gould, 1984). Briefly, mammary or thyroid tissue is removed from donor rats, minced and enzymatically dispersed. Serial dilutions containing known numbers of monodispersed "morphologically intact" cells are prepared, and known cell numbers are inoculated into subcutaneous fat pads of syngeneic recipient animals. The recipients in most cases are hormonally manipulated to elevate prolactin or thyrotropin in mammary and thyroid experiments, respectively. The cells may be irradiated *in situ* in the donor rat before removal and dispersion, *in vitro* after suspension, or *in vivo* in the graft site in the recipient rat.

Under appropriate hormonal stimulation, clonogenic cells in the graft inocula proliferate to form morphologically and functionally normal mammary alveolar units (AU) or thyroid follicular units (FU). The presence of one or more AU or FU at the cell inoculation site is used as the end point in terminal dilution assays for clonogen survival. The fraction of sites with AU or FU is related to the number of cells inoculated according to the model of Porter et al. (1973):

$$\log M = \log K + S \log Z$$

where M is the average number of clonogens in inocula of a given suspension, Z is the average number of morphologically intact cells inoculated per site, S is a measure of slope and K is the clonogenic fraction. A computerized iterative method is used to estimate K and S. K and S in turn are used to calculate AD50 or FD50 values, i.e., the number of inoculated morphologically intact cells required to produce one or more AU or FU in 50% of the graft sites. It is of importance to note that the value of S in individual experiments rarely deviates significantly from 1.0 in accord with the clonogenic origin of AU and FU.

Clonogen survival is then calculated from the ratios of unirradiated to irradiated AD50 or FD50 values, and survival curve parameters are determined according to the multitarget-single hit relationship. Survival curve parameters for mammary and thyroid epithelial clonogens irradiated in vivo and with and without time allowed for completion of ISR are summarized in Table 1. For comparison, the hepatocyte data of Jirtle et al. (1984) are included.

CARCINOGENESIS RISK PER CLONOGEN

The designs of two experiments, one recently published and one still in progress, are summarized in Tables 2 and 3. Essentially the same methodology was employed in setting up these experiments as in the clonogen survival studies except that the inocula contained greater numbers of clonogens chosen on the basis of the survival information. In addition, in the mammary study, elevated prolactin levels were induced by implantation of a pituitary gland adjacent to a silastic capsule containing estrone in the spleen of the recipient rats (Clifton, Barnes and Haning, unpublished), and the animals were adrenalectomized to further potentiate the appearance of the mammary cancer (Clifton and Crowley, 1978). Recipient rats in the thyroid study were thyroidectomized and maintained on an iodine-deficient diet (Mulcahy et al., 1984).

The most striking finding in the first experiment is the high neoplastic initiation frequency as indicated by both the total tumor incidence and the carcinoma incidence at the lowest inoculated cell number. For example, observations were made on 29 sites, each of which had been grafted with 26 clonogenic cells which had received 5 Gy in vitro (Table 2). This experimental

Table 1. Radiation dose-clonogen survival parameters as assayed by clonogen transplant procedures (Clifton and Gould, 1984)

	X-ray survival			14 MeV neutron survival	
Tissue	D_o, rad	n(No ISR)	n(ISR)	D_o, rad	n
Mammary	130	5	17	97	1
Thyroid	200	3	10	--	1
Liver	280	1	2	170	1

Table 2. Thyroid neoplasms from known numbers of irradiated and unirradiated clonogenic cells (Mulcahy, Gould and Clifton, 1984).

Group[a]	Number grafted/site		Number	Number of sites[c]			Neoplasms/10^3 clonogens	
	Cells	Clonogens	of rats[b]	Total	Tumor	Carcinoma	Tumor	Carcinoma
C1	59,200[d]	411[e]	8	38	8	0	0.51	--
C2	28,600	205	10	47	9	2	0.93	0.21
C3	14,800	103	11	54	12	5	2.16	0.90
C4	7,400	52	12	57	11	2	3.71	0.67
C5	3,700	26	10	45	12	2	10.26	1.71
Totals or mean			51	241	52	11	1.49	0.31
E1	200,000[f]	411[g]	9	43	27	11	1.52	0.62
E2	100,000	205	12	59	29	12	2.40	0.99
E3	50,000	103	9	43	20	9	4.52	2.03
E4	25,000	52	7	31	14	5	8.68	3.10
E5	12,500	26	6	29	12	3	15.92	3.98
Totals or mean			43	205	102	40	2.79	1.09

[a]"C" indicates recipients of unirradiated thyroid cell grafts; "E" indicates recipients of thyroid cells which were exposed to 5 Gy x-rays in vitro before transplantation. Each rat received 5 grafts. All recipients were thyroidectomized and maintained throughout on an iodine-deficient diet.

[b]Number of rats autopsied.

[c]"Total" indicates number of graft sites examined; "tumor" includes both carcinomas and adenomas.

[d]Number of unirradiated morphologically intact thyroid cells grafted per site.

[e]Number of clonogens grafted per site assuming 1 clonogen per 144 morphologically intact cells.

[f]Number of irradiated morphologically intact cells grafted per site.

[g]Number of surviving irradiated clonogens assuming a survival fraction of 0.296 after 5 Gy in vitro.

group thus tested for initiation in a total of 26 x 29 or 754 clonogenic cells. Twelve tumors were observed of which 3 were carcinomas. The frequency of neoplasms was thus 16 tumors including 4 carcinomas per 10^3 clonogens tested.

The "spontaneous" frequency of tumors in the comparable groups which had received 26 unirradiated clonogens per graft site, was lower but significant. The total tumor incidence was 10 per 10^3 clonogens, and the carcinoma incidence was 1.7 per 10^3 clonogens (Table 2). The net incidence attributable to the 5 Gy radiation exposure was thus 6 tumors including 2.3 carcinomas per 10^3 clonogenic cells.

The high frequency of initiation per clonogenic cell reflected in these data has already been confirmed in the study of grafted mammary clonogens still in progress (Table 3). In group A

Table 3. Experimental design (in progress): Neoplasia in grafted mammary clonogens (Clifton and Gould, unpublished).

Group[a]	A	B	C
Number of rats:			
total[b]	150	144	145
alive 3/30/84	61	32	69
autopsied	89	112	76
Total cells/rat[c]	4×10^6	4×10^6	4×10^6
Irradiation:			
dose	7 Gy	7 Gy	0 Gy
day	-1	+31	--
Autopsied rats with:			
carcinoma	13	36	2
fibroadenoma	2	2	2

[a]Group A: Mammary cells were irradiated with 7 Gy ^{137}Cs gamma rays in vivo in the donor rat 1 day before transplantation; Group B: The graft site in the recipient rat was irradiated with 7 Gy 250 kVp x-rays 31 days after mammary cell grafts; Group C: Not irradiated.

[b]Initial number of graft recipient rats corrected for 4 rats lost to autopsy. All recipients were adrenalectomized and intrasplenically implanted with a pituitary and silastic capsule containing estrone.

[c]Morphologically intact monodispersed mammary cells per inoculum per rat.

of this experiment, the mammary donors were irradiated with 7 Gy ^{137}Cs gamma rays. The mammary glands were not removed for transplantation until 24 hours later to allow completion of ISR. The graft sites in groups B and C were transplanted with the same number of morphologically intact unirradiated mammary cells as group A. The graft sites of group B were irradiated with 7 Gy 31 days after transplantation; the grafted mammary cells of group C remained unirradiated (Table 3). As significant and different numbers of animals are still alive in the three groups, the analysis is presented primarily for illustration.

Carcinoma frequency thus far is highest in group B, and significantly higher in group A than in group C. The relative radiogenic cancer risk over all animals was 6 fold in group A and 18 fold in group B (Table 3). Expressed in terms of only those animals autopsied, the relative risks were somewhat decreased but are still manyfold in both irradiated groups.

Data from group B cannot be analyzed on an absolute risk per clonogen basis. By 31 days after grafting when the transplant sites of group B were irradiated, each clonogen would be expected to have given rise to a multi-cellular mammary structure. The fraction of cells in such structures which retain clonogenic potential is not known.

The cancer data from groups A and C have been analyzed in terms of the transplanted surviving clonogen numbers (Table 4). The cancer risk per irradiated group A clonogen was about 1 per 1000 cells; in contrast, the risk in group C was about 1 per hundred thousand clonogens. Thus, the relative risk of cancer in the irradiated cells of group A was 80-90 fold that in the unirradiated cells (Table 4).

At the time of the radiation exposure of group B, the graft sites of groups B and C would be expected to have the same unknown number of clonogens. Assuming a surviving clonogen fraction of 0.066 after irradiation of group B, a relative risk per surviving clonogen can be calculated. The relative risk per surviving clonogen in group B was 185-275 times that in group C (Table 4).

These relative mammary cancer risk figures should be interpreted with caution because of the high probability of intercellular interactions which may suppress the appearance of cancer more at the high surviving clonogen numbers in group C than the lower surviving cell numbers of groups A or B. Indeed, the second major finding in the thyroid carcinogenesis experiment is the nature of the relationship between tumor incidence and grafted clonogen number (Table 2). We had naively expected that the fraction of graft sites which developed neoplasms would increase with increasing numbers of grafted initiated cells in a relationship

Table 4. Mammary carcinoma risk per surviving clonogenic cell (see Table 3).

Group	A	B	C
Number of rats:			
total	150	144	145
autopsied	89	112	76
Number of clonogens/rat:[a]			
grafted	1.8×10^3	1.8×10^3	1.8×10^3
surviving	1.2×10^2	--	1.8×10^3
Total surviving clonogens:			
total	1.8×10^4	--	2.6×10^5
autopsied	1.1×10^4	--	1.4×10^5
Cancer:			
observed	13	36	2
per total clonogens	0.7×10^{-3}	--	0.8×10^{-5}
per autopsied clonogens	1.2×10^{-3}	--	1.4×10^{-5}
Relative cancer risk:			
per total clonogens	87.5	275[b]	1.0
per autopsied clonogens	80.0	185	1.0

[a]Assuming 1 clonogen per 2200 morphologically intact cells, and a surviving fraction of 0.066 after 7 Gy irradiation.
[b]Assuming at time of irradiation, groups B and C had the same unknown number of cells at risk per graft site, and that 0.066 of these survived 7 Gy irradiation in group B.

similar to the dependence of the fraction of FU-positive graft sites on the number of grafted cells in the survival studies. This did not occur. Over a 16-fold increase in irradiated clonogen number per graft site, the incidence of radiation-related neoplasms only increased 2-3 fold (Table 2). Furthermore, the incidence of "spontaneous" thyroid neoplasms did not increase significantly over the same clonogen dose range. Total neoplasm and carcinoma frequencies per clonogen markedly increased in both groups as grafted cell numbers decreased (Table 2). Clearly, the efficiency of the expression of initiation in overt neoplasia is suppressed at the higher inoculated cell doses, and the nature of this interaction requires further study.

THE CLONOGENS

An underlying hypothesis in the design and development of these techniques has been that the cells from which radiogenic neoplasms arise are members of the same population which is necessary for tissue repair and repopulation following irradiation or other insult. It follows that knowledge of the relative numbers and nature of these cells in normal tissues is critical to quantitative consideration of neoplasia.

As noted above, the AU and FU which arise from grafted mammary and thyroid clonogens are morphologically and physiologically normal structures (Clifton and Gould, 1984). But what of the frequency of such clonogenic cells in normal glands? And how do they respond to physiologic change?

The case of the thyroid is straightforward; a very high percentage of thyroid epithelial cells are clonogenic under appropriate hormonal stimulation. In our experiments, in which 4-5 week old male rats have been used as both thyroid cell donors and recipients, FD50 values have ranged between 50 and 100 morphologically intact cells in the thyroidectomized recipients, and the S values of the transplant equation (see above) were about 1.0 (Mulcahy et al., 1984). Assuming an FD50 of 100 cells per graft inoculum and an S value of 1.0, it can be calculated that on the average 1 clonogenic cell from an inoculum of 144 morphologically intact cells would survive at the graft site.

Studies with tritiated thymidine pre-labeled thyroid cell inocula show loss of about 90% of the label at the graft site by 5 days after grafting (Mulcahy et al., 1980). If this figure is reflective of cell loss from the graft site in general, then after inoculation of 144 morphologically intact cells on the average, 14 would remain at the graft site, one of which would be clonogenic. Now if some fraction, say 40%, of the morphologically intact cells in the inoculum were fibroblasts, then on the average the single clonogen is 1 of about 8 thyroid epithelial cells remaining at the graft site. Thus, thyroid clonogens are common.

In contrast, similar calculations from an AD50 of 1550 morphologically intact mammary cells from a young adult female rat (Table 5) indicates one transplantable clonogen per approximately 2200 morphologically intact mammary cells. Even with considerable cell loss after grafting, mammary clonogens are much less common than thyroid clonogens. The relative sparsity of mammary clonogens thus required that large numbers of cells be transplanted in the carcinogenesis study (Table 3). This beclouds interpretation of the data in that survival at the graft site may not be the same at low and very high grafted cell numbers, and cell-cell interactions are clearly cell-number dependent. For these reasons, and as a first step in further characterization of the cell population from which radiogenic cancer is presumed to arise, we are applying both biological and physical methods to enrich the clonogenic mammary cell concentration and to gain information on the effects of physiologic changes on their concentration.

The effect of a variety of physiologic conditions on the AD50 values of mammary cells grafted in young adult female recipients co-grafted with prolactin-secreting MtT are summarized in Table 5. As compared to cells from untreated young adult females, the AD50 value of cells from rats in which lactation had been induced by mammotropic-secreting pituitary tumor (MtT) grafts were markedly increased, i.e., prolactin induced proliferation and secretion markedly reduced the concentration of clonogens. Similarly, male mammary tissue contained a low clonogen concentration. Assay of mammary cells from multiparous non-lactating females yielded an AD50 value twice that of glands from young adult virgins, and glands of older virgins also have a somewhat lower clonogen concentration. A similar decrease in the clonogen concentration in older rats has been observed in thyroid cells (Watanabe et al., 1983). In contrast, the relative clonogen concentration in 14-day-old female glands is at least as great as in young adults, perhaps higher.

Ovariectomy had no detectable effect on the AD50 value, whereas adrenalectomy significantly reduced it. The adrenalectomy

Table 5. Influence of donor age and physiologic condition on AD50 values (Kamiya and Clifton, unpublished).

Sex	Age (Days)	Condition[a]	AD50 (95% c.l.)
F	50-55	virgin	1,552 (1,082-2,101)
F	36-41	virgin	1,331 (939-1,792)
F	50-55	MtT-lactating	15,720 (11,136-21,264)
F	>180	virgin	2,164 (1,519-2,876)
F	>180	multiparous	3,694 (2,572-4,948)
M	50-55	untreated	11,311 (7,931-15,712)
F	50-55	ovariectomized	1,792 (1,264-2,388)
F	14	untreated	879 (602-1,181)
F	14	untreated	1,360 (884-1,913)
F	50-55	adrenalectomized	677 (285-1,052)
F	50-55	adrenalectomized	730 (447-1,067)

[a]Surgical treatments and MtT grafts performed 14 days before grafting.

effect may be related to the glucocorticoid-induced inhibition of mammary differentiation which we previously observed (Clifton and Furth, 1960). In any event, it appears that some increase in the concentration of mammary clonogens may be achieved by donor adrenalectomy.

The age-related differences in relative mammary clonogen concentrations, if they occur in girls and women, may be related to the marked and progressive decrease in radiogenic breast cancer risk of Japanese women with increasing age at the time of exposure to the atomic bomb (Tokunaga et al., 1984). The further reduction in clonogen concentration in mammary glands of multiparous females is consistent with the well-known cancer preventive effect of early pregnancy in women (McMahon et al., 1973).

We plan to apply both physiologic and physical methods to further concentrate the mammary clonogens, and ultimately to characterize them morphologically and immunohistologically. Finally, clonogen concentration will greatly aid in mammary carcinogenesis studies aimed at defining LET and dose-rate effects at the cellular level.

INITIATION AND PROGRESSION: SOME SPECULATIONS

The high frequency of radiogenic initiation seen in these experiments suggests either an initiation target per clonogen so large as to be inconceivable, or a large number of small targets, hits in any one of which results in initiation (Gould, 1984). The comparably high frequency of initiation in polycyclic hydrocarbon-treated tracheal epithelium observed by Terzaghi et al. (1983), and the high transformation frequency reported by Kennedy et al. (1980) in irradiated 10T $^1/_2$ cells in culture show that our findings are not unique. These results indicate that radiogenic initiation, and probably initiation by other agents, is not the rare rate-limiting mutational event often presumed (Mole, 1984).

It is not necessary to conclude that all or any of the postulated multiple targets for radiogenic initiation are the same as those for initiation by other carcinogenic agents or conditions, nor that the sequence of events which follow initiation during progression to malignancy is in all cases similar. Given that the final product of carcinogenesis is a cell population which is heritably refractory to the multiple mechanisms which control cell number and differentiation, and that genetic alterations ranging from point mutation through altered gene regulation to incorporated foreign genetic material is characteristic of such cancer cells, what may be the factors common to initiation events in any of a large number of sensitive targets? We suggest that these data and considerations are most consistent with an old and unoriginal hypothesis that initiation is any intracellular event which results in instability in the genetic apparatus. The nature of the instability determines the range of types and likelihood of subsequent progressive changes. Accordingly, progression is viewed as due to clonal selection from among more neoplastic variants. What is on the cellular level within limits a random process, on the cell population or tissue level appears orderly.

It is nearly certain that initiation by polycyclic mutagenic hydrocarbons is the result of chemical interaction of the carcinogen with DNA and consequent gene alteration (Miller and Miller, 1974). In contrast, the high frequency in a spectrum of human and animal cancers of abnormalities of chromosome structure and/or number (Wake et al., 1981; Sasaki, 1982) suggests that initiation may also result in genetic instability at the chromosomal level. It is of interest to note that chromosomal abnormalities have been observed early in the progression of three types of tumors induced by hormonal imbalance in the absence of other known carcinogens: mouse Leydig cell tumors induced by estrogen (Hellstrom, 1961), estrogen-induced rat pituitary tumors (Waelbroeck-VanGaver, 1969), and thyroid tumors in rats resulting from chronic exposure to iodine deficiency (Beierwaltes and Al Saadi, 1968). In the latter case, cells of the hyperplastic and adenomatous areas which preceded cancer had a high frequency of diverse aneuploidies and translocations. In contrast, carcinomas were characterized by a high frequency of a single translocation and deletion.

In our experiments described above, we employed specific endocrine imbalances to potentiate the expression of radiogenic initiation, i.e. to hasten progression. As expected, neoplasms developed from unirradiated control cell grafts in some animals. Hormone imbalance-induced hyperplasias and adenomas characteristically regress following correction of the endocrine imbalance and frank carcinomas are a very late phenomenon (Clifton, 1959; Clifton and Sridharan, 1975). We suggest that in addition to hastening progression, endocrine imbalance imposes an environmental condition which causes the cell to act as if initiated. The genetic instability brought about in the hormonally driven cells leads slowly to the accumulation and selection of changes comparable to those which occur more rapidly during progression after initiation, but in the absence of initiation. The process is interrupted or slowed on correction of the hormonal imbalance.

Just as we speculate that the targets for radiogenic and hydrocarbon initiation may not all be the same in a cell of a given type, given the diversity of cell life cycles and of cellular control mechanisms, the spectrum and size of the target population initiable by a given carcinogen might differ from cell type to cell type, and hence partially govern the tissue specificity of carcinogen action.

Intercellular interactions play an important role in post-irradiation repair of potentially lethal damage *in vitro* and *in vivo*, Durand and Sutherland (1972) described a significant increase in the repair capacity of CHO cells irradiated in close contact in spheroids as opposed to monolayers in culture. As noted above, ISR has been described in mammary, thyroid and liver cells after exposure *in vivo* (c.f. Clifton and Gould, 1984). Evidence is also accumulating that cell-cell interactions play a role in the expression of initiation in overt cancer. Medina et al. (1978) and DeOme et al. (1978) reported that the presence of normal mammary cells in close association with mMTV-initiated mouse mammary nodule cells suppressed the growth of the latter. And we here observed suppression in tumor formation per initiated clonogenic thyroid cell as the number of cells per graft site was increased. Whether these findings reflect repair or reversal of initiation, selective lethality among initiated cells, or suppression of progression is under investigation.

CLOSING COMMENTS

The emerging picture of radiation carcinogenesis, and perhaps carcinogenesis by other agents, seems both good news and bad news. The bad news is that initiation is far more frequent than has usually been recognized, and that we know perhaps less about it than we believed. The good news is that normal cell populations are a good deal more capable of suppressing expression of initiation in overt cancer than has been recognized, though the mechanisms remain unclear. Finally, if progression is cancer-rate limiting, more time is available to alter the outcome of a presumably initiating exposure by prophylactic intervention.

As the intercellular interactions are clarified and quantitated, the clonogen transplantation-hormonal stimulation techniques will be employed to investigate and compare initiation by low and high LET radiations at low and high dose rates. For example, in principle, it should ultimately be possible to test biophysical hypotheses of radiation action at low dose rates by transplantation and observation of neoplasia from appropriate admixtures of high dose-rate irradiated and unirradiated cells.

REFERENCES

Beierwaltes, W.H., and Al Saadi, A.A., 1968, Sequential cytogenetic changes in development of metastatic thyroid carcinoma, *in*: "Thyroid Neoplasia," S. Young and D.R. Inman, eds., Academic Press, New York.

Clifton, K.H., 1959, Problems in experimental tumorigenesis of the pituitary gland, gonads, adrenal cortices and mammary glands: A review, *Cancer Res.*, 19: 2.

Clifton, K.H., and Crowley, J.J., 1978, Effects of radiation type and role of glucocorticoids, gonadectomy and thyroidectomy in mammary tumor induction in MtT-grafted rats, Cancer Res., 38:1507.

Clifton, K.H., and Furth, J., 1960, Ducto-alveolar growth in mammary glands of adreno-gonadectomized male rats bearing mammotropic pituitary tumors, Endocrinology, 66:893.

Clifton, K.H., and Gould, M.N., 1984, Clonogen transplantation assay of mammary and thyroid epithelial cells, in: "Clonal Regeneration Techniques," C.S. Potten, and J.H. Henry, eds., Edinburgh, Churchill Livingstone, in press.

Clifton, K.H., and Sridharan, B.N., 1975, Endocrine factors and tumor growth, in: "Cancer, a Comprehensive Treatise," p. 249, vol. 3, F.F. Becker, ed., Plenum Press, New York.

DeOme, K.B., Miyamoto, M.J., Osborn, R.C., Guzman, R.C., and Lum, K., 1978, Detection of inapparent nodule-transformed cells in the mammary gland tissues of virgin BALB/cfC3H mice, Cancer Res., 38:2103.

Durand, R., and Sutherland, R., 1972, Effects of intercellular contact on repair of radiation damage, Exp. Cell Res., 71:75.

Gould, M.N., 1984, Radiation initiation of carcinogenesis in vivo: A rare or common cellular event, in: "Radiation Carcinogenesis: Epidemiology and Biological Significance," p. 347, J.D. Boice and J.F. Fraumeni, eds., Raven Press, New York.

Hellstrom, K.E., 1961, Chromosomal studies on diethylstilbestrol-induced testicular tumors in mice, J. Natl. Cancer Inst., 26:707.

Jirtle, R.L., Michelopoulos, G., McLuin, J.R., and Crowley, J., 1981, Transplantation system for determining the clonogenic survival of parenchymal hepatocytes exposed to ionizing radiation, Cancer Res., 41:3512.

Jirtle, R.L., Michelopoulos, G., Strom, S.C., DeLuca, P.M., and Gould, M.N., 1984, The survival of parenchymal hepatocytes irradiated with low and high LET radiation, Br. J. Cancer, 49 (Suppl. VI):197.

Kennedy, A.K., Fox, M., Murphy, G., and Little, J.B., 1980, On the relation between x-ray exposure and malignant transformation in C3H 10T1/2 cells, Proc. Natl. Acad. Sci. USA, 77:7262.

McMahon, B., Cole, P., and Brown, J., 1973, Etiology of human breast cancer: A review, J. Natl. Cancer Inst., 50:21.

Medina, D., Shepherd, F., and Gropp, T., 1978, Enhancement of the tumorigenicity of preneoplastic mammary nodule lines by enzymatic dissociation, J. Natl. Cancer Inst., 60:1121.

Miller, E.C., and Miller, J.A., 1974, Biochemical mechanisms of chemical carcinogenesis, in: "The Molecular Biology of Cancer," p. 377, H. Busch, ed., Academic Press, New York.

Mole, R.H., 1984, Dose-response relationships, in: "Radiation Carcinogenesis: Epidemiology and Biological Significance," p. 403, J.D. Boice, and J.F. Fraumeni, eds., Raven Press, New York.

Mulcahy, R.T., Gould, M.N., and Clifton, K.H., 1984, Radiation initition of thyroid cancer: A common cellular event, Int. J. Radiat. Biol., 45: 419

Mulcahy, R.T., Rose, D.P., Mitchen, J.M., and Clifton, K.H., 1980, Hormonal effects on the quantitative transplantation of monodispersed rat thyroid cells, Endocrinology, 106:1769.

Porter, E.H., Hewitt, H.B., and Blake, E.R., 1973, The transplantation kinetics of tumor cells, Br. J. Cancer, 27:55.

Sasaki, M., 1982, Current status of cytogenetic studies in animal tumors with special reference to non-random chromosome changes, Cancer Genet. Cytogenet., 5:153.

Terzaghi, M., Klein-Szanto, A. and Nettesheim, P., 1983, Effect of the promoter 12-O-tetradecanoylphorbol-13-acetate on the evolution of carcinogen-altered cell populations in tracheas initiated with 7,12-dimethylbenz[a]anthracene, Cancer Res., 43:1461.

Tokunaga, M., Land, C.E., Yamamoto, T., Asano, M., Tokuoka, S., Ezaki, H., Nishimori, I., and Fujikura, T., 1984, Breast cancer among atomic bomb survivors, in: "Radiation Carcinogenesis: Epidemiology and Biological Significance," p. 45, J.C. Boice and J.F. Fraumeni, eds., Raven Press, New York.

Waelbroeck-Van Gaver, C., 1969, Tumerus hypophysaires induites par les aestrogens chez le rat. II. Etude cytogenetique, Eur. J. Cancer, 5:119.

Wake, N., Slocum, H.K., Rustum, Y.M., Matsui, S., and Sandberg, A.A., 1981, Chromosomes and causation of human cancer and leukemia, KLIV. A method for chromosome analysis of solid tumors, Cancer Genet. Cytogenet., 3:1.

Watanabe, H., Gould, M.N., Mahler, P.A., Mulcahy, R.T., and Clifton, K.H., 1983, The influence of donor and recipient age on the quantitative transplantation of monodispersed rat thyroid cells, Endocrinology, 112:172.

DISCUSSION

Mendelsohn: In the mammary gland, age or hormonal status may change the proportion of cells that are clonogenic by changing the number of fully differentiated cells. This might have no effect on the total number of clonogenic cells in the gland. Thus, I have difficulty seeing the relationship between AD(50) and cancer risk.

Clifton: We are, in fact, measuring the number of clonogenic cells, both as a fraction and as a total number. If you know the total number of cells required to produce a structure, and you know that the structure you are scoring can arise from a single clonogenic cell, then from this you get the number of clonogenic cells as a fraction of the total number of morphologically intact cells. From this you know the total number of clonogens that you had in your test tubes, but not the total number of clonogens in the gland. But one might derive this number by looking at the total number of cells one can get out of the gland. The values, in our experiments with the young male rat thyroid gland, for example, turns out to be on the order of $5x10^5$ cells.

Mendelsohn: Does the fraction of clonogenic cells vary with age, or after thyroidectomy?

Clifton: Oh yes. We have found a higher fraction of clonogenic cells in young animals. When you treat such young animals with a carcinogen, then the initiated clonogen population is likely amplified with age. Thus, you may have fewer total cells in the gland of a young animal, but if 25% of the cells are clonogenic, for example, as opposed to 5% in an older animal, then by the time the animal has five times as many cells, it presumably has five times as many initiated cells.

Sonneblick: Is the hormonal instability reversible?

Clifton: Yes, if you correct the hormonal imbalance, in most cases the process of progression subsides.

Sonneblick: Are we to assume that the genetic instability to which you referred is irreversible? Are the cells on their way to malignancy?

Clifton: That is a very good point. When I said the process is reversible, I should have said it becomes progressively less reversible as time goes on, so that you do finally have a cell that is no longer hormone-responsive and is fully malignant.

Sonneblick: Could you discuss further the stages between initiation and full malignancy with regard to hormonal status or other modulating factors?

Clifton: The process of hormonal carcinogenesis is frequently characterized by the production of tumors which themselves will regress on correction of the hormonal imbalance. But if you keep stimulating them, then you get carcinomas which do not. After regression of hormone dependent tumors, if you reintroduce the hormonal imbalance, the tumors come back very quickly. In other words, you have a cell with a "memory".

Sonneblick: It has been reported in the literature that there are only about 200 or so credible cases of spontaneous regression of human tumors. Could you give some reasons other than errors in the original diagnoses or misses in detection for this small number?

Clifton: I always think of the lady who was coming in for treatment at Wisconsin in her 22nd year with known metastatic breast cancer. She had premenopausal breast cancer that had been surgically removed. Metastases had been found at the time and she had spent 22 years chasing later metastases. I think it is very likely she had a very hormonally responsive tumor to begin with and that it was very slow growing after menopause.

Fry: Would it be worth developing an alternative method of increasing the thyrotrophin levels and would you expect any difference in the carcinoma incidence?

Clifton: Do you mean designing a method of raising thyrotropin levels other than by iodine deficiency? Unfortunately, the rat does not have thyrotropic tumors, and I do not like the use of modified uracil drugs to block T3 and T4 production because I am not all that confident that they themselves might not have some biochemical effects, other than the effect on the iodine concentrating system. So, we have used iodine deficiency as a means of maintaining chronic levels of TSH. Some of our data suggested that for optimum differentiation of unirradiated clonogenic cells into thryoid follicles some iodine is needed. We have therefore considered adding a suboptimal level of iodine back to the low iodine diet, or putting animals on an iodine/iodine deficiency cycle that might speed up the entire process of carcinogenesis. It may be that the thyroid hormone deficiency affects the proliferation of the cells.

Shellabarger: Is the hormone-imbalance of the host selected to optimize the stimulation of potential cancer cells?

Clifton: Yes, within the limits of our experience to date. Your group and ours, as well as others, have done a lot of work on hormone imbalance in whole animals that have been exposed to radiation. The hormone combinations that we have employed here are ones that highly potentiate the appearance of carcinomas following radiation exposure of the whole animal.

INDUCTION AND MANIFESTATION OF HEREDITARY CATARACTS

Udo H. Ehling

Institut für Genetik
Gesellschaft für Strahlen- und Umweltforschung
D-8042 Neuherberg, Federal Republic of Germany

INTRODUCTION

The lens is a transparent, highly refractive structure located between the pupillary portion of the iris and the vitreous. A cataract is an opacity of the lens causing a reduction of visual function. Ehling (1963) pointed out that morphologically comparable cataracts in mammalian species have very often the same mode of inheritance.

The direct comparability of the genetic endpoint of mice and man was one aspect which lead us to initiate a systematic investigation of the induction of dominant cataracts in the mouse (Ehling, 1983; Kratochvilova and Ehling, 1979). The other aspect was that Ehling (1976) developed a concept for the direct estimation of the risk of radiation-induced genetic damage to the human population expressed in the first generation, based on the induction of dominant mutations in mice. In addition, the dominant cataract mutation test was developed for the systematic comparison of the induced mutation rates to dominant alleles causing cataracts and to recessive visible alleles at specific loci in the mouse. The results are used to determine if factors equally influence the induced mutation rates to recessive and dominant alleles.

The lens is enclosed by a capsule to which the zonular fibers are attached. Blood vessels, lymphatics and nerves are not found in the lens, which continues to increase in volume, weight, and size throughout life. All cells are retained in the lens, the older fibers are found toward the nucleus and the younger ones toward the cortex. Thus at any age the internalized concentric layers of the lens reflect the process of fiber cell differentiation, maturation, and aging of the lens (Maisel et al., 1981). This unique development of the lens makes this organ an ideal object for biochemical and molecular investigations. The aim of these studies is the investigation of the nature of radiation or chemically induced dominant mutations resulting in cataractogenesis.

DOMINANT CATARACT MUTATIONS IN MICE

The dominant cataract test (Kratochvilova and Ehling, 1979) screens for mutations causing an opacity of the lens. To detect dominant cataract mutations, the F_1-offspring were examined biomicroscopically with the aid of a slit lamp for lens opacities at 4 - 6 weeks of age. Pupils were dilated with a 1% atropine solution and the lenses of mice observed at 42x magnification with a narrow-beam slit lamp illumination at a 20° - 30° angle from the direction of observation.

Presumed mutant individuals exhibiting a lens opacity were outcrossed to normal mice, and at least 20 offspring were examined to confirm the genetic nature of the cataract. When, among 20 offspring, no individual exhibited the phenotype, it was concluded that the lens opacity was not due to a dominant mutation with a penetrance value equal to or greater than 0.32 (Favor, 1982). For those F_1 individuals that produced offspring with the lens opacity phenotype, the presumed mutant was considered confirmed, and mutant lines were established to determine the penetrance and the expressivity of the gene.

In a series of papers the induction of radiation-induced dominant cataracts (Ehling et al., 1982; Kratochvilova, 1981; Kratochvilova and Ehling, 1979) and ethylnitrosourea-induced dominant cataracts (Ehling et al., 1982; Favor, 1983a,b) were reported.

SPECIFIC LOCUS MUTATIONS IN MICE

A specific locus test is conducted by mating treated mice homozygous wild type at a set of marker loci to untreated animals homozygous recessive at the marker loci. Resultant offspring are expected to be heterozygotes at the marker loci. In the event of a mutation at one of the marker loci in the treated wild type animal, the resulting offspring individual will express the recessive phenotype characteristic for the locus. The specific locus stock extensively used has the following markers: a, non-agouti; b, brown; c^{ch}, chinchilla; d, dilute; p, pink-eyed dilution; s, piebald; se, short ear. The coat color is affected by 6 markers and 1 marker affects the size of the external ear (se, short ear). The d and se loci are closely linked (recombination, 0.16%). A double d-se mutation may represent a deletion involving both loci.

Genetic changes that are detectable by the method include lesions both within the marker locus and external to it. Any alteration that leads to a change in the gene product resulting in an altered phenotype is scored by the specific locus method. Intermediate alleles, mutations that do not cause complete absence of the gene product, are very likely to be the result of intragenic changes. The most common grosser changes are small deficiencies involving the locus in question. The length of a deficiency that is compatible with viability of heterozygotes probably depends on the content of specific chromosome regions and may be as great as 7 centimorgans (Russell and Matter, 1980). The resulting mutant phenotypes are normally characteristic for the marker locus. Although this reduces the need for genetic confirmation of the presumed mutations, genetic tests of allelism are routinely done and combined with characterization of effects of the mutation on viability.

The specific locus method has been extensively used and results of radiation (Russell, 1972; Searle, 1975) and chemical (Ehling, 1978, 1981; Russell et al., 1981) mutagenicity experiments have been reviewed.

COMBINED EXPERIMENTS FOR THE DETECTION OF DOMINANT CATARACT AND SPECIFIC LOCUS MUTATIONS

The detection of dominant cataract mutations was combined with the scoring of specific locus mutations. The advantage of a combined investigation of dominant cataract mutations and specific locus mutations in mice is at least threefold:

1. The number of scorable mutations is increased in comparison with a simple specific locus experiment by a factor of four to five.

2. The combined investigation allows the comparison of the mutation frequency of selected and unselected loci.

3. In the same experiment the frequency of mutations with a dominant and a recessive mode of inheritance can be compared.

Groups of (101/ElxC3H/El)F_1 hybrid male mice, 11 weeks old, were exposed to 3 doses of ^{137}Cs γ-rays. In one experiment, a dose of 4.55 + 4.55 Gy (0.55 Gy/min) with a 24-h fractionation interval, and in the others, a single dose of 5.34 or 6.00 Gy (0.53 Gy/min) were used. During irradiation the males were placed in a lucite container and the anterior part of the body was shielded with lead. The control males were handled in the same way but sham irradiated.

Ethylnitrosourea (ENU), Serva, Heidelberg, was dissolved in 0.07 M phosphate buffer (pH 6.0) immediately before use. The mutagenicity of ENU was tested in hybrid male mice (101/ElxC3H/El)F_1, 10-12 weeks old. Single doses of 80 mg/kg, 160 mg/kg and 250 mg/kg of the test compound in a volume of 0.5 ml were injected intraperitoneally. All injections were completed within 40 min after the chemical was dissolved. The weights of the hybrid males ranged from 25 to 29 g. The amount of mutagen injected did not vary from the nominal value for the particular animal by more than 5%. The highest dose of ENU had no effect on the survival of the treated animals. The control males received an equal volume of the solvent.

Immediately after treatment, each male was caged with an untreated test-stock female, 10-13 weeks old. The offspring were counted, sexed and carefully examined externally at birth for any variant phenotype. The litters were examined again when cages were changed, the final examination for specific locus mutations being at 19-21 days of age. At that time the offspring were ledgered and ear punched. Specific locus mutations were confirmed by an allelism test.

After they had been scored for specific locus mutations, all F_1 offspring were examined biomicroscopically with the aid of a slit lamp for lens opacities. Pupils were dilated by treatment with atropine. Presumed mutant individuals exhibiting a lens opacity were outcrossed to strain-101 mice for genetic confirmation.

The results of the radiation experiments were discussed recently in detail (Ehling et al., 1982). A summary of the data is given in Table 1.

Table 1. Recessive and Dominant Mutations Induced in Mice by γ-Rays (Ehling et al., 1982)

Dose (Gy)	Dose rate (R/min)	Germ cell stage treated	Number of F_1 offspring	Number of mutations at 7 specific loci	Mutation rate per locus x 10^5	Number of dominant cataract mutations	Mutation rate per gamete x 10^5	Presumed mutation rate per locus x 10^5 (a)
0	-	-	103 218	6[b]	0.8	-	-	-
0	-	-	8 174	2	3.5	0	0	0
5.34	53		1 721	3	24.9	1	58.1	1.9
6.00	53	postsperma-togonia	865	3	49.5	1	115.6	3.9
4.55+4.55	55		272	2	105.0	1	367.6	12.3
5.34	53		10 212	7	9.8	3	29.4	1.0
6.00	53	spermatogonia	11 095	14	18.0	3	27.0	0.9
4.55+4.55	55		5 231	9[c]	24.6	6	114.7	3.8

[a]Based on the assumption that 30 loci can be scored.

[b]Untreated historical control for the laboratory.

[c]A simulataneous d-se mutation included, which is caused by double non-disjunction.

A total of 38 specific locus mutations was observed in 29 396 offspring. The radiation-induced frequency of specific locus mutations agrees well with the results observed in other laboratories (Russell, 1972; Searle, 1974). The induction of specific locus mutations served as a positive control for the induction of dominant cataract mutations. A total of 15 dominant cataract mutations was observed in the same number of offspring. The pooled frequency of dominant cataracts induced in postspermatogonia (P = 0.04) was significantly different from the observed control frequency. A comparison of the overall frequency of induced specific locus mutations and cataract mutations in postspermatogonia and spermatogonia after exposure showed that there were 2.5 times more recessive mutations induced by γ-radiation than dominant mutations. Taking into account that in humans, according to McKusick (1983), 23 dominant hereditary diseases are known which develop a cataract and approximately 10 additional cases, which are not so well established, then it is very likely that at least 30 loci coding for dominant cataracts are scored in this experiment. If 4 times more dominant loci are scored in these experiments than recessive loci and the ratio between induced specific locus mutations and dominant cataracts in spermatogonia is 2.5, then we can conclude that the per locus radiation-induced mutation rate is approximately 10 times greater for recessive specific locus mutations than for dominant cataract mutations.

The frequency of chemically induced recessive and dominant mutations is summarized in Table 2. For the induction of dominant cataracts a proportional increase is observed in the dose range of 80 and 160 mg/kg of ENU, with no subsequent increase of the observed mutation frequency for 250 mg/kg of ENU (Favor 1983a,b). The different shapes of the dose-effect curves for recessive and dominant mutations are not due to the distinct genetic endpoints scored, but to the difference in the mutation spectrum of ENU-induced specific locus mutations (Ehling, 1983). The d-locus mutates 2 times more often in the high dose group than in the low dose group. A still more pronounced difference exists for the p-locus. For the determination of the ratio of recessive to dominant mutations it is necessary to take into account the different shapes of the dose-effect curve for both genetic endpoints for the highest group. Therefore, the comparison has to rely on the dose points for 80 and 160 mg/kg. The ratio of the rates per locus for dominant cataract mutations to specific locus mutations is 1 : 7-8. A similar ratio was observed in the radiation experiments. The spontaneous ratio was likewise not used for the comparison because just one spontaneous dominant mutation was observed in the control group. For the comparison it is necessary to have a more reliable mutation rate.

Table 2. Ethylnitrosourea-Induced Recessive and Dominant Mutations in Spermatogonia of Mice (Favor 1983a,b)

Dose (mg/kg)	Number of F_1 offspring	Number of mutations at 7 specific loci [(a)]	Frequency per locus x 10^5	Number of F_1 offspring	Number of dominant cataract mutations	Mutation rate per gamete x 10^5	Presumed mutation rate per locus x 10^5 [(b)]
0	227 805	19 (13)	1.2	21 643	1	4.6	0.2
80	13 274	20	21.5	5 090	5	98.2	3.3
160	8 658	35 (32)	57.8	6 435	14	217.6	7.3
250	9 766	64 (57)	93.6	9 352	17	181.8	6.1

[a]Number of independent mutations in parenthesis.

[b]Based on the assumption that 30 loci can be scored.

ESTIMATION OF THE GENETIC RISK

In using the data from the mouse to arrive at quantitative estimates of the radiation genetic risks for humans, 3 general assumptions are made, according to Sankaranarayanan (1982), unless there is evidence to the contrary:

1. The amount of genetic damage induced by a given type of radiation under a given set of conditions is the same in the germ cells of humans and those of the test species which serves as a model.

2. The various biological and physical factors affect the magnitude of the damage in similar ways and to similar extents in the mouse and in humans.

3. At low doses and at low dose rates of low LET irradiation, there is a linear relationship between dose and frequency of genetic effects studied.

Using these assumptions and a concept developed by Ehling (1976) one can quantify the genetic damage expressed in the first generation (Ehling, 1984a). It should be emphasized that cataract mutations observed in the mouse can be directly compared with the manifestation of cataracts in man. Most of the dominant cataracts have a severe effect. Most of the human dominant cataracts are of clinical importance. The cataract data can be directly used for the estimation of human ill health. For the estimation of the genetic risk we make the following assumptions:

1. The dose-effect curve for the induction of dominant cataract mutations is linear.

2. Dominant cataract mutation rates are representative for all dominant mutations.

3. According to McKusick (1983), the number of well established dominant mutations in man is 934. The number of well established dominant mutations with a cataract in man is 23. The ratio of all dominant mutations to cataracts is the same in man and mouse. This ratio of 41 can be used to convert the induced mutation rate of dominant cataracts to estimate the overall dominant mutation rate.

The estimated effect of 1 Gy per generation of high intensity exposure of spermatogonia for 1 million liveborn individuals is summarized in Table 3.

Table 3. Estimated Effect of 1 Gy per Generation of High Intensity Exposure of Spermatogonia for 1 Million Liveborn Individuals

$0.45 - 0.55 \times 10^{-4}$	Cataract mutations/gamete/Gy
x 41	Multiplication factor for the overall dominant mutation rate
x 1.25 - 1.5	Correction factor for low penetrance
= 2 300 - 3 400	Expected cases of dominant diseases

Based on experiments by Kratochvilova (1981) one can conclude that the average penetrance for the induced mutations in heterozygote conditions is 80%. In order not to underestimate the frequency of expected cases with dominant mutations one has to multiply the frequency by 1.25. The correction factor for 1.25 is the lower limit, based on the observed average penetrance of 80%. The upper limit, due to the fact that we miss in these experiments some induced mutations with low penetrance can be only estimated. I assume it can be as high as 1.5. For acute exposure with high intensity radiation of spermatogonia of man with 1 Sv, one can expect 2 300 - 3 400 induced mutations of clinical importance in the first generation in 1 million liveborn individuals.

No experimental data are available for the induction of dominant cataract mutations by low intensity radiation in male mice or for the mutation rate in female mice. Therefore, an estimation of the population risk is only possible with reservations. Such an estimation can only be based on the generalization of results obtained with the specific locus method.

In general, for exposure with low intensity we expect only one-third the frequency of mutations as with high intensity exposure. Therefore, we multiply the expected cases by 0.3. In addition, the UNSCEAR Report used a factor of 0 - 0.44 for the induction in females. The UNSCEAR Report (United Nations, 1982) estimated the risk from the induction of mutations having dominant effects in the progeny to lie in the range of 1 000 - 2 000 cases of affected individuals per million born per Gy of low LET, low dose-rate exposure of males. The lower limit of expected cases is based on the cataract data, the upper limit on skeletal data. For the exposure of females the estimate of risk under similar conditions is 0 - 900 cases per million births.

One advantage of the direct estimation of the genetic risk is that the results based on the induction of dominant cataracts can be compared with the data based on dominant hereditary disorders of another system, for example, on the induction of dominant skeletal mutations. In a series of experiments Ehling (1966) and Selby (1982) established the radiation-induced rate of dominant mutations affecting the skeleton in offspring of exposed mice. Considering the difference between the quantification of the genetic risk based on skeletal and cataract mutations and the sample size of these experiments, the estimations are in good agreement.

However, in contrast to the determination of the multiplication factor for cataracts, it is difficult to determine the multiplication factor for the conversion of the mutation rate for dominant skeletal defects into the estimation of the overall dominant mutation rate. These difficulties are well documented by the UNSCEAR Report (United Nations, 1977) and the BEIR III Report (National Research Council, 1980). Another advantage of the dominant cataract method is that the observed variants are directly subjected to a genetic confirmation test. The essential differentiation between variants and mutants was discussed recently (Ehling, 1984b). In addition, until now we have no detailed information for the pentrance of the induced skeletal mutations. Therefore, the comparison of the estimation of the genetic risk based on the direct method with the observation obtained in Hiroshima and Nagasaki will be limited to the results of radiation-induced cataracts.

DIRECT ESTIMATION OF THE GENETIC EFFECTS OF THE ATOMIC BOMBS

According to the WHO Report "Effects of Nuclear War on Health and Health Services" (1984), the genetically significant population dose in Hiroshima and Nagasaki was based on the estimation that the mean parental dose for approximately 19 000 children with one or both parents irradiated within 2 500 m was

117 rem. In a calculation model the size of the genetically significant population exposed can be estimated on the basis of the number of 19 000 children born to parents of which one or both has been irradiated. On the assumption of two children per pair of parents, the genetically significant population dose in Hiroshima and Nagasaki becomes 19 000 x 117 x 0.5 = 1.1×10^6 man-rem, absorbed at high dose rate. The doses are at present under revision (Oftedal, 1984). The estimation of the number of dominant mutations expected in 19 000 born after parental exposure in Hiroshima and Nagasaki are summarized in Table 4.

Table 4. Estimation of Dominant Mutations Expected in 19 000 Offspring Born after Parental Exposure in Hiroshima and Nagasaki

$0.45 - 0.55 \times 10^{-6}$	Cataract mutations/gamete/R
$\times 1.1 \times 10^6$	man-rem (genetically significant population dose)
= 0.5 - 0.6	Expected cases of dominant cataracts
x 41	Multiplication factor for the overall dominant mutation rate
= 20 - 25	Total number of expected dominant mutations

For the estimation of the number of dominant cataract mutations in 19 000 born after parental exposure in Hiroshima and Nagasaki the mutation rate of $0.45 - 0.55 \times 10^{-6}$ cataract mutations/gamete/R has to be multiplied with the genetically significant population dose of 1.1×10^6 man-rem. The correction factor for low penetrance is not used for the comparison, because both sets of data were collected under similar conditions. According to the estimation of Table 4 one would expect less than 1 additional dominant cataract due to the fact that the parents were exposed in Hiroshima or Nagasaki. The expected overall frequency of dominant mutations in the first generation would be 20 - 25.

These figures can be compared with the data of Schull et al. (1981) on 4 indicators of genetic effects from studies of children born to survivors of atomic bombings of Hiroshima and Nagasaki. The indicators are frequency of untoward pregnancy outcomes (stillbirth, major congenital defect, death during first postnatal week); occurrence of death in live-born children, through an average life expectancy of 17 years; frequency of children with sex chromosome aneuploidy; and frequency of children with mutation resulting in an electrophoretic variant.

With these criteria only a fraction of the 25 expected induced dominant mutations could be recovered. Therefore, the conclusion of the authors, "in no instance is there a statistically significant effect of parental exposure; but for all indicators the observed effect is in the direction suggested by the hypothesis that genetic damage resulted from the exposure", is in good agreement with the direct estimation of genetic risk of the children of Hiroshima and Nagasaki based on mouse experiments. However, due to the uncertainties of the human data, I would not accept the other conclusion of the authors that for humans the doubling dose for the first three indicators is 156 rem. "This is some four times higher than the results from experimental studies on the mouse with comparable radiation sources."

The per locus ratio of radiation-induced dominant to recessive mutations induced in spermatogonia of the mouse is approximately 1 : 10 (Table 1). If this ratio is representative for all dominant and recessive mutations one can use this ratio for the estimation of induced recessive mutations. In this case one would expect not only 25 dominant mutations but also 250 recessive mutations induced in 19 000 offspring born to parents of which one or both were exposed in Hiroshima or Nagasaki, based on the assumption that the number of dominant and recessive loci are equal.

Similar to the quantification of the radiation-induced genetic damage of the first generation the data of Table 2 can be used for the estimation of the expected number of dominant mutations in the first generation after ENU-exposure.

It is until now not possible to consider the total burden of chemical mutagens. Therefore, it is necessary to establish standards for the exposure limits of specific chemical mutagens. This limit can be expressed as a fraction of an increase in the spontaneous mutation rate (10^{-5} or less) or as a number of accepted mutations. For example, for a single compound the allowable level of risk for all dominant mutations could be 10^{-6}. These figures are debatable and depend on the number of chemical mutagens a society will permit. Independent of the detail of these exposure limits for a chemical mutagen it is necessary to discuss these problems and find an acceptable and responsible solution.

MANIFESTATION OF HEREDITARY CATARACTS

A recessive cataract mutation in rabbits (kat) in the homozygous state disturbs the metabolism of the lens and permits the penetration of water into the lens. This absorption of water, depending on the state of hydration of the animal during the 2nd month of life, underlies the evolution to complete opacification. Two groups of rabbits with sutural cataracts, one receiving a diet poor in water to cause dehydration of the animals, the other receiving a diet rich in water were compared. Total cataract developed in 16% of the first as against 85% in the second group (Ehling, 1957). Investigations of hereditary cataracts in mice indicated that one initiating factor of caractogenesis, the hydration of the lens, is associated with an increase in the intralenticular concentration of Na^+ and a decrease in the concentration of K^+. Examples of osmotic cataracts with an elevated Na^+/K^+ ratio are the Nakano and the Philly mouse (Shinohara and Piatigorsky, 1980).

The Philly mouse, a derivative of the Swiss-Webster strain, develops an osmotic cataract during the fourth postnatal week. The Philly cataract progresses from an initial faint subcapsular opacity to a dense nuclear cataract in about 1 month. Crystallin synthesis is severely depressed in the fiber cells of the Philly cataract. This appears to be caused, at least in part, by ionic changes within the lens that interfere with the translation of crystallin messenger RNA's. Carper et al. (1982) observed that the messenger RNA for a β-crystallin polypeptide with a molecular size of 27 kilodaltons, first detected 5 to 10 days after birth in the normal mouse lens and the Nakano mouse cataract, was absent in the Philly mouse cataract. The heterozygous Philly lens had intermediate levels of the 27-kilodalton β-crystallin polypeptide and exhibited delayed onset of the cataract. The deficiency of functional 27-kilodalton β-crystallin messenger RNA is the earliest lesion reported yet for the Philly lens and points to a transcriptional or posttranscriptional developmental defect causing this hereditary cataract.

The mutants detected in the experiments with ionizing radiation and chemical mutagens open the possibility to study various metabolic processes on different levels (protein, DNA) during the formation of the cataract. The systematic investigation of recovered cataract mutations in mice will be an ideal source to study the genesis of hereditary cataracts. These studies could be a model for the elucidation of the manifestation of cataracts in man.

SUMMARY

A cataract is an opacity of the lens causing a reduction of visual function. The organogenesis of the lens in various mammals is similar. Therefore, a cataract mutation in mice can be directly compared with cataracts in man. Screening for dominant cataracts in mice was combined with the scoring of specific locus mutations. This combination increases the number of scorable loci, allows the comparison of unselected and selected loci and makes possible a systematic comparison of dominant and recessive mutations. In a combined experiment, dominant cataract and specific locus mutations were scored in the same offspring of mice after treatment of spermatogonia. In radiation experiments the induced frequency of dominant cataracts in spermatogonia was after single exposure 4.5 - 5.5 x 10^{-5} mutations/gamete/Gy and for specific locus mutations 1.6 - 2.8 x 10^{-5} mutations/locus/Gy. In experiments with ethylnitrosourea (ENU) the induced frequency of dominant cataracts was 0.7 - 1.3 x 10^{-5} (mutations/gamete)/(mg ENU/kg body weight) and for specific locus mutations 2.6 - 3.3 x 10^{-6} (mutations/locus)/(mg ENU/kg body weight).

The radiation-induced mutation rate can be used for the direct estimation of the genetic risk in humans. The genetically significant population dose for 19 000 offspring in Hiroshima and Nagasaki was estimated to be 1.1 x 10^{6} man-rem, absorbed at high dose rate. For the 19 000 offspring one would expect less than 1 radiation-induced dominant cataract and a total of 20 - 25 dominant mutations. If the number of dominant and recessive loci are equal one would expect in addition the induction of 250 recessive mutations in this population.

Chemically induced dominant cataract mutations could be used to determine the allowable level of exposure for a single compound.

The genetically characterized mutations will be an ideal source for studies in the field of developmental genetics. The investigation of α-, β-, and γ-crystallins by electrophoretic methods, the activity determinations of enzymes and the systematic use of cDNA hybridization may lead to an understanding of the genesis of dominant cataracts in mice and man.

ACKNOWLEDGMENT

Dominant cataract studies were supported by contract No. 305-81-BIO D, and the specific locus experiments by 136-77-1 ENV D of the Commision of the European Communities.

REFERENCES

Carper, D., Shinohara, T., Piatigorsky, J., and Kinoshita, J.H., 1982, Deficiency of functional messenger RNA for a developmentally regulated γ-crystallin polypeptide in a hereditary cataract, Science, 217:463.

Ehling, U.H., 1957, Untersuchungen zur kausalen Genese erblicher Katarakte beim Kaninchen, Z. menschl. Vererb.-Konstitutionsl., 34:77.

Ehling, U.H., 1963, Vererbung von Augenleiden im Tierreich, in: "Bericht über die 65. Zusammenkunft der Deutschen Ophthalmologischen Gesellschaft in Heidelberg," pp. 228-238, Bergmann-Verlag, München.

Ehling, U.H., 1966, Dominant mutations affecting the skeleton in offspring of X-irradiated male mice, Genetics, 54:1381.

Ehling, U.H., 1976, Die Gefährdung der menschlichen Erbanlagen im technischen Zeitalter, Fortschr. Geb. Röntgenstr., 124:166.

Ehling, U.H., 1978, Specific locus mutations in mice, in: "Chemical Mutagens," Vol. 5, pp. 233-256, A. Hollaender and F.J. de Serres, eds., Plenum Publishing Corporation, New York-London.

Ehling, U.H., 1981, Mutagenicity of selected chemicals in induction of specific locus mutations in mice, in: "Comparative Chemical Mutagenesis," pp. 729-742, F.J. de Serres and M.D. Shelby, eds., Plenum Publishing Corporation, New York-London.

Ehling, U.H., 1983, Cataracts - Indicators for dominant mutations in mice and man, in: "Utilization of Mammalian Specific Locus Studies in Hazard Evaluation and Estimation of Genetic Risk," pp. 169-190, F.J. de Serres and W. Sheridan, eds., Plenum Publishing Corporation, New York-London.

Ehling, U.H., 1984a, Methods to estimate the genetic risk, in: "Mutations in Man," pp. 292-318, G. Obe, ed., Springer-Verlag, Berlin-Heidelberg.

Ehling, U.H., 1984b, Variants and mutants, Mutat. Res., (in press).

Ehling, U.H., Favor, J., Kratochvilova, J., and Neuhäuser-Klaus, A., 1982, Dominant cataract mutations and specific locus mutations in mice induced by radiation or ethylnitrosourea, Mutat. Res., 92:181.

Favor, J., 1982, The penetrance value tested of a presumed dominant mutation heterozygote in a genetic confirmation test for a given number of offspring observed, Mutat. Res., 92:192.

Favor, J., 1983a, A comparison of the dominant cataract and recessive specific locus mutation rates induced by treatment of male mice with ethylnitrosourea, Mutat. Res., 110:367.

Favor, J., 1983b, Studies on ethylnitrosourea-induced dominant cataract mutations in mice, in: "Symposium on Mutagenesis: Basic and Applied," pp. 11-12, L.N. Mithila University Press, Kameshwaranagar, Darbhanga, India, December 22-23, 1983.

Kratochvilova, J., 1981, Dominant cataract mutations detected in offspring of gamma-irradiated male mice, J. Hered., 72:302.

Kratochvilova, J., and Ehling, U.H., 1979, Dominant cataract mutations induced by γ-irradiation of male mice, Mutat. Res., 63:221.

Maisel, H., Harding, C.V., Alcalá, J.R., Kuszak, J., and Bradley, R., 1981, The morphology of the lens, in: "Molecular and Cellular Biology of the Eye Lens," pp. 49-84, H. Bloemendal, ed., John Wiley & Sons, New York.

McKusick, V.A., 1983, "Mendelian Inheritance in Man," The Johns Hopkins University Press, Sixth Edition, Baltimore-London.

National Research Council, Committee on the Biological Effects of Ionizing Radiations, 1980, "The Effects on Populations of Exposure to Low Levels of Ionizing Radiation," National Academy of Sciences, Washington, D.C.

Oftedal, P., 1984, Genetic damage following the nuclear war, in: "Effects of Nuclear War on Health and Health Services," pp. 163-174, World Health Organization, Geneva, Switzerland.

Russell, L.B., and Matter, B.E., 1980, Whole-mammal mutagenicity tests: Evaluation of five methods, Mutat. Res., 75:279.

Russell, L.B., Selby, P.B., Halle, E.v., Sheridan, W., and Valcovic, L., 1981, The mouse specific locus test with agents other than radiations. Interpretation of data and recommendations for future work, Mutat. Res., 86:329.

Russell, W.L., 1972, The genetic effects of radiation, in: "Peaceful Uses of the Atomic Energy," Vol. 13, pp. 487-500, United Nations, New York, International Atomic Energy Agency, Vienna.

Sankaranarayanan, K., 1982, "Genetic Effects of Ionizing Radiation in Multicellular Eukaryotes and the Assessment of Genetic Radiation Hazards in Man," Elsevier Biomedical Press, Amsterdam.

Schull, W.J., Otake, M., and Neel, J.V., 1981, Genetic effects of the atomic bombs: a reappraisal, Science, 213:1220 (Quotation p. 1220).

Searle, A.G., 1974, Mutation induction in mice, in: "Advances in Radiation Biology," Vol. 4, pp. 131-207, J.T. Lett, H. Adler, and M. Zelle, eds., Academic Press, New York-London.

Searle, A.G., 1975, The specific locus test in the mouse, Mutat. Res., 31:277.

Selby, P.B., 1982, Induced mutations in mice and genetic risk assessment in humans, in: "Progress in Mutation Research," Vol. 3, pp. 275-288, K.C. Bora et al., eds., Elsevier Biomedical Press, Amsterdam.

Shinohara, T., and Piatigorsky, J., 1980, Persistence of crystallin messenger RNA's with reduced translation in hereditary cataracts in mice, Science, 210:914.

United Nations Scientific Committee on the Effects of Atomic Radiation, 1977, "Ionizing Radiation: Sources and Biological Effects," United Nations Publication A/32/40, United Nations, New York.

United Nations Scientific Committee on the Effects of Atomic Radiation, 1982, "Ionizing Radiation: Sources and Biological Effects," United Nations Publication A/36/49, United Nations, New York.

World Health Organization, 1984, "Effect of Nuclear War on Health and Health Services," Report of the International Committee of Experts in Medical Sciences and Public Health to Implement Resolution WHA 34.38, World Health Organization, Geneva, Switzerland.

DISCUSSION

Abrahamson: With respect to your chemical cataract mutations, have you tested them from F_1 to F_2 and demonstrated transmission? The reason I ask this, is the word drifts around the Society of Geneticists that many of these things are not reappearing in the second generation and are not showing up as transmitted mutations. I would like to get this point on record.

Ehling: You many soon see in Mutation Research, a short letter about "Variants and Mutants" addressing this point. In our laboratory it is a law that we call something a variant if it is just a deviation from the normal phenotype, and only call it a mutant if we have proof that it is transmitted. At least 20-30 offspring resulting from an outcross of the original suspected mutant are examined to determine if we have a dominant mutation (Ehling et al., 1982). There was, of course, a problem with skeletal mutations, which are more difficult to assess since we have to breed the F_2 generation and then sacrifice and examine the F_1 to decide whether we have a mutation or not. If the F_1 is normal, we kill the offspring; if we have a presumed mutation, we check it in the F_2. This is one reason why I rely more on the cataract data. There is another problem that some inbred strains of mice have a very high spontaneous background of cataracts. You need to work with inbred strains or outbred strains where the spontaneous background is low.

Abrahamson: With respect to the Environmental Academy Report on chemical mutagens, we have no problem when you have positive results with the mouse cataract assays, but questions arise when there is a negative. The committee, in fact, recommended that the cataract and skeletal test be used not only for the mouse, but be tested in other mammalian species.

Ehling: In a textbook of human genetics (Neel and Schull, 1958), the authors make the excellent point that you cannot compare radiation-induced mutations in mice or other organisms, with chemically induced mutations because the gonads are very well buffered. In our first experiment at Oak Ridge with a super mutagen, MNNG, we had negative results. We now get mutations with other compounds and there is no problem anymore. MNNG is very active at injection site, but it does not reach the gonads in an active form. You may have overwhelming evidence of its mutagenicity in yeast, but the nature of the mammalian system is such that the gonads are very well protected. I think a very essential point is that until now there has been relatively little evaluation of the effects in the germinal system of the mouse of any substance mutagenic in other systems, although there has been excellent evaluation for carcinogenicity. Until we have this information, how can we evaluate the negative findings? Perhaps for a long time you could say "forget the mouse tests", but when you have seen the figures, for example for ENU, which is very mutagenic in our test, then I think we also have to accept the findings from a whole series of negative experiments where we have found no mutation. I think there are many reasons for these results, such as the blood/testis barrier. I recognize how expensive the mouse germinal-cell mutation test is and therefore we are trying to develop the so-called parallelogram approach (Sobel, 1984). As an alternative, for instance, I advocate the possibility of using the doubling dose for different end points.

Neel: I will reserve most of my comments for this afternoon, but there are two points worth making now. The first is that from the beginning, the expectation of a demonstrable genetic effect from the Hiroshima-Nagasaki findings was very low. In view of the magnitude and cost of the studies, I asked the National Academy to convene a committee that would confirm my judgment. A note in Science (Genetics Conference, 1947) made it clear that from everything we then knew of dose, the returns would be small, but we could not hold back from undertaking the study. My second point is a familiar complaint. You have made some very broad generalizations, and once again I would plead that you put limits on these very global extrapolations.

Ehling: I think it is extremely important to have the data of Hiroshima-Nagasaki because of one believes in extrapolation then in this respect the data fit very well with that we would expect. Your second point is very debatable. I have used a range (Ehling, 1976) also the BEIR-Report used a range. But one question is, what incidence are we using for the doubling dose, for instance? This value will affect our estimates. Stevenson (1959) gives a figure of 10,000 recessive and 10,000 dominant mutations; other estimates show that we have less than 1000 dominant genes. I think it is important to state clearly at the start the assumptions that are made in the extrapolation. For instance, the BEIR-Report assumes males and females are equal in sensitivity. However, they then used the mutation rate for the male and zero mutation rate for the female. So basically they assumed no induction of mutations in female mice. Now this has been changed. We should be aware of the many assumptions we have to make simply because there are no data available. If one gives a range of mutation rates, this implies a security which no one should attach to these numbers; however, if the order of magnitude is correct, I think we have achieved something.

Lee: How was the number of cataract loci determined?

Ehling: They are based upon data from V.A. McKusick, "Mendelian Inheritance in Man", 6th ed., The Johns Hopkins University Press, Baltimore, 1983.

Lee: I am referring to your mouse data.

Ehling: I have assumed that the number of loci in the mouse is similar to that in humans; there are 25,000 genes in the mouse and 30,000 genes for humans, a factor of 1.2. You can argue about these, but the important point is that we are using the ratio, and if I consider the number of cataracts we found in these experiments, and the number of cataracts reported in the literature, I am sure that we have at least 30 different dominant cataracts.

Lee: Have you done genetic tests for allelism?

Ehling: Yes! We will have to localize the cataract mutation, and we are presently doing such tests.

Lee: You also mentioned that some of these cataract mutants have been cloned. Which ones were cloned?

Ehling: The work I am referring to was done in the National Institutes of Health.

Neel: In response to the question concerning humans, the classic test for allelism has not been carried out. The differentiation between mutants is largely based on the detailed morphology of the cataracts. So we are making an assumption which is not necessarily correct. This point, of course, brings up the question about the distinct morphological types that occurred in your mouse experiments, because in order to make this type of comparison with human data, we do need to know whether you have distinct morphological types that are breeding true.

Ehling: I will answer with a quotation in German; Vogt et al. (1940) has written "Es gehort zu den groβten Wundern des Erbgeschehens, mit welcher Exaktheit innerhaib des klaren Liniengewebes die angeborenen Erbleiden lokalisiert und geformt sind und sich gegen die klare Substanz abrenzen... Es gehort zum Erstaunlichsten in der Erbpathologie, daβ auf einem scheinbar so kleinen Gebiete wie der Linse das Keimplasma mit einer Genauigkeit differenziert, die an chemische Reaktionen erinnert." Cataracts are really very beautiful. We have described the different types in Journal of Heredity (Kratochivola, 1981); time did not allow me to discuss these. Where we have tested, we have shown that each distinct type represents a separate mutation.

Evans: A number of groups of researchers have now cloned some of the crystallin mutants. I think it is very important that you actually have a look at your mutants using probes to see precisely what you have got.

Ehling: We have been in contact with NIH and we have obtained their clones; we are just initiating a program in my laboratory because I think that there is a possibility to determine the nature of the type of mutations we have induced.

Evans: Let me ask you another question rather divorced from this, and look at the seven specific locus experiments. Since the original work done by Russell at Oak Ridge (1965) and Lyon (1974) how much more information have we on other recessive loci in the mouse, in terms of dose-response? What are the shapes of the dose-response curves in many of these recessive tests. Are they dose-squared rather than linear?

Ehling: This is a long-standing controversy. First, there was a set of six loci tested in Harwell. When we initiated our program to compare the mouse and hamster, we tested five of the specific loci for inbred strains BALB/c and four loci for strain DBA/2. Knowing the literature on mutation rate in yeast and *Neurospora* we expected tremendous differences. The fact is that the mutation rate is basically equal, independent of the genetic background. I do not think that the radiation-induced mutation rate, the data which we have known for 20 years, will answer that question. For chemical mutagens, the situation is different. For procarbazine we have a strictly linear dose-effect relationship. For MMS-induced mutations at the postspermatogonial stages, we have an indication of a linear-quadratic dose-effect curve. It is almost impossible to prove this, because we get data at 20 mg/kg, 40 mg/kg, and then, of course, when we go to 80 mg/kg there we have only dominant lethal mutations. But, my point is that perhaps the shape of the curve is not so essential. The spontaneous dominant lethal mutation rate in our conventional animal assays was 14%, and now in our new facilities it is 9%. For procarbazine and ENU 25% of the specific locus mutations are homozygous lethals, for MMS it is 75%, and 66% for radiation-induced mutations. There are biological reasons suggesting that you can induce intragenic or point mutations with radiation, but there is a sizeable fraction that are probably due to two hits.

Evans: If you take the radiation experiments and look for double mutations, I think they turn up fairly frequently. What happens with the chemicals?

Ehling: After exposure of spermatogonia to gamma rays we observed 10% double mutants (d-se)(Ehling et al., 1982). We have recovered close to 30 mutants induced by chemicals and found no double mutants. There is one other point, Lyon *et al.*, (1972) have given the mutation spectrum for high dose-rate and for low dose-rate; for low dose-rate the frequency is not as high, exactly as you expect. I cannot say now exactly what the frequency of specific locus mutations was, nor their effects in the condition following low dose-rate treatment, but it was substantially lower than the 2/3 we have obtained at high dose-rate.

Abrahamson: Are you talking about the female?

Ehling: No, that is the male (Lyon et al., 1972).

Sonneblick: When you had zero irradiation in your mice, did you get any cataracts?

Ehling: We have had one spontaneous cataract in over 22,000 offspring.

Andrew: Do you have information on whether all of the dominant cataract mutations you observed are mutations of the genes coding for the crystallins themselves, or whether some are other determinants of lens structure or function?

Ehling: I would be glad if I could give you an answer, but I cannot yet say anything about that.

REFERENCES

Ehling, U.H., 1976, Die Gefahrdurg der menschlichen Erbanlagen in technischen Zeitalter, Fortschr. Rontgenstr., 124:166.

Ehling, U.H., and Neuhauser-Klaus, A., 1972, Procarbazine-induced specific locus mutations in male mice, Mut. Res., 15:185.

Ehling, U.H., Favor, J., Kratochvilova, J., and Neuhauser-Klaus, A., 1982, Dominant cataract mutations and specific-locus mutations in mice induced by radiation or ethylnitrosourea, Mut. Res., 92:181.

Genetics Conference, 1947, Genetic effects of the atomic bombs in Hiroshima and Nagasaki, Science, 106:331.

Kratochvilova, J., 1981, Dominant cataract mutations detected in offspring of gamma-irradiated male mice, Journ. Hered., 72:302.

Lyon, M.F., 1974, Mutation induction in mice, in Advances in Radiation Biology, 4:pp 131-207, Lett, G.T., Adler, H.I., and Zelle, M., eds., Academic Press, New York and London.

Lyon, M.F., Phillips, J.S., and Bailey, H.J., 1972, Mutagenic effects of repeated small radiation doses to mouse spermatogonia. 1. Specific-locus mutation rates, Mut. Res., 15:185.

McKusick, V.A., 1983, "Mendelian Inheritance in Man", The Johns Hopkins University Press, Sixth Edition, Baltimore-London.

Neel, J.V., and Schull, W.J., 1958, Human Heredity, The University of Chicago Press.

Russell, W.L., 1965, Studies in mammalian radiation genetics, Nucleonics, 23:53.

Sobel, F.H., 1984, Problems and perspectives in genetic toxicology, pp 1-19, in Mutations in Man, ed. G. Obe, Springer-Verlag, Berlin-Heidelberg.

Stevenson, A.C., 1959, The load of hereditary defects in human populations, Radiat. Res., Suppl. 1:306.

Vogt, R., Wagner, H., and Schlapfer, H., 1940, Erbbiologie und Erb-pathologie des Auges, Handbuch der Erbbiologie des Menschen Bd 3, Hrsg. G. Just, Berlin.

EXTRAPOLATION FROM LARGE-SCALE RADIATION EXPOSURES: CANCER

Charles E. Land

Radiation Epidemiology Branch
National Cancer Institute
Landow 3A-22
Bethesda, MD 20205

INTRODUCTION

The risk of cancer following exposure to ionizing radiation is a minor public health problem (less than 3% of all cancer deaths plausibly can be attributed to radiation, including natural background (Jablon and Bailar, 1980), vs. about 30% for cigarette smoking (US Surgeon General, 1982)), about which we know a great deal and for which we are being required to give exceedingly specific estimates (Public Law 97-414, 1983). How has this come about? There is and has been great public concern about radiation hazards, certainly, and much misinformation is abroad that requires correction. But fundamentally, I believe, the present situation results from a history of highly productive research into the biological effects of ionizing radiation. The first-order results of this research, that there are health hazards associated with exposure and that these hazards can be serious for high enough exposures, have naturally led to demands for second- and third-order inferences: how great are the hazards from low-level exposure, how is the hazard from any level of exposure distributed over time after exposure, how does sensitivity vary by age, sex, and other host parameters, and how does radiation interact with other risk factors?

There is a practical (or worldly) side to the problem, that parallels, but does not always coincide with, the scientific urge to learn more about a natural phenomenon. The practical point of view converges on individual cases: How has a particular exposure history affected the likelihood that a given individual will get cancer, or what is the likelihood that a given individual's cancer was caused by his or her exposure? It may be objected that we do not know, nor can we expect ever to know, enough to characterize an individual's cancer risk from radiation exposure. All our information comes from observations of groups of exposed people, or by analogy with experimental observations on groups of animals or cells, and we are more comfortable, as scientists, with inferences stated in terms of averages, or in general terms. Realistically, however, it is difficult to see how inferences can be made about individual cases without generalizing from other experience. Even scientists, when faced with an individual case about which a decision must be made, will generalize from inferences about populations or experimental results; in other words, they will "play the odds." It seems, therefore, that the main scientific responsibility for questions involving individual cases is not to highlight the difference between inferences for a population and for an individual (this is understood), but rather, to ensure the quality of the inferences which must be generalized, and the method of generalization.

It is hazardous to estimate, or to devise algorithms for estimating, risk or the allocation of blame in individual cases where there may be a causal association between radiation exposure and cancer. One hazard is that the job may not be done properly - that objectivity may be lost, that scientific errors may be committed, or that arbitrary assumptions may be treated as fact, either explicitly or implicitly. A very different kind of hazard is that, in the interests of scientific objectivity, scientists may so temporize, and inferences may be so qualified, that the information provided cannot easily be incorporated into the making of societal decisions. One of the two major themes of this presentation is that one can act to minimize both of these hazards simultaneously.

My second major theme is that going through the exercise of applying what we know about radiation and cancer risk to individual cases can be scientifically beneficial. I know of no better way to gain an appreciation of the essential things that we do not know well enough, and of how various bits of information fit together.

INFERENCES ABOUT GROUPS VS. INFERENCES ABOUT INDIVIDUALS

Because we cannot distinguish a radiation-induced cancer from one with a different cause, we can study the risk associated with radiation exposure only by observing cancer frequency in groups of people or experimental animals exposed to ionizing radiation. Given a sufficiently long period of observation on a sufficiently large population, it is a simple matter to estimate the excess risk associated with exposure: the estimate is the number of cancers observed, less the number expected based on the experience of an otherwise similar, but non-exposed, population. More elaborate analyses are required to extrapolate risk from the observed population to another group, differently distributed with respect to radiation dose, age, sex, baseline risk, length of follow-up, and other characteristics.

Extrapolation of risk to an individual is not much different from extrapolation to a group that is homogeneous with respect to dose, age at exposure, sex, and possibly a few other factors. One difference is that a group may be distributed comparably to the observed population with respect to factors other than (say) dose, age, and sex, whereas an individual can correspond to only a single value for any relevant factor. Thus, assuming that an elaborate analysis has yielded a risk estimate specific to dose, age, and sex, but not other factors, extrapolation to a group may require the assumption that the group be "otherwise similar" to the original population, that is, similarly distributed with respect to other factors. Extrapolation to an individual, on the other hand, can require only that the individual be otherwise "typical" of the original population. Operationally, however, it is difficult to see how this condition could be interpreted otherwise than in a group context: the individual can be thought of as having been randomly selected from some group to which inferences based on the observed population can be applied.

If, as I have argued above, there is no operational difference between making inferences about closely specified groups and about individuals with respect to radiation exposure and cancer risk, there may be a psychological difference. We may feel more accountable for the statements we make if their application to individual cases, involving actual people, is clear than if the statements are general and less immediately applicable to particular cases. Given a social and political environment in which related decisions are being made, however, and in which all relevant information is, or should be, brought to bear on these decisions, there is no avoiding accountablility, and there is a considerable responsibility for ensuring that the information we provide be incorporated properly.

Perhaps the most detailed and specific societal application of information about radiation exposure and cancer risk is a determination of the extent to which a documented radiation exposure may have caused a particular cancer, diagnosed in a given individual at a given time. The determination involves an assessment of the likelihood of that kind of cancer occurring at that time, given the exposure relative to the likelihood in the absence of that exposure. It depends, therefore, upon age-specific baseline risk, the estimated excess risk for the given exposure to a person of the specified sex and age at exposure at the specified time after exposure, and upon modifications in either baseline or excess risk, or both, that might be associated with the patient's personal characteristics or history of exposure to other carcinogens or risk-modifying agents. Applications of this type actually are being made, and result in societal decisions based upon whatever information, good or bad, is available to the decision makers. Poor information may be used in preference to better information if it is more accessible to decision makers, and in particular, if it is in a form that is easier to apply to the problem at hand (Bond, 1983; Jablon, 1983).

THE INFORMATIONAL BASIS FOR ESTIMATING RISK

The basis for estimating cancer risk in human populations following exposure to ionizing radiation is the aggregated epidemiological data on cancer risk in irradiated human populations, together with whatever information we can bring to bear on the problem from experimental and theoretical radiobiology and medicine. That is, risk estimates are derived from epidemiological data, but how the estimates are calculated, the mathematical structure that specifies what one observation or set of observations has to do with another, derives in large part from other sources.

We know that ionizing radiation can cause cancer in man because studies of different populations with documented exposures to high radiation levels (hundreds or thousands of times natural background) have consistently found higher cancer rates than those in comparable, non-exposed populations (Boice and Land, 1982). Radiation was used to cure or alleviate the symptoms of disease long before its carcinogenic potential was fully appreciated, and even now it is the treatment of choice for some diseases (including cancer) for which the potential benefit outweighs the risk of subsequent cancer. Studies of patient populations given radiotherapy or extensive diagnostic radiology constitute much of the epidemiological basis for our knowledge of radiation carcinogenesis (Boice, 1981). Information on the effects of radiation from ingested or inhaled materials comes from studies of workers who swallowed radium while painting instrument dials with luminous paint (Rowland and Lucas, 1984), and from studies of uranium and other hard rock miners working in atmospheres heavily contaminated with radon (BEIR, 1980). Finally, the survivors of the atomic bombs dropped on Hiroshima and Nagasaki in 1945, and natives of the Marshall Islands who in 1954 were exposed to radioactive fallout from a Pacific nuclear weapons test, have been extensively studied (Kato and Schull, 1982; Conard et al., 1980).

From these studies it appears that ionizing radiation can, at some level of exposure, increase the risk for many, and perhaps most, of the types of cancer that occur in man. It is also clear that radiation does not create unique forms of cancer, but merely increases the risk of tumors that occur naturally. For the acute forms of leukemia, chronic myeloid leukemia, and female breast cancer, the association is so strong as to appear certain (BEIR, 1980). It is generally accepted that ionizing radiation does not cause chronic lymphatic leukemia (BEIR, 1980; United Nations, 1977). Other forms of cancer lie somewhere between these two extremes, although for many the position on this scale is uncertain because there is very little information.

For any given cancer site, the credibility of a presumed association between risk and radiation exposure depends upon several factors, including the following:

(1) Statistical significance - this is a combination of the total number of cancer cases observed and the apparent size of the excess relative to normal risk. For a fixed number of cases the strength of the association is greater if the excess risk is relatively large, and for a fixed relative excess the association is more credible if the evidence is based on many cases.

(2) Specificity - how certain is it that the association was not due to something else? Credibility is helped by the existence of a non-exposed population, otherwise similar to the exposed population, in which an excess risk was not observed. It is especially important to satisfy this requirement in studies of medically exposed populations, for which the conditions leading to exposure may themselves be related to the risk of subsequent cancer.

(3) Dose response - does the level of risk appear to increase with increasing radiation dose to the tissue of interest? Each of us is exposed to natural background radiation at the very least, and so the concept of increased risk from additional exposure involves the assumption of a gradient of risk with increasing exposure.

(4) Consistency - is the association seen in a number of exposed populations, and are the apparent excesses similar when such factors as dose, age, and period of observation have been taken into account? Spurious associations can arise by chance between cancer risk and practically anything, or exposure can be fortuitously related to a true risk factor. But such spurious associations are very unlikely to arise in many different populations, exposed under different circumstances. We therefore tend to place most credence in associations that turn up frequently and under diverse circumstances of exposure, while distrusting isolated reports not verified by other experience.

ADDRESSING SPECIFIC ELEMENTS OF CANCER RISK

The effect of radiation exposure on cancer risk in an exposed population can be addressed in the following ways: For a fixed exposure (dose, radiation quality, and temporal distribution of exposure), and a fixed period of follow-up, how does risk vary by such personal characteristics as sex, age at exposure, and personal history? For a homogeneous group of people and a fixed follow-up interval, how does risk vary according to exposure characteristics? For a homogeneous group, and a fixed exposure, how does risk vary over time after exposure? Each of the above questions addresses a different determinant of risk, and is stated conditionally on fixed values of all other determinants.

In fact, however, such questions only rarely can be asked without making fairly stringent assumptions about the way in which the answer to the question varies according to the values of other parameters, such as that the influence of sex on excess risk is independent of radiation dose, or that the shape of the dose-response curve is the same for risk 10 years after exposure as it is for risk 20 years after exposure. Only by making such assumptions (not necessarily the ones given above) is it possible to bring enough data to bear upon a question of interest to have a reasonable chance of getting a clear answer. But such an approach involves asking, not the questions we really want answered, but approximations to them - not, "How does the excess risk associated with a 10-rad exposure depend upon sex?" but, "On the average, for different dose levels, how does excess risk compare between males and females?" Of course, if excess risk among males is a constant proportion of that among females, for all dose levels, the two questions have the same answer; moreover, it may be possible to test the hypothesis of a constant proportion, and thereby learn something new about radiation carcinogenesis.

EXAMPLE: DOSE RESPONSE

For compelling statistical and methodological reasons, it is hopeless to observe risk in a population whose excess risk is only a few percentage points above normal, and to expect to estimate that excess with any precision (Land, 1980). It is only when the excess is fairly large, at least 50% of normal risk and preferably much greater, that we can estimate it with any real confidence even for a very large study population. Thus most of what we know about radiation-related cancer in man is based on studies of populations exposed to radiation doses high enough to yield substantial fractions of radiogenic cancers.

If we are interested in the risk from lower-dose exposures it is necessary to have an extrapolation rule by which the estimated high-dose risk determines lower-dose risk estimates. A fairly simple rule is to assume that excess risk is proportional to the number of ionizations produced in the tissue at risk, that is, proportional to dose. This rule is equivalent to drawing a straight line from a point representing zero excess risk at zero dose to the point representing the excess risk estimated at whatever dose was received by the population studied (the so-called linear model). Rules expressing risk as a weighted sum of dose and dose-squared (quadratic models), and more complicated functions that allow for competing effects of cell inactivation, are suggested by theoretical and experimental radiobiology (Upton, 1977).

Data from simple systems, like pink stamen hair mutations in tradescantia, suggest a quadratic (linear-quadratic) dose-response relationship at the low end of the dose scale, with equality for contributions proportional to dose and to dose-squared (crossover dose) at between about 50 and 200 rads of low-LET radiation and a related reduction in effect for fractionated or protracted exposures as compared to acute exposures (NCRP, 1980). Experimental carcinogenesis is less straightforward; extremely diverse dose-response patterns have been observed for the single class of reticular-tissue tumors in the mouse for example (Kohn and Fry, 1984). Overall, however, the main body of radiobiological opinion, as reflected in publications by the BEIR Committee (1980), the United Nations (1977), the NCRP (1980), and the ICRP (1977), favors a curvilinear dose-response relationship for low-LET radiation consistent with a dose-rate reduction factor of about 2.5 at dose levels in the 150-200 rad range.

Epidemiological data are not very informative about the shape of the extrapolation model that should be used, for the same reasons that low-dose data tend not to be informative about excess risk (Land, 1980). In general, the available data are consistent with linearity, but also consistent with other models (BEIR, 1980; Land, 1980). The breast cancer data are more strongly suggestive of linearity (Land et al., 1980; Tokunaga et al., 1984).

When fitted to general dose-response models with several free parameters, epidemiological data tend to yield imprecise parameter estimates that cause the risk at low-dose levels to be poorly defined. More restrictive dose-response models, such as models in which the crossover dose is fixed, yield estimates with smaller standard deviations, but such estimates do not reflect uncertainties about the assumptions made.

It must be frustrating, to someone charged with making societal decisions based on estimated risk, when an expert committee appointed to produce such estimates provides not one estimate with an associated measure of uncertainty but several, each conditional on a different dose-response model. Unless all the estimates agree broadly (e. g., that the risk is very low, or that it is very high) the decision maker must somehow devise a rule that takes all the estimates into account, and this is a more complicated task than basing a rule on a single estimate.

The BEIR Committee's report (1980) gave cancer risk estimates (all sites except leukemia and bone cancer) for low-LET radiation corresponding to three different dose-response models: a linear model in which excess risk was proportional to dose, a pure quadratic model in which risk was proportional to the square of dose, and a "preferred" linear-quadratic model in which risk was proportional to dose plus dose-squared/116. In other words, the committee, or a majority of it, felt that the most likely dose response was one for which the linear term dominated below a so-called "crossover" dose of about 116 rad, and the dose-squared term dominated above that value, but expressed its uncertainty by also giving estimates corresponding to the limiting forms of the general linear-quadratic model, as the crossover dose became infinitely large (the linear model) or infinitely small (the pure quadratic model).

It is interesting that the BEIR Committee felt able to provide risk estimates using a particular, closely-specified dose-response model (excess risk depends upon only one free parameter when crossover dose is fixed), but exhibited considerable uncertainty about which model to use (i.e., uncertainty about the value of the crossover dose). The available data were not sufficient, however, to define estimates based upon a more general model in which the crossover dose itself was a free parameter to be estimated from the data. The crossover value in the "preferred" model was selected for two reasons: first, 116 rad was the value estimated when the general linear-quadratic model was applied to dose-response data for leukemia incidence; second, that value was about in the middle of the range 50-200 rad of values obtained from experimental models. Thus, the BEIR Committee relied upon analogy, first with leukemia and second with experimental models. If the committee had quantified its uncertainty with respect to the crossover value, it could have presented a single estimate which would have incorporated additional information, and uncertainty, about the crossover dose. Such an estimate would have been more easily used for regulatory and other purposes, and it would have expressed more explicitly the nature and range of uncertainty that the committee attached to its estimates.

Incorporation of subjective uncertainty, which is often uncertainty that is based upon data but is poorly articulated, into risk estimates is a common aspect of Bayesian treatments of statistical decision procedures (Ferguson, 1967). As applied to the present problem, the approach is to quantify subjective uncertainty (and information) about the value of a parameter by representing it by an artificial data set whose central estimate and uncertainty measure correspond to the collective opinions of the expert committee. Such a data set can then be combined with actual data to define a risk estimate using standard statistical methods. Table 1, derived from a more detailed discussion of the problem (Land and Pierce, 1983), illustrates how the BEIR Committee might have approached the problem by postulating data on the crossover value corresponding to a lognormal distribution with a median value of 100 rad and with uncertainty represented by various values for the standard deviation in the log scale. The analysis incorporates this prior information with data on leukemia risk and risk of other cancer, under the assumption of a common crossover value for both classes of cancer,

Table 1. Example of Bayesian analyses combining A-bomb survivor leukemia incidence data, 1950-1971 (Ichimaru et al., 1978), with mortality data for cancers of other sites, 1950-1974 (Beebe et al., 1978), using a general linear-quadratic dose-response model with subjective lognormal prior distributions for the crossover dose. Adapted from Land and Pierce (1983, Table 6).

Crossover Dose Estimate (90% CI)		Est. Excess Risk at 1 Rad, ± SD	
Prior	Posterior*	Leukemia	Other Cancer
100 (0, ∞)	186 (12, 2800)	1.12 ± .94	1.63 ± 1.47
100 (.05, 200000)	174 (14, 2200)	1.08 ± .89	1.57 ± 1.40
100 (.16, 64000)	170 (14, 2100)	1.07 ± .87	1.55 ± 1.38
100 (.67, 15000)	162 (15, 1720)	1.05 ± .84	1.51 ± 1.33
100 (1.9, 5200)	153 (16, 1400)	1.01 ± .79	1.47 ± 1.26
100 (5.2, 1900)	140 (19, 1100)	0.97 ± .72	1.40 ± 1.16
100 (16, 610)	121 (27, 550)	0.89 ± .59	1.28 ± 0.95
100 (32, 310)	110 (38, 220)	0.84 ± .47	1.20 ± 0.79
100 (51, 195)	104 (53, 200)	0.81 ± .38	1.15 ± 0.67
100 (74, 135)	100 (86, 117)	0.79 ± .32	1.12 ± 0.59
100 (85, 117)	100 (92, 108)	0.79 ± .32	1.12 ± 0.59

*Estimate obtained after analysis, incorporating information from both data and prior distribution.

but with different values for risk. As the prior information on crossover dose is made to appear more nearly certain, the estimates for leukemia and other cancer risk per rad at low-dose levels approaches the values that would be obtained by postulating a crossover value of 100 rad. More uncertainty, however, yields crossover dose estimates somewhat higher, and higher risk estimates which are themselves more uncertain, as reflected in their estimated standard errors.

EXAMPLE: TIME TO TUMOR

One marked difference between making inferences about groups and individuals is the immediacy in the latter case of questions about time after exposure. For example, in the case of a compensation claim for a cancer diagnosed at age 50, 24 years after a radiation exposure at age 26, the issue concerns only baseline and excess risk at the time of diagnosis; risk at other times is important only to the extent that it bears on risk 24 years after exposure.

We cannot directly observe time between exposure and diagnosis of a cancer caused by that exposure, because we cannot distinguish between radiation-induced cancers and others. We can, however, compare the distribution of cancer diagnoses over time following exposure in a population subgroup for which it is clear that most of the cancers are radiogenic, with the temporal distribution of cancers in a similar population subgroup for which it is clear that very few of the observed cancers were caused by radiation. The A-bomb survivor data are particularly suitable for such comparisons, because everyone was exposed at the same time and has been followed for subsequent cancer risk in exactly the same way (Land and Tokunaga, 1984). Also, there is a large number with high-dose exposures, with consequently high proportions of certain types of cancer that were caused by radiation, and even larger numbers of survivors whose exposures were much smaller and among whom the percentage of radiation-induced cancers has been correspondingly small.

Comparisons with respect to time to diagnosis between cancers seen among high-dose and low-dose members of the same exposed population, when standardized by exposure age and made specific to cancer site, can have crucial implications. If the two distributions are different, as is the case for leukemia, and especially, for chronic granulocytic leukemia or for acute leukemia following exposure at a young age, the next step is to characterize this difference with respect to possible effects of exposure age, radiation dose, or other factors. This is a particularly difficult kind of analysis, and it has been only partially successful for leukemia. The data suggest a wave-like distribution, like the lognormal, whose characteristics may vary according to histological type and exposure age (Fig. 1).

If, on the other hand, the two distributions are remarkably similar, as is clearly the case for breast cancer and only somewhat less clearly so for lung cancer (Land and Tokunaga, 1984), the implications for risk estimation are straightforward. If radiation-induced cancers have the same, or about the same, time-to-diagnosis distribution as cancers not caused by radiation, then excess risk as a proportion of baseline risk (relative excess) must be constant over time following exposure. Except for a period of transition during which the relative excess must increase from zero at the time of exposure to its eventual constant value, the likelihood that a particular cancer was caused by a particular exposure must not depend upon how long after the exposure the diagnosis was made. This conclusion is quite different from that implied by the leukemia data; the likelihood of a radiation etiology appears to depend strongly upon time of leukemia diagnosis.

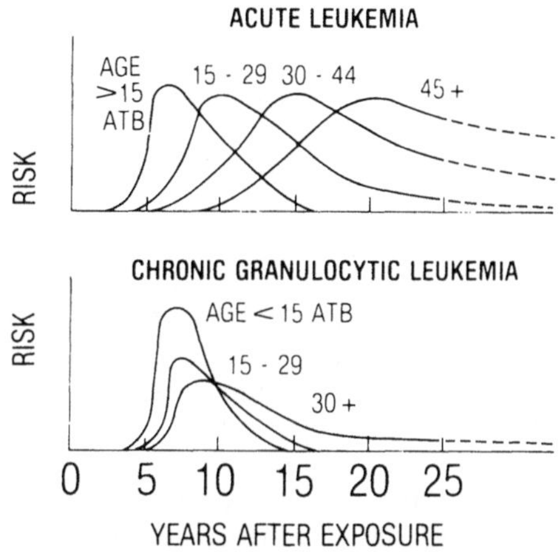

Fig. 1. Schematic representations of induction period for radiation -induced leukemia: Influence of age at exposure and calendar time on the leukemogenic effect of radiation. (Redrawn from Ichimaru and Ishimaru, 1975.)

Time-to-diagnosis modelling also has profound implications for the application of risk estimates to groups, but refined treatment is not always required. The excess risk over a time interval corresponding to an actual period of observation in a studied population can be estimated without reference to the temporal distribution of risk, as long as finer detail is not required. Extrapolation beyond the period of observation, on the other hand, for example to the end of life, is strongly model dependent; the 1980 BEIR Committee found that, for cancers other than leukemia and bone cancer, considered as a group, and for a single, brief exposure to a general population, projected lifetime risk assuming constant relative risk was about 4 times as high as it was under the so-called constant absolute risk model (BEIR, 1980). In the latter model excess risk, in absolute terms, is assumed to be invariant over time, following a minimal latent period.

The BEIR Committee did not favor one temporal extrapolation model over the other; both are time-honored and computationally simple models, obvious approaches to be used in the absence of information. More recent analyses, however, which have focused on the variation of risk over time within successively longer observation periods, have tended to yield results consistent with a constant relative excess and inconsistent with a constant absolute excess over time (Kato and Schull, 1982; Land et al., 1980; Land and Tokunaga, 1984), and it seems reasonable to expect the pattern to continue with increasing follow-up, or at least to change only slowly over time. A possible exception, other than leukemia and bone cancer, is cancer of the thyroid, for which some analyses suggest a constant, or even decreasing, absolute excess with increasing follow-up (Shore, 1984). Thyroid cancer is normally an indolent disease, however, and it is entirely possible that observed patterns of risk over time in populations known to be exposed to radiation may predominantly reflect temporal patterns in the level of diagnostic effort applied to these populations.

Because the constant relative excess model is consistent with the data available for a number of cancers, notably those of the female breast and the lung but also for a number of others (Kato and Schull, 1982), its usefulness as a practical rule of thumb seems assured, especially in the absence of a strong competitor. But the model is biologically naive, having arisen initially as a computationally simple, first-order model. Do the predictions of the model agree with those of current theories of carcinogenesis?

In a very general sense, the constant relative excess model is consistent with the widely accepted multi-stage theory of carcinogenesis (see, e. g., Day and Brown, 1980), in that radiation exposure may cause certain irreversible cellular changes that in themselves do not suffice for the development of a cancer, but that, if followed by other changes, may result in one or more transformed cells capable of relatively unrestrained replication. The rather prompt "wave" of increased leukemia and bone cancer risk following irradiation (BEIR, 1980) suggests that for these sites subsequent changes are not required or occur soon after exposure. For certain other cancers (female breast, lung, and stomach) observed conformance to the constant relative excess model suggests that the necessary changes subsequent to irradiation are strongly related to age, and that non-radiogenic cancers result when substantially the same events occur following cellular changes similar to those caused by ionizing radiation.

In the context of the multi-stage theory the constant relative excess model should hold if the initial changes needed for carcinogenesis, whether by radiation or other agents, occur mainly at younger ages and the subsequent changes mainly at older ages. In that case the proportion of radiation-initiated cells relative to all initially modified cells should remain constant throughout the period of subsequent change. If, on the other hand, initiation by factors other than radiation should be significant at older ages, then given an exposure, and possible initiation, by radiation at a young age and the possibility of initiation by other agents at older ages, the relative importance of the radiation exposure with respect to any subsequent cancer should decline with increasing follow-up. Interestingly, the constant relative excess model appears to fit very well for radiation-induced breast cancer, and the epidemiology of breast cancer in both irradiated and non-irradiated populations suggests that sensitivity to carcinogenic agents is high early in life and decreases markedly with increasing age (Tokunaga et al., 1984; Cole, 1980). For lung cancer, on the other hand, there is fairly strong evidence that radiation exposure at ages over 50 can increase risk, and if the same is true for other carcinogens one might expect excess risk associated with radiation exposure to decline with increasing follow-up. No such decline has been observed (Kato and Schull, 1982; Land and Tokunaga, 1984), but the data do not rule one out.

The carcinogenic process results in a transformed cell with the capacity to replicate uncontrollably. More time is required before sufficient cell divisions (about 30 generations) result in a detectable tumor. It is this consideration that leads us to believe that there must be a minimum time to tumor detection. Tumor doubling times for human tumors have been observed to correspond closely to lognormal distributions (Steel, 1977). In a study of doubling times for 780 metastatic and primary tumors in human patients, Steel (1977) confirmed this general finding and estimated means and variances of times in the logarithmic scale. Estimated median doubling times were extremely long for primary adenocarcinomas of the colon and rectum (632 days), but on the order of three months for primary carcinomas of the breast and lung and a little over 2 months for primary bone sarcomas. These estimates correspond to median times from transformation to diagnosis of 5-8 years for primary breast, lung, and bone cancers. The measurements were made fairly late in tumor development, when the tumors were already large enough to be seen radiographically. Moolgavkar et al. (1982) have argued that tumor doubling times are probably shorter when the tumor is smaller. Time to diagnosis data for bone cancer in patients injected with 224-Ra (Mays and Spiess, 1983), on the other hand, agree closely with a lognormal distribution with median 10 years, roughly twice the time estimated by Steel for bone cancer.

Tumor development occurs only at the end of the carcinogenic process, and there is no reason to believe that radiation-induced cancers grow more quickly or slowly than other cancers of the same tissues. Tumor growth is important to the constant relative excess model, therefore, mainly in the years immediately following irradiation when, epidemiologically speaking, cells recently initiated by radiation are competing not only with cells initiated by other agents but not yet transformed at the time of irradiation but also with cells farther along in the multi-stage process and, in particular, those already transformed and replicating. Because a radiation exposure cannot result immediately in a detectable cancer, excess risk must start at zero. After some number of years, when all, or practically all, transformed cells existing at the time of irradiation have been diagnosed as cancers, cells initiated by the radiation exposure should be competing only with cells initiated but not yet transformed at the time of irradiation and with cells initiated subsequent to irradiation. At that point excess risk should be either steady or declining relative to background. In the interim, however, the relative risk should increase, and its rate of increase corresponds to the distribution of the time needed for a transformed cell to grow into a readily detectable tumor.

Steel's (1977) analysis of doubling time data for primary cancers of the breast and lung suggests that time in years from transformation to diagnosis is approximately lognormally distributed with mean 2.1 and standard deviation 0.7 on the logarithmic (base e) scale. Thus scarcely any (< 1%) transformed cells could form detectable cancers within 2 years after transformation, about 10% would do so within 3.5 years, about 25% within 5 years, and about half within 8 years. The distribution is strongly skewed to the right: 13 years would be required for 75% of all transformed cells to develop into detectable tumors, and 20 years would be needed for the percentage to reach 90%.

In a population with no detectable tumors at the time of irradiation, any pool of transformed cells and developing tumors existing immediately prior to irradiation should be substantially exhausted (i. e., have developed into observable tumors) by 20 years post-irradiation. That is, after 20 years the variation over time of the relative frequency of diagnosed cancers due to radiation exposure vs. background initiation should be determined by the number of initiated, but not transformed, cells at the time of irradiation and variations in the levels of initiation and promotion subsequent to irradiation. During the first 10 years after irradiation the relative importance of radiation and background initiation should be dominated by the depletion of the pool of transformed cells existing at the time of irradiation, while during the second 10 years, when depletion of the pool is proceeding more slowly, the later pattern gradually should prevail.

AN ILLUSTRATION

Let p(a) represent the probability of an early-stage carcinogenic event at age a. Let q(a') denote the probability, given early-stage events at some prior age, that subsequent changes will result in a transformed cell at age a', capable of developing into a detectable tumor. Finally, let f(a',y) denote the probability that a transformed cell at age a' will lead to cancer diagnosis y years later. Then the probability of cancer diagnosis at age a", given an early-stage event at age a, is approximately

$$P(a''|a) = \sum_{a'=a}^{a''} q(a')f(a',a''-a')$$

The unconditional probability of diagnosis at age a" is

$$P(a'') = \sum_{x=0}^{a''} p(x)P(a''|x).$$

Now suppose that the probability of an early-stage change is mainly a matter of age-specific sensitivity to carcinogenic agents whose prevalence does not depend on age. Assuming that sensitivity varies by age in about the same way for all early-stage carcinogens, p(a) ought to be approximately proportional to the relative excess cancer risk for a radiation exposure at age a. Suppose further that the likelihood of cancer diagnosis at age a' in the general population is roughly proportional to the frequency of late-stage changes, assuming a background level of early-stage changes. Then q(a') should be roughly proportional to the population baseline cancer rate. Finally, suppose that the probability distribution of the time from transformation to tumor diagnosis is lognormal with mean 2.1 and variance 0.7 on the logarithmic scale. That is, f(a',y) is independent of a' and is specified by Steel's estimate for tumor doubling times.

Given a radiation exposure at age a, the probability of an early-stage change is some constant times p(a). The relative excess risk at age a", therefore, is proportional to the ratio

$$R(a'',a) = P(a''|a)/P(a'').$$

Figure 2 shows variations in the pattern of R over age at diagnosis for different exposure ages a, when the above assumptions are applied to cancers of the female breast, lung, and the digestive tract (esophagus, stomach, and colon), and for a hypothetical cancer for which both p(a) and q(a') are constant. For each of the actual cancer sites, the values p(a) were derived from the 1980 BEIR report (1980, Table V-14) and q(a') was derived from the 1973-1977

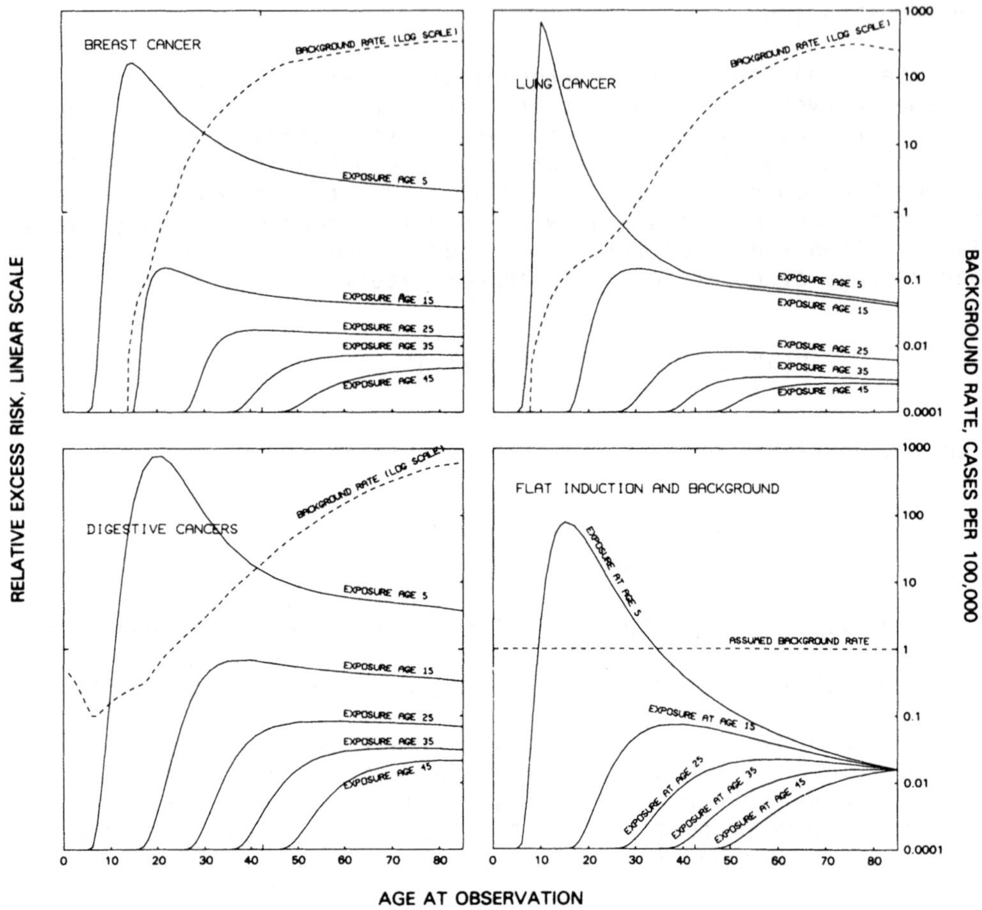

Fig. 2. Relative excess risk as a function of age at irradiation and age at diagnosis, using a two-stage model for changes leading to a transformed cell and a lognormal model for time from transformation to cancer diagnosis. For each cancer the frequency of initial changes has been assumed to be proportional to the relative excess risk associated with irradiation at each age, and the frequency of second-stage changes has been assumed to be proportional to baseline risk. In the lower right-hand panel, both frequencies have been assumed not to depend upon age.

SEER rates (Young et al., 1981). The same exposure factor was applied at each exposure age. Relative excess risk is shown on an arbitrary linear scale; age-specific baseline risk, on a semilogarithmic scale, is shown superimposed on each graph as an aid in interpreting the curves for R.

For each of the three cancer sites considered, baseline risk increases steeply with age, while the relative risk of radiation-induced cancer decreases with age at exposure. For each site the graphs predict a peak in relative excess risk following exposure at an early age, but this peak occurs at ages for which baseline risk is exceedingly low. If true, therefore, it would be unlikely for the peak to be detected, because it would be reflected in few or no cases. At ages where baseline risk levels would predict more cases, the relative excess is flat or slightly decreasing. For exposures at older ages the increase in R is more gradual. Overall, the graphs do not correspond exactly to the constant relative risk model, but the agreement is close enough, at ages for which appreciable numbers of cases might be expected and allowing for a 10-15 year period dominated by tumor growth, for the model to be an adequate approximation.

The lower right-hand panel of Figure 2 shows that the constant relative risk model would not be expected to hold, even approximately, if baseline risk did not increase, or if sensitivity to early-stage changes did not decrease, with increasing age. Thus the model is consistent with multi-stage carcinogenesis only under certain conditions, which nevertheless appear to be fairly common.

SUMMARY

Even though much is known about cancer risk associated with exposure to ionizing radiation, societal demands for detailed risk assessment go far beyond our ability to satisfy them according to customary standards of scientific accuracy. Society's requirement, however, is for the best information available, however good it may be, and not necessarily for the definitive solution to the problem posed. Bayesian methods may be useful for presenting incomplete and varied data and informed scientific opinion in a form suitable for use in societal decision making, while at the same time providing a disciplinary framework for incorporating opinion into scientific recommendations.

Inferences about individual cases and their relationship to risk of radiation carcinogenesis provide an especially severe test of the completeness of our understanding of the relationship between exposure and risk. This is particularly true with respect to the distribution of excess risk over time following exposure. Recent work suggests that a standard model often used for projection of risk forward in time, the constant relative excess model, may give a surprisingly accurate picture of time to tumor for a number of cancer sites.

REFERENCES

Beebe, G. W., Kato, H., and Land, C. E., 1978, Studies of the mortality of A-bomb survivors. 6. Mortality and radiation dose, 1950-1974, Radiat, Res., 75:138.

BEIR Report, 1980, Committee on the Biological Effects of Ionizing Radiation, "The Effects on Populations of Exposure to Low Levels of Ionizing Radiation," National Academy Press, Washington, D. C.

Boice, J. D., Jr., 1981, Cancer following medical radiation, Cancer, 47:1081.

Boice, J. D., Jr., and Land, C. E., 1982, in: "Cancer Epidemiology and Prevention," D. Schottenfeld and J. F. Fraumeni, Jr., eds., W. B. Saunders Company, Philadelphia.

Bond, V. P., 1983, Radiation exposure and cancer: The probability of causation, in: "Proceedings of the Seventh International Congress of Radiation Research, Sessions C: Somatic and Genetic Effects," J. J. Broese, G. W. Barendsen, H. B. Kal, and A. J. van der Kogel, eds., Martinus Nijhof, Amsterdam.

Cole, P., 1980, Major aspects of the epidemiology of breast cancer, Cancer, 46:865.

Conard, R. A., Paglia, D. E., Larsen, P. R., Sutow, W. W., Dobyns, B. M., Robbins, J., Krotosky, W. A., Field, J. B., Rall, J.E., and Wolff, J., 1980, "Review of Medical Findings in a Marshallese Population Twenty-Six Years After Accidental Exposure to Radioactive Fallout," BNL 51261, Brookhaven National Laboratory, Upton, New York.

Day, N. E., and Brown, C. C., 1980, Multistage models and primary prevention of cancer, J. Nat. Cancer Inst., 64:977.

Ferguson, T. S., 1967, "Mathematical Statistics: A Decision-Theoretic Approach," Academic Press, New York.

Ichimaru, M., and Ishimaru, T., 1975, Leukemia and related disorders, J. Radiat. Res., 16 (Supl):89.

Ichimaru, M., Ishimaru, T., and Belsky, J. L., 1978, Incidence of leukemia in atomic bomb survivors belonging to a fixed cohort in Hiroshima and Nagasaki, 1950-71. Radiation dose, years after exposure, age at exposure, and type of leukemia, J. Radiat. Res., 19:262.

Internation Commission on Radiological Protection, 1977, "ICRP Publication 26: Recommendations of the International Commission on Radiological Protection," Pergamon Press, Oxford.

Jablon, S., 1983, Probability of causation - practical problems, in: "Proceedings of the Seventh International Congress of Radiation Research, Sessions C: Somatic and Genetic Effects," J. J. Broese, G. W. Barendsen, H. B. Kal, and A. J. van der Kogel, ed., Martinus Nijhof, Amsterdam.

Jablon, S., and Bailar, J. C., 1980, The contribution of ionizing radiation to cancer mortality in the United States. Prev. Med. 9:219.

Kato, H., and Schull, W. J., 1982, Studies of the mortality of a-bomb survivors. 7. Mortality, 1950-1978: Part 1. Cancer mortality, Radiat. Res., 90:395.

Kohn, H. I., and Fry, R. J. M., 1984, Radiation carcinogenesis, New Engl. J. Med., 310:504.

Land, C. E., 1980, Estimating cancer risks from low doses of ionizing radiation, Science, 209:1197.

Land, C. E., Boice, J. D., Jr., Shore, R. E., Norman, J. E., and Tokunaga, M., 1980, Breast cancer risk from low-dose exposures to ionizing radiation: Results of a parallel analysis of three exposed populations of women, J. Nat. Cancer Inst., 65:353.

Land, C. E., and Pierce, D. A., 1983, Some statistical considerations related to the estimation of cancer risk from low-dose exposures to ionizing radiation, in: "Epidemiology Applied to Health Physics, Proceedings of the 16th Midyear Topical Meeting of the Health Physics Society," CONF-83010, Nat. Technical Information Service, Springfield, VA.

Land, C. E., and Tokunaga, M., 1984, Induction period, in: "Radiation Carcinogenesis: Epidemiology and Biological Significance," J. D. Boice, Jr., and J. F. Fraumeni, Jr., ed., Raven Press, New York.

Mays, C. W., and Spiess, H., 1983, Epidemiological studies of German patients with Ra-224, in: "Epidemiology Applied to Health Physics, Proceedings of the 16th Midyear Topical Meeting of the Health Physics Society," CONF-83010, Nat. Technical Information Service, Springfield, VA.

Moolgavkar, S. H., Day, N. E., and Stevens, R. G., 1982, Two-stage model for carcinogenesis: Epidemiology of breast cancer in females, J. Nat. Cancer Inst., 65:559.

National Council on Radiation Protection and Measurements, 1980,

"Influence of Dose and its Distribution in Time on Dose-Response Relationships for Low-LET Radiation," NCRP Report No. 64, National Council on Radiation Protection and Measurements, Washington, D. C.

Public Law 97-414 Section 7(b), 1983, U. S. Govt. Printing Office, Washington, D. C.

Rowland, R. E., and Lucas, H. F., Jr., 1984, Radium dial workers, in: "Radiation Carcinogenesis: Epidemiology and Biological Significance," J. D. Boice, Jr., and J. F. Fraumeni, Jr., ed., Raven Press, New York.

Shore, R. E., 1984, Radiation-induced thyroid cancer, in: "Radiation Carcinogenesis: Epidemiology and Biological Significance," J. D. Boice, Jr., and J. F. Fraumeni, Jr., ed., Raven Press, New York.

Steel, G. G., 1977, "Growth Kinetics of Tumors. Cell Population Kinetics in Relation to Growth and Treatment of Cancer," Clarendon Press, Oxford.

Tokunaga, M., Land, C. E., Yamamoto, T., Asano, M., Tokuoka, S., Ezaki, E., and Nishimori, I., 1984, Breast cancer among atomic bomb survivors, in: "Radiation Carcinogenesis: Epidemiology and Biological Significance," J. D. Boice, Jr., and J. F. Fraumeni, Jr., ed., Raven Press, New York.

United Nations Scientific Committee on the Effects of Atomic Radiation, 1977, "Sources and Effects of Ionizing Radiation." United Nations, New York (Report E.77.IX.1).

Upton, A. C., 1977, Radiobiological effects of low doses: Implications for radiological protection, Radiat. Res., 71:51-74.

U. S. Department of Health and Human Services, 1980, "Cancer: The Health Consequences of Smoking: A Report from the Surgeon General," U. S. Govt. Printing Office, Washington, D. C.

Young, J. L., Jr., Percy, C. L., and Asire, A. J. (ed.), 1981, "Surveillance, Epidemiology, and End Results: Incidence and Mortality Data, 1973-1977," National Cancer Institute Monograph 57, NIH Publ. No. 81-2330, U. S. Govt. Printing Office, Washington, D. C.

DISCUSSION

Jablon: The linear-quadratic response model with cell-killing was developed on the basis of laboratory experimentation with animals marked by considerable genetic homogeneity. Humans, however, are marked by considerable genetic heterogeneity. What effect would this have, do you suppose, on the nature of the response function or its form?

Land: I think it would depend on how much variability there was. If there were some exquisitely sensitive elements of the population that saturated at very low doses, you could get almost any kind of a shape of curve. Such elements would make the curve steeper at low doses and flatten off at higher doses, but since radiation doesn't seem to be unique in its effects, these people would have a very high background level of risk. However, we have not identified anything that creates such unique susceptibilities to radiation carcinogenesis.

Clifton: The linear-quadratic model was introduced to radiobiology as a description of acute biophysical events in cells. In its application to cancer dose-response curves, we are dealing with the endpoint of a series of complex events initiated by radiation. The linear-quadratic model may well fit genetic data and other one- or two-step effects, and may be useful in handling some cancer data. It should not, however, be concluded that initiation by radiation either follows or does not follow linear-quadratic dose-response relationships on the basis of whether cancer incidence does or does not follow such a relationship.

Land: If the events happening after radiation do not have anything to do with radiation dose, then it seems that this will not affect the dose-response curve. The question is, how many of the initiated cells develop into cancers? Has this anything to do with the dose? If not, it seems to me that there is no problem. You are just dividing by some constant, aren't you?

Clifton: It seems to me that you have made the assumption that every initiated cell, or a fixed proportion of them, gives rise to cancer, and that you are looking for an individual to develop only one cancer. Now that individual may have many, many initiated cells which may or may not develop into other cancers that you will see; this may affect your whole consideration. Overt cancer is unlikely to be a simple function of initiated cell number.

Land: I do not see how it could affect the dose-response curve, provided that the data are based on comparisons of similar groups of people exposed at the same time to different doses and given substantially the same lifetime experiences after exposure.

Jablon: If I could jump into this, I would say I am very discouraged. We have been told earlier in this workshop that statistical models alone would not provide answers, and that we had to get our information out of biologists. And now Dr. Land is trying to use some biological information to supplement the very weak human data and we are saying that we are not sure that this can be done. It seems to me that we are getting a recipe for going nowhere.

Clifton: No, I do not think so. But, I do not think we can make mechanistic assumptions from the fact that cancer incidence dose-response curves follow a certain pattern on the basis of current biological information. But this does not mean that we should not try to find out what the relationships are, for those relationships will be extremely important in risk calculation. The mechanistic assumptions that are inherent in the alpha-D, beta-D squared model are inappropriate for mechanistic analysis of cancer induction because we are talking about a much more complex process.

Abrahamson: Ignoring dosimetry of A-bomb survivors, which we all know is undergoing revision, I have heard that there is some uncertainty with respect to how the fluoroscopy dosimetry was calculated which could have altered the shape of the dose-response curve--perhaps constraining it to becoming more linear.

Land: I usually hear that argument in terms of, "did the exposed people really get as great a dose or as little dose as they said?" But this is another fundamental radiobiological problem--if you have highly fractionated doses, the dose-response curve just about <u>has</u> to be linear? How can it be anything else?

Abrahamson: I agree, but if you are averaging a whole series of doses, and averaging a high-dose point and a low-dose point you will drop the slope of the curve considerably.

Jablon: Twelve years ago, I became interested in the question of dosimetry errors as in the data from Hiroshima-Nagasaki. Given that the dose estimates for survivors of Hiroshima-Nagasaki cannot be accurate, the question is, what effect will those inaccuracies have on not only the doses, but on the shape of the apparent dose-response curve. Now if you assume (as seems to be generally true) that dose estimates tend to be lognormally distributed about true values, it follows that the average estimate will be biased high. How much of a high bias you get depends upon what the variances are. Taking some reasonable estimates of variances, I convinced myself that if the response was truly linear, the dose-response curve would show a "bendover", because the people experiencing a true dose of 150 would appear to have had 175 rad, and so on. The other side of that coin is a response curve, which should appear as a quadratic, would tend to have its far end pushed down.

Land: I can support Dr. Abrahamson; I went through the same calculations and then discovered that he had already done it. The fluoroscopy data are not, by themselves, strong enough to allow us to choose between linearity or non-linearity. It seems to me that the inference of linearity comes from two sources. One source is the strong A-bomb survivor data themselves, which on the basis of curve-fitting, argue against very strong non-linearity. The second source of inference is that the risks from these highly fractionated exposures seem to agree well not only with A-bomb survivors who were exposed all at once, but also from mastitis patients, who had fairly high but well-documented doses in three, four, or five fractions. If there is no fractionation effect, then there is not appreciable non-linearity. If, in fact, the doses from the fluoroscopy patients were wrong, and they were much higher than supposed, then we should abandon that part of the argument.

THE FEASIBILITY AND URGENCY OF MONITORING HUMAN POPULATIONS FOR THE GENETIC EFFECTS OF RADIATION: THE HIROSHIMA-NAGASAKI EXPERIENCE

James V. Neel

Department of Human Genetics
University of Michigan Medical School
Ann Arbor, MI 48109

INTRODUCTION

By actual account, this is the eighth symposium devoted to some aspect of risk assessment with reference to ionizing radiation and chemical agents in which I have participated in the past 5 years. All of these have resulted in impressive-appearing publications, as I presume, will this one. We are doing a tremendous amount of talking. I wish I felt the progress was proportional to the verbiage.

Much of what I will say today about the studies in Japan will not be new to those of you who have attended these or still other workshops and, yes, I do see a lot of familiar faces here. I do, however, propose to go well beyond the simple facts of those studies, using them as a point of departure to develop a position which I would like to put right up front, lest it get buried in what looks like just another presentation of the Hiroshima-Nagasaki data.

This position, very simply stated, is that the mess created by current public uncertainties and fears concerning the genetic effects of low-level pollutants, radiation or chemical, as the case may be, has now reached such proportions that it is appropriate for concerned scientists, of whom there are a good many here, to attempt to reach a reasonable consensus of what needs doing and then present this to the responsible and involved agencies. The need is pressing for the formulation of a coherent research policy by some central advisory group. To the extent that this policy will be implemented by the necessity to coordinate not only the efforts EPA--there will be the necessity to coordinate not only the efforts of the agencies involved, but also the efforts of the laboratories maintained by these agencies. There will also be the need to coordinate the efforts of Universities and private laboratories. I know the situation with respect to the laboratories of the DOE best. If some significant fraction of their technical capabilities were brought to bear in a coordinated fashion on some of the research needs I will be mentioning, I am convinced rapid progress could be made in our ability to evaluate the impact of low-level exposures. Such a coordinated program might, incidentally, help meet recurrent questions concerning the missions of the laboratories. But it won't happen without a little of the gentle persuasion that I'm sure some individual scientists will find distasteful.

This theme, of the need for these studies, as you will recognize, is in fact really neither original or new. I remind you that the National Science Foundation, in its first Congressionally mandated effort to provide the Congress with a 5-year projection of the nation's research needs, published in 1980, listed as one of its five high priority items an improvement of the scientific and technological base for risk assessment. As far as I can see, neither the responsible federal agencies nor the scientific community have responded to that opportunity on anything like an appropriate scale. To be sure, this area does not yield the quick returns of the recombinant DNA or new immunological technologies, and clearly we are in a country oriented to quick returns, but just maybe the magnitude of the issue has become sufficiently clear and the technologies are sufficiently in sight that now is the time to move it.

As a final opening point, I suggest that the genetic data most urgently needed are not those creating the presumption of genetic damage--new demonstrations of induced mutations in animal models or increases in sister chromatid exchanges, DNA adducts, chromosomal damage, or abnormally shaped sperm in exposed persons--but epidemiological studies on the offspring of individuals with unusual, potentially mutagenic experiences. While, however, arguing the urgent need for data on transmitted effects from human populations, I will later argue for a second round of studies with the mouse model, these studies now utilizing the same new genetic end points it is becoming possible to employ in human studies.

SOME LEGAL DEVELOPMENTS INVOLVING THE PERCEIVED GENETIC EFFECTS OF RADIATION EXPOSURE

My personal involvement in the practical implications of knowledge--or lack thereof--of radiation effects seems to be cylical--like the seven year itch. This past year I have been drawn into three current attempts which beautifully illustrate the complexity of the issues upon us. The first and second of these involve the efforts of the Environmental Protection Agency to promulgate standards of exposure for the general population from the proposed geologic repositories for high-level radioactive waste and from radionuclides. References to the appropriate documents, and some of the attending legal actions, will be found elsewhere (Neel in press, a,b). The fact is that the laws of the land are forcing EPA to attempt to predict the biological effects on humans of radiation exposures which are a fraction of background levels. These extrapolations extend three orders of magnitude beyond the boundaries of the experimental data, and are about 90% based on observations on another species, the mouse. With this audience, I do not have to belabor the possibility for error in these extrapolations. We will all agree to the need to be properly cautious in our approach to this situation. On the other hand, the financial implications of these regulations are very great; in my opinion, never was the data base for a proper risk-benefit analysis weaker.

The third very active problem area in which I find myself involved is the issue of the somatic and genetic sequelae of the exposures sustained by veterans who participated in the various atomic bomb tests. The Congress has now mandated (PL 98-160) what will be a rather expensive effort to determine the short- and long-term consequences of these exposures, which average less than 1 rem of surface radiation as the situation is now understood. The approach will of necessity rely heavily on a questionnaire. If 40 questions concerning the ill effects of an agent are asked of a population with no exposure whatsoever to that agent, on average one question will yield a "significant" positive effect at the 5 percent level, and one a "significant" negative effect at the same level. The latter will be lost sight of but the former seized upon. Thus, although I understand the concerns of these veterans, I am very impressed by the potential for "false positives" in a study like this.

It is this third development which in particular has convinced me of the need for a carefully coordinated program on the part of the scientific community. Better we take the initiative in the most appropriate studies the situation permits, of which more later--imperfect though the opportunities and prospects be for a decisive outcome--than that we face a collection of mandated studies on groups who are epidemiological nightmares. Perhaps if the scientific community had taken the lead in developing and applying better technologies some years ago, we would not face the mandate of this study on the veterans.

HUMAN DATA: THE STUDIES IN HIROSHIMA AND NAGASAKI

With this effort to establish the proper Zeitgeist in which to consider studies of the presumed results of a mutagenic experience, let us turn to a specific example--not so much as a model as a learning experience. The first step towards studies on the genetic effects of the atomic bombs in Hiroshima and Nagasaki were taken in 1946; after a year and a half of planning, a comprehensive program which continues down to the present time was initiated in 1948. The program has been described and the findings presented on numerous occasions (e.g., Neel and Schull, 1956; Schull and Neel, 1958; Kato et al., 1966; Schull et al, 1966, 1981; Neel et al., 1982). I will not at this time attempt one more complete recapitulation but comment only on the essential. The majority of the study population in the two cities consists of a cohort of children defined at the fifth foetal month, when their mothers registered the fact of their pregnancy for ration purposes. In both cities that cohort has been subdivided into the children of the distally exposed, who were beyond the zone of the radiation delivered by the explosion, and the children one or both of whose parents were proximally exposed, this defined as within 2000 meters of the hypocenter of the explosions. The original studies on the cohorts were primarily morphological, but in 1968 cytogenetic studies were initiated, and in 1976, biochemical studies, consisting both of an effort to detect mutations altering the electrophoretic mobility of a battery of proteins and of quantitative studies of an enzymatic subset of these proteins, to detect mutations causing loss of function. These latter two studies have involved for the most part subsets of the original cohort. Thus the great majority of our observations can be related to a <u>prospectively</u> defined cohort, a fact we consider of cardinal importance in studies of this type.

The most recent analysis of the accumulated data, which included an effort to generate an estimate of the genetic doubling dose of radiation, was published in 1981 (Schull et al, 1981). (The genetic doubling dose is defined as that amount of radiation which will produce the same frequency of mutation that occurs spontaneously.) The analysis dealt with four qualitative-type indicators: 1) frequency of untoward pregnancy outcomes (congenital defect and/or stillbirth and/or death during first postnatal week), 2) death among liveborn infants prior to an average age of 17, 3) frequency of sex-chromosome aneuploids, and 4) frequency of mutations altering the electrophoretic mobility of proteins. In no instance were there significant differences between the children of so-called proximally exposed parents, who received from an estimated 1 <u>rem</u> up to the maximum compatible with survival, and the children of the distally exposed, who received essentially no excess radiation. However, the findings were all in the direction of the hypothesis of a genetic effect of the exposure. Let us assume that the exposure had indeed had a genetic effect, i.e., that humans are no different in principle than every other properly studied plant and animal species. To derive from these data an estimate of the doubling dose, it was then necessary to estimate to what extent "spontaneous" mutation in the distally exposed parents contributed to the first two of the above-mentioned indicators under the conditions of post-war Japan. With what seemed to be reasonable assumptions, we then, averaging all the results together, estimated a doubling dose of 156 rem. The error attached to that estimate is undoubtedly large and somewhat indeterminate. Although at face value that doubling dose estimate is some four times higher than that commonly quoted for the mouse, our estimate of the mutational component in congenital defect and early death in Japan during the period covered by this study is substantially higher than that employed in most of the extrapolations from mouse to man. The result is that our direct estimate of the absolute impact of the exposure on the next generation is rather similar to that yielded by some of the extrapolations in current use.

Unfortunately, just as that manuscript was going to press, major questions were raised concerning the adequacy of the radiation exposure estimates. These questions concern at least four different aspects of the dosage question (radiation spectrum of the bombs, absorbance of radiation by moisture in the air, shielding properties of Japanese roof tile, attenuation of gonadal dose by body tissues). As the focus of attention has shifted from one aspect of the dosage question to the next, dose estimates have gone up and down like a yo-yo, but mostly down. My quasi-informed guess is that the new estimates will be about 75% of the old; the only certain conclusion is that the doubling dose estimate of 156 rem is no longer valid.

The continuation since 1980 of the various aspects of the program just described has resulted in substantial accretions to this data base. Specifically, there will be additional data with respect to survival, the frequency of sex-chromosome aneuploids, and, especially, the frequency of electrophoretic mutations. In addition, two types of data which have been accumulating will become available for analysis, namely, data on the frequency of mutation yielding reciprocal translocations, and, on a small scale, data on the frequency of mutation yielding loss of activity for a battery of 11 enzymes. We propose, as soon as the new dosage estimates are in, to bring the entire study up to date, at which time the question of the error to be attached to the estimate must be addressed as definitively as possible.

HUMAN DATA: THE NEW TECHNOLOGIES

Even with the expected accretions to the Hiroshima-Nagasaki data just described, given the numbers to be available, and the imprecision inherent in some of the assumptions that enter into the doubling dose calculation, the error surrounding any estimate of doubling dose remains undesirably large. In this section we briefly consider two technologies that may make very significant contributions to improving this estimate. Both technologies have the advantage that, unlike congenital defects/chromosomal abnormalities of certain types, the indicator traits are apparently not usually the object of strong selection nor subject to biased reporting, so that they can be studied some years after the mutagenic exposure. On the other hand, it is presumed these effects serve as a kind of "litmus paper" for the occurrence of more serious genetic events. Furthermore, since I regard it as certain that studies will be undertaken on other populations with lesser exposures and smaller numbers of children than in the Japanese cohorts, not only are there other applications for these technologies but they can be so standardized (and are so standardizable) that the results of various studies can be readily combined.

Of the two new approaches on the horizon, one is closer than the other. The first makes use of the ability with two-dimensional polyacrylamide gel electrophoresis of complex protein mixtures, coupled with very sensitive polypeptide stains, to visualize hundreds of proteins from a cell type, tissue, or body fluid. Time does not permit a description of the present state of our efforts to build a monitoring system around this technology (cf. Neel et al., 1984). A major aspect of our program has been the development of computer algorithms which will detect a polypeptide which is present in a child but not in its parents (Skolnick, 1982; Skolnick et al., 1982; Skolnick and Neel, in press).

At first glance a 2-D gel projects formidable complexity. Since, however, in the first dimension the separation of the polypeptides is on the basis of their pI, and in the second dimension primarily on the basis of molecular weight, one can conceptualize a 2-D gel as an array of one-dimensional electrophoretograms ordered from the "top" to the "bottom" of the gel by molecular weight. The reading of these polypeptides for variation in the pI corresponds in many respects to the reading of 1-D gels for variants. The experience in Japan has demonstrated the feasibility of studying mutation altering electrophoretic mobility; 2-D gels represent a much more efficient approach to doing just that.

It is important not to be perceived as suggesting that in the present state of 2-D PAGE, the millenium has arrived. Far from it. We are having lots of problems. A research-quality gel for the study of mutation demands that some 30 steps be implemented with precision. There is an urgent need for technical improvements resulting in more reproducible gels. There is also a need to increase the amount of reliable information which can be extracted from a gel or blood sample. Our own group is pushing in two directions. One is towards better quantification of the stained polypeptides. If this can be done with high accuracy, it should permit the identification of variants (and mutations) in which the normal gene product is not produced. There is reason to believe that mutation resulting in these nulls is much more frequent than mutation resulting in electromorphs (e.g., Mukai and Cockerham, 1977; Johnson and Lewis, 1981; Johnson et al., 1981). This development would thus greatly extend the usefulness of the method. The other push is towards developing the ability to extract from these gels a sufficient quantity of both normal and variant polypeptides to permit the kind of characterization necessary for rigorous biochemical genetics. Incidentally, at the far end of the spectrum of possibilities raised by these gels is the synthesis of an oligonucleotide probe patterned after a suitable amino acid sequence in a polypeptide isolated from a gel, after which, by any of several procedures, it should be possible to establish the position of the corresponding structural gene with reference to a chromosome. It boggles my mind that we can be mapping a newly arisen mutation affecting a protein to a chromosomal position before we have any idea what the protein does!

The other new approach on the horizon to the study of mutation involves the evolving ability to characterize DNA rapidly and detect nucleotide differences between individuals. In principle, the study of mutation at the DNA level is the same as at the polypeptide level: One must develop efficient technologies to detect attributes of a child's DNA not present in either parent. My failure to devote more time to this possible development reflects a lack of time rather than of interest, plus the fact the 2-D technology is somewhat further along. The challenge is of course to develop the most efficient approach to the detection of DNA sequences (or deficiencies or duplications) in a child which are not present in either parent. There are a number of forms this approach could assume; I expect only experience will permit a choice between them.

One of these new technologies, the 2-D gel approach, is in the process of being incorporated on a pilot basis into the study in Japan, and the second will be as soon as the technology is a little further along. The experience there should be extremely valuable to any future studies on other populations.

THE NUMERICAL REQUIREMENTS OF GENETIC STUDIES: TWO VIEWS

At this juncture it is appropriate that we consider the numerical requirements of genetic studies of this type. Mutation is a rare event. For instance, mutation resulting in electrophoretic variants of proteins (presumably for the most part due to nucleotide substitutions) occurs with an approximate frequency of 3-4 x 10^{-6}/gene/locus/generation. One thus needs relatively large numbers to detect modest increases (or decreases) in mutation rates. There are two rather different approaches one can take to estimating the necessary magnitude of any proposed study. On the one hand, one can take the position that a mutational effect should not be inferred or its magnitude estimated until there is a statistically significant difference between the study and the control population. We have on several different occasions examined the numbers necessary for what might be termed "a conclusive result," under a variety of assumptions as to the baseline mutation rate, the magnitude of the Type I and Type II errors one will accept, and the minimal magnitude of the effect one wishes to demonstrate (Neel, 1971, 1980). The numbers are indeed formidable. An example we have used frequently requires two samples of approximately 7,000,000 observations each, given conventional assumptions concerning Type I and Type II errors and baseline mutation rates, to demonstrate a 50% difference in the mutation rates of the two populations from which the samples are drawn.

Before you dismiss such numbers as absolutely preposterous, recall that with respect, for instance, to 2-D gels, each spot usually represents the product of two alleles, and a single type of preparation might yield 50 spots unique to that preparation. With, say, four convenient sources of preparations per individual human or experimental animal, a single organism could yield data on mutation at 400 loci. The 7,000,000 observations thus reduce to about 18,000 samples.

The other approach, much more applicable in exposures to radiation than in exposures to chemicals, is to assume that one is in an estimation rather than a hypothesis-testing mode. This approach is appropriate to radiation exposures because of the certainty that radiation penetrates to produce mutation and the relative ease with which gonad exposures can be calculated--neither true for the chemical mutagens. On the other hand, given the public concerns which are upon us, perhaps, just perhaps, the course with respect to the chemical mutagens is to proceed with studies where the basic motive is "damage control," not hypothesis testing. Every set of findings permits some estimation of treatment effect--the larger the numbers, the more accurate the estimate. In this approach, the error attached to the observation can be used to set an upper limit to the magnitude of the possible effect, even when the observation itself is counter-hypothesis.

If within the next decade, several additional genetic studies were to be conducted on the most appropriate human groups that can be identified, it would be important they be designed so that the results can be combined not only _inter se_ but with the

Japanese data. I suggest it will also be important, even when the event is as rare as mutation, to attempt to subdivide exposure categories, so that one can take advantage of the greater statistical power of a regression-type analysis of effect on dose (rather than a simple 2 x 2 contrast).

WHY WE CANNOT JUST EXTRAPOLATE FROM ANIMAL MODELS TO HUMAN RISK

Given the obvious difficulties in developing genetic risk estimates from studies of the sequelae of human exposures, it would indeed be convenient if we could extrapolate with the requisite intellectual rigor from animal models to the human situation. I would like to suggest that at this point, the internal inconsistencies in the data from the chief human surrogate, the mouse, have reached the point where the uncertainties in such extrapolations are very clear. At the outset of this critique, I express my admiration for the rigor which has accompanied the effort to build up a corpus of data on mammalian radiation genetics. Furthermore, in the 30 years following World War II, the data on human populations were clearly inadequate to do more than confirm in a general way the order-of-magnitude accuracy of the projections from mice. It is, indeed, the very breadth of this effort in mammalian radiation genetics and the care with which it has been conducted, that has brought to light some of the problem areas on which a number of geneticists have commented recently (Denniston, 1982; Kohn, 1983; Lyon, 1983; Neel, 1983). In this section I will mention briefly five problem areas in the field:

1. Effect of Radiation on Litter Size, Preweaning Mortality, Weight at Weaning, Frequency of Malformations, and Other General Indicators of Fitness in Experimental Mammalian Populations

Green (1968) has reviewed in detail the numerous experiments (at least a dozen) undertaken in the 1950s and 1960s, involving mice, rats, and swine, treated with relatively large amounts of radiation for from several to as many as 30 generations, in which the indicator traits were the kinds of outcomes just listed. There is a near-absence of the anticipated strain deterioration, leading Green to write: "Attempts to measure the effects of presumptive new mutations on the fitness and other traits of heterozygotes, either after one generation of parental exposure or after several generations in which the mutations may accumulate have, in general, been unsuccessful" (p. 115). Although explanations such as the relatively small size of an individual experiment or close inbreeding in some series (leading to the rapid elimination of induced mutations) have been invoked to explain the paradox, this scarcely seems an adequate explanation for the total body of data. Given the importance of these and related indicators in the studies of humans, this situation deserves clarification.

2. Studies on Murine h-2 Locus

As summarized by Kohn and Melvold (1976), a series of studies on X-ray-induced mutation rates at the 40 or more genes screened for mutation at the h-2 locus by the "cross-compatibility" technique employed has yielded an induced rate some 2% of that yielded by the 7-locus test which sets the standard in mouse radiation genetics. While attempts have been made to explain this discrepancy on technical grounds, I do not find the arguments entirely convincing.

3. The Relatively Low Rate of Radiation-Induced Lethals in Mice Yielded by Roderick's Experiments

Using a pericentric inversion large enough to "tie up" genetic segments proximal to the inversion in both chromosomal arms, Roderick (1983) has recently developed an estimate of the sensitivity of the mouse genome to radiation-induced lethal mutations which is approximately two orders of magnitude below the estimate yielded by the 7-locus method of Russell. Writing that "this discrepancy of one hundred fold is perplexing and deserves comments," Roderick seems to favor the different character of the loci in the two test systems as the most likely explanation.

4. Some Emerging Questions Concerning the Radiation-Induced Dominant Skeletal Mutations of the Selbys'

In recent years, the demonstration that radiation (and ENU) induces dominantly inherited mutations affecting the eyes and skeleton of mice (Ehling, 1966; Selby and Selby, 1977; Kratochvilova and Ehling, 1979) has been used to extrapolate to the first-generation phenotype impact of the radiation of humans (e.g., Selby, 1983; Ehling, 1983), an extrapolation adopted by several national and international committees. Recently, Lovell, Leverton, and Johnson (in press) find that in a different strain of ENU-treated mice, in which skeletal variation is assessed by a different approach, induced inherited skeletal malformations have a negligible frequency. While this observation, of course, in no way invalidates the earlier finding, it does at the very least suggest the need for caution in the acceptance of the extrapolations which have been made. A repetition of the findings of the Selbys in several other strains is highly desirable.

It is true that one cannot use "negative data" to predict the deleterious effects of radiation (although such data can be used to set upper limits), but it does seem as if this corpus of data indicating a lesser sensitivity of the mouse to mutagens than commonly assumed, if confirmed, has now reached a level such that in the future it cannot be ignored. At the same time, a theoretical basis for regarding mouse-man extrapolations with added caution is emerging. This constitutes our fifth point.

5. Spontaneous Mutation Rates in Mice and Humans.

Now that comparable indicator traits--homologous proteins--are available for mutation rate studies in mice and humans, it is striking (in "confirmation" of what has long been suggested by the morphological approach) how similar spontaneous rates appear to be in mice and humans (Neel, 1983). Since, however, the human breeding cycle is some 25-fold longer than that of the mouse, with the attendant increased exposures to mutagens, one interpretation of this observation is that the similarity is due to the development of superior genetic repair mechanisms in the hominid line of evolution. There are thus theoretical reasons for presuming humans may be less sensitive to exogenous genetic insults than mice.

I would suggest that in the field of mammalian radiation genetics, the administrative and scientific question which is now of paramount importance is the degree to which it is desirable to concentrate our effort on resolving some of these discrepancies and improving the basis for extrapolation from mouse to man with current techniques, and the degree to which human studies are emphasized, the assumption being made that it is not either/or but that there will continue to be a mix of both. In reaching this decision, we may have to be guided in part by the obvious reluctance of the public to accept some of our extrapolations: The public wants people data where possible.

TOWARD A UNIFIED PROGRAM

I would now like to make a proposal which I do not expect will find easy acceptance. It is that an appropriate share of the enormous technical and intellectual resources of the National Laboratories, plus appropriate nonlab collaboration, be coordinated in a push to bring on line all aspects of a program to monitor mutation rates at the nucleic acid and protein levels. This effort must include an appropriate experimental component, but the primary objective should be to define human populations for study and initiate these studies (even though the populations are less than optimal) before further studies of even less appropriate populations than the veterans who participated in the atomic bomb tests are mandated.

In both the proposed animal and human studies, I urge that indicators of genetic damage in the somatic cells of the subjects be pursued at the same time that indicators of transmitted genetic damage in their offspring are studied. This would constitute the initial step towards building the intellectual bridges and "conversion factors" which some day may render it possible to predict germ-line genetic damage from studies of appropriate somatic cell indicators.

With respect to the human studies, I suggest that in addition to a continuation of the genetic studies in Japan, along the lines sketched out, there should be at least one study initiated here in the U.S. just as soon as possible. A compilation of groups with unusual exposures to radiation has been sponsored by the Nuclear Regulatory Commission (U.S. NUREG, 1980). A group of unusual interest, although small and highly dispersed, is composed of the children of individuals receiving abdomino-pelvic radiation and/or chemotherapy for malignant disease. The recent radiation accident in Juarez, Mexico (Marshall, 1984) also defines a group of interest, but it is imperative that if any collaborative study be proposed, great care must be taken to observe the very proper Mexican sensitivities.

It would not seem unreasonable to suggest that some properly constituted group, such as the Committee on Federal Research into the Biological Effects of Ionizing Radiation, buttressed by appropriate advice from individuals familiar with the groups under consideration, as well as individuals familiar with the details of large-scale field surveys, could play an important role in designating U.S. populations for study. Given the strong correlation between mutagenesis and carcinogenesis, a tip-off as to the suitability for genetic studies of a population exposed to a potential mutagen would be an increased frequency of malignancy in the population. However, in view of the 20-year lag period between a carcinogenic exposure and the appearance of an increased frequency of most solid tissue tumors, reliance on an indicator with a shorter lag period, such as myelogenous leukemia, seems appropriate.

With respect to the animal studies, I would suggest considerable economy of effort could be achieved by using the same experimental population for both the studies on proteins and the studies on DNA markers. This does not imply that all the work be done in one laboratory; there are ample opportunities for inter-laboratory sharing.

Lest I seem to be advocating a large piece of applied research, I remind you that both the proposed human and laboratory studies contain a large component of very basic science. To outline all the ramifications is beyond the scope of this presentation, but the opportunities for "basic" data range from a better understanding of genetic variation at these levels (and the discrepancy between the frequency of DNA and protein polymorphisms) to the fundamental contribution to evolutionary theory which would result from establishing how frequently deletion- and duplication-type mutations of DNA occur.

THE PRACTICAL IMPLICATIONS/APPLICATIONS OF AN IMPROVED DATA BASE

In addition to its contribution to radiation biomedicine, what are the practical implications of such a program. They would seem to fall into three categories.

Firstly, improved data would clearly be useful to regulatory agencies. One has only to return to the derivation of the EPA protocol for the disposal of high - level radioactive waste to appreciate the need for such data.

Secondly, improved data should go far to forestall mandated studies on populations which by all standards have been subjected to some de minimus increased risk, such as the veterans involved in the nuclear bombs test shots. But for these better data to discharge that function, they must be collected in such a fashion that there will be public confidence in their validity, an issue to which I return shortly.

Thirdly, such data would clearly assist in the appropriate legal approach to the individual who has sustained an unusual mutagenic exposure and has reason to believe his children have suffered in consequence. In this context, I would like to endorse, for genetic purposes, the concept of "proportionate risk" in whose development Bond (1959, 1981, 1982) has played such an important role. You will recall that this concept requires indemnification for the development of a malignancy following a carcinogenic exposure, partial indemnification to begin at some arbitrary level of exposure, this increasing with increasing exposure up to full indemnification. The concept will be much more difficult to apply to mutagenesis than to carcinogenesis, because the genetic end points are so many and varied. The alternative to attempting to proceed in this manner will unfortunately be the continuation of an adversary situation with no clear guidelines.

CREDIBILITY

We come finally to the difficult issue of credibility. This is one that the laboratory scientist does not face, at least to nearly the same extent as the genetic epidemiologist. The public may not readily accept the results of extrapolations from mice to humans, but they will seldom doubt the sincerity of the attempt. But with respect to the results of studies on human populations, a deep-seated cynicism and skepticism has become all too entrenched. As long as the Department of Energy is perceived as a proponent of nuclear energy, the results of any studies it conducts on the effects of radiation will unfortunately be suspect to many. Studies by the Department of Health and Human Services will be somewhat less suspect, but, for instance, when Land et al. (1984)--quite correctly, in my opinion--critiqued the study of Lyon et al. (1979) on leukemia in the southern 17 counties of Utah between 1951 and 1958, the major period of aboveground nuclear weapons testing at the Nevada Test Site, I suspect this was perceived by the inhabitants of that area as Big Government again moving to quash the courageous local investigator. The situation is only slightly less confusing when OTA criticizes a study by the NAS.

Recently the House has been considering a resolution (H.R. 2350) which would establish a Presidential Commission on Human Applications of Genetic Engineering. Assuming that a Presidential Commission is the most sacrosanct body our political system currently provides, would it be appropriate to urge the creation of such a Commission to consider the entire issue of pollution effects? This suggestion is not meant to be critical of the epidemiological studies of the past, nor to suggest any revisions with respect to the funding of the studies; it pertains solely to creating a mechanism for rendering the findings of these studies more acceptable.

REFERENCES

Bond, V. P., 1959, The medical effects of radiation, in: "Proceedings, Thirteenth Annual Convention, National Association Claimants' Counsel of America," p. 117, W.H. Anderson, Cincinnatti.

Bond, V. P., 1981, The cancer risk to radiation exposure: Some practical problems, Hlth. Phys., 40:108.

Bond, V. P., 1982, Statement. "Hearings on Radiation Exposure Compensation Act of 1981 (S. 1483)", p. 242, GPO, Washington.

Denniston, C., 1982, Low-level radiation and genetic risk estimation in man, Ann. Rev. Genet., 16:329.

Ehling, U. H., 1966, Dominant mutations affecting the skeleton in offspring of X-irradiated male mice, Genetics, 54:1381.

Ehling, U. H., 1983, Cataracts--indicators for dominant mutations in mice and man, in: "Utilization of Mammalian Specific Locus Studies in Hazard Evaluation and Estimation of Genetic Risk," p. 169, F. de Serres, ed., Plenum Press, New York.

Green, E. L., 1968, Genetic effects of radiation on mammalian populations, Ann. Rev. Genet., 2:87.

Johnson, F. M., and Lewis, S. E., 1981, Electrophoretically detected germinal mutations induced in the mouse by ethylnitrosourea, Proc. Natl. Acad. Sci. USA, 78:3138.

Johnson, F. M., Roberts, G. T., Sharma, R. K., Chasalow, F., Zweidinger, R., Morgan, A., Hendren, R. W., and Lewis, S. E., 1981, The detection of mutants in mice by electrophoresis: Results of a model induction experiment with procarbazine, Genetics, 97:113.

Kato, H., Schull, W. J., and Neel, J. V., 1966, A cohort-type study of survival in the children of parents exposed to atomic bombings, Amer. J. Hum. Genet., 18:339.

Kohn, H. I., 1983, Radiation genetics: The mouse's view, Radiat. Res., 94:1.

Kohn, H. I., and Melvold, R. W., 1976, Divergent X-ray induced mutation rates in the mouse for H and "7-locus" groups of loci, Nature (Lond.), 259:209.

Kratochvilova, J., and Ehling, U. H., 1979, Dominant cataract mutations induced by λ-irradiation of male mice, Mut. Res., 63:221.

Land, C. E., McKay, F. W., and Machado, S. G., 1984, Childhood leukemia and fallout from the Nevada nuclear tests, Science, 223:139.

Lovell, D. P., Leverton, D., and Johnson, F. M., Lack of evidence for skeletal abnormalities in offspring of mice exposed to ethylnitrosourea (ENU), Proc. Natl. Acad. Sci. USA, in press.

Lyon, M. F., 1983, Problems in extrapolation of animal data to humans, in: "Utilization of Mammalian Specific Locus Studies in Hazard Evaluation and Estimation of Genetic Risk," p. 289, F. J. de Serres, and W. Sheridan, eds., Plenum Press, New York.

Lyon, J. L., Klauber, M. R., Gardner, J. W., and Udall, K. S., 1979, Childhood leukemias associated with fallout from nuclear testing, N. Eng. J. Med., 300:397.

Marshall, E., 1984, Juarez: An unprecedented radiation accident. News and comment, Science, 223:1152.

Mukai, T., and Cockerham, C. C., 1977, Spontaneous mutation rates at enzyme loci in Drosophila melanogaster, Proc. Natl. Acad. Sci. USA, 74:2514.

National Science Foundation, 1980, "The Five-Year Outlook: Problems, Opportunities and Constraints in Science and Technology," Vol. I and II, U. S. Government Printing Office, Washington.

Neel, J. V., 1971, The detection of increased mutation rates in human populations, Perspect. Biol. Med., 14:522.

Neel, J. V., 1980, Some considerations pertinent to monitoring human populations for changing mutation rates, 1980, in: Proceedings, XIV International Congress of Genetics "Well Being of Mankind and Genetics," p. 225, MIR Publishers, Moscow.

Neel, J. V., 1983, Frequency of spontaneous and induced "point" mutations in higher eukaryotes, J. Hered., 74:2.

Neel, J. V., in press, a, Some aspects of quantifying and acting upon the risks of radiation exposures, in: "Proceedings, Banbury Center Conference on Risk Quantitation and Regulatory Policy," M. Shodell, ed., Cold Spring Harbor Press, New York.

Neel, J. V., in press, b, How can we best evaluate, and compensate for genetic hazards in the environment and work place, in: "Genetics and the Law: Third National Symposium," A. Milunsky, ed., Plenum Press, New York.

Neel, J. V., and Schull, W. J., 1951, "The Effect of Exposure to the Atomic Bombs on Pregnancy Termination in Hiroshima and Nagasaki," National Academy of Sciences-National Research Council Publication 461, Washington, D. C.

Neel, J. V., Schull, W. J., and Otake, M., 1982, Current status of genetic follow-up studies in Hiroshima and Nagasaki, p. 39, in: "Progress in Mutation Research," K. C. Bora, G. R. Douglas, and E. R. Nestmann, eds., Elsevier Biomedical Press, Amsterdam.

Neel, J. V., Rosenblum, B. B., Sing, C. F., Skolnick, M. M., Hanash, S. M., and Sternberg, S., 1984, Adapting two-dimensional gel electrophoresis to the study of human germ-line mutation rates, p. 259, in: "Two Dimensional Gel Electrophoresis of Proteins," J. E. Celis, and R. Bravo, eds., Academic Press, New York.

Roderick, T. H., 1983, Using inversions to detect and study recessive lethals and detrimentals in mice, p. 135, in: "Utilization of Mammalian Specific Locus Studies in Hazard Evaluation and Estimation of Genetic Risk, F. de Serres, and W. Sheridan, eds., Plenum Press, New York.

Schull, W. J., and Neel, J. V., 1958, Radiation and the sex ratio in man, Science, 128:343.

Schull, W. J., Neel, J. V., and Hashizumi, A., 1966, Some further observations on the sex ratio among infants born to survivors of the atomic bombings of Hiroshima and Nagasaki, Am. J. Hum. Genet., 18:328.

Schull, W. J., Otake, M., and Neel, J. V., 1981, Hiroshima and Nagasaki: A reassessment of the mutagenic effect of exposure to ionizing radiation, in: "Human Mutation: Biological and Population Aspects," p. 277, E. Hook, ed., Academic Press, New York.

Selby, P. B., 1983, Applications in genetic risk estimation of data on the induction of dominant skeletal mutations in mice, in: "Utilization of Mammalian Specific Locus Studies in Hazard Evaluation and Estimation of Genetic Risk," p. 191, F. de Serres, and W. Sheridan, eds., Plenum Press, New York.

Selby, P. B., and Selby, P. R., 1977, Gamma-ray-induced dominant mutations that cause skeletal abnormalities in mice. I. Plan, summary of results, and discussion. Mut. Res., 43:357.

Skolnick, M. M., 1982, An approach to completely automated comparison of two-dimensional electrophoresis gels, Clin. Chem., 28: 979.

Skolnick, M. M., Sternberg, S. R., and Neel, J. V., 1982, Computer program for adapting two-dimensional gels to the study of mutation, Clin. Chem. 28:969.

Skolnick, M. M., and Neel, J. V., in press, An algorithm for comparing 2-D electrophoretic gels, with particular reference to the study of mutation, Adv. Hum. Genet.

U. S. Nuclear Regulatory Commission, 1980, "The Feasibility of Epidemiological Investigations of the Health Effects of Low-Level Ionizing Radiation. Final Report, (NUREG/CR-1728)," Division of Technical Information and Document Control, U. S. Nuclear Regulatory Commission, Washington, D. C.

DISCUSSION

Ehling: You compared the paper of Lovell and Johnson (1983) with the approach made by Paul Selby (1977). One cannot do this because Johnson and Lovell (1983) used methods where they took measurements of bones that clearly identify inbred strains of mice; these are quantitative characters but not gene mutations. Therefore, the outcome of their study is not relevant to the results of Paul Selby. In regard to the cataracts, these experiments are being made by J. West at Harwell and I think they will be published soon.

Neel: I agree with your latter point, but not with your former point.

Ehling: It is such a completely different approach. You could then quote Gruneberg with his studies (1983) in which he shows that minor skeletal variants have nothing to do with mutations, but depend on many other factors.

Neel: My comment has two sides. First, we do need a repitition of the Selby's work in other strains. There are horrendous extrapolations being made on the basis of work with one mouse strain to the total load of human morbidity from radiation exposures. Secondly, I do not readily accept your point that mutation should have no affect on the metrics that Lovell and colleagues measured.

Ehling: I did not say that, but I think that you cannot compare two completely different approaches and say that one approach does not work, therefore, the other approach is wrong. The two experiments are quite different.

Neel: I hold that one approach had not confirmed the other approach, but not that either is wrong. We have a difference of opinion here that I do not think we will settle too readily.

Brown: Are the human cohort studies proposed of sufficient power to detect the highest potential rates of mutation in the population?

Neel: The numbers game, of course, has troubled us for many, many years, and here, I would take the position that we're going to close in on truth slowly and by degrees. I happen to have a slide that illustrates just one of many exercises we've gone through. We assumed ° errors of 0.05 and β errors of 0.2, which are the usual types of assumptions one makes; we also assumed a detectable spontaneous rate of 10^{-5} locus/generation, and then asked how many individual observations do we need to detect differences in mutation rates of various magnitudes between two samples. For instance, at the 50% point, it would require two bodies of 7 million observtion each to yield a statistically significant difference. On the other hand, if you take the position, as we do in radiation, that mutations are produced by any radiation exposure, you can proceed somewhat differently. As you are no longer testing a hypothesis, you can now generate the best estimate possible from the data. Seven million sounds unrealistic, but I think there is a fighting chance such numbers may be achieved with the new technologies. For instance, there are in blood samples, about 200 proteins we can score with high accuracy (in plasma erythrocyte/lysate, platelet/lymphocyte preparations). So from each person we could obtain 400 locus tests. It is a question of whether society wants to put into the research enterprise something like one-thousandth of the money that will be spent on regulation and litigation in the next ten years.

Abrahamson: Would you care to comment on the far greater uncertainty associated with the non-biological issues of high level radioactive waste disposal, such as effluents, transport, metrology relative to the uncertainty of the biological effects.

Neel: I hope I have made it clear that the uncertainties I am speaking to are due to strain and locus differences which would not have been anticipated at the beginning of this research. In arriving at some of the risk evaluations, certainly there are many other factors to be taken into account besides the straight genetics. But right now, I want to clean up our part of the problem.

Thilly: It would seem to me that a test to check for genetic change in a population would pick up those chemical expsoure which are, presumably, the strongest in our environment. This meeting started with Richard Peto pointing out that a large portion of the population is exposed daily to a large variety of nitrosamines, polycyclic aromatics, etc., via the lungs and gastrointestinal tract. Do you think that the electrophoretic approach or the multi-enzymic approaches would be capable of detecting somatic genetic changes resulting from continuous smoking?

Neel: As you know from the work you presented, there are a lot of intriguing ways to look at genetic damage in somatic cells, ranging from DNA adducts to sister chromatid exchanges and chromosome breaks. Right now, we have no idea of how to relate to those observations to transmitted genetic damage. I have suggested that any future study in mouse or humans should encompass those somatic cell indicators, so that we can build the necessary intellectual bridges between somatic genetic damage and transmittable genetic damage. Once we have made these connections, we may be able to avoid the very laborious genetic studies and trust the somatic observations.

Thilly: Could you reflect on the idea perhaps we might be able to get far larger number of germinal mutations by examining sperm samples from a much smaller number of individuals exposed to smoking, occupational hazards or chemotherapy, for example? Would this information not give a value for germinal mutations that we could then expand into the kind of analysis you're discussing?

Neel: The only data I trust completely is based on transmitted damage, as reflected in the children of exposed persons. If I were waving the magic wand, I would say we should go for the worst cases that we could identify around the world. For instance, in Japan, our group has access to a Japanese population of former workers from the installation that made sulphur mustard gas during World War II (Neel et al., in press). The frequency of bronchogenic carcinoma in this group is very high, and there is probably an increase also in primary hepatomas and gastric cancer. This is probably the most highly mutagenized human population in the world. The population is small and is disappearing rapidly, either through the death of the workers or from their children moving out of the area. In another five years, we will have largely lost the opportunity to study this group. We are beginning a pilot study with 2-D gels to examine a variety of protein markers for evidences of mutation in Hiroshima. I would like to reach out from Hiroshima to Okujima, which is only forty miles away; in fact, we are already doing that with one-dimensional protein electrophoresis. We should consider other groups that ought to be brought into our surveys.

Ricci: I wish to state, for the record, that I find it extremely upsetting that we have questioned the propriety of individuals having recourse to the court system for redress. Insofar as there is an under-representation here of lawyers, economists, social scientists, this is an unfair statement to make. I would like to leave you with the following message. It is a necessary condition for humans to exist in an organized society in which we do have redress at individual levels through the court system. I also suggest that the decision handed down by the Circuit Court in Utah on radiation exposure is a document of about 500 pages, and the decision is a fairly well reasoned one. We have to sue each other left and right, nevertheless, individually, we must have recourse should we feel that we were not properly represented (a) through the political system, and (b) to our employers.

Neel: Let me come back very strongly on that. I have not drawn that inference, for I firmly believe that justice should be done, but on the basis of knowledge, not guess work.

Ricci: Whenever the courts are faced with scientific information only, they take that information as advisory. For example, when the courts last year looked upon the impact of formaldehyde on nasal carcinoma, the uncertainties which were given to them (however inaccurate the studies turned out to be later on) were such that the court could not do anything with it.

REFERENCES

Gruneberg, H., 1963, The Pathology of Development; A study of inherited skeletal disorders in animals, Wiley, New York.

Johnson, F.M., and Lovell, D.P., 1983, Dominant skeletal mutations are not induced by ethylnitrosourea in mouse spermatogonia, Environ. Mutagen., 5:496.

Lovell, D.P., and Johnson, F.M., 1983, A search for skeletal mutations induced by ethyl nitroso urea (EtNu) using morphometric methods, Environ. Mutagen., 5:246.

Selby, P.P., and Selby, P.R., 1977, Gamma-ray-induced dominant mutations that cause skeletal abnormalities in mice, Mutat. Res., 43:357.

References

PROSPECTS FOR CELLULAR MUTATIONAL ASSAYS IN HUMAN POPULATIONS

Mortimer L. Mendelsohn

Biomedical Sciences Division,
Lawrence Livermore National Laboratory
Livermore, CA, 94550

INTRODUCTION

Practical, sensitive, effective, human cellular assays for detecting somatic and germinal mutations would have great value in environmental mutagenesis and carcinogenesis. When available, such assays should allow us to fill the void between human mutagenicity and the data that exist from short-term tests and from mutagenicity in other species. We will be able to validate the role of somatic mutation in carcinogenesis, to identify environmental factors that affect human germ cells, to integrate the effects of complex mixtures and the environment in the human subject, and to identify people who are hypersusceptible to genetic injury. Human cellular mutational assays, particularly when combined with cytogenetic and heritable mutational tests, promise to play pivotal roles in estimating the risk from low-dose radiation and chemical exposures. These combined methods avoid extrapolations of dose and from species to species, and may be sensitive enough and credible enough to permit politically, socially and scientifically acceptable risk management.

Cytogenetic methods already have wide applicability to the human (Evans, this volume), and heritable mutational measurements are undergoing an extensive epidemiological trial in Japan (Neel, this volume). However, we presently have no validated, generally accepted assays for human somatic or germ cell mutations. Bleak as this sounds, I believe the technology is almost available and that we are on the verge of having several promising cellular mutational methods for human application.

Somatic and germinal assays have one enormous advantage over heritable assays. As is clear from the studies of the Hiroshima and Nagasaki populations, observable heritable genetic changes at any particular locus occur with great rarity (Neel, this volume). To find sufficient mutational events for a measurement of rate, one must study millions of offspring, millions of loci, or some combination of the two. The expectation is that somatic and germinal mutation rates will be similarly small, but for these measurements any one person can provide the millions or billions of cells needed to estimate a mutation rate. The challenge is to detect the rare, one-in-a-million events at the cellular level and to have good assurance that the events represent mutational phenomena.

Detection strategies at the cellular level can be based on four types of changes:

- alteration in behavioral phenotype, such as drug resistance.
- alteration in gene product, such as a modified protein.
- alteration or loss of a messenger RNA.
- alteration or loss of a DNA sequence.

The farther down the list one goes, the closer one is in principle to a verifiable mutational event; however, to my knowledge the RNA- and DNA-based methods are not yet available for single-cell application. To identify RNA and DNA changes would require methods that detect loss of a single sequence or gain of a minimally deviant single-copy sequence in a single cell. Assays based on changes in behavioral phenotype and in gene product seem more feasible with present technology and are discussed below.

HPRT SOMATIC CELL MUTATION BASED ON 6-THIOGUANINE RESISTANCE

Albertini has pioneered the development of assays in human peripheral blood lymphocytes based on resistance to 6-thioguanine. In the original assay (Strauss and Albertini, 1979), freshly drawn lymphocytes were exposed to phytohemagglutinin (a mitogen), 6-thioguanine (a purine analogue that is cytotoxic to cells attempting DNA synthesis in the presence of effective levels of the enzyme HPRT (hypoxanthine phosphoribosyltransferase)), and tritiated thymidine (a label to detect those cells successfully carrying out DNA synthesis).

Autoradiography was used to count both the rare, resistant, thymidine-labelled cells, and the cells at risk (based on labelled cells in cultures not exposed to 6-thioguanine).

Encouraged by early results, Albertini and his colleagues collected a sizable body of data using this assay (Albertini, 1982). They observed background levels that did not change with age, increased levels in cancer patients before and after cancer therapy, and increased levels after PUVA therapy. However, instability of background levels and consistently high variant frequencies with occasional values as high as one variant per 100 cells at risk led to the suspicion that the assay was detecting phenocopies as well as mutants. One source of phenocopies seemed to be circulating cells committed to DNA synthesis before exposure to mitogen. These cells and the excessively high variant frequencies were eliminated by introducing a cryopreservation step into the assay.

At this point in the evolution of the method, sufficient doubt had been cast on the mutational validity of the autoradiographic endpoint to make Albertini refocus his efforts toward a clonogenic version of the assay (Albertini et al., 1982). He was able to clone single cells in microwell cultures at an efficiency up to 50%, using a crude T-cell growth factor. This step led to a dilution assay which could be used both to estimate the frequency of thioguanine-resistant cells, and to support the mutational nature of the phenotype by demonstrating the corresponding HPRT enzyme deficiency in the resistant clones. Albertini is now in a position to validate the modified autoradiographic assay or to use the clonogenic assay, and to continue on with his epidemiologic applications of the methods.

Several other laboratories are pursuing these approaches with mixed success. Amneus et al.(1982) at the University of Upsalla have incorporated into the original Strauss and Albertini procedure the method of flow cytometric sorting to concentrate late S and G2 cells prior to autoradiography. This facilitates counting and eliminates some sources of phenocopies. Their recent results suggest that the variant cells found early after a mutagenic insult are predominantly due to DNA lesions which interfere with enzyme production (Amneus, Mattson, and Fellner-Feldegg, personal communication). As the lesions are repaired, these variants disappear and are replaced by less frequent, presumably stable mutants. This group feels that the short-term autoradiographic assay in human peripheral blood lymphocytes may be useful for estimating DNA damage but not mutation. On the other hand, Morley et al.(1982, 1983) have described autoradiographic and clonogenic assays which they believe successfully detect HPRT mutations in human peripheral blood lymphocytes. Evans and colleagues (this volume) have recently produced credible radiation dose-responses in vitro in human lymphocytes using a clonogenic version of the HPRT assay. Similarly, Jones et al.(1984) at Livermore have a clonogenic assay that is working well with mouse splenocytes mutagenized in vivo.

In spite of the difficulties, it seems likely that some form of HPRT assay will be available for general application in the near future. Assays for other genes involving drug resistance are also possible, although this approach is inherently limited for human application to those genes functioning effectively as single copies (that is, only to genes on the X chromosome) or to dominant selectable markers, such as diptheria resistance.

HEMOGLOBIN SOMATIC CELL MUTATION ASSAY

Papayannopoulou et al. (1976) at the University of Washington developed a monospecific polyclonal antibody to Hb S (sickle hemoglobin); they showed that the fluorescent antibody could be used to identify sickle-trait erythrocytes in smears, and raised the possibility of using this method as an assay for sickle-like somatic mutations in erythrocyte precursors. This general strategy of using known mutated gene products to generate immunologically specific reagents to detect new cellular mutants has broad potential applicability in somatic mutagenesis.

The same group developed a variety of similar immunologic probes to hemoglobins with single amino acid substitutions, such as Hb C (Papayannopoulou et al.,1977), as well as probes to hemoglobin frame-shift variants, such as Hb Wayne and Hb Cranston (Stamatoyannopoulos et al., 1980). They showed that the rate of appearance of Hb S- and C-bearing erythrocytes in normal adults was on the order of one in 10 million, and that this rate increased after cancer chemotherapy. However, making such determinations manually by fluorescence microscopy was extremely arduous, requiring one person-month to count a single sample.

In collaboration with Papayannopoulou and associates, a team from the Lawrence Livermore National Laboratory attempted to automate the Hb S method using flow cytometry (Bigbee et al., 1981). They developed a way to stain erythrocytes in suspension, and to examine one million cells per second. However, direct application was prevented by false-positive events occurring at roughly one per one hundred thousand erythrocytes. By using flow sorting to concentrate fluorescent cells, they were able to confirm the original frequencies of occurrence of variants, but found the results too variable for quantitative or epidemiologic application.

Continuing independently, the Livermore group (Bigbee et al., 1983; Jensen et al., 1984) has developed a suite of monoclonal, monospecific antibodies that recognize a variety of single base change, single amino acid substitutions in human, mouse and monkey hemoglobins. They have also developed improved methods of fixation that provide better antibody staining, more durable cells, and cells suitable for flow cytometric counting by scattered light. The problem of false-positive signals continues to plague the method but is now reduced by a factor of 10. With the present methods, in reconstruction experiments it is possible by flow sorting to retrieve 88 antibody labelled cells (coefficient of variation, 10%) from a mixture of 100 labelled and one billion unlabelled cells. Further developments using human, mouse and monkey blood samples are underway using the partially automated sorting method, as are attempts to automate the procedure fully.

An important limitation of this method has to do with its incredible resolution. Much as one may want the flexibility to measure single, specific base changes (i.e., the ultimate resolution of a genetic method), certain mutagens, such as ionizing radiation, may be very inefficient at producing such small lesions. In addition, the smaller the target size, the rarer the expected outcome. As mutant frequencies drop one or more orders of magnitude below one per million, it becomes increasingly difficult to detect the rare events or to have a large enough population of precursor cells at risk to measure the frequency of variants properly.

GLYCOPHORIN SOMATIC CELL MUTATION ASSAY

An alternative to the limitations of the hemoglobin-based assay would be an immunological assay that recognizes gene-loss mutations at the cellular level. Prior attempts at such assays (Atwood and Petter, 1961) are thought to have failed to detect true mutational events perhaps due to the high prevalence of phenocopies mimicking the desired genetic change. This problem can apparently be avoided in a setting where two alleles are codominantly expressed and can be separately detected in the same cell. The loss of either allele can then be detected in cells in which the other allele is still functioning normally and serves as a physiological control. Glycophorin-A in erythrocytes appears to be just such a system.

Human glycophorin-A is a 131-amino acid, trans-membrane sialoglycoprotein that codes for the M and N serotypes. M individuals have a serine and glycine respectively at the 1 and 5 position, counting from the amino-terminal external end. N individuals have a leucine and glutamic acid at these same positions. Otherwise, the two molecular types are identical. Each allele on average produces roughly 250,000 copies per erythrocyte whether in a homozygous or heterozygous configuration. The principle of the somatic mutation method is to label M and N independently with monoclonal antibodies carrying different fluorophores. In heterozygous individuals, every normal erythrocyte should show both fluorescent colors. A mutation inactivating one allele would result in erythrocytes lacking the M or N antigen. Such cells will fail to bind one of the monoclonal antibodies while expressing the second normally. Symmetry is expected. Thus the rate of variants seen for one color should be similar to that seen for the second.

Bigbee et al.(1983, 1984) have produced a series of non-cross-reacting monoclonal antibodies against the M or N glycophorins, as well as several antibodies that recognize both forms equally. When the antibodies are fluorescently labelled, they show high affinities to glycophorin in fixed erythrocytes in suspension. The flow cytometric signals of such erythrocytes typically have coefficients of variation of 12%, with peak means that vary from one person to the next by a coefficient of variation of as little as 3%. The Livermore studies have confirmed that the expression of the two alleles is quantitatively independent.

To date, putative mutants have been sought using only half of the assay, that is, an anti-M antibody has been used in conjunction with one of the anti-M,N antibodies. In this limited form, the method works well, finding roughly 8 variants per million erythrocytes in normal individuals. A preliminary trial on patients undergoing chemotherapy found average rates increased by 100%, with one individual increased by 800%. The full assay with both antibodies is now being tested and hopefully will provide confirmation through symmetry that the variants are indeed mutants.

One of the Livermore anti-M antibodies is able to recognize glycophorin-Mc, a rare allele which differs from N only in the presence of a serine at the 1-position. This antibody is thus discriminating M from N by the difference in the single terminal amino acid. It could be used to detect single amino acid substitutions in NN homozygotes, and should be comparable in resolution and performance to the hemoglobin reagents. Thus, with the set of antibodies now in hand, it may be possible to measure both point mutation and gene loss at the glycophorin locus.

LDH-X SPERM CELL MUTATION ASSAY

The same techniques described for somatic cells can in principle be applied to germinal cells, although presently only mature sperm can be readily sampled in the human. Ansari, Baig and Malling (1980) have described what may be the prototype for such methods. They immunized mice with rat LDH-X, a form of lactate dehydrogenase that is localized specifically in germinal cells. Highly specific polyclonal antibodies were produced, which recognized rat sperm, but also had the ability to recognize roughly one per million mouse sperm. The presumption is that the recognizable mouse sperm were expressing an LDH-X gene which mutated to the rat genotype at a single site. The sperm were identified by the staining reaction or fluorescence of their midpiece. A dose-response study in mice treated with procarbazine, a known heritable mutagen, gave very convincing results.

Unfortunately Malling has been unable to reproduce these results and has exhausted his supply of the active antibody. Nevertheless, the potential remains to apply similar techniques to human sperm. Whether such germinal methods will be predictive of heritable mutation remains to be seen, but whatever the outcome, cellular methods in sperm should provide an important insight into agents capable of reaching the testis and causing the expression of abnormal gene products.

DISCUSSION

None of the methods discussed above is available today for application to human testing, yet I hope it is clear why I am optimistic that such methods should soon be at hand. Obviously they will need extensive elaboration before general application. Validation will be especially important since the approaches for the most part do not measure mutations directly. Validation will presumably be made by a variety of consistency tests in human subjects and model systems. In some situations it may be possible to clone cells and verify directly; in others, such as the methods involving anucleate erythrocytes, cloning is out of the question. For the erythrocytic methods, the best one can hope for is that isolation of the variant cells will allow biochemical confirmation of the expected electrophoretic or other mutational changes in the involved proteins.

The full extension of these methods requires a great deal more than the few isolated examples given in this presentation. Ideally one would want methods that are sensitive to all types of DNA lesions, including base substitutions in all four bases, frame-shift lesions, and small and large deletions, inversions or translocations. A fully representative spectrum of genes should be included, as well as a broad and relevant sampling of tissues and cells. For somatic mutation there is an expectation that an elevated rate would be predictive of cancer risk. But would a high rate in erythrocytes predict only polycythemia? Or would erythrocytic mutations also predict cancer in a variety of other tissues and organs?

We must understand well the kinetics of how mutants appear and disappear in cell populations. These kinetics will involve turnover of cell compartments, the maturity of the initially mutated cell, and possible selective disadvantage or advantage for the mutant phenotype. A test system with short duration (i.e. short memory or short integration time) would be best suited for studying acute exposure situations. Intermediate- to long-term memory would be best for studying the effects of occupations, life styles, or geographical factors, or for searching out repair defective individuals. Finally it will be important to understand better what the time domain means to the rate calculation. Is the relevant parameter mutations per cell cycle, mutations per hour, mutations per generation, or some combination of these? How differently is this expressed in species with different lifetimes and different body sizes?

Clearly much remains to be done to develop the methods and to gain some understanding of their application. In spite of this, direct testing in humans, in my view, remains the single most tractable way to approach the low-dose risk problem for carcinogenesis and mutagenesis. I hope the resources and scientific strengths will be available to make it happen quickly and effectively.

ACKNOWLEDGEMENTS

Work performed under the auspices of the U.S. Department of Energy by the Lawrence Livermore National Laboratory under contract number W-7405-ENG-48.

REFERENCES

Albertini, R.J., 1982, Studies with T-lymphocytes: An approach to human mutagenicity monitoring, in: "Banbury Report 13, Indicators of Genotoxic Exposure," pp. 393-412, B.A.Bridges, B.E.Butterworth, and I.B.Weinstein, eds., Cold Spring Harbor Lab.

Albertini, R.J., Castle, K.L., and Borcherding, W.R., 1982, T-cell cloning to detect the mutant 6-thioguanine-resistant lymphocytes present in human peripheral blood, Proc. Natl. Acad. Sci. USA, 79:6617.

Amneus, H., Matsson, P., and Zetterberg, G., 1982, Human lymphocytes resistant to 6-thioguanine: Restrictions in the use of a test for somatic mutations arising in vivo studied by flow-cytometric enrichment of resistant cell nuclei, Mutat. Res., 106:163.

Ansari, A.A., Baig, M.A., and Malling, H.V., 1980, In vivo germinal mutation detection with "monospecific" antibody against lactate dehydrogenase-x, Proc. Natl. Acad. Sci. USA, 77:7352.

Atwood K.C., and Petter, F.J., 1961, Erythrocyte automosaicism in some persons of known genotype, Science, 134:2100.

Bigbee, W.L., Branscomb, E.W., Weintraub, H.B., Papayannopoulou, Th., and Stamatoyannopoulos, G., 1981, Cell sorter immunofluorescence detection of human erythrocytes labeled in suspension with antibodies specific for hemoglobin S and C, J. Immunol. Methods, 45:117.

Bigbee, W.L., Langlois, R.G., Vanderlaan, M., and Jensen, R.H., 1984, Binding specificities of eight monoclonal antibodies to human glycophorin A: Studies using M^cM and $M^kEn(UK)$ variant human erythrocytes and M- and MN^V-type chimpanzee erythrocytes, J. Immunol. (in press).

Bigbee, W.L., Vanderlaan, M., Fong, S.S.N., and Jensen, R.H., 1983, Monclonal antibodies specific for the M- and N-forms of human glycophorin A, Mol. Immunol., 20:1353.

Jensen, R.H., Bigbee, W.L., and Branscomb, E.W., 1984, Somatic mutations detected by immunofluorescence and flow cytometry, in: "Biological Dosimetry", pp. 161-170, W. Eisert and M.L. Mendelsohn, eds., Springer-Verlag, Heidelberg.

Jones, I.M., Burkhart-Schultz, K., and Carrano, A.V., 1984, Cloning of thioguanine resistant lymphocytes for measurement of *in vivo* mutation in the mouse (in preparation).

Morley, A.A., Cox, S., Wigmore, D., Seshadri, R., and Dempsey, J.L., 1982, Enumeration of thioguanine-resistant lymphocytes using autoradiography, *Mutat. Res.* 95:363.

Morley, A.A., Trainor, K.J., Seshadri, R., and Ryall, R.G., 1983, Measurement of *in vivo* mutations in human lymphocytes, *Nature (London)*, 302:155.

Papayannopoulou, Th., Lim, G., McGuire, T.C., Ahern, V., Nute, P.E., and Stamatoyannopoulos, G., 1977, Use of specific fluorescent antibodies for the identification of hemoglobin C in erythrocytes, *Am. J. Hematol.*, 2:105.

Papayannopoulou, Th, McGuire, T.C., Lim, G., Garzel, E., Nute, P.E., and Stamatoyannopoulos, G., 1976, Identification of haemoglobin S in red cells and normoblasts, using fluorescent anti-Hb S antibodies, *Br. J. Haematol.*, 34:25.

Stamatoyannopoulos, G., Nute, P.E., Papayannopoulou, Th., McGuire T.C., Lim, G., Bunn, H.F., and Rucknagel, D., 1980, Development of a somatic mutation screening system using Hb mutants. IV. Successful detection of red cells containing the human frameshift mutants Hb Wayne and Hb Cranston using monospecific fluorescent antibodies, *Am. J. Hum. Genet.*, 32:484.

Strauss, G.H., and Albertini, R.J., 1979, Enumeration of 6-Thioguanine resistant peripheral blood lymphocytes in man as a potential test for somatic cell mutation arising *in vivo*, *Mutat. Res.*, 61:353.

DISCUSSION

Painter: Could a spectrofluometric assay be made on the whole hemoglobin or red blood cell sample, rather than on a sorted one?

Mendelsohn: You can take hemoglobin and ask how much of it is sickle hemoglobin.

Painter: First you would have to wash the cells to get rid of the antibody, then it may be feasible.

Mendelsohn: Let me answer your question in a slightly roundabout way to avoid any misunderstanding about the sickle cell situation. There are three ways that a person can have sickle cells. The most obvious is for the person to inherit a sickle gene from a parent who likewise has sickle cells. In such a heterozygous sickle person every red cell will contain from 30 to 50% sickle hemoglobin. The second way is for the person to inherit a sickle gene from parents who do not have sickle red cells. The mechanism here would be a heritable mutation in a parental germ cell leading to sickle cell trait in the affected progeny. Judging from somatic rates, this is a rare event, occurring perhaps once in every 10 million individuals. Again every red cell in the person would partially express sickle hemoglobin. The third way is for the person to undergo a somatic mutation in which case only a small fraction of the red cells would be affected. In addition to these mechanisms, molecules of sickle hemoglobin can arise because of transcriptional and translational errors. Perhaps one in every million hemoglobin molecules will be so affected and will be distributed randomly among all cells. This level of molecular error is well below the detection limit of our cell-based method, but it can be detected by a radioimmune assay of hemoglobin in solution. Such an assay reflects the sum of the genetic and the processing errors.

Neel: It will require individual judgment to decide whether the background is too high. However, this may not be a problem because you can count so many cells. How do you decide whether there is too much noise in the system?

Mendelsohn: We are going to have to use common sense to evaluate noise or to decide whether a variant is indeed a mutant. One possibility for the latter is to do single cell electrophoresis to confirm that we really have the variant protein that the assay was searching for. I do not have a definitive answer to your question, but it is obviously going to be an important issue when we are ready to promulgate these techniques.

Painter: If you have an event that causes a mutation, would you expect that there would be a burst of mutation because of amplification from the one mutant cell?

Mendelsohn: The yield and its kinetics will depend on which cell mutated. An erythrocytic precursor cell at or near its last division will express the mutation rapidly but in only a few daughter cells. The evidence of mutation will have little permanence and will last roughly for the 120-day survival span of a mature red cell. At the other extreme, an early stem cell undergoing the mutation may express many mutated erythrocytes over an extended period of time, perhaps many years. An even greater amplification theoretically could come in early embryogenesis when a handful of cells can be the precursor of the entire adult bone marrow and can be expected to express throughout the individual's lifetime. One predicts that the more the amplification, the smaller the number of cells at risk, and the larger the time constant. All this is too complex to work out at this stage of the assay, but should become clearer as the methods are put to use.

Longfellow: In the context of risk extrapolation to low dose exposure, let me play devil's advocate. Perhaps the wide range of error possible with mathematical extrapolations based on assumptions gleaned with high-dose experiments are safer than having an assay for somatic mutation screening without an assessment of the actual consequence of having such abnormal cells.

Mendelsohn: Obviously the presence of a few mutant red cells in one's blood has no direct medical consequence. Its importance is indirect, and is driven by the somatic mutation theory of carcinogenesis and by other potential health impacts of genetic damage, such as the high-leverage effects of heritable mutation. It is not known how important base-substitutions or gene losses are in cancer induction, nor is there firm information on the comparative dose-responses of genetic damage and carcinogenesis. But given what little mechanistic information we have and the intractability of low-dose assessment, I believe the measurement of somatic mutation in people should be given high priority. Once the methods are in the hands of the epidemiologists, we should soon find out just how predictive and powerful they are. At the moment I see few attractive alternatives.

CYTOGENETIC AND ALLIED STUDIES IN POPULATIONS EXPOSED TO RADIATIONS AND CHEMICAL AGENTS

H.J. Evans

Medical Research Council
Clinical and Population Cytogenetics Unit
Western General Hospital
Edinburgh, U.K.

Genetic hazards to man following exposure to environmental mutagens are in many people's minds associated with mutational changes induced in germ cells and then transmitted to offspring. These mutations, whether spontaneous or induced, can be broadly classified into one of two categories: (a) chromosomal mutations, which involve either a change in the structure or a change in the number of chromosomes and which are changes that can be readily visualised in chromosome preparations analysed under the microscope, and (b) gene mutations, which encompass a range of changes from single base pair substitutions in DNA to frame shifts involving single base or much larger deletions and including also the loss of very much larger DNA segments encompassing one or more genes from a given chromosome region. There is considerable overlap between these two categories of chromosomal and gene mutations and it is quite evident that many things that we see inherited as single phenotypic changes, that is apparently single gene effects - for example some of the thalassaemias - are really a mixed bag and may encompass quite large deletions which can sometimes be visualised under the microscope.

Now when such mutations occur in human germ cells, their effects if deleterious, are readily ascertained if they result in a congenital abnormality or in an inherited disease or predisposition to such disease. However, because of the nature of our genetic organization and the mechanisms involved in both spontaneous and induced mutagenesis, exactly the same mutations that occur in germ cells or germ cell precursors also occur in somatic cells, but here their consequences, although certainly important, are often far less obvious. For example, a mutation to loss of function of a cluster of immunoglobulin genes that occurs in a germ cell will produce an immunologically incompetent child with severe and usually lethal inherited disease, whereas exactly the same mutation occurring in a somatic skin cell, or a cell lining the gut, or a nerve cell in the brain, may have no untoward consequences whatsoever towards the individual in which this mutation has occurred. On the other hand if this somatic mutation occurred in a primitive cell in the bone marrow that was responsible for producing a large proportion of the individual's immunologically competent cells, or if it occurred early in embryonic development, then this somatic mutation could have very serious somatic consequences.

In terms of risk assessment we are especially interested in induced mutations in man but before going on to discuss the detection of such mutations, particularly in somatic cells, I should first like to remind you that spontaneous mutations in man are not that infrequent in both somatic and germ cells. In germ cells they contribute to our not insubstantial genetic burden and in somatic cells to a variety of forms of ill-health. Let me by way of background then first consider spontaneous heritable germ line mutations.

1. Spontaneous heritable germ line mutations in man

(a) Chromosomal mutations

It is just about 25 years ago that Peter Nowell and David Hungerford described a basically simple technique that enabled us to readily examine the chromosome constitution of an individual. The technique essentially involves the short term - 2 day - culture of the lymphocytes in one or two drops of human peripheral blood and a 1 ml blood sample provides us with 10^6 lymphocytes and many hundreds of mitotic metaphase chromosome spreads. More recently we have been able to directly examine chromosomes in human sperm, but the technique cannot be easily applied to all individuals and we have no simple routine method for visualizing chromosome mutations in human oocytes. Chromosome mutations in human germ cells are therefore not easy to vizualise, but they are readily detected as constitutional chromosome changes in blood cells of live newborn, still births, or from cultured fibroblasts from the products of conception that are lethal early or later on in development *in utero*.

If we examine lymphocyte chromosomes from populations of consecutive live newborn, it turns out that the average incidence of chromosome anomalies is around 1 in 150 (0.6%). The bulk of these anomalies are new mutations and some two-thirds of them represent babies having an abnormal number of chromosomes, usually an additional chromosome, i.e. 47 not 46, and responsible for conditions such as Down's and Klinefelter's syndromes, the remainder involving various chromosome structural changes. In still births the

Anomaly	Live births (n=43558)	Perinatal deaths (n=500)	Spontaneous abortions <28wks (n=941)	<18wks (n=255)	<12wks (n=1498)
Polyploidy	0.13	2.00	53.13	101.96	160.21
Autosomal Trisomy	1.24	28.00	151.98	250.98	330.44
Monosomy X	0.046	2.00	72.26	156.86	93.46
Other	4.43	24.00	27.62	39.22	30.70
TOTALS	**6.26**	**56.00**	**304.99**	**549.02**	**614.82**

Fig. 1. Chromosome anomalies in spontaneous abortions - perinatal deaths and live births (all per 1,000).

incidence of chromosome anomalies is around 5%, increasing to 30% in spontaneous abortions occurring in the second trimester and 60% in the first trimester (Fig. 1). The bulk of these abortuses are embryos with an additional or missing chromosome and, taking into account viable pregnancies, these data imply that around 7% of all human conceptions involve a chromosomally abnormal gamete. Moreover, in many cases we have good evidence for an increasing chromosomal mutation frequency with increasing parental age for these germ cell mutations.

One of the main points that I wish to emphasize here is that the germ cell chromosomal mutation frequency is high, so that the background 'noise level' is high. This high noise level, together with the fact that running chromosome analyses on large populations of newborn, or on carefully defined abortus material, make it difficult and expensive, but not impossible, for us to detect induced germ cell chromosomal mutations in man.

(b) Gene mutations

Single gene mutations in human germ cells are ascertained through their physical or clinical consequences in offspring. A glance at McKusick's catalogue of Mendelian inheritance in man shows a list of some 3,000 inherited characters many of which are associated with some form of ill-health and many of which are believed to be inherited as a single gene. All of these represent mutations that have arisen at one time or another and they encompass a range of phenotypic effects.

It turns out that something like 1-2% of the live newborn inherit a single gene that is responsible for some form of ill-health, and around 9% have other abnormalities as a consequence of the interaction of a number of inherited genes. However, only a proportion of these abnormal genes are new mutations. For some genes the mutation rate is very low, for example most cases of Huntington's chorea are not due to fresh, or new, mutations and the mutation frequency here is of the order of 1 in 10^6-10^7, that is one mutation in every 1-10 million sperm. For other genes the mutation rate is much higher, for instance the recessive X-linked gene for non-specific mental retardation that is responsible for the presence of at least half of the non-Down's syndrome males in institutions for the mentally retarded, has a mutation rate of around 7×10^{-4} per gamete per generation, i.e. almost 1 in every 1,000 sperm. Other relatively highly mutable loci are those involved in Duchenne muscular dystrophy and neurofibromatosis where the mutation rate per gamete per generation is of the order of 5-10×10^{-6}.

As in the case of the chromosome mutations in germ cells, single gene mutations also appear to increase in frequency with increasing age of parents, and whether this is a consequence of continued exposure to environmental agents, or to an aging effect on DNA repair processes, or on DNA replication, is unknown and will not be discussed here. What I would say here, however, is that although it is theoretically possible for us to detect elevated single gene mutation frequencies in our populations as a consequence of exposure to mutagens, there are many practical difficulties. I do not wish to spend time discussing these, but the problems of obtaining accurate background frequencies of a number of different relatively rare events in a population, and of complete screening and of effective and accurate ascertainment (diagnosis) for monitoring or for post-exposure sampling are fairly obvious.

Let us now turn to consider somatic mutations that occur spontaneously as well as those that are induced when we expose our cells, or ourselves, to mutagens.

2. Spontaneous and induced somatic mutations in man

First, a general point. I have already emphasized that the principles of genetics and direct studies on experimental animals tell us, and show, that the kinds of spontaneous and induced mutations that occur in germ cells also occur in somatic cells. Our interest in detecting, measuring and studying somatic mutations in man then follows for four reasons:

(i) Looking for mutations in germ cells in populations of people involves looking at a large number of people, whereas large numbers of somatic cells can be obtained from a single person.

(ii) Exposure of a population to a mutagen will result in

increasing the incidence of somatic as well as germ cell mutations and analysis of chromosomes from blood lymphocytes or skin gives us an immediate measure of the mutational consequences of exposure in somatic cells and an indirect measure of effect in germ cells. In this context we can view our somatic cells as providing a window to show what is happening to our germ cells.

(iii) It is a simple matter to obtain repeated samples of blood cells so that we are able to monitor exposures of single individuals and groups of individuals and obtain information on whether an individual has been exposed to a mutagen and also information on the level of exposure - or dose.

(iv) Somatic mutations may have health consequences to the individual in just the same way as germ cell mutations may have health consequences to his or her offspring.

(a) Spontaneous and induced chromosomal mutations

The incidence of acquired somatic chromosome mutations, as opposed to inherited and hence constitutional mutations, can be readily established by analysing chromosome preparations from short term lymphocyte cultures and counting the numbers of cells having abnormal numbers of chromosomes, or having abnormally structured chromosomes that may be partially deleted, duplicated or involved in translocation. These aberrations are readily visible and countable. How frequently are they found?

If we examine blood chromosomes from a population of healthy, young, non-cigarette smoking adults then we find that the frequency of cells that are missing or have gained a chromosome, i.e. have 45 or 47 chromosomes, is of the order of 1%. If we examine the cells in detail then we find that around 1% have chromosome breakage or other structural rearrangements and for a specific type of rearrangement such as a dicentric the incidence is around 1 in 1-2,000 cells. If we examine cells from older, healthy individuals not unduly exposed to known mutagens, then these frequencies are increased and particularly so of the cells with chromosome gains or losses which may be five or six times higher in 60 year olds relative to 20 year olds.

From what I have said it is evident that the background frequencies of chromosomal mutations in our blood cells are relatively high and easy to measure. Although I should mention that a detailed analysis of a metaphase cell, to identify and count aberrations, takes time and skill and it is almost a week's work for a trained technician to score around 1,000 cells. This ease of obtaining blood cells and of scoring chromosome aberrations, coupled with the fact that exposure to most human mutagens and all ionizing radiations results in dose-related increases in aberration frequencies, has led

to the use of chromosomal mutation assays in blood cells as the only practical measure to date of monitoring exposure of people to mutagens, and particularly to ionizing radiations, and to provide us with some kind of biological dosimeter. Let us then turn to consider the response of our chromosomes to ionizing radiations.

If we expose peripheral blood lymphocytes to X-rays, or other ionizing radiations *in vitro*, then we observe an increased aberration frequency which is dose dependent. The shape of the dose-response curve depends on radiation quality - linear with very high LET, almost dose-squared with X-rays. An important fact is that with X-rays, for example, 20 rads gives us approximately 1 dicentric per 100 cells. Since we only see 1 dicentric in 1,000 to 2,000 cells normally, this would imply that our chromosomes are sensitive and that we should, and indeed we can, detect effects at lower exposures. Now these are results from experiments undertaken *in vitro*, what happens *in vivo*?

Some early studies in my laboratory were undertaken on blood lymphocytes from cancer patients exposed to whole body X-irradiation at doses of between 25 and 50 rads. Some blood cells from these patients were simultaneously exposed to the same doses *in vitro* and the results showed the same response to X-rays for chromosome aberration induction whether the cells were exposed *in vivo* or *in vitro*, with a dose of 25 rads of X-rays giving an 18-fold increase in the incidence of dicentrics plus rings per cell (Fig. 2). Radiation exposures are, however, rarely whole body and over the years we've been studying a group of patients in Edinburgh who have ankylosing spondylitis and who have been treated with a course of X-rays delivered as a series of fractions to a limited area of the spine. Blood cells sampled 24 h after a single initial exposure of

Case No.	Age (yrs)	Before Irradiation			After Irradiation (25R)		
		No. of Cells	Dic.	Rings	No. of Cells	Dic.	Rings
1	50	100	1	0	200	3	2
2	61	100	0	0	200	1	1
3	79	100	0	0	200	8	1
4	60	100	0	0	200	5	1
5	53	100	0	0	200	8	0
6	40	100	0	0	200	5	1
Total		600	1	0	1,200	30	6

Fig. 2. Dicentric and ring aberrations in peripheral blood lymphocyte chromosomes of 6 cancer patients before and after whole-body irradiation with 25 rads of 2-MeV X-rays.

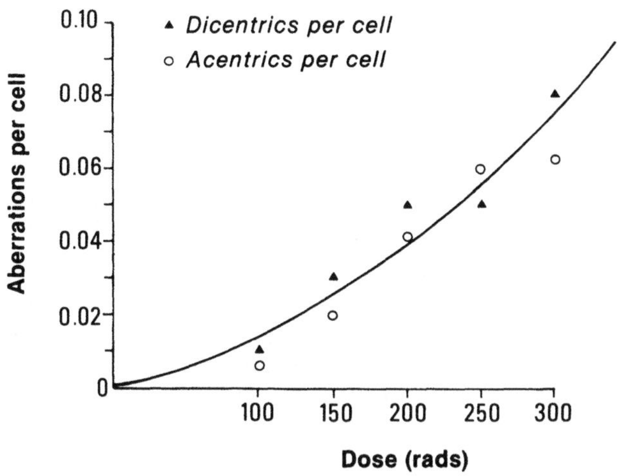

Fig. 3. Chromosome aberrations in blood lymphocytes of 5 different patients with ankylosing spondylitis sampled 24 h after a single partial body exposure to X-rays.

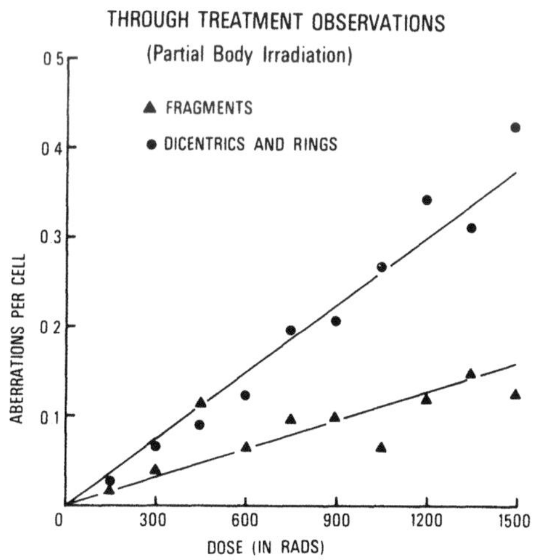

Fig. 4. Mean aberration frequencies after a series of successive fractionated partial body X-ray exposures to 5 patients (see Fig. 3).

from 100-300 r to each of five individuals show a clear dose-response relationship for aberration production (Fig. 3). Repeated samples taken after each fraction from the same individuals show a cumulative, approximately linear response (Fig. 4).

Now a proportion of the lymphocytes that are circulating in the peripheral blood are very long-lived cells and we have been sampling blood cells from patients exposed to radiations many months or years previously. These time studies show two things (Fig. 5). First that the frequencies of unstable aberrations, that is dicentrics, rings and fragments decline with time after exposure. They decline as a consequence of cell division, turnover and cell death. They nevertheless remain at an elevated frequency for 20 to 30 years after exposure. Second, the stable rearrangements (these are translocations within and between chromosomes that do not involve the formation of dicentric structures or fragments from broken chromosomes) remain at very high levels.

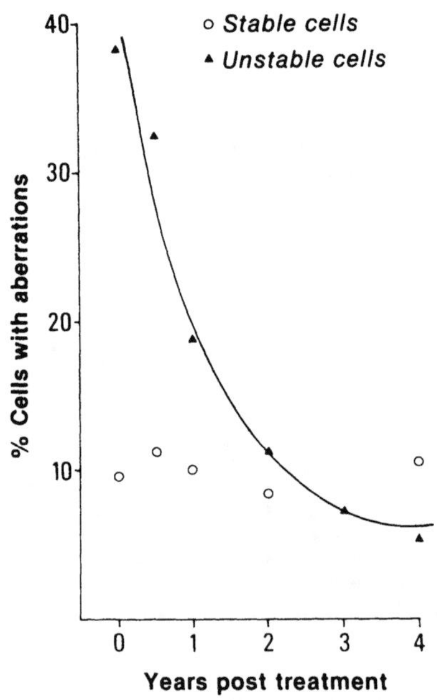

Fig. 5. Frequencies of cells with stable (e.g. translocations) and unstable (e.g. dicentrics and rings) aberrations with time following partial body X-ray exposure of a series of patients with ankylosing spondylitis.

The patients that I have referred to received high doses of radiation, but I've already said that we can detect the effect of exposures of 10-20 rads *in vivo* and certainly 5-10 rads *in vitro*. In view of this, and of our knowledge of the effects of time post exposure on aberration yield, around 12 years ago we set up a large experiment on a group of occupationally exposed dockyard workers who were refuelling nuclear submarines and who were exposed to gamma rays at levels within the occupational limits of 5 rads per annum. We took blood samples from over 200 men before they were ever occupationally exposed to radiation, and then sampled blood cells from them at 6 monthly intervals over a period of 10 years. All the slides were coded and scored blind and the total analysis was not performed until the experiment had run for 10 years.

We scored a very large number of cells and the results show that with increasing accumulated dose there is a detectable increase in aberration yield (Fig. 6) - although no workers were exposed above the occupational limits. If we consider just the dicentrics and rings we see (i) that there is no threshold, (ii) that there is an approximately linear dose response and (iii) that we observe significant increases in aberration frequency at doses clearly well below the occupational limits. The frequency of aberrations we know declines with increasing time post exposure and so we analysed our data in terms of total dose, early dose (up until 18 months prior to

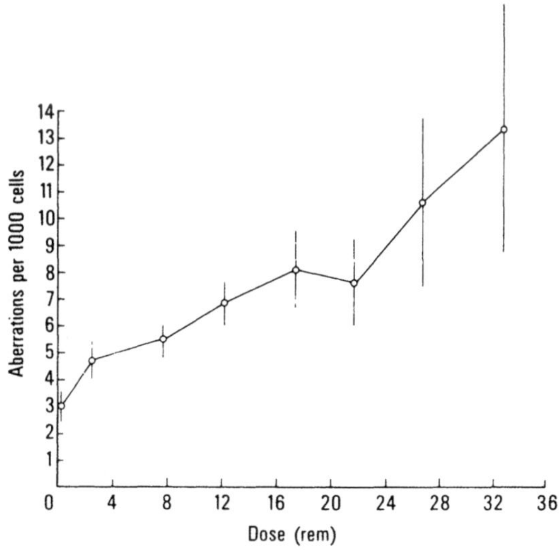

Fig. 6. Frequency of chromosome aberrations in peripheral blood leucocyte chromosomes of nuclear dockyard workers exposed to cumulative doses of gamma rays over a ten-year period.

final sampling) and late dose (dose accumulated) over the last 18 months). This analysis showed that the rate of increase of all aberrations against total dose is around 3/cell/rem x 10^4, but if we consider only recent dose it is around 4.4/cell/rem x 10^4. I mention these figures since Lloyd and colleagues at Harwell repeated our study, but on Atomic Energy Authority personnel, and their results published last year confirm our findings and show a closely similar dose response.

Radiations are, of course, but one class of mutagenic hazards and there are vast numbers of mutagenic and carcinogenic chemical agents in the environment which we would expect would induce chromosomal and gene mutations in man. One such agent is cigarette smoke. Cigarette smoking may be responsible for almost one-third of induced cancers in developed countries and a few years ago we and others showed that cigarette smoke condensate is a potent mutagen to human cells exposed in culture and its potency is such that we would expect to see effects on human chromosomes exposed *in vivo*.

In a double blind study we analysed chromosomes from matched smokers and non-smokers, with the smokers smoking on average 10 cigarettes a day. Our findings (Fig. 7) show a highly significant increase in aberration frequencies in these blood cells taken from light smokers and larger increases have been observed in heavier smokers. Light smokers would appear to show a level of genetic damage that is roughly equivalent to what we would expect of a radiation worker receiving ~2 rads per year.

In addition to populations of cigarette smokers, various populations occupationally exposed to known or suspected mutagens have been studied and Fig. 8 lists a number where significant increases

	Non-smokers	Smokers
Total subjects	41	55
Mean age (yrs.)	45.8	48.2
Total cells	4,100	5,500
gaps	92 (2.24)	171 (3.11)
deletions	36 (0.87)	56 (1.02)
exchanges	14 (0.34)	46 (0.84)
Cells with damaged chromosomes	91 (2.21)	176 (3.20)

Fig. 7. Incidence of chromosomal aberrations in blood lymphocytes of cigarette smokers and non-smokers.

AGENT	POPULATION SIZE
X or γ-rays	>1000
Arsenic	33
Benzene	>190
Chloromethylether	12
Chloroprene	>50
Epichlorhydrin	>100
Organophosphates	>180
Pesticide/Herbicide	>40
Styrene	>50
Vinyl chloride	>500
Ziram	9

Fig. 8. Some human populations occupationally exposed to mutagenic agents and showing increased chromosome aberration frequencies in the blood lymphocytes.

in aberration frequencies have been observed in exposed workers as compared with controls. This list is incomplete and does not include, for example, populations exposed to ethylene oxide, but I think it emphasizes the point that cytogenetic assays for chromosomal mutations do allow us to detect and indeed monitor exposure of individuals to a wide range of mutagens. Our major problem here is the effort required to count sufficient cells to obtain meaningful results and this I shall return to later.

(b) Spontaneous and induced gene mutations

Detecting spontaneous or induced somatic gene mutations in man presents quite a different problem from detecting chromosomal mutations. We must remember that there must be at least some 40,000 or more different genes in the human genome and in any mutational assay we are restricted to assaying for an alteration of a cell's phenotype with respect to the product of one - or at most a few - genes. With an assay for chromosomal mutations we are essentially screening the whole genome for macrochanges; with a gene assay we are screening perhaps less than 100,000th of the genome for any change which involves a change in gene expression.

A mutation at any gene locus is a relatively rare event, but *in vitro* culture systems have been developed which allow certain specific mutant cells a selective advantage over normal cells in terms of viability and growth. These selective assays enable us to detect the presence of a single mutant cell in a population of a million or more cells, but they have a number of disadvantages including: (i) first they are few in number; (ii) they are bedevilled by the problem that man is a diploid organism so that many mutations may not be expressed unless the cell or person is already heterozygous for that mutation; (iii) most of the assays involve the use of specific fibroblast cell strains; and (iv) mutant cells can only be readily detected if they are allowed to proliferate in culture for a period of around 14 days. These assays have, however, been used to measure mutation rates at certain loci in human somatic cells in culture and they show that mutation frequencies of 1 in 10^6 or 1 in 10^7 cells are not unusual.

In the last few years there have been enormous strides in our understanding of the different classes of lymphocyte present in human peripheral blood and one important development has been the discovery that T lymphocytes, which are the ones that respond to PHA in culture and which can proliferate for a few days in culture, can, by the addition of T cell growth factor, be made to proliferate for periods of weeks or even months in culture. Dick Albertini and colleagues were the first to make use of this finding to develop a system for measuring mutations in human lymphocytes and we have further developed and extended this approach.

The mutation that we have been using involves the loss of activity of the HPRT gene, which is located at the end of the X-chromosome so that one copy only is functional in both males and females. Inactivation, or loss, of this locus results in the mutated lymphocyte being resistant to the toxic effects of culture in the presence of the purine analogue 6-thioguanine. Cells with the enzyme metabolize thioguanine and die; cells without survive and proliferate.

To determine mutation frequencies in human peripheral blood lymphocytes we culture separated blood lymphocytes in paired 96 well culture plates, one of which contains culture medium with phytohaemagglutinin and 6-thioguanine and the other culture medium and PHA but no 6-thioguanine. The cells are first cultured for three days as bulk cultures, to allow expression, and five cells are placed in each well of plates with culture medium minus 6-thioguanine, to give us a plating efficiency, and 5,000 cells are put into each well containing 6-TG. The cells are then cultured for 14 days and we count the numbers of colonies produced in the various wells. When we do this we find that the mutation frequency is of the order of 1 in 10^5 to 10^6. Of interest is the fact that if we do this for a range of people of different ages then we find that the frequency of mutations increases with age (Fig. 9). The rate of increase in mutation frequency with age is approximately three mutant cells per 10^7 cells per year and the increase would appear to be linear with

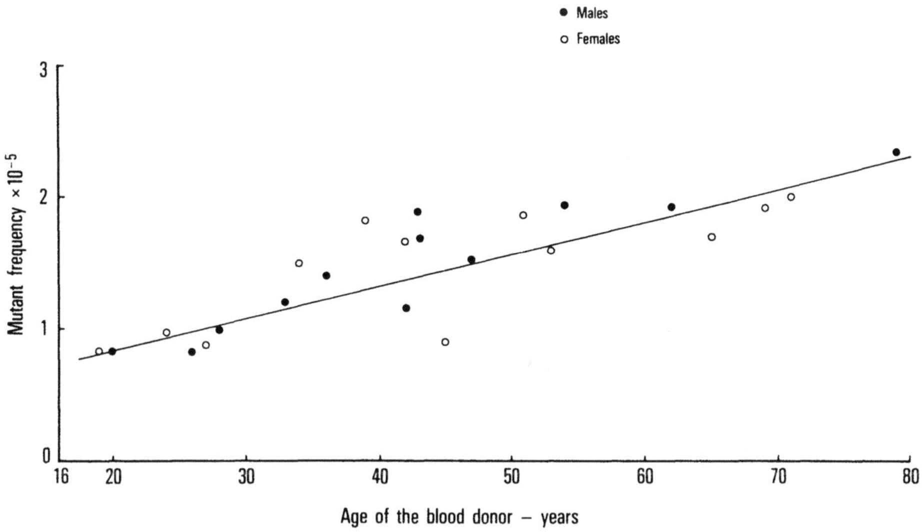

Fig. 9. Frequency of 6-TG⁻ (HPRT⁻) lymphocytes in peripheral blood samples of a series of healthy individuals in relation to age.

time. I should emphasize here that we are assaying for loss of activity, so that the phenotypic expression of resistance may reflect any kind of mutational change ranging from a single base substitution making the gene product inactive, to a loss of a large chunk, or even the whole, of the chromosome containing the gene.

Although we are still engaged in characterizing the cells and optimizing culture conditions, we have also begun to use this system to assay the response of the human genome to mutagen exposure *in vitro*. Exposure of G_0 lymphocytes to X-rays can be readily detected at doses of greater than 50 rads and the dose response is curvilinear suggesting that many of these mutations are due to chromosome aberrations (Fig. 10). We have also studied a number of chemical mutagens and are relating the incidence of mutation to the incidence of induced SCEs in these cells (Fig. 11). The system that we use here is partly automated and needs to be further refined; moreover we could do with other dominant selectable marker genes. However, we have just started to look for mutations in people exposed to mutagens and the system holds promise that we shall be able to do this, but we have only just started this work.

3. Other approaches to detect induced somatic mutations in man

I think it is evident that at the present time the only practical approach that we have for detecting *in vivo* induced somatic mutation in man is through analysing chromosomes in peripheral blood cells of

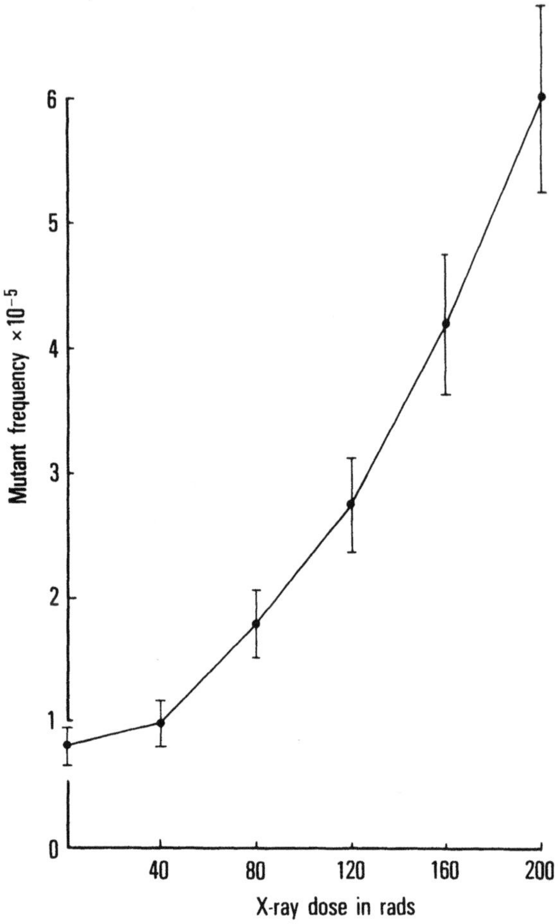

Fig. 10. Frequencies of HPRT⁻ mutant lymphocytes following various X-irradiation exposures *in vitro*.

exposed individuals. The system is in world-wide use as a monitoring system for radiation over-exposure, but, in terms of its more general applicability it has two drawbacks: (i) not all mutations are chromosomal mutations and the spectrum of mutation may differ between different mutagens. This drawback is virtually irrelevant for radiation exposures, but may be of importance for certain chemical mutagens. (ii) Chromosome analysis is time consuming, tedious and requires the involvement of highly trained personnel, for large numbers of cells have to be scored. A trained individual can analyse at most some 200-300 metaphase cells per day and many hundreds or even thousands of cells may need to be analysed from any one individual. Can we get around this? There are at least two

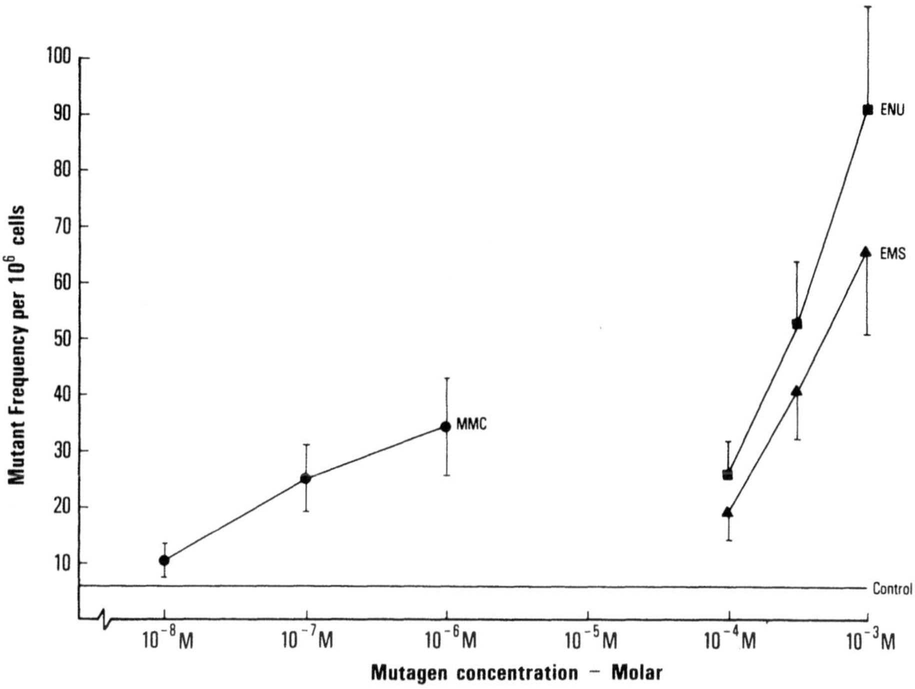

Fig. 11. Frequencies of $HPRT^-$ mutant lymphocytes following exposure *in vitro* to various concentrations of ethyl nitrosourea (30 min) and to ethyl methane sulfonate and mitomycin C (both 5 h).

possibilities. First, instead of scoring metaphase chromosomes we could score interphase micronuclei which arise from chromosome fragments that are lost from nuclei at the time of cell division. This is a much simpler, but less sensitive, analysis that detects only a fraction of the total chromosome damage and I am not going to consider it here. Second, we could use a fast throughput machine to analyze our chromosomes for us. We have been pursuing this second approach and I would like to quickly present some results. One obvious approach is to use a Fluorescence Activated Cell Sorter. The principles of such machines will be known to all as will the fact that it is possible to use them to count and distinguish chromosomes in soups as well as cells in suspension as first shown by the Livermore group in the USA. Chromosome soups are made from cultured peripheral blood lymphocytes and then forced down a fine vibrating nozzle. On the way down the fluorochrome-stained chromosomes pass through a laser beam, light is excited and scattered and measured so that the mass of each chromosome is essentially determined as it flashes down in single file down the pipe. Single chromosomes end up in drops at the end of the stream and one can direct selected drops by charging them and in this way collect specific chromosomes.

The flow machine measures, classifies and counts each chromosome as it passes down the pipe and the first important point is that it does this at a rate of around 2,000 chromosomes per second. Second it produces a chromosome profile which differs between different individuals - just like a fingerprint - but is constant for any sample taken from each individual (Fig. 12). Now we would expect that if an individual is exposed to a radiation dose that results in a significant amount of chromosome breakage, then his flow chromosome profile should change. The expectation would be that because of breakage there would be relatively fewer large chromosomes with more smaller chromosomes and a poorer definition of peaks and valleys in the flow profile - or flow karyotype.

We have therefore undertaken quite a large number of experiments in which we've taken a normal blood sample, divided it into aliquots giving each aliquot a dose of radiation of, say, from zero to 400 rads,

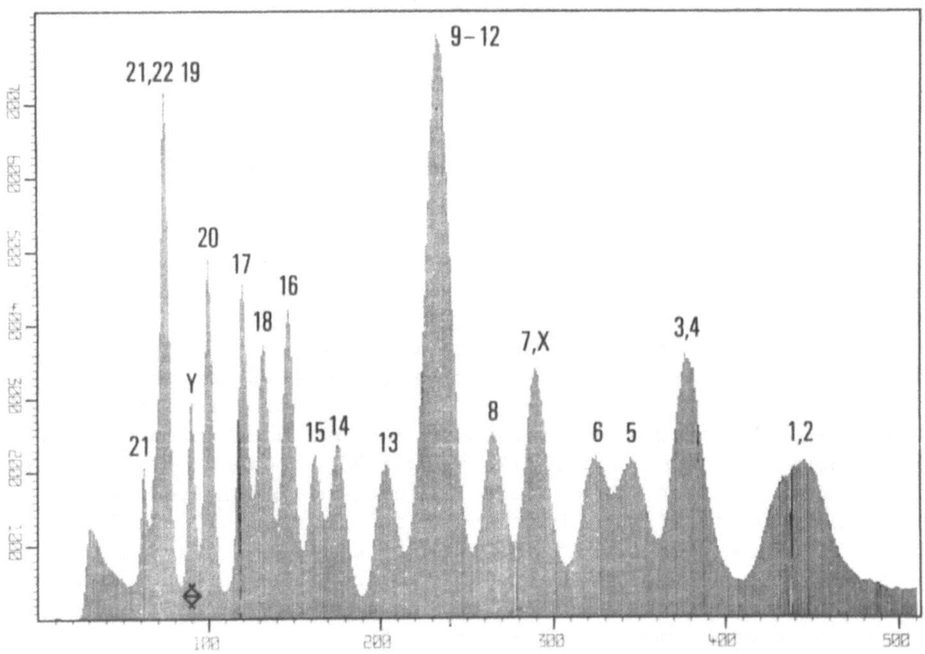

Fig. 12. Flow profile from lymphocyte chromosomes stained with Hoechst 33258. The sample is from a normal male and the numbers indicate the chromosome numbers in the individual peaks.

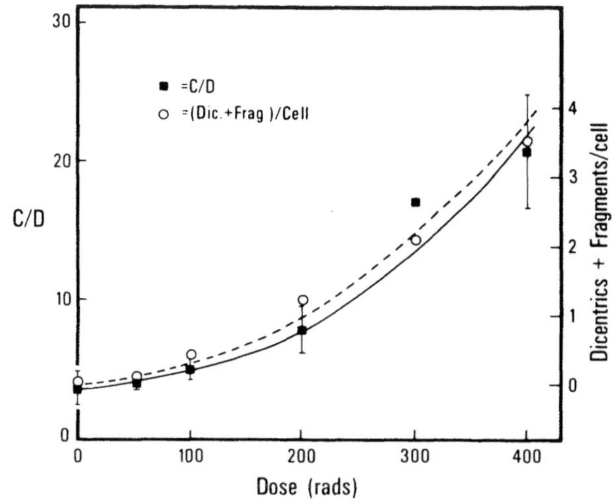

Fig. 13. Frequencies of dicentrics and fragments observed from the scored microscope preparations, and changes in flow profile of chromosomes from the same original population of mitotic lymphocytes, all after exposure of cells to a graded range of X-ray dose.

cultured the samples, ran half of each sample through the flow machine and with the other half made standard chromosome slide preparations. An observer then scores the slides for chromosome aberrations (dicentrics and fragments) and another observer analyzes the flow profile - all done blind without knowledge of dose, etc. We find that we can readily relate change in flow profile to aberration frequency (Fig. 13) and with refinements of technique we are now able, with some difficulty, to detect exposures down to 25 rads. We should with this system be able to develop it so that we can detect exposures of 10 rads, and if certain other refinements work, then possibly the effects of lower exposures - but that is another story. The essential fact here is that we can score very large numbers of chromosomes very rapidly and so possibly circumvent the long labor involved in scoring slides manually.

Another approach to a more rapid method of detecting aberrations is to develop a machine which will automatically, or semi-automatically, enable us to analyze chromosome preparations on slides. We have been developing such a machine in our Unit. This machine is referred to by us as FIP - Fast Interval Processor - and it has been taken over by the company Shandon Ltd. which is shortly going to

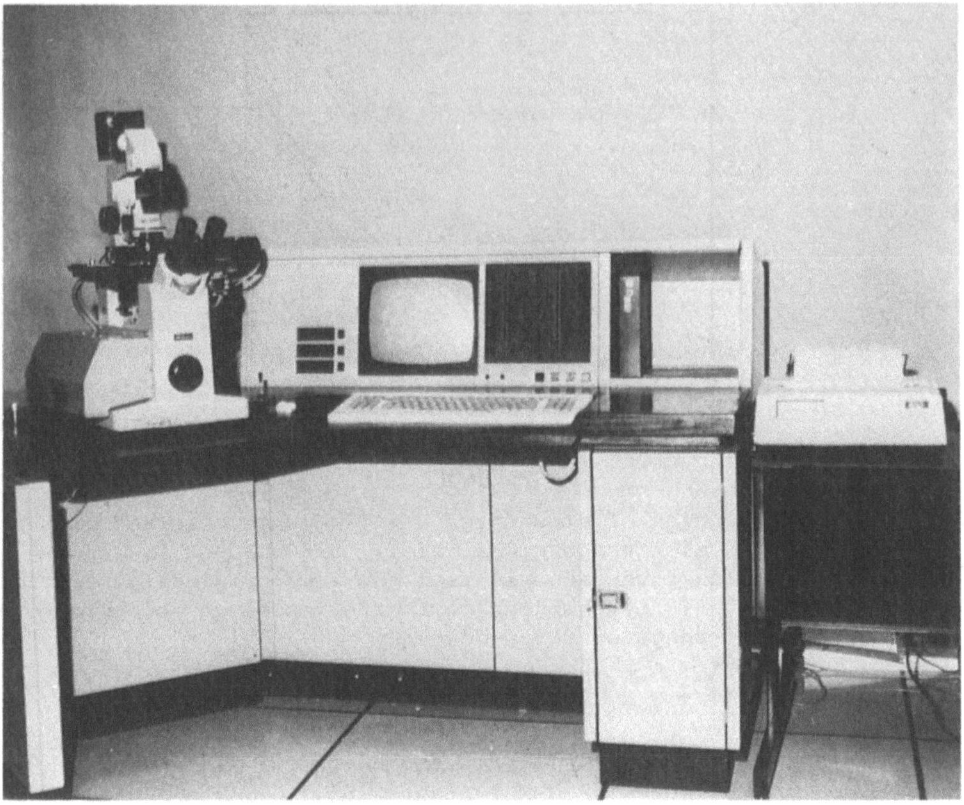

Fig. 14. The FIP slide scanner used for rapid metaphase finding and relocation, and rapid (1,000 cells/second) measurement of cell DNA contents.

produce it on a commercial scale (Fig. 14). FIP can measure the DNA contents of nuclei as it were on the fly and can count, and give DNA measurements, on just over 1,000 cells per second - i.e. around one million in 15 minutes. At the present time the machine is used for cervical smear and DNA densitometry screening. In our laboratory it is also used for finding and locating metaphases for aberration scoring and although our final scoring is done manually, the use of FIP just about doubles our rate of manual scoring. We have almost completed some front-end programming so that the machine will be a semi-automatic karyotyper and when this is done it will be a very useful chromosome aberration scoring device.

In addition to scoring for chromosome aberrations, the two kinds of machine that I have talked about can also be used to detect and count mutant cells if we have specific antibodies which can react with a mutant protein present in single mutant cells. Antibodies for certain types of mutation are already in existence and this is an area where further developments are certainly to be expected. In the case of chemical mutagens, antibodies can be produced against specific chemical adducts in DNA and this is also an important approach which does not of course detect mutations, but rather DNA damage some of which may result in mutations.

4. Concluding comment

In conclusion, I think that the prospects for detecting and measuring the effects on human peripheral blood lymphocyte chromosomes of low-level exposure of populations to mutagens are good. At the present time we can, with the use of expensive labor, provide useful and practical information on populations exposed to radiations, or to a range of chemical agents, and with the development and application of new techniques it should be possible to minimize the labor and maximize the efficiency and resolution of detection of induced somatic mutation in man.

DISCUSSION

Neel: Apropos the relationship between cigarette smoking and mutation: we want to approach this problem in the data from Hiroshima and Nagasaki. At the end of our study, we will have a handful of mutations: we now have seven, but perhaps there will be ten. We shall then get smoking histories, and set up paired controls, to explore the relationship.

Evans: That would be a good thing to do.

Painter: One of the things your data from control populations showed, and something that always struck me, is that we are carrying around a fairly high load of chromosome aberrations. The question then arises, that if we can see an increase on this background of 6%, how do we know the difference in the increase between the deleterious aberrations and ones that have no effect (the ones that we have been carrying should not be any problem).

Evans: If you are carrying 6% aberrations, you are in a relatively bad way; for that value is an average for someone who is aged 60-70 years or so. It is very hard trying to equate an increase in a wide spectrum of chromosome aberrations in somatic cells to an increased risk of a specific disease, but one thing is very clear - that many of the aberrations are lethal to cells. So, for example, your B cell production in response to an immune challenge is diminished if aberrations are present in B cell precursors. Some cells are killed as a consequence of the aberrations they possess, and you also have viable cells with chromosome aberrations which are mutations: overall your net output of efficient cells is reduced. We now know that there are at least 20 different oncogenes in humans that have been mapped in the human chromosome complement. At least three of the oncogenes are at sites which are involved in chromosomal rearrangements in specific cancers. So, certain tumors in humans are clearly associated with certain types of aberrations. Thus we have some rather obvious effects of chromosome changes which result in (or are associated with) malignancy. I think we tend to neglect the possible consequences of somatic mutations, which are important in their own right. For example, if you have a mutation which affects chromosome 14 and the heavy chain immunoglobulin complex in the cells in which that gene complex is not allelically excluded, you no longer can manufacture immunoglobulin satisfactorily from these cells, and you may suffer some consequences. If there are aberrations involving functional genes in tissues where those functions are essential, then the efficiency of the tissue/organ will be diminished. Now, it is very hard to obtain hard and fast data to demonstrate the deleterious effects of such mutations, but I think they must occur and they must be deleterious. So, aberrations are bad for you, but how bad, I cannot tell you.

Painter: I think that I might suggest the idea that as long as aberrations kill cells, they probably are not bad. Concerning the load we carry, I think we have a lot of reserves.

Evans: I am not at all convinced that we have all these reserves. One of the arguments for the increased aberrational frequency seen in people as they age is that their proliferating tissues have to resort to make use of 'poorer' cells. It is not simply a question of whether your repair capacity diminishes as you get older, or that these cells have simply been exposed for a longer time period to environmental mutagens. It may also be a reflection of the reduced size of the pool of precursor cells that is available as one ages. This argument has been used to explain the increased frequency of non-disjunction in aging human females; eggs that have been in the ovary for forty years, (that is, the poor eggs or the "bottom of the barrel"), are finally used up. It is the increased frequency of non-disjunction in these old eggs that results in the increased frequency of Down's syndrome in children of older women. Let me make one further point. We looked very hard of course at our nuclear dockyard workers for cancer incidence, but in a population of a few hundred people, there is no hope of seeing any possible radiation effect, even if such an effect existed. But now that we have everyone who has been occupationally exposed to radiation in the UK listed on computer, so we shall eventually get information on the frequency of neoplasms in people exposed at different levels. All that the aberration frequency in cells of a given individual tells me is that this person was or was not exposed to radiation, and it will tell me how much of a dose the person may have received. The dose may be two rads of radiation or ten cigarettes a day; they are the same, as the aberration frequencies are similar.

Longfellow: Let us come back to the question made, and misunderstood, before. My contention is that when you and Dr. Mendelsohn set out to validate your assays you take populations known to be high risk populations. Those people are aware that they have been sampled, and, if any subsequently get cancer, an association will be made between the cancer and their exposure. How do you protect yourself against such an extrapolation before you are ready to use it as a scientific tool?

Evans: Whatever you do in life, someone will misuse it. There is no question about that, but in our case we do have controls both within and outside the sampled populations.

Thilly: There are a few parts of your presentation which I find difficult to understand. For example, it is reasonably demonstrated that in human cells there are certain chemicals which are good clastogens, but they are not so effective at causing gene mutations, and vice versa. It troubled me that in the HGPRT assays you confounded the concept of clastogenesis with gene mutations, asserting, without providing evidence, that the HGPRT locus is so close to the end of the X-chromosome that the loss of such a fragment would be nonlethal. It is, in fact, some distance from the end of the X-chromosome. It has been well documented in a number of human cell systems that HGPRT-deficient cells are produced by acridines, such as ICR-191 which under appropriate conditions do not cause clastogenesis. It has also been demonstrated with a wide variety of chemicals, such as ethylmethanesulphanate that when HGPRT cells are induced, and they can also be reverted. Such changes are not consistent with a clastogenic event. Finally, Capecchi et al., (1974) showed some years ago with hamster cells that about 50% of HGPRT cells induced by a wide variety of chemicals give cross-reacting material for a mutant HGPRT protein. Furthermore, it is my understanding that there has been no demonstration of any loss of chromosomal fragments in HGPRT deficient cells. Could you therefore clarify your remarks further.

Evans: The HGPRT locus maps at Xq26-27, that is, at the terminal positive band at the end of the long arm of the X chromosome, so that the locus is about five-sixths of the way down the long arm and is almost, but not quite, at the end of the long arm. So 80% or so of the long arm is on the centromere side of the locus. The bulk of aberrations involving breakage and loss of part of the X-chromosome are therefore going to involve loss of that gene, of that there is no shadow of any doubt. The second answer is very simple. You can get a "loss" mutation by either having the gene turned off; from a product that does not work; from the loss of a segment of chromosome containing that gene; or from the loss of a whole chromosome. Dr. Mendelsohn has been discussing in detail the kind of point mutational changes you are referring to, and I wished to emphasize the chromosomal aspects. As for evidence for X-chromosome aberration involvement in HPRT mutation, I would refer to the work of Cox and Masson (1978) who exposed human fibroblasts to X-rays, using the standard 6-TG selective system, and showed that 40% of the mutant cells that grew in the presence of TG could be seen to have chromosome structural rearrangements involving the long arm of the X chromosome. With ultraviolet light induced $HPRT^-$ mutants I understand that in contrast to X-rays most mutants are not chromosomal deletions or chromosomal anomalies (unpublished work). There is indeed clear evidence that $HPRT^-$ mutants produced by exposure of fibroblasts to certain chemical mutagens make a protein that cross-reacts with HPRT, but is non-functional; and the fact that such point mutations can be induced is not disputed. However, there is plenty of evidence for loss mutations as a consequence of loss of a chromosome segment and I think that many X-ray-induced mutations may be loss mutations of this sort. In our lymphocyte system the dose-squared response is very much in line with what we would expect for a mutation that is a consequence of a chromosomal aberration. Now that we can maintain relatively long-term cultures of cloned human lymphocytes we shall be able to define our mutants in more detail and we shall be assaying for products that cross-react with HPRT after exposure of lymphocytes to a range of mutagens.

REFERENCES

Capecchi, M.R., Capecchi, N.E., Hughes, S.H., and Wahl, G.M., 1974, Selective degradation of abnormal proteins in mammalian tissue culture cells, Proc. Natl. Acad. Sci. USA, 71:4732.

Cox, R., and Masson, W.K., 1978, Do radiation-induced thioguanine-resistant mutants of cultured mammalian cells arise by HGPRT gene mutation or X-chromosome rearrangements, Nature (London), 276:629.

ROUND TABLE DISCUSSION: WHERE THE FUTURE?

Chair: A. Hollaender
The Council for Research Planning in Biological Sciences, Inc.
1717 Massachusetts Avenue, N.W.
Washington, DC 20036

I am very pleased to have a chance to talk about the future of the development of radiation biology and chemical mutagenesis. I am reminded of when I moved to Oak Ridge National Laboratory in 1946, and Eugene Wigner asked me, "What would you like to do?" "Well," I said, and I proceeded to outline what I thought needed to be done. I am very proud to say that I can see from the discussions we have had today that I was right. At first, concentration on genetics and nucleic acids was not a very popular subject. As a matter of fact, the people at the Atomic Energy Commission in Washington asked what has nucleic acid to do with radiation. I do not have to explain it here! And as you know it has worked out very well.

For this evening's discussion, I would like to let our imagination go as we discuss together what you would like to see done, and how you see radiation biology and the whole field of chemical mutagenesis developing in the next twenty to thirty years. I think we should bring in new ideas. Already we have done a lot of mouse experiments. I was reminded of researches at Oak Ridge, where the mouse genetic studies involved about 30 to 40 million mice, and cost us about $30 million. But these experiments gave us an answer of a kind which we needed, even though we do not have a good understanding of the basic problems that are involved. Please do not hesitate to use your imagination in our discussion. I would like to call on the first speaker, Dr. Haynes. who will open the evening's meeting.

WHERE THE FUTURE?

PART I - ROUND TABLE DISCUSSION

Robert H. Haynes

Biology Department
York University
Toronto, Canada

INTRODUCTION

Thank you very much, Dr. Hollaender. I think that it is a marvelous opportunity for all of us to be at this session, which is chaired by Alexander Hollaender, who is certainly one of the monumental figures in American and international science. This year, Alex has received the Enrico Fermi Award which is the highest honor of the U.S. Department of Energy, indeed it is one of the highest scientific honors that can be bestowed in this country. I cannot image that anyone is more deserving of this award than Alex, and he joins the illustrious company of previous award winners, which included people like John von Neumann and many of the great luminaries of atomic energy research in America. Alex, we are all incredibly impressed at your vitality. I think what has impressed me about you over the years is that your eyes are on the future, and I think that no higher compliment can be paid to a scientist than that. I am certain that Alex knows the answer to the question: Where the future? The only accurate statement that I can make about the future is that it lies ahead.

Prophecy, especially for scientists, is a far riskier business than going over Niagara Falls in a barrel. Ergo, caveat lector! Since I make no pretense to clairvoyance, I offer only my prejudices on issues whose resolution I consider important if we are to deal effectively with the health hazards of environmental chemicals and radiation.

Initially, I thought that I should summarize the ideas presented by other speakers; however, I accumulated such a long list that this has become impracticable. From the topics discussed at this meeting, it would appear that cancer is considered to be the most significant health hazard associated with human exposure to low doses of radiation (ionizing and ultraviolet) and mutagenic chemicals. However, I would urge, particularly in discussions with government officials and the public, that we emphasize strongly the mutagenicity of these agents, and not leave the impression that it is only their carcinogenicity that is of concern. Newly induced mutations can be deleterious in many ways, not the least of which may be to predispose people to a wide variety of diseases of which cancer is only one. Having made this point, I will, in fact, concentrate on carcinogenicity in my remarks, even though I believe that the more general problem of mutagenicity should be our primary concern.

POTENTIAL VERSUS ACTUAL SOURCES OF RISK

In assessing both individual and national research priorities, we should keep in mind the distinction between the various ways in which mutations and cancer can arise, and how they most commonly do arise. These are two very different matters. Basic research has enabled us to identify many potential sources and mechanisms of mutational change and neoplastic transformation. However, with a few important exceptions, we have not had similar success in singling out the most important actual causes of cancer and genetic disease, whether in individuals, occupational groups, or entire populations. Still, the epidemiological and laboratory identification of potential causative agents does justify the consideration of specific preventive or regulatory measures to protect public health. It is essential for governments to have the political will and ability to act effectively, even on the basis of imperfect knowledge, if any health benefit is to be derived from research in this difficult and contentious area.

The epidemiological studies reviewed by Richard Peto makes it clear that certain so-called lifestyle factors are far more important sources of human cancer than the low doses of radiation and industrial pollutants to which most of us are exposed. Cigarette smoking, high-fat diets, certain sexual behaviors, alcohol consumption, and excessive exposure to intense sunlight are major contributors to cancer incidence in industrialized populations. More research certainly is needed on the molecular and other mechanisms whereby these factors cause cancer. However, a much higher priority should be assigned to political, economic, sociological, and psychological studies on how to wean people away from the hazardous practices that increase cancer risk. At the same time, incentives must be developed to encourage those in the tobacco and alcoholic beverage industries, and their suppliers, to diversify their activities into other, more benign, areas of enterprise. This is not to say that regulation and control of human exposure to a wide variety of industrial chemicals is not needed. However, efforts in this area are unlikely to have a marked effect on total cancer incidence except among the relatively small groups who are occupationally exposed to such agents. By comparison with the adverse health effects of cigarette smoking, the problem of population exposure to low doses of radiation from the nuclear industry simply pales into insignificance.

Prospective epidemiological studies and human population monitoring are slow and expensive enterprises. However, much can be achieved by computer-assisted record-linkage studies of existing health and occupational records. We are now at a stage where greatly increased collaborative efforts between experimenters and epidemiologists would be especially fruitful in assessing the significance of new sources of mutational change that have been identified in the laboratory. For example, folate deprivation in yeast and other cells brought about by antifolate and sulfa drugs greatly enhances the frequency of mitotic recombination and other genetic exchange events (Barclay et al., 1982). These effects arise from drug-induced deoxyribonucleotide pool imbalances, a condition that is capable of provoking all known types of genetic change in organisms ranging from viruses to humans (for a review, see Kunz, 1982). This finding would suggest that cancer incidence should be monitored in people suffering prolonged, untreated folate deficiencies or those treated over long periods of time with sulfa drugs to determine whether they are at a higher risk of cancer or if they suffer elevated levels of chromosome aberrations. I give this example to illustrate how laboratory scientists and epidemiologists might cooperate in putting new ideas to the test in the human content; I am sure that many other such possibilities exist.

Even if one has reasonable dose-response data, it is clear that there exist no generally acceptable extrapolation models for calculating risks for exposure to low levels of mutagens. We seem to have here a classic Alphonse-Gaston situation: the experimenters defer to the modelers, who, in turn, defer to the experimenters, each waiting on the other for some insight to resolve the problem. Unfortunately, statistical uncertainties associated with biological variations will be a continuing bane of both theory and experiment. As a result, I doubt that the problem of low-dose extrapolation in a risk assessment context will ever be resolved in a fully satisfactory way. Thus, we will continue to be faced with a large number of potential hazards whose risks cannot be calculated except within very broad (I am speaking of orders of magnitude) limits. With the extrapolation problem seemingly intractable, there are risks in risk assessment. The fact that numbers can be generated might lull laymen into the belief that some precision has been achieved, but we should remember that in this field there are such things as good and bad numbers!

In areas where little is known for certain, it is difficult to distinguish between major and minor issues. However, high mutagen doses are more hazardous than low doses, so research and regulatory efforts should be devoted to identifying and controlling those mutagenic materials to which people are exposed at substantial levels. Such materials generally will occur as complex mixtures that contain thousands of discrete compounds. Synergistic or antagonistic interactions may occur among these ingredients, and some of the chemicals present might even be antimutagens. The amount of any one ingredient is likely to be small and its biological behavior in pure form may not be related in a simple way to its contribution to the mixture. Thus, I feel that much more effort should be devoted to testing and monitoring the net mutagenicity of the complex mixtures of materials that occur in human diets, as elements of the work-place, and as pollutants in the environment. As part of this effort, it is essential to learn more about the interactions that can occur among different mutagens in various relative concentrations, and to identify those materials that have antimutagenic activity. The amount of field and laboratory work entailed in a project of this kind is truly enormous, and furthermore, much of it does not have the scientific appeal of analyzing the mutagenic properties of pure, well-defined, and chemically interesting substances. However, I believe that such practical field-related work is essential if we are to gain an accurate insight into the mutagenicity of our environments.

Thus, our greatest priority is to get our priorities right. We should concentrate our regulatory, political, and research efforts on those factors that, on epidemiological grounds, are known to cause human cancer and other mutation-associated diseases. And we should acknowledge that there comes a point of diminishing returns in pursuing meaningful data from studies employing ever smaller doses of radiation and chemicals.

The Problem of Mutagen Burden

Alexander Hollaender repeatedly has called attention to the importance of estimating the mutagen burden on individuals, occupational groups, and populations. The problem for chemicals is much more difficult than that for ionizing radiation. However, its resolution has great significance for public policy, and the public health benefits to be expected from regulatory measures are enormous.

The mutagen burden on an individual or population at present is not a precisely measurable quantity. It can be envisioned as the cumulative mutagen exposure to any genotoxic agent, of either endogenous or exogenous origin. The total mutagen burden would be some weighted summation of the burdens for all mutagens. This burden will depend in an important way on the lifestyle, diet, place(s) of residence, and occupation of the people concerned. At present, we have no practical way of defining precisely such a quantity, let alone measuring it.

For purposes of discussion, I find useful to consider the total burden as being divisible into "avoidable" and "unavoidable" components. One component of the unavoidable mutagen burden is natural background radiation which will vary geographically. Medical and dental sources of radiation are avoidable in principle, but, in practice, are generally unavoidable, as few would be prepared to forego their immediate benefits. As for chemicals, an unavoidable component would be those reactive species that arise endogenously in the course of normal metabolism, for example, oxygen radicals (Ames, 1983). Also, chemical mutagens are widespread in human foods and beverages, and, although this contribution to the burden might be reduced by dietary changes, it is unlikely to be eliminated completely because we must eat and drink something.

In the case of ionizing radiation, the unavoidable background dose to humans is small. Thus, even minor absolute increases in population exposure must be considered significant simply because the background _is_ so low. The adoption of sensible regulatory measures in health physics has served to protect both workers and the public from hazards that otherwise would have arisen from the irresponsible use of ionizing radiation and the proliferation of radioactive sources.

In the case of chemical mutagens, we simply have no idea whether the total burden should be considered "large" or "small", nor do we know how it might be distributed between "avoidable" and "unavoidable" components. Thus, regulatory agencies and the public are faced with the dilemma illustrated in Fig. 1. If the unavoidable mutagen burden at present is small and remains so for the foreseeable future, then appropriate regulatory measures would not only prevent future increases, thereby maintaining the _status quo_, but also, if they are sufficiently rigorous, would reduce the present level of avoidable burden. In the latter situation, regulatory measures would be expected to make a significant impact on the incidence of cancer and mutation-associated diseases (Fig. 1, left). On the other hand, if the mutagen burden already is high, and the avoidable component is small, then even the most Draconian regulatory measures will make only a slight impact on public health beyond maintaining the _status quo_

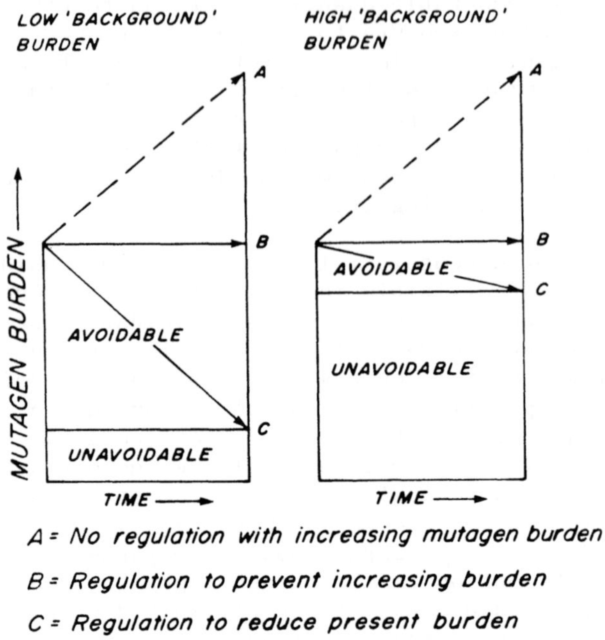

Fig. 1. Schematic representation of the mutagen burden question. The vertical axis represents the total mutagen burden on an individual or group. The left-hand diagram represents the case in which the total burden is largely "avoidable." There is only a low background of "unavoidable" mutagens to which people are exposed. The upper dashed line represents the change in this burden with time: here it is assumed that the avoidable burden increases at a substantial rate as years pass, but the unavoidable burden remains unchanged. The right-hand diagram represents the case in which the total burden is largely "unavoidable" background, but the avoidable burden, as in the left diagram, is increasing with time.

(Fig. 1, right). The worst possible scenario, from the standpoint of enthusiasts for the vigorous regulation of environmental chemicals, would be that in which the avoidable mutagen burden is a very small fraction of the total, and the effective rate of change in net mutagen exposure also is small. In such a dismal situation, regulatory measures would have no significant effect on public health.

Unfortunately, we do not know what is the actual mutagen burden on the human population. Most of us have our prejudices, but we should recognize such opinions for what they are. However, I am convinced that one of our top priorities must be to develop data that will help to decide first, whether the chemical mutagen burden is high or low, both on the general population and specific groups, and second, how this burden might reasonably be apportioned between avoidable and unavoidable components. Again, such a project would be massive in scope and the results probably hard to interpret. Population monitoring of the mutagenicity of all accessible body fluids would seem to be an obvious brute-force approach. If such studies were accompanied by cytogenetic analyses and follow-up of the individuals surveyed, it might yet be possible to determine whether the observed incidence of cancer and allied diseases is induced as a result of our immersion in the infamous "sea of mutagens." Further, we might learn whether this burden could be significantly reduced by lifestyle or regulatory decisions.

The foregoing discussion is conditioned by two major assumptions. First, that most of the changes involved in human cancer are induced somatically by mutagenic agents, and the contribution of "spontaneous" mutations to human disease is minor. Second, that for all practical purposes, cell populations in human tissues are homogeneous with respect to mutagen sensitivity and mutation rates. I feel, however, that these premises need to be re-examined critically in light of recent work on DNA repair in simple eukaryotes, and by the knowledge that large numbers of genetic loci may be involved in the maintenance of genetic stability in higher organisms.

The Complex Biochemical Basis of Genetic Stability and Change

Classical genetics established that mutations are rare, sudden and discrete events that presumably cause genes to pass from one state to another. Early searches for chemical mutagens gave negative or equivocal results, and, by the mid-thirties, only the high energy quanta of ionizing and ultraviolet radiations were established mutagens. These observations led Max Delbrück, in 1935, to propose that the remarkable stability of genes and the surprisingly high thermal Q_{10} for mutagenesis could be explained physically on the basis of the Polanyi-Wigner theory of molecular fluctuations. It would be necessary to assume only that the molecules of which genes are made have unusually high energy thresholds for the appropriate isomeric transitions. Such transitions could be stimulated by high energy radiation but not in ordinary chemical reactions. This view of the molecular basis of energy stability and change is illustrated schematically in Fig. 2 (left). Because of the alleged physical stability of genes, their presumably accurate mode of duplication, and the

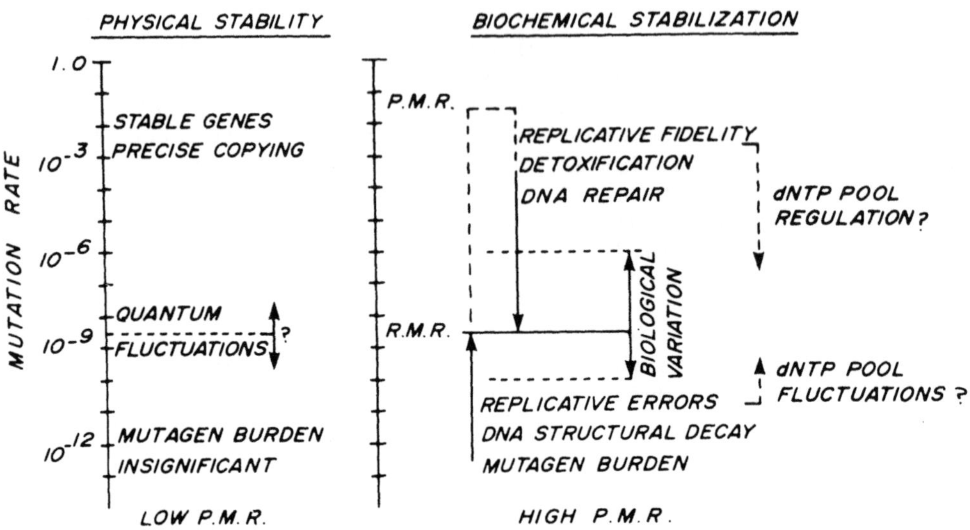

Fig. 2: Schematic diagram contrasting physical and biochemical mechanisms for the molecular basis of genetic stability and change. Left: Induced mutations would be rare since normally they could arise only from exposure to the low environmental levels of ionizing and ultraviolet radiations. Right: The potential mutation rate (P.M.R.) is very high, but various biochemical devices suppress it to low residual mutation rates (R.M.R.). There can be considerable biological variation in mutation rates, especially when mutations affecting the many genes involved in the maintenance of genetic stability are taken into account.

apparent lack of any significant mutagen burden, it was concluded that the potential mutation rate of cells (P.M.R.) was intrinsically very low. Mutations were assumed to arise from rare, quantum-statistical fluctuations in the isomeric states of the genetic molecules.

With subsequent discoveries of chemical mutagens and the many biological factors that can affect mutation rates, H. J. Muller concluded in 1952 that mutation is the result of some "biochemical disorganization in which processes normally tending to hold mutation frequencies in check are to some extent interfered with." Subsequent work on the structure, replication and repair of DNA has made it clear that organisms have high potential mutation rates and that their genetic stability is maintained dynamically by many complex biochemical processes.

Genetic "noise" can arise from spontaneous physiochemical "decay" of DNA structure, from attack on DNA by endogenous and exogenous mutagens, and from replication errors. The latter may be associated with altered DNA templates, dysfunctional DNA replication or repair complexes, and improper concentrations of deoxyribonucleotides (dNTP) and of the divalent cation Mg^{++}. Were it not for the existence of a complex array of molecular and biochemical mechanisms that counteract these natural tendencies toward error and decay, cells would have much higher mutation rates than actually observed. Indeed, it would have been impossible for genomes to have evolved much beyond a few hundred nucleotides in length (Eigen and Schuster, 1979). These stabilizing processes include various enzymes for the detoxification of mutagens, for DNA repair, and for maintaining replicational fidelity (for example, exonuclease "proofreading" during DNA replication and post-replicational mismatch correction). In addition, there is evidence that deoxyribonucleotide precursor pools are stringently regulated so as to minimize genetic change (Kunz, 1982; Haynes, 1984).

This more complex picture of the biochemical basis of genetic stability and change is summarized in Fig. 2 (right). It provides a broader view of mutagenesis than the well-known DNA damage-repair hypothesis (Haynes et al., 1984). Because most of the processes that promote genetic stability are enzymatic in nature, a large number of genetic loci must be involved in the synthesis and regulation of the necessary biochemical machinery. Even in yeast almost 100 loci are known to be involved in various modes of DNA repair, and many more undoubtedly remain to be discovered (Haynes and Kunz, 1981). When one adds to repair processes the proteins involved in mutagen detoxification, DNA replication, dNTP biosynthesis and pool regulation, it becomes clear that organisms probably devote a significant fraction of their genomes to encoding mechanisms that protect against genetic change. If this is the case, cells must possess a rather large number of loci which, when mutated, could give rise to clones that exhibit some form of impaired genetic stability. Work involving a variety of organisms has shown that mutations in loci involved in DNA replication, repair, and dNTP biosynthesis may confer mutator and hyper-recombinational phenotypes, as well as sensitivity to physical and chemical mutagens (Haynes and Kunz, 1981; Kunz and Haynes, 1981; Meuth, 1984). Although relatively few such genes have been identified so far in humans, there is no reason to think that mammals have any fewer loci involved in the maintenance of genetic stability than does yeast.

Such considerations raise several important questions for future research. First, how many human genes are involved in maintaining genetic stability? Second, how heterozygous is the human population with respect to these genes, and how deleterious are their mutant alleles in the heterozygous state? Third, are allelic frequencies for these genes sufficiently large to render the human population significantly heterogenous for spontaneous mutation rates and mutagenic sensitivity? These questions have considerable practical importance. If the human population is very heterogenous with respect to mutagen sensitivity, then an important policy question arises immediately. Should industrial standards for protection against the genetic hazards of radiations and chemicals be set to protect the most sensitive individual in the population, or only those of median sensitivity? I will not even attempt to discuss the question here, but it has implications not only for the cost of protection measures, but also for individual rights to employment in a safe environment.

Induced and Spontaneous Somatic Mutation in Carcinogenesis

An interesting theoretical question arises from the possible existence of a large number of human loci involved in maintaining genetic stability. The question relates to the relative significance of induced, as opposed to spontaneous, genetic changes in neoplastic transformation.

It is generally agreed that carcinogenesis is a multi-step process and that several somatic mutations or chromosomal rearrangements are involved in the progression from a normal cell to a fully developed malignant neoplasm. Recent work on cellular oncogenes (Land et al., 1983) and tumor cytogenetics (Yunis, 1983) indicates that some of these changes are highly specific and involve only a small number of genes. In principle, any one of these alterations could be induced by mutagenic agents, or could arise spontaneously via any of the routes to genetic change summarized in Fig. 2 (right). Thus, the set of tumorigenic events could be all induced, all spontaneous, or some mixture of both. Unfortunately, we do not know how the observed cancer incidences are divided among these three theoretical categories for their origin. It is important to get some quantitative insight into this issue because only that fraction of tumorigenic alterations induced by avoidable mutagens can be reduced by personal decision or public health control measures.

It is clear from the epidemiological data, especially for cigarette smoking, that a substantial fraction of lung cancer incidence is indeed induced and the same appears to be true for other types of tumors as well. However, if there really are about one hundred or more human genetic loci involved in the maintenance of genetic stability, then it is possible that "spontaneous" genetic changes might play a more important role in carcinogenesis than is commonly thought. To simplify the discussion I will designate these loci by the letter "s" (for "stability").

Even to contemplate such a large number of s-loci may seem unrealistic. It can be argued that although a large number of genes have been identified in yeast on the basis of their sensitivity to radiation and chemical mutagens, some of them might, in fact, be allelic (allelism tests have not yet been carried out for all of them); furthermore, some might not have any direct effect on the maintenance of genetic stability. On statistical grounds, however, it is clear that all the relevant loci, even for DNA repair, have not been identified; in any case, there exist many more such loci than was imagined only few years ago. I would like to play devil's advocate for a moment and explore the consequences of the possible existence of, let us say, one to two hundred s-loci in the human genome. On such an assumption and current multi-step, somatic mutation models of carcinogenesis, we can demonstrate that spontaneous genetic changes may play an important role in the origin of human cancer.

Consider a normal human epithelial cell containing a large number of s-loci, some fraction of which are heterozygous, though the cell has essentially a normal resistance to mutagens: there would be a large number of ways in which the genetic stability of cells could be impaired. Thus, spontaneous mutation rates, even though low on a per locus basis, could become significant because the number of relevant loci (and cells at risk) is large, and most of the DNA lesions induced at low mutagen doses would be subject to efficient repair. However, cells containing spontaneously mutated s-genes (certainly if homozygous and possibly if heterozygous) would then generate clones with reduced genetic stability. Such clones are likely to combine mutator, hyper-recombinator and mutagen-sensitive phenotypes. These clones would have a substantially enhanced chance of undergoing the further, more specific, genetic changes involved in carcinogenesis. Such changes could arise both spontaneously (because of the mutator or hyper-rec phenotypes) or be more readily induced by low mutagen doses (because of the enhanced mutagen sensitivity of such unstable clones). It is plausible to imagine that spontaneous mutation, including mitotic recombination and other genetic exchange events, might play a rather important role in tumorigenesis, although tumors in which all of the necessary genetic alterations arose from spontaneous events might be small. However, many spontaneous events might be included in the large class of tumors whose development involved both induced and spontaneous events. If such a view has merit, effective control of environmental mutagens would have its major health benefit in the younger age groups. Older people would accumulate the changes necessary for tumorigenesis primarily by spontaneous mutagenesis, a phenomenon that might not be subject to ameliorative measures.

CONCLUSIONS

I fear that the general tone of my remarks has been rather more pessimistic than I would have liked. This is not because I feel that little can be done to reduce the incidence of cancer and other diseases associated with human exposure to mutagenic agents. It derives rather from the very difficult, but not necessarily insurmountable, problems that face us in this area. If the epidemiological data are accepted at face value, and I see no reason to reject them, then a large fraction of neoplastic disease arises from lifestyle factors. But how does one persuade people to change their personal habits in order to reduce some future risk to health? In an increasingly competitive economic world, how much protection against relatively small risks can we afford? Clearly, there is no reason to allow the human mutagen burden to increase wherever reasonable regulatory measures would hold it in check. However, until we learn much about the magnitude of this burden, what proportion of it is avoidable, and what fraction of the genetic changes in tumors is induced rather than spontaneous, it will be difficult to give legislators and the public a quantitatively accurate assessment of the expected public health benefits from the control of environmental mutagens.

As always, future challenges for basic research are large and exciting. Discovery and analysis of the action of oncogenes promise to provide at long last a clear insight into the molecular mechanisms of carcinogenesis. The realization that genetic stability is maintained by an astonishing variety of complex biochemical mechanisms opens new research horizons in DNA repair, replication and recombination. It is a pity that Delbrück's purely physical theory of genetic stability and change is wrong. With biochemical stabilization there are an unfortunately large number of ways in which the complex genetic machinery can be damaged or go wrong. However, if this is the price organisms have had to pay for genomic evolution beyond the 100-nucleotide level, I suppose, on balance, it was worth it.

ACKNOWLEDGMENTS

I owe a debt of gratitude to my colleagues and students, especially Friederike Eckardt, Bernard Kunz, and Evan McIntosh, for many helpful discussions on the subject of this paper. Work in my laboratory, for many years, has been supported by grants from the Natural Sciences and Engineering Research Council of Canada.

REFERENCES

Ames, B.N., 1983, Dietary carcinogens and anti-carcinogens, Science, 221:1256.

Barclay, B.J., Kunz, B.A., Little, J.G., and Haynes, R.H., 1982, Genetic and biochemical consequences of thymidylate stress, Can. J. Biochem., 60:172.

Delbrück, M., 1935, Atomphysikalisches Modell der Genmutation, Part 3 of Timoféeff-Ressovsky, N.W., Zimmer, K.G. and Delbrück, M., Über die Natur den Genmutation und der Genstruktur, Nachr. Ges. Wiss. Göttingen, 1:189.

Eigen, M., and Schuster, P., 1979, "The Hyper Cycle: A Principal of Natural Self-organization," Springer-Verlag, Berlin.

Haynes, R.H., 1984, Molecular mechanisms in genetic stability and change: the role of deoxyribonucleotide pool balance, in: "Genetic Consequences of Nucleotide Pool Imbalance," F.J. de Serres, ed., Plenum Publishing Corporation, New York (in press).

Haynes, R.H., Eckardt, F., and Kunz, B.A., 1984, The DNA damage-repair hypothesis in radiation biology: comparison with classical hit theory, Brit. J. Cancer, 49, Suppl. VI:81.

Haynes, R.H., and Kunz, B.A., 1981, DNA repair and mutagenesis in yeast, in: "Molecular Biology of the Yeast Saccharomyces: Life Cycle and Inheritance," pp. 371-414, J.N. Strathern, E.W. Jones, and J.R. Broach, eds., Cold Spring Harbor Laboratory, Cold Spring Harbor, N.Y.

Kunz, B.A., 1982, Genetic effects of deoxyribonucleotide pool imbalance, Environ. Mutagen., 4:695.

Kunz, B.A., and Haynes, R.H., 1981, Phenomenology and genetic control of mitotic recombination in yeast, Ann. Rev. Genet., 15:57.

Land, H., Parada, L.F., and Weinberg, R.A., 1983, Cellular oncogenes and multistep carcinogenesis, Science, 222:771.

Meuth, M., 1984, The genetic consequences of nucleotide precursor pool imbalance in mammalian cells, Mutat. Res., 126:107.

Muller, H.J., 1954, The nature of the genetic effects produced by radiation, in: "Radiation Biology," Vol. I, Pt. 1., "High Energy Radiation," p. 417, A. Hollaender, ed., McGraw-Hill Book Co., Inc., New York.

Yunis, J.J., 1983, The chromosomal basis of human neoplasia, Science, 221:227.

WHERE THE FUTURE?

PART II - ROUND TABLE DISCUSSION

Marvin A. Schneiderman

Uniformed Services University of the Health Sciences
Department of Preventive Medicine and Biometrics
4301 Jones Bridge Road, Bethesda, MD 20814-4799

During the course of this meeting while trying to find some meaningful way to summarize what had been said, I found myself writing several outlines for these remarks. As I wrote each one, I found upon listening to another two or three stimulating and provocative speakers that the outline had to be put aside - and a new one written. As a consequence, these remarks constitute a "no latent period" paper: the exposure lasted for three days, and this response comes immediately thereafter. The toxicologists will recognize what we have here as an acute disease-type response rather than a chronic disease-type response. That means there should be place for substantial afterthoughts, after these thoughts.

In the course of my writing the outlines, I recalled several quotes that might have done by way of introduction. Having listened to the molecular biologists and the difficulties that they saw, and hearing toxicologists describe problems of age, sex, race, different kinds of exposure, multiple exposures, and the statistician-epidemiologists recount all the problems that humans get into, I realized that perhaps the appropriate quotation came from Hamlet: "There are more things on heaven and earth, Horatio, than are dreamed of in your philosophies." There was also a second quote that came to mind, one from my wife's grandmother, whose command of English was not as good as Shakespeare's (well, it was not her native language), but who could have expressed the same idea. What grandmother might have said having listened for these three days was "Oi!" accompanied by a lift of the shoulders and a slight turn of the head. Yes, life, and hence biology, is complicated. But complicated or not, it can be lived with - with, perhaps, a little humility. I too, am convinced that there are more things on heaven and earth - and I also shake my head and say "Oi!" But this reaction does not deter me from adding my support to some of the points raised by Dr. Haynes. What we have been trying to do here is to reconcile three things: first, modern science that shows us once again the enormous complexity and diversity of living organisms; second, the gross "black box" effect that we see when we look at the exposure-response relationships stripped of details and nuances; and finally, the need of the environmental health administrators to develop decisions that are so reasonable and so well worked out that they will withstand challenge or, even better, discourage challenge. I saw and heard lots of problems, but I do not think we have set ourselves an impossible task. Hard? Yes. Impossible? No.

We were talking this week both about science and the interface between science and policy. I do not think that science and policy can be fully separated. In the issue of Science, May 25, 1984, the first book review was of a publication on science policy. The reviewer starts out with the statement, "Science policy decides what research is to be done." It would appear that we have talked about for the last several days is what research has been done and what can be done, so that information can be transmitted to the people who are concerned with protecting our health. What can we develop out of the science so that we can help the administrators? Like the Science reviewer, I think science affects policy and policy affects science. As policy affects the science that we will do, so too does the science drive the policy by saying, "this science is worth doing; this science is worth following up."

While balancing precariously at the interface of science and policy, I heard (and experienced, first hand) evidence this week of problems in language and communication. Clearly, as scientists and technicians, we have problems in talking with the administrators. We also have problems in talking with each other and in hearing each other, especially the latter. At times, in the course of this meeting, I found myself failing to "hear" what was being said, and I then began to feel great sympathy for the administrators who have had to listen to me from time to time. It was as if I were watching a motion picture that had been made in Brazil; the language in the film was colloquial Portuguese and the subtitles were French.

I think we, as scientists, have to consider ourselves sometimes as translators. And, in doing so, we must be aware that with good translators something may be gained in the translation, and not lost. John Bailar talked about the scientist and the technician in their roles as scientists, as employees (of the federal government, industry or the university), and as citizens.

He made it clear that we, as citizens, must become involved in talking to the people who have to make the decisions and furthermore in talking to them in a way that they understand. That involvement means we also have to listen to what they say, to find out what it is that they want to know, and what they do not understand. Dr. Bridges asked the question, "How do we get the best people to talk to the administrators?" I am not sure I know what he means by "best," and I hope he would not want to cut off communication if the "best" were reticent. We must become aware that we will not necessarily get the best people to talk to the administrators, but rather we will get the people who have volunteered, those who want to do it. From what was said here it is obvious to me that more and more of us seem to want to talk to the administrators (in this company we should have a few "best" people). We are going to have troubles and difficulties at times, but to be effective, we have to recognize we must want to talk to the administrators. A great many of them, the lawyers and the politicians, really do wish to hear us. I recall Senator Tsongas of Massachusetts remarking (in effect) during his first year in the Senate, "Oh my, what do I do about all these issues that I am confronted with that I do not know anything about; however, I come from Massachusetts--greatest university state in the United States--surely there are people there who can help me." (He may even have meant you and me.) He quickly discovered that he could say to academics from these universities, "You know I have gotten involved in these technical issues. I do not understand them. Who at the University can and wants to help me?" He was overwhelmed with good help and advice. And he is not the only Senator who wants scientific advice. The key is that in talking to the non-scientist administrators and legislators we should be willing to do it essentially on their terms and when it fits their schedules.

This then brings me to a set of questions that we might talk to administrators about. Models for risk assessment? What is the matter with the models? They have defects that should make the administrators cautious (as if they were not already very cautious),but defects or not, there is no question that dose-response models will be used because they are required to be used. There is an Executive Order (No. 12291, I believe) from the President of the United States requiring the regulatory agencies to consider regulations from the point of view of cost and benefit. Any regulation that would lead to costs of more than $100 million a year has to be justified on the basis of a cost-benefit analysis. Cost-benefit analysis means that risk assessment must be made. The administrators of regulatory agencies <u>are</u> worried about the models, about which are the appropriate models and how to use them particularly in cancer. In recent risk computation on the effects of perchlorethylene, I saw that risk assessment had been computed using five different models and one off-stage specter lurking in the shadows. The models were the standard multi-stage, the one-hit, the multi-hit, the Weibull, the probit, etc. (the "etc." is the specter model that still needs to be developed, which encompasses all the good and none of the bad features of the others). Two of these models gave absolutely wild answers. They indicated that the probability of cancer was unity - certainty - at quite low doses.

Models such as these the regulator had to reject out of hand no matter what we said to him. What should we, as scientists, try to do to help? We are going to have to develop better knowledge of biology and promote the better science, probably along the lines of some of the things talked about here this week, so that we will eventually be able to help the regulators choose from among the models those that are closer to our new understanding of biology. We need to tell the legislators that we have major defects in the models now. We have susceptibility models, and mechanistic models. And is it that never-the-twain shall meet? Mechanism is extremely important. Susceptibility is extremely important. We must develop models using knowledge of mechanisms upon which are superimposed the variability found in human populations. I don't quite see where James Trosko's ideas will fit - but I think they will. We need to develop models in carcinogenesis that take into account problems of promotion; none of the present mathematical models handles the issue of promotion at all well. We must also develop models for defects other than cancer, such as pregnancy-wastage, developmental defects and deficits, and for illnesses in which there are clear thresholds. For them we still use the "safety factor" approach and so far it seems to have worked well (perhaps we have just been lucky). But now, people are trying to be far more quantitative than in the past with respect to regulation, and they really want to make valid computations based on sound biology. They know there is a limited amount of money that can be spent on regulation and on the various regulatory activities, and rational regulators must balance the need to take care of the most important illnesses first, with the existing, or soon to be developed, ability to do positive and useful things in preventing many illnesses.

The next question I want to ask the administrators about is "How much basic biology should we do?" The more we learn about the biology, the more we seem to learn about individual differences. The more we learn about individuals who are likely to develop the disease, the more we should be able to regulate wisely.

This now brings me to some of Richard Peto's opening remarks. When talking about causes of cancer I find myself agreeing with Peto substantially on the importance of factors such as cigarette smoking. That is based on hard data. I think the range of 10-70%

he put on dietary factors as causal agents shows the softness of those data. When Peto says the 2-8% of cancers are related to industrial exposure, recall that he has also said that the restraints and regulations that must occur with respect to industrial exposure must be very much greater proportionately than the proportion of illnesses these exposures cause. Industrial exposures are much more easily modified than are personal behavior. Five percent of all cancers related to industrial exposure implies a high risk for exposed persons. If 5% of all cancers are associated with industrial exposure, then, for exposed workers who are mostly men, about 25-30% of their cancers will be occupationally related. And by and large, society (i.e., industry and government) can do something to limit or prevent these exposures. It may very well be true that 10-70% of cancers are related to diet, but, if we cannot do very much about diet other than through education and persuasion, which takes a long time, I would urge that we now go on to take action where we can prevent disease now.

I think it is useful and interesting that these sessions were opened by Peto, a statistician and epidemiologist with active political leanings, and are being closed by myself, a statistician and epidemiologist with some (less active, I must admit) slight political leanings. The two of us are action-oriented. Is there some obvious action to recommend for the future? What is the future likely to bring? I have heard that an important function of an epidemiologist is to forecast what is going to happen to humans. I can therefore attempt to do some forecasting on the question of what is going to happen to cancer incidence because changes in cancer funds will lead to changes in regulatory policy, and we all believe research should lead to lowered cancer rates. For even being willing to try to make this prediction I am indebted to Morton Mendelsohn who invited me to the Livermore Laboratories about a year or so ago to think and talk about this problem. Here is what I see: By the year 2000 there will be a decline in age-adjusted cancer rates. Cancer rates are already declining for the youngest age group (persons under 45) for many forms of cancer. And they are declining a little less in the next oldest age group, and perhaps still increasing in the next oldest age group, and in the oldest age groups. Women are going to do better then men; cancer rates for women are lower than they are for men for almost every primary site that is not sex specific. There are still substantial differences among cancer rates in different geographic areas, which, however, are getting smaller. There are also substantial differences in this country in cancer rates among the races.

I think it is extremely important that we look at these differences to help gain understanding of both the biology and the environment. Black males in the U.S., who smoke, start smoking on the average at a later age than do whites; they also smoke fewer cigarettes but have lung cancer rates that are 40% higher than white males. There must be something else going on other than cigarette smoking. There are several other diseases in which there are substantial black/white differences that ought to provide us with clues to etiology and perhaps better means of prevention. What does this imply about appropriate areas for future research?

Many of the papers at this meeting detailed new work and new developments in molecular biology, and new techniques that will enable us to make far better measurements than ever before. At the moment, these developments seem to be on the micro-scale approach in biology. There are going to be important things for us to do on a macro scale, too. We are going to have to work very much more closely with the social scientists. As the social scientists have pointed out to us, at times we have really stuck our foot in our mouth when we have tried to pursue pure science with human subjects. I recommend the book Love Canal, published last year by Adeline Levine, a sociologist from State University of NY at Buffalo. Dr. Levine has written about the alienation of the population that lived at Love Canal by the scientists who came to work there. For anybody who is going out to work with human populations, it is imperative to remember that human populations are not research resources, not experimental animals. Human populations are people who have serious concerns about what you and I are measuring and what it means, and they really want to know what to make of the results of our tests. Our findings may mean publications to us, but they mean life (or may mean life) to the people on whom we make our measurements. So we are going to have to listen to the sociologists and the social scientists in order to learn how to work with exposed people as people, and not look upon them as "subjects" or "cases." Dr. Bailar brought us some of the thoughts and the points of view of the political scientists with whom we will have to interact more closely if our science is to be reflected in meaningful changes in public behavior.

We have been much involved this week with the problems that used to be considered the concern the pathologists, that is, the "splitters" versus the "lumpers." Should regulation proceed from the "black box" concept, where we do not know what is going on in the black box, and we assert that we don't need to know all the details (the lumpers approach)? Or do we reject this idea and take apart the black box, sometimes in so much detail that we lose sight of the fact that, if we are going to develop the film to give us a recognizable picture (i.e., take any reasonable actions), dark is needed so that the film is not spoilt. I am inclined to try to be a synthesizer (rather than a lumper or a splitter) and I depend on people concerned with detail to give me a basis for synthesis.

Dr. Haynes mentioned several other things that I would like to underline. He remarked that we have to be very careful about the numbers we generate. We must make clear to people that 3×10^{-6} does not mean 3.0000×10^{-6}. The General Accounting Office, in April 1984, published a report evaluating EPA's risk-benefit/cost-benefit operations. The GAO were fully in favor of doing cost-benefit computations but not in carving the results in granite. Their major recommendation was that EPA include estimates of the uncertainty of their results. Noting that the General Accounting Office has recognized that there are uncertainties in estimates being made makes me think that the scientists have had some effect.

However, the GAO seemed to express little of the uneasiness about risk assessment that the statisticians and others expressed here. One way of handling the uncertainties in the past has been to be "conservative." This approach has led to making a higher estimate of the risk than really might occur (i.e., an upper confidence limit). I think as our knowledge develops, our estimates will become less "conservative." We are going to get closer and closer to an estimate that we can all agree on, and the ranges of our estimates will not be as broad. If you are public-health oriented, the estimates of risk will come down from above. A health-of-the-economy view would place more emphasis on avoiding false positives. These two points of view, both reasonable, are in conflict. Let me illustrate this possible conflict with a military example. When I learned aircraft identification in the Army it was in a class room. We learned from slides how you might identify one aircraft from the other, but when we went out into the field viewing aircraft under not too ideal conditions, doubts turned up. If you were part of an anti-aircraft crew and a plane overhead went "brrr," the safe thing to do when in doubt was to shoot at it. Think of false positives and false negatives (in identifying the aircraft) from the point of view of the crew on the ground who, if they are skilled anti-aircraft gunners, are essentially in charge of whether that airplane is going to continue to fly overhead or not. A false negative is dangerous to the ground crew. A false positive is very much more important (and dangerous) to the aircraft crew. When in doubt the ground crew says, "Oh, that's possibly an enemy." That's a positive, and they shoot at it. A false positive becomes overwhelmingly important to the air crew who may be shot down by friendly fire. Shooting down one of our own aircraft could happen because it is safer for the ground people to make this false-positive decision. False positives and false negatives are not equal; they do not carry equal weight nor equal costs. They carry different importance to different people. To that anti-aircraft crew, the false negative carried high weight; to the air crew, the false positive carried even more weight. We have to keep false positives and false negatives in mind when we look at testing systems and their possible consequences. Whether the regulator looks upon himself as a member of the anti-aircraft crew or the aircraft crew may depend much on the tenor of the times. In the past I think we had many more anti-aircraft crew members involved in regulation than we do now.

What do I see in terms of our future testing, given the possibilities of false positives and false negatives? I think we are going to continue doing animal testing, including long-term feeding studies. I was a little surprised by Richard Peto, who, after estimating that 10-70% of all cancers were probably related to diet, then seemed to say, "Let's not do feeding studies; they didn't find asbestos, or cigarette smoking, et cetera as carcinogens." (There was a time when some epidemiologists argued that no carcinogens that affected humans were ever found in the laboratory. I have not heard this claim recently.) But how do you test dietary components if you do not do feeding studies? I think we are going to do feeding studies but in a somewhat more sophisticated way than before. At later stages, the full-scale animal test will be run in parallel with other assays that may help to define mechanisms. I think we may even learn ways to do such experiments that are not so outrageously expensive. As a way of defining applications of the research there will also be a place for the NAS/NRC kinds of committee, which consists of people who are from many different disciplines, who will try to give us a coherent view of the place of risk analysis. Pointing out where risk assessment may fail without supplying some alternatives is not a fully useful process, however. The usefulness of having people from the different disciplines, once they get to know each other, is that they will be able to say things like, "Fred, I do not understand a word of what you said. Would you please explain it to me again, and tell me why this work of yours is important or useful?" With some struggle, an explanation will be forthcoming that Fred (and others) will be able to write in language that nonspecialists will understand, and which will enable the nonspecialist, nonscientist, to visualize the science better so as to regulate better.

I want to conclude here. Largely, I think we have to learn to listen better, too, in order to regulate better. I must add that I have not defined "better," but with your help perhaps I can try. As for my not understanding Fred (or several Freds) - and Fred not understanding me or Charles Land or John Van Ryzen or Charles Brown, could we meet outside after work and start telling each other what we think our respective disciplines have to offer to improve the health and well-being of our fellow citizens? If we improve each other's understanding (please remember, I am a layman in your field) perhaps we can transmit this understanding to the regulators and administrators. Add to that if we hear some of things being said by people in the other disciplines that we have thought are not so important to the regulatory process, such as sociology, political science, law, perhaps we will learn to say better what we need to say. While all of us may not be the best people to talk to the regulators and the legislators, when we do talk with them we can still be useful and helpful.

DISCUSSION

Rybicka: The report of Dr. Peto showed that 35% of cancer incidence is due to cigarette smoking and 30% to the diet. It seems that these two factors could be affected to reduce cancer but there is a surprising lack of awareness among the general public. People know that smoking is not good for their health but they are not aware of the recent steep increase in the lung cancer among smokers. Information about the role of diet is less popular. It seems that the widely spread advertisement of the role of diet would help to reduce the role of cancer incidence.

Schneiderman: I think cigarette smoking is the most important single cause of chronic disease in the United States today. In 1978 I published a paper suggesting how this might be handled, with new proposals for labels on the cigarette packages, proposals on taxation and I also suggested monetary arrangements so that the tobacco farmers would not face bankruptcy (Schneiderman, 1978). These ideas have not been accepted yet. Do you remember that after Richard Peto gave his numbers for mortality from cigarette smoking he suggested that dietary factors were important contributors to cancer. He gave the range of 10% to 70%, and his "best" estimate was 35%. He asked, "What can we do about this in terms of what we know about diet?" He then surprised me by saying that he believed that we might be able to reduce the incidence by 1 or 2% with the appropriate dietary measures. Peto made reference to some conflicting data recently developed. Dr. Lee indicated that there may be an increased risk of cancer associated with the intake of polyunsaturated fats. Yet that is the type of fat that we are recommending people eat. Thus it seems that increasing the proportion of polyunsaturated fat, while reducing the risk for heart disease, might also increase the risk for cancer. Further, when we look at the range of cholesterol levels in this country, we discover that the persons with the lowest cholesterol levels have higher cancer rates than people with higher cholesterol levels; there are similar reports from Norway (U.S. Department of Health, Education and Welfare, 1979). I think the National Academy's recommendations that we cut down our fat intake and increase our intake of cruciferous vegetables are excellent (National Research Counsel/National Academy of Science, 1982). The report was a breakthrough in that it said, "Let us do something before we have definitive data." The Committee pointed out that if we had taken steps with respect to cigarette smoking before we had definitive data, several million people might not now be dead - or, at least, might have lived longer.

Holtzman: A couple of years ago, I was fortunate enough to be loaned out to a regulatory agency in science administration. We took an uncohesive program and turned it into a risk analysis program for evaluation of specific industries. One of the earliest things tht I was told was to go out to the laboratories, talk to the people doing the research, and find out what they were doing. I was also expected to explain to them our concerns and needs. One of the mandates of the risk analysis program was to determine the gaps in our information, and to obtain the appropriate data from the people who were doing the supporting research. Unfortunately, the Office of Management and Budget (OMB) came along and cut out the money for supporting research to get this information. Two years later, I am here at Brookhaven National Laboratory, and I find out that the same thing is happening to us now, and not only here, but around the country. How do we face something like this? How do we deal with OMB and the administration?

Starr: There are a couple of points I want to make. One is that I think Dr. Schneiderman is expressing his own personal bias with respect to the value, or cost, of false positives relative to false negatives. I think his example unfairly weighs the loss of two lives in a fighter aircraft against the entire population of the aircraft carrier. In fact, the cost of false positives may be every bit as great as the cost of false negatives, when you consider the lost benefits to society at large from the uses of a chemical in terms of quality of life. The other point is with respect to Dr. Haynes' remarks regarding the distinction between avoidable and unavoidable mutagens, or mutagenic sources. If we regard the mutagens as avoidable and unavoidable on a societal or national level, and voluntary and involuntary on a personal level, we set a 2 x 2 classification. The involuntary and unavoidable risks can't be resolved and the voluntary and avoidable ones such as smoking seem to be an individual problem. As long as those of you who are smoking pipes and cigarettes at this meeting are informed of the consequences, it is not a national problem. The involuntary and avoidable ones present the real difficulty, because the perception of risk on a personal level is nonlinear, especially for rare events in the tails. These rare events remain very important on a personal level, even though they are of little consequence on a societal level.

Lee: I want to take issue on one point. Once something becomes a mutagen of genetic concern, it is no longer an individual issue.

Schneiderman: First of all I would like to say that Dr. Starr is absolutely right. There is no doubt in my mind that I am biased; I am biased in favor of my survival, and I would consider anyone who is not as suicidal. I apologize if I led Dr. Starr to believe that I equated false positives with false negatives. A false positive or a false negative is more or less important to you depending on where you sit. If, for example, I was in the aircraft, a false positive would be worse than a false negative. If, on the other hand, I was the carrier or on the ground, then false negatives have worse consequences for me than false positives.

DeMarco: If we are not concerned with our survival, do you suppose we are suicidal? It seems to me when one talks to young people today one confronts that very issue. We face the realization that they are either (1) uninformed, therefore frivolous about the decision, or (2) they are depressed and they mean it.

Schneiderman: Or (3) they think that they are immortal, and that is why the young do not pay attention.

Longfellow: I want to pick-up on what Dr. Haynes was saying about the psychosocial aspect. I think that there is a tendency for scientists to play the ostrich. They fail to see the full impact of their work and the continuity of the risk assessment process. One major aspect is our failure to be diplomatic and to even attempt to explain to the non-scientist the full spectrum of our knowledge, particularly by putting the scientific findings into terms that the layman can understand (for instance, the relation between smoking and mortality). Our second failing is to make believe that the legislators in Washington are a breed apart and that we are without influence over our fate (e.g., research funding). They are a part of the public-at-large, as much as your next door neighbor or the folks that you see at a cocktail party. When the public better understands the need for research funding and the consequence of failing to provide it, they can then place appropriate pressures on the Congress and the Senate to allocate funds. In the 60's and 70's there was a tremendous increase in government funding of cancer research. In that atmosphere there was little tendency to scrutinize the research properties of other programs and other scientists as long as they were not impacting on the availability of research funds which you needed. Now, however, as funding has become tighter, there is a greater awareness of the finiteness of the available pie. The slice that you wanted is also the one I wanted, and by rights am entitled to. Instead of making a case for a more generous allocation of pie, we are reduced to jealous infighting and protection of our turf. I think that funding could be increased significantly by going back to the Congress, and convincing our representatives that our work has relevancy and import, particularly for the long-term future of mankind. I think it behooves us as scientists to become a little less like ostriches and to take a part in shaping our future.

Abrahamson: You are talking here about scientists representing science to government, to lawyers, and to judges, but what you have to remember is that you are not the only scientist doing this. Someone else may be representing the other side in any given issue, and he or she will have the same credentials that you have. Between the two opposing scientific views, who will be believed?

Neel: I think Dr. Haynes put it very well. The administrator has also to have confidence in you as a person. You are going to him not only as a representative of science, but as an individual.

Do you remember Dr. Wilbur Cohen, who was for a time Secretary of Health and then went back to Michigan to be Dean of the School of Education? Some time ago, when the money situation began to get tight, our young researchers set up a symposium to discuss how they could influence the decision-making process. Dr. Cohen was one of the scheduled speakers. What he said, very simply, was that the walls of Jericho are not going to fall down immediately simply because we scientists think tht we have a message for the legislature. We have first to establish channels of communication and trust, that you build up over time.

Clifton: And this process does take time. People have to have confidence in you, and essentially, to know you as a person.

Hollaender: Radiation biology was put on a solid base when we were able to determine the background radiation correctly. Why can we not at least attempt to do something like this in regard to chemicals, especially unavoidable chemicals?

Haynes: I agree that this requirement really is an item of the highest possible priority. There are no easy technical answers. It is hard enough to measure background radioactivity if you are taking into account all the different radioisotopes. But to measure background with chemicals is quite another matter. Building a large-scale data base on the mutagenic responses of human body fluids and using this in epidemiological studies of human populations might be of some help. We might perhaps use the techniques that Dr. Neel and Dr. Mendelsohn have been suggesting.

Borg: I would like to draw your attention to a BNL document from 1978 to show that initiatives dealing with risk assessment in public policymaking systematically and on a firm scientific basis is hardly new. It is entitled "Quantification of the effects of environmental carcinogens/mutagens on human beings. Critical evaluation of present knowledge" (Borg and Setlow, 1978). I have time to read only the first sentence, "Now is the time for critical assessment of three aspects of the quantification of environmental carcinogenic and mutagenic risks - thresholds, extrapolation, and background - that directly affect policy making and legislation." In 1979 we solicited support for the work proposed from Dr. Schneiderman (then at NCI), among others.

Hollaender: Last week I attended a meeting in Ovana on plasmids and bacteria where we heard some exciting developments, which 20 years ago would have been thought impossible. Perhaps we can follow them into this area, where you can take a little piece of plasmid and insert it into a chromosome. I hope when we have another meeting of this type that we can explore this; it is not an impossible idea, but it will take time and a lot of basic work before the technique can be successfully developed. I think our time is up now and I want to thank the speakers for their contributions.

Setlow: On behalf of the Medical Department and the Biology Department, I wish to thank all the participants for being so witty and intelligent, and for transferring so much new information to us.

Hollaender: The volume which we shall publish will be dedicated to Dr. Charles Dunham, who was a Director of the Division of Biology and Medicine, U.S. Atomic Energy Commission which later became the Department of Energy.

REFERENCES

Borg, D.C., and Setlow, R.B., 1978, Quantification of the Effects of Environmental carcinogens/mutagens on human beings: critical evaluation of present knowledge, BNL internal document, Oct. 1978 (revised April, 1979).

National Research Council/National Academy of Science, 1982, Diet, Nutrition, and Cancer, National Academy Press, Washington, D.C.

Schneiderman, M.A., 1978, Legislative possibilities to reduce the impact of cancer, Prev. Med., 7:424.

U.S. Department of Health, Education, and Welfare, 1979, Proceedings of the Conference on the Decline in Coronary Heart Disease Mortality, R.J. Havlik and M. Feinleib, eds., National Institutes of Health Publication No. 79-1610, Washington, D.C.

THE AGENT CARRIER AND TRANSFER APPROACH TO RADIOBIOLOGICAL RESPONSES

Victor P. Bond* and Matesh N. Varma

Brookhaven National Laboratory

Upton, NY 11973

INTRODUCTION AND GENERAL APPROACH

A conceptually close analogue for evaluating different approaches to the assessment of the average risk to cells exposed to "low-level radiation" can be found in physics. Here a physical target(s) of atomic or nuclear dimensions is exposed in a specified field of incident particles, and the probability of a particle-target encounter, and thus the risk of target transformation, is determined by observing the number of encounters that occur during a given exposure time t_E. This approach is described by the simple equation,

$$N_H = N_E \Phi \bar{\sigma}, \text{ or} \tag{1}$$

$$I_H = \Phi \bar{\sigma} , \tag{2}$$

where: N_E and N_H are the number per cm^3 of exposed and damaged targets, respectively;
I_H is N_H/N_E for a given t_E;
Φ is the exposure expressed in terms of the interactive fluence (that is, particles capable of interacting with a target), in particles/cm^2, during exposure time t_E;
σ can be viewed as either the "risk coefficient", the risk per unit fluence of a physical target sustaining damage as

*Victor P. Bond played a major role in instigating this Workshop, and we had planned that his presentation would lead off the meetings. Unfortunately, Dr. Bond became ill and was unable to attend. We are honored to include his paper in this volume. The Editors.

a result of the particle-target interaction, or alternatively, as the average value of the apparent "cross section" of the target, in units of cm^2/target.

At low exposures the interactions are stochastic, and because the number density of particles relative to that of the targets is so small as to virtually exclude more than one hit on the same target (that is, single-hit kinetics), only the number or proportion of hit and damaged targets, as opposed to the amount of energy transferred per hit target, can increase with exposure. Therefore, I_H is proportional to the independent variable Φ, the product of the fluence rate ϕ and the exposure time t_E. This gives rise to the simple linear, no threshold relationships shown in Fig. 1, in which I_H includes all particle-target interactions.

Consider the quantification of the risk to similarly exposed biological cellular systems using the physics analogy. The biologically active agent is energy, which can be transferred from the agent carriers (charged particles) only in a stochastic encounter with a cell target(s). Thus, the physics involved is the same as that described above, in that the target is of molecular dimensions. The particle fluence remains conceptually identical, but is limited to the charged-particle agent carriers or "interactive fluence"[1] Φ. This is because uncharged particles or electromagnetic radiations mostly transfer energy to cells or organisms indirectly through the production of charged particles. That only I_H, and not the mean energy transfer per hit cell target, will increase with exposure can be assured by limiting considerations to low-level exposure (LLE).[2] I_H cannot be determined directly in the living cell, but can be estimated indirectly by means of an instrumented surrogate or phantom target (see below).

Of more direct interest in risk assessment is I_q, the small fraction of cells in I_H with irreversible injury, showing cell transformation or, more generally, any spontaneously irreversible change of state termed a "quantal" effect to include chromosome abnormalities and cell inactivation (Finney, 1964). The value of I_q, and

---

[1]"Interactive" only in the sense of being able to interact with a cell target, and not in the sense of each particle having actually interacted.

[2]LLE is defined here as exposure (to radiation) in a field so weak that the occurrence of an additional hit on a previously hit and damaged cell target is negligibly small. Then, it is only the number or incidence of damaged cell targets per exposed target that can increase with exposure. Because repair of "sub-effect" damage occurs, reasonably large exposures spread over long periods of time also meet these criteria.

therefore of the relevant risk coefficient $\bar{\sigma}_q$, must be determined either by scoring the incidence of transformed cells directly or, indirectly, by scoring the number or fraction of descendant clones of transformed cells. Because the processes involved are also stochastic, I_q will be a constant fraction of I_H, and the curves for I_H or I_q against exposure Φ are linear and without threshold (Fig. 1).

One can obtain the slope and thus the risk coefficient $\bar{\sigma}_q$ for any quantal effect and radiation field. However, the disadvantages of this strictly empirical approach are severe. The risk coefficient $\bar{\sigma}_q$ for biological targets in any cell system and for any type of quantal response is highly dependent on the energy, mass, charge and velocity of the charged particle(s), that is, on the radiation quality. This leads to myriad values of $\bar{\sigma}_q$ for any specified cell system and quantal response. The multiplicity of risk coefficients severely limits the predictability of $\bar{\sigma}_q$ because of its applicability to only one of numerous situations encountered with different mixtures of particles, charges and velocities. A more coherent and and simplifying approach is needed.

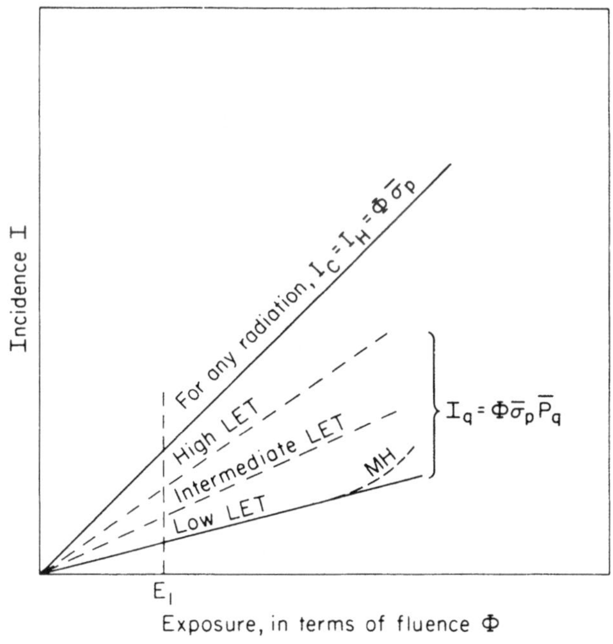

Fig. 1. Schematic curves for the incidence or number of hits per exposed targets, as a function of the particle fluence. Although the incidence of hit targets I_H applies to either physical or biological targets, I_q refers only to biological targets.

Such a new approach is obtained if each charged particle is viewed as simply a carrier of a common and potentially harmful agent, energy, rather than being regarded as a separate agent capable of causing biological damage and thus requiring a separate exposure-response function. Thus, each type of particle of a specified velocity carries a different total amount of energy, only a portion of which is transferred to the detector, the cell target, as a result of an interaction. Since the interactions are stochastic, the range of possible energy transfers, or "hit sizes,"[3] must vary widely from near zero to an upper limit, determined by the maximum amount of energy carried by the particle. No matter how monoenergetic any particle beam may be, any given amount of exposure to that or any other beam of charged particles must result in a wide spectrum of possible hit sizes.

From the general principles of pharmacology and toxicology, the probability of a quantal cellular response would be expected to be highly dependent on hit size and independent of the quality of the radiation. This expectation is augmented by the fact that not all target damage is effective in causing a quantal response, and that such subeffective damage to the target can be repaired rapidly. These observations suggest the existence of a function such as that shown in Fig. 2, a "hit size effectiveness function" (HSEF), in which the fraction of cells responding quantally is shown as a function of the amount of the common agent, energy, transferred to a cell. This function adds new dimensions in at least two distinctly separate ways. Viewed as a "cell sensitivity function" (CSF), it is a true cell analogue of the pharmacological organ "dose-quantal response" curve. It is in terms of this function that the absolute or relative sensitivities of different cell systems should be assessed.

Viewed as a "hit size effectiveness function" (HSEF), however, the new function permits one to weight the spectrum of hit sizes for any amount of LLE to a radiation of any quality, on the basis of the effectiveness of each hit size in causing a quantal cell response. The incidence of quantally responding cells for that exposure can be determined as shown in Fig. 3.

---

[3]The "hit size" is the analogue for stochastic agent transfer or "dose" for deterministic agent delivery. Although in principal the biological response to a given amount of agent is independent of mode of agent delivery, "hit size" is used to preserve the traditional use of "dose" for the deterministic mode of agent delivery to an individual(s) specified in advance. Also, both practically and as a result of less reproducibility of agent distribution among the prospectively unspecified individuals hit in the stochastic mode, the variance in biological response will generally be greater.

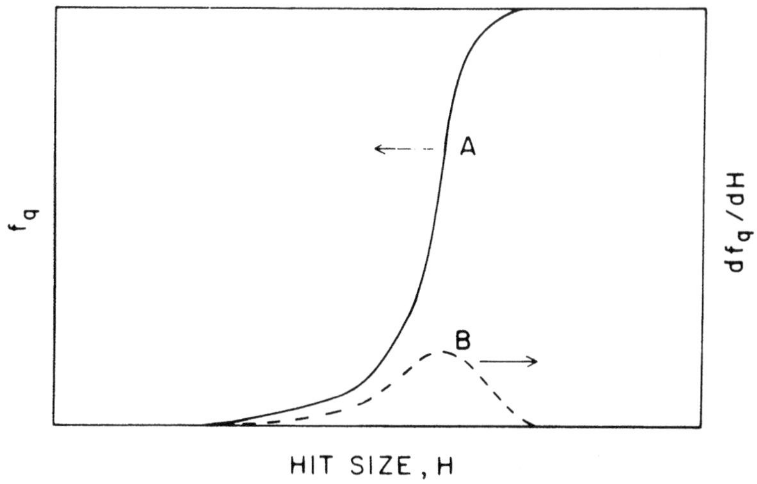

Fig. 2. Schematic of the cell sensitivity function, or, alternatively, the hit size effectiveness function. Curve A is the analogue for the cell of the dose-response function for an organ. Curve B, obtainable only by differentiating the observed function A, yields an estimate of the distribution of cell target sensitivities.

If the HSEF is generally applicable, it would alter drastically the manner in which risk assessment and radiation quality are now viewed conceptually and handled practically. For instance, it would eliminate the need for any one of the multiplicity of exposure-response functions exemplified by the curves labeled I_q in Fig. 1, and replace them by a single function, the HSEF. In so doing, it could obviate entirely the need for a standard radiation, relative biological effectiveness (RBE), quality factors (Q), dose equivalent and rem. It should then be possible to assess the expected excess incidence of single-cell quantal effects in a given cell system or in the host organism for any amount of LLE to any quality of radiation or mixture. It would simplify greatly risk assessment involving LLE for different radiations or mixtures of radiations.

In this paper, we will elaborate our thesis focusing upon why the approach and its application in radiation risk assessment is necessary with LLE. However, because it is so different from present approaches, we will first discuss relationships between the proposed and present approaches. Preliminary accounts of the overall approach have been published (Bond, 1979, 1982, 1984; Bond and Varma, 1983.

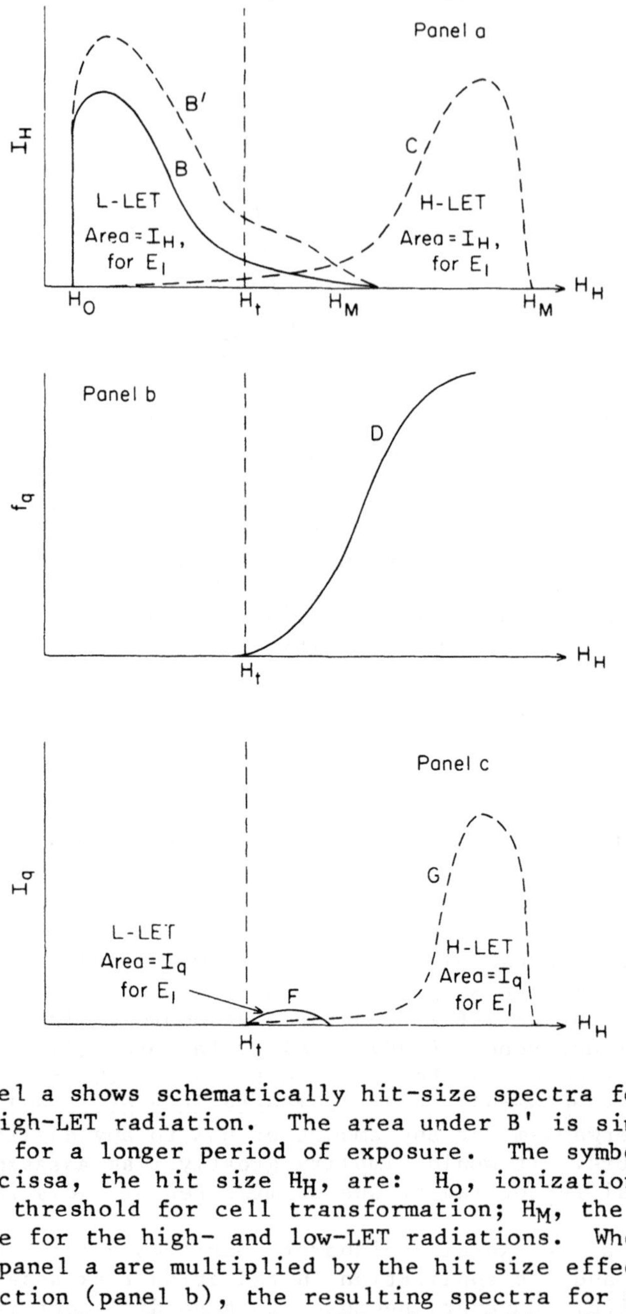

Fig. 3. Panel a shows schematically hit-size spectra for a low- and a high-LET radiation. The area under B' is similar to B, but for a longer period of exposure. The symbols on the abscissa, the hit size H_H, are: H_o, ionization potential; H_t, threshold for cell transformation; H_M, the maximum hit size for the high- and low-LET radiations. When the spectra in panel a are multiplied by the hit size effectiveness function (panel b), the resulting spectra for the incidence of transformed cells as a function of hit size, are shown in panel c. For a given total exposure, the area under each spectrum corresponds respectively to the incidence of transformed cells for the high- and low-LET radiations.

DOSE, EXPOSURE AND THRESHOLD VERSUS NONTHRESHOLD FUNCTIONS

Absorbed Dose

Although Φ is the conceptually correct independent variable that leads to a "linear, no-threshold" function for single-cell effects in the context of LLE, "absorbed dose" is now used "officially" and almost exclusively for that purpose (ICRU 33, 1980). It is therefore necessary to show that absorbed dose becomes effectively N_H or the dimensionless ratio I_H, and that its use with LLE and single-cell effects is conceptually inappropriate and misleading. Conversely, it must also be shown that, if the actual hit size or cell dose is used as the abscissa instead of absorbed dose, the cellular response curve effectively has a threshold. In other words, it is a matter of choice whether a given set of data is to be be displayed as a threshold function (that is, the fraction of quantal responders of a single hit size versus the hit size), or as a linear, no-threshold function (the number or proportion of individuals hit, without regard to hit size and including a preponderance of zero hit sizes versus the number or incidence of injured individuals, without regard to the severity of injury, which spans a wide range, including zero).

With respect to the first point, this can be shown most easily by writing the formula for the mean absorbed dose to a population of cells in situ or in tissue culture medium and subjected to LLE to radiation,

$$\bar{D}_A = \frac{H_1 + H_2 ---}{N_E} = \frac{H_1 + H_2 ---}{N_H} \cdot \frac{N_H}{N_E} \qquad (3)$$

where: $\bar{D}_A$ = the mean absorbed dose to the cells;
H = the hit size, or the amount of energy transferred to the individual cell;
N_H = number of hit cells;
N_E = number of exposed cells, with $N_H << N_E$. Eq. 3 can be extended,

$$\bar{D}_A = \bar{H}\frac{N_H}{N_E} = kN_H = k'\Phi \ , \qquad (4)$$

where: Φ = charged particle fluence, determined microdosimetrically (N_H/σ);
$k = \frac{\bar{H}}{N_E}$ (ergs/gm);
$\bar{H}$ is the mean hit size to cells that are hit;
$k' = k\sigma$ (ergs/gm)cm^2, where σ = cross sectional area of the sphere simulating the cell apparent target diameter.

It is then apparent from Eq. 4 that, when N_H is much smaller than N_E, the mean hit size $\bar{H}$ to the cells becomes constant, and it is only N_H (or I_H), proportional to Φ, that can increase with exposure.

The same phenomena seen in Eq. 4 above are shown schematically in Fig. 4. With LLE to radiation, $\bar{H}$ remains constant and only N_H or I_H increase with increasing exposure. However, with increasing Φ, eventually all targets are hit at least once. Under these conditions the total energy deposited per cell target, or the total hit size, continues to increase with increasing exposure. Thus, because of multiple energy depositions in each cell, the mean hit size to the cell targets approaches as its limit the average energy per gram to the organ or culture medium. Therefore absorbed dose, a dependent variable of Φ in the LLE range in the moderate to large exposure region, becomes a usable independent variable (Fig. 4).

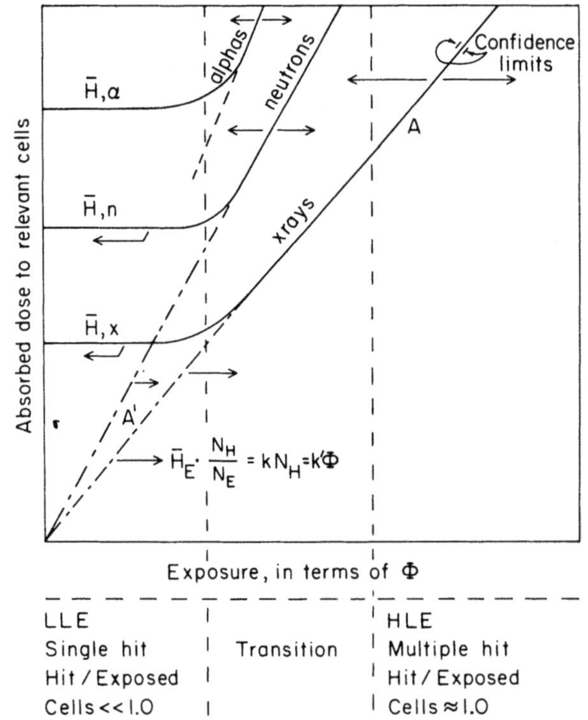

Fig. 4. Plot showing the relationships among Φ, absorbed dose to an organ, and absorbed dose to a group of cells in an organ or culture medium. The confidence limit for x-rays, shown as very small for curve A, becomes very large as this curve extends into the LLE region and becomes curve $\bar{H}$,x.

Figure 5 shows the consequences of this "Janus quality" of absorbed dose. It is usable as the independent variable for HLE where the probability of quantal organ effects are of primary consideration, it is of marginal value in the transition zone, but it is seriously misleading in the LLE region, where effects of single-cell origin are of primary concern. The fluence Φ is appropriate in all exposure regions.

Definitions of Exposure, HLE versus LLE

Although it is desirable to express the dose as the amount of radiation involved in an exposure in terms of a physical quantity related most directly to the severity of the biological effect, usually it is necessary to measure, and frequently to express, that amount in terms of its proximate vicinity to the biological object, i.e., in terms of exposure to agent carriers. With HLE the biologi-

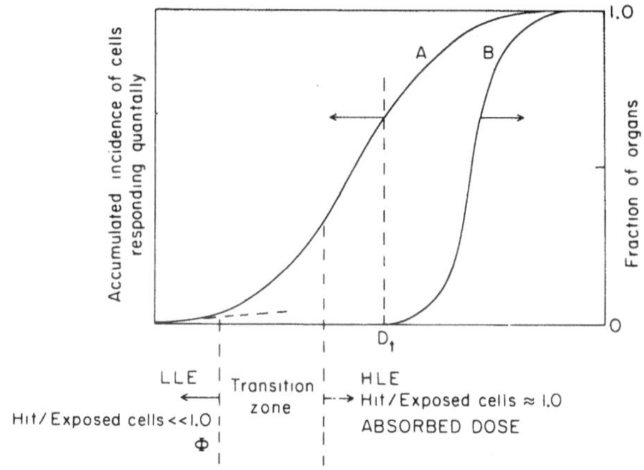

Fig. 5. Schematic plots, curve A for the incidence of cell death versus the exposure; and curve B for the fraction of like-dosed failed or transformed organs due to organ cell population depletion from cell death as a function of the absorbed dose of radiation. The transition zone extends from above the region of single-cell injury or death, through the region of cell population depletion and thus organ injury, to the region of organ failure and death. In this transition zone exposure, expressed in ICRU terms as the energy per unit mass of air in the environs of the organ or its cells, begins to approximate effectively both the mean cell "dose" and mean organ absorbed dose. This is because all cells are hit at least once, and thus added hits per cell can only increase the mean dose per cell and to the organ.

cal effect is on the individual organ (or tumor), and the severity of effect is believed to be related most closely to the amount of energy deposited per unit mass of tissue, that is, the object-oriented quantity, absorbed dose. Because the absorbed dose, the total energy concentration per unit tissue mass from multiple stochastic interactions, is proportional to the exposure Φ (Eqs. 3 and 4, Fig. 4), the amount of agent in the tissue (or the dose) is readily determined from that in the proximate environment, the exposure, Φ. To the degree that the exposure is measured in the actual tissue or organ, the more accurate will be the estimation of absorbed dose. However, with LLE and single-cell effects, energy is transferred stochastically so that the amount deposited per cell is not proportional to, and cannot be derived from the exposure Φ. Even were the exposure measured in tissues in close proximity to the cells of interest, it still would not measure the hit size or the mean dose to the cells.

To be exposed to low-level radiation is to be subjected to the possibility of an undesirable single-cell effect resulting from agent transfer due to a stochastic interaction or collision with a target. Therefore, the amount or the quantity exposure must be expressed in units of a constant multiplied by the number of possibilities for a collision. This is given by the product of the number of agent source "pass bys" per unit time, the exposure time t_E, and the constant σ. As noted above, this product, Φ, is the total exposure.

The total impact of a given exposure Φ on a population may be quantified in terms of N_H, the observed number of hit and injured individuals, without regard to the severity of injury. This same value, N_H, also expresses the expected excess number, N_H, for any given exposure Φ, in the observed population.

If N_H is normalized to per exposed person, N_H/N_E, the resulting expected incidence I_H provides also an estimate of the total risk, $\bar{R}_H$, to the statistically average exposed individual, therefore the average risk is a synonym for I_H.

If the severity of the impact on the population is taken into account, then I_H in Eq. 1 must be replaced by I_q. Thus,

$$\bar{\sigma}_q = \bar{\sigma}_p \, \bar{P}_b = \left(\frac{N_H}{N_E}\right)\left(\frac{N_q}{N_H}\right)\frac{1}{\Phi} = \frac{I_q}{\Phi}\ , \tag{5}$$

where: N_H/N_E determines the mean physical probability that an exposed cell will be hit;
N_q/N_H determines the mean biological probability that a hit cell (among many cells with different hit sizes) will respond quantally;
I_q is the product of N_H/N_E and N_q/N_H;

I_q/Φ determines $\bar{\sigma}_q$, the expected excess incidence of quantal cell effects per unit Φ, or the risk coefficient central to any biological risk assessment involving late effects of a single cell origin;
$\bar{P}_b$, the mean biological probability, is the weighted mean for the entire hit-size spectrum, corresponding to one value of Φ.

In summary, the difference between exposure and dose is that, with exposure, Φ is the independent variable for both HLE and LLE. With HLE, Φ is significantly dissipated through multiple collisions so that some large fraction of the beam energy is transferred to the tissue. Thus the absorbed dose is proportional to, and therefore readily predictable from, Φ. However, with LLE, the agent is transferred only stochastically from agent source to unselected individual(s), in amounts that are also stochastic, so that the hit size or dose to cells is not predictable from the exposure. Because the majority of exposed individuals are not hit, it is only the number of singly hit individuals that can increase proportionately with exposure. Thus curves for Φ versus I_H, I_i, or I_q are linear and without threshold.

Threshold versus Nonthreshold Functions

The representation of a complete set of data in the response of a population to exposure to an agent as either a threshold or a nonthreshold and linear function is purely a matter of choice rather than a requirement dictated by intrinsic properties of either the biological system or the agent. That this statement is true is shown most easily using organ quantal effects. Consider, as an example, a uniform population of 200 mice, divided randomly into 10 groups of 20 each, and given doses of increasing size of any chemical, biological or physical agent. If the amount of injury to an individual organ and therefore to the organism (the amount of organ tissue damaged or killed, or the severity of organ functional impairment) is measured as a function of dose, a quasi-threshold and curvilinear function, a dose-effect curve is obtained such as that marked A in Fig. 6a. Below the threshold D_T the system is "elastic," so that injury is spontaneously reversible. The focus is on the severity of the impact on the individual, as a function of dose. Groups of organs are exposed at each selected dose level only to improve the estimate of the response of the average individual.

As indicated by the S-shaped curve B in Fig. 6b, the fraction f_q of quantal responders in a group or population of "like-dosed" individuals, increases with dose. The derivative of this dose-response curve, shown as df_q/dD on the right-hand ordinate, yields an estimate of the distribution of individual organ sensitivities, which cannot be determined prospectively for any individual. The figure shows that the severity of the quantal effect on the organ or

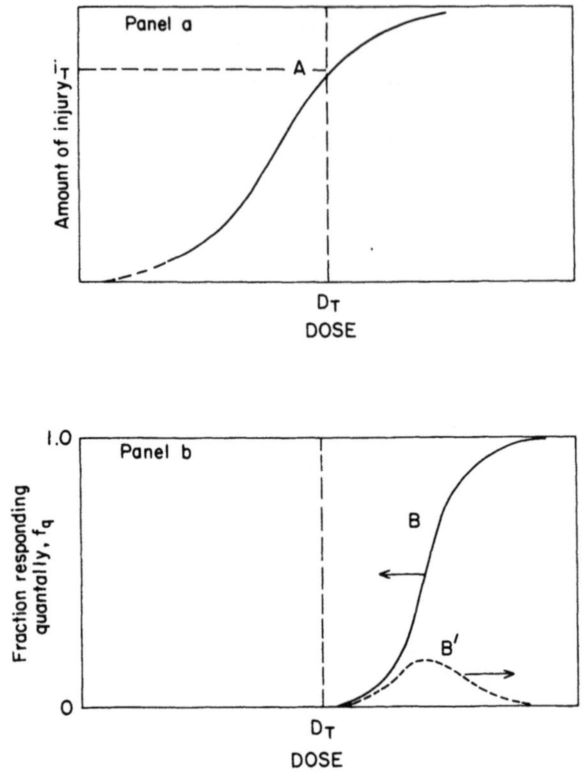

Fig. 6. Panel a. A schematic dose-effect curve for the mean severity of injury to "like-dosed" organs (individuals), as a function of the amount of agent, the dose. Severity can be measured in terms of the amount of organ tissue or number of cells damaged or killed, or in terms of the amount of functional organ impairment. There is a quasi-threshold, dependent in part on the sensitivity of the detection of injury. The curve reflects the severity of impact on the individual, and not on a population. Above the threshold dose D_T, the organ may undergo an all-or-none phase transition from effectively functional to effectively non-functional. This effect, which is usually irreversible in the absence of corrective intervention, is termed a "quantal effect."

Panel b. A schematic dose-response curve in which the severity of the health impact on a group of individuals is plotted against dose. The impact is measured as the fraction of "like-dosed" individuals who show a quantal response. Note that the severity of health impact on the individuals with a quantal response does not change above the threshold, and all that can increase with increasing dose is the fraction of individuals in a group who show the quantal response.

individual does not change with increasing dose. It is therefore only the severity of the health impact of a given dose on a population of "like-dosed" individuals, expressed as f_q, that can increase with dose. Further, the figure shows that of the total of 200 exposed mice, approximately 100 received doses above the threshold, of which about 50 showed the quantal response, death. In order to transform the threshold function shown in Fig. 6 into a proportional or linear, no threshold function, one need only use the identical set of physical and biological data used in constructing Fig. 6b, and simply change the independent variable, from dose to the sequential number of doses delivered randomly as shown in Fig. 7a, and then to the number of dosed individuals, as shown in Fig. 7b.

The linear, no-threshold curves in Fig. 7b can be converted to an exposure-incidence curve by using a different but identical population of mice, and by 1) keeping the number of mice to be hit N_H at 200, but adding perhaps 800 mice that are to receive zero dose, for a total of 1000 exposed mice and 2) selecting animals to be exposed by means of a quasi-stochastic process, and keeping all 1,000 animals in the pool so that each has a chance of being selected and dosed twice or more. In this quasi-stochastic process within the limits of allowed doses, the 1,000 animals are drawn randomly from a common pool, placed in an exposure container, and returned to the pool afterwards. The radiation source is programmed so that 4 out of 5 times that a mouse is in the exposure container, the beam is turned on but does not intersect the exposure container (that is, zero dose is delivered to the mouse). One out of 5 times, one of the 10 doses used previously is delivered, in random order, so that a total of 200 mice is dosed at least once and a few more than once. For consistency in terminology, and in order to restrict the meaning of dose to agent delivery by ordered processes, all dosed mice, under these conditions of random delivery will be called hit, and all doses will be termed hit sizes. Also, the series of beam pulses that either encompass or miss entirely the exposure container can be termed a fluence of radiation sources, since collectively the mice exposed in the container "see" a number of beams per unit area directed towards them, with different agent transfer capabilities ranging from zero to the largest hit size possible.

We now plot the increase in the incidence of hit animals, that is, number hit without regard to hit size divided by the number exposed (1,000 total of non-hit plus hit animals), against the number of agent carrier beams per unit area in order to obtain curve I_i (Fig. 8). From this we proceed similarly to obtain the curve I_q. The initial segment of both curves is linear and without threshold, but with slopes smaller than those in Fig. 7 by the dilution factor of unhit individuals, that is, by a factor of 5. These curves for stochastically hit organs are the analogues of exposure response curves for cells hit stochastically as the result of LLE to radiation. The start of a multihit component to the curve

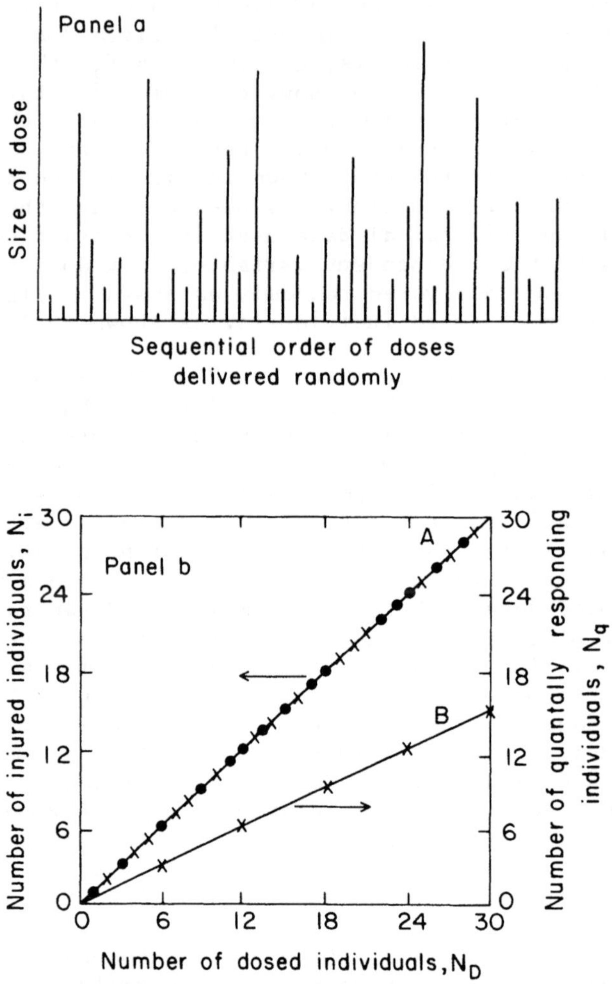

Fig. 7. Panel a. The size of hit plotted against sequential order of hits delivered stochastically. Note the random pattern of hit sizes, ranging from the smallest to the largest possible.

Panel b. The number of individuals injured N_i as a function of number of individuals injured without regard to severity of injury N_H. Also shown is the number of individuals injured quantally, N_q. Note that panel b is correlated with panel a, in which we see that a large hit can occur as the first hit, the last or anywhere intermediate. Thus, although only those individuals who received an above threshold hit size responded quantally, such responses are seen at any time during exposure. The no threshold is given in terms of the number of hit and injured individuals, and not in terms of the amount of agent to an individual.

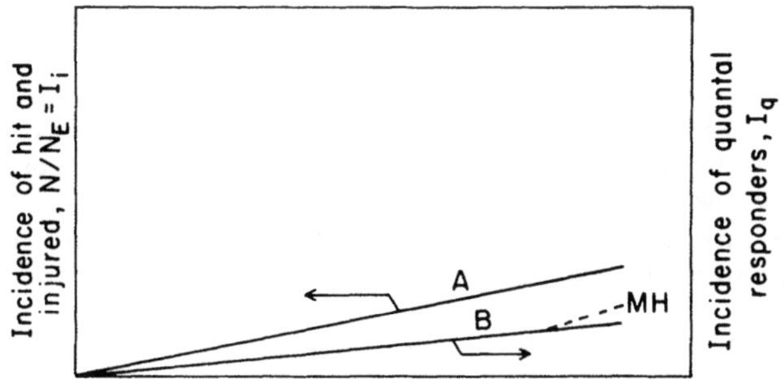

Fig. 8. Schematic exposure-response curves, in which are shown both the incidence, I_H, of hit and injured individuals, and I_q, the incidence of quantally responding individuals. These curves are identical to those in Fig. 7, except that the slope of the curve is reduced by a factor of 5. This factor is equal to N_H/N_E, the ratio of hit and injured, to the total exposed (that is, the 200 individuals hit, divided by total of 1,000).

I_q, labeled MH, is due to the same mouse being hit more than once. Thus the increase in I_q becomes greater than proportionality because the total hit size of a few mice is increased. The appearance of curve MH illustrates a requirement that if single-hit kinetics and therefore independence of action is to be maintained, then N_H must be much less than N_E.

QUANTIFICATION OF I_H AND Φ; THE HSEF

It is not possible to detect a hit on a living cell that has not responded quantally, therefore the total number or incidence of hit cell targets can be quantified only by the use of a physical surrogate for the cell target, a cell phantom. Such phantoms, although not specifically identified as such, have been developed extensively as microdosimeters (Kellerer, 1976; Rossi, 1968, 1979). These phantoms are spherical proportional counters filled with tissue-equivalent gas and operated at reduced pressures, such that the response per particle in principle is the same as that in the living cell target. Thus the instrument, with appropriate scaling for any given LLE, measures the number of singly hit simulated cell targets per exposed target, or the incidence of singly hit cells.

Analogous to the cross section σ_p in physics, $\bar{\sigma}_q$ (Eq. 5) represents only the apparent target size. This can exceed appreciably the actual or geometrical target size of macromolecular dimensions because there may be more than one target, and because physical or chemical influences can extend well beyond the physical cross section of a charged particle. Thus σ_q is regarded here purely as a parameter of the system, namely the risk coefficient I_H/Φ, and not as an accurate estimate of target(s) size. In addition to measuring I_H when viewed as a phantom cell, the proportional counter also measures Φ when viewed as a charged particle detector of known cross section. The instrument also yields the spectrum of hit sizes for any given exposure in terms of the pulse height.

Even though it implies energy transfer, the hit size, H, is used here because it is noncommittal with respect to which physical parameter is most relevant to the amount of target damage and thus the probability of a quantal cell response. As with macrocollisions, the most relevant impact parameter is not known. Although energy is most easily measured, it could as well be the imparted momentum, as has been suggested in another context (Turner and Hollister, 1965), or the impact velocity (Richmond et al., 1961). If energy is used, it could be expressed as either the total energy imparted per target, the energy density along the particle track, or energy density in the real or assumed target volume. In practice, any of the microdosimetric quantities, ε, y, or z could be used (Rossi, 1968).

The estimation of the HSEF is described in detail elsewhere (Varma and Bond, 1983), and will be discussed here only briefly. Because no quantally responding cell can be identified with any given hit size, the fraction of cells with the same hit size that respond quantally cannot be determined directly. Therefore one needs multiple independent inputs to the effectiveness of a given hit size, in terms of producing a quantal response. Highly quantitative microdosimetric hit-size spectra and cell quantal response data for several radiations of different qualities fill this requirement, not only because the spectra are different, but because the total effectiveness of different hit-size spectra can be compared in terms of the RBE of the different quality radiations. Alternatively, the radiation quality, the interaction and agent transfer characteristics of a given size spectrum, can be varied by use of multiple hits to increase the total energy transferred per cell, to simulate the larger hit sizes achievable alternatively by using higher LET radiations. The procedure carries the assumption that the probability of a quantal response depends only on the total hit size even if given in increments, provided that the time between increments is short in comparison to that for repair of cellular injury. Therefore, high exposure rates must be used. Only the lower reaches of the multihit portion of the curve are used in order to minimize complicating influences, such as cell death. In either approach, σ_q must be among the independent parameters employed in obtaining the best fit HSEF by computer analysis of a given set of biological and microdosimetric data (Varma and Bond, 1983).

RADIOBIOLOGICAL FUNCTIONS, EXPECTED AND OBSERVED

We note from the above discussions that energy can be delivered only stochastically to individual cells in amounts also determined stochastically. Further, since it is only the severity of the population health impact that can be quantified in terms of the number or incidence of quantally responding cells, one would expect to observe only exposure-response functions, that is, cell exposure-quantal response functions that are initially linear and without threshold.

That the expected curves are actually observed is illustrated by the examples shown in Figs. 9 and 10. Additional examples and the fact that the range of initial and limited single-hit kinetics vary with low-LET radiation, can be extended if larger exposures are given at reduced exposure rates to permit maximum repair of sub-effective cell target damage are shown in Figs. 11 and 12, and de-

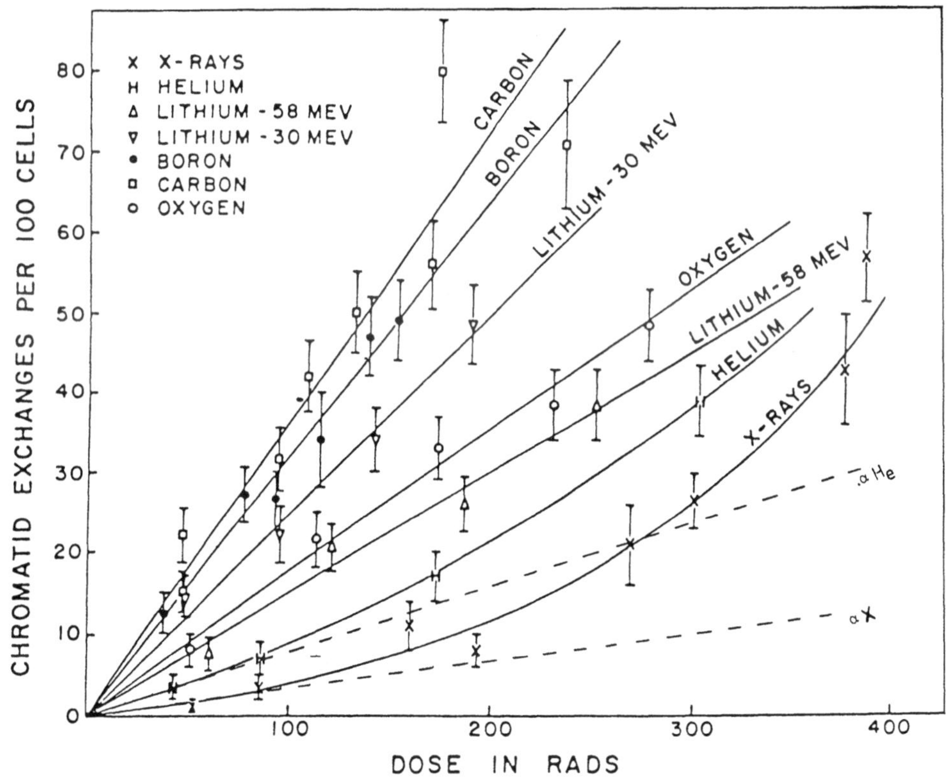

Fig. 9. A series of linear, no-threshold curves for the single-cell quantal effect, chromosome abnormalities, for a number of radiations of different quality or LET.

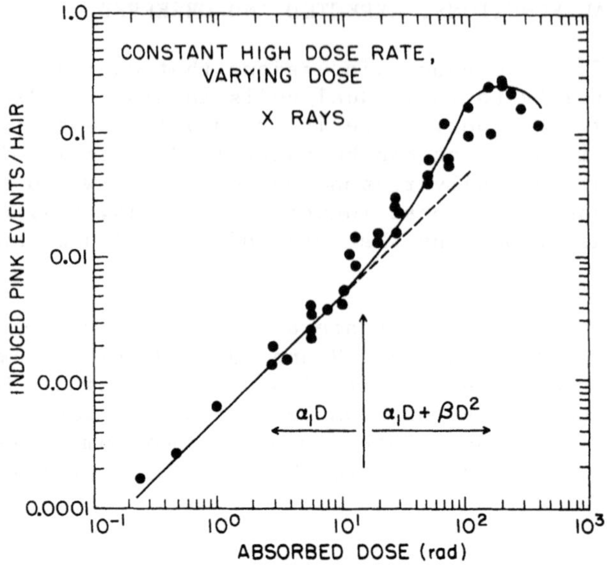

Fig. 10. X-ray dose-response curve for induced pink mutations in Tradescantia, on a log log plot to show detail in the low-dose range. The solid circles indicate experimental points at high dose rate. The low-dose portion of the solid curve and its dashed-line extrapolation have a slope of unity corresponding to a linear, no threshold dose-effect relationship. The increased slope at higher doses indicates that the response involves a higher exponent of dose. For explanation of the linear "$\alpha_1 D$" and quadratic "$\alpha_1 D + \beta D^2$" portions of the curve see text from Sparrow et al., 1972; Underbrink and Sparrow, 1974; Nauman et al., 1975; Underbrink et al., 1976.

scribed in detail elsewhere (Bond, 1979; NCRP, 1980). Bond (1982) has also discussed the clonal and thus probably single-cell origin of many human malignancies, and the fact that the full expression of a malignancy is probably a multistage process that involves promoters, inhibitors or other factors. These factors are nearly always endogenous, and thus exposure of the individual could interfere with their normal role. However, a significant effect would be expected only with HLE where cell death, and thus organ damage, is prevalent. This influence can be sharply reduced or eliminated by limiting observations to the LLE range. Thus, while the theory and observations outlined have not been shown to apply to other than single-cell autonomous systems and cannot be defended rigorously as applicable to the induction of human cancers, they do appear to describe well what is observed in single-cell systems. Our ideas do not appear to be seriously contradicted by animal or human data based on either radiation mutagenesis or carcinogenesis.

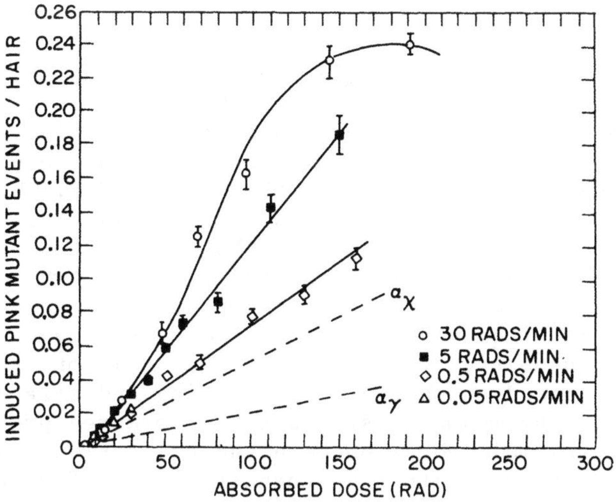

Fig. 11. Dose-response curves for pink mutant events/hair after x-irradiation at 0.05 and 0.5 rad min^{-1} (combined), and at 5 and 30 rad min^{-1} (from Nauman et al., 1975). The dashed lines represent the alpha terms in Equation 3.2 for x-rays and gamma rays.

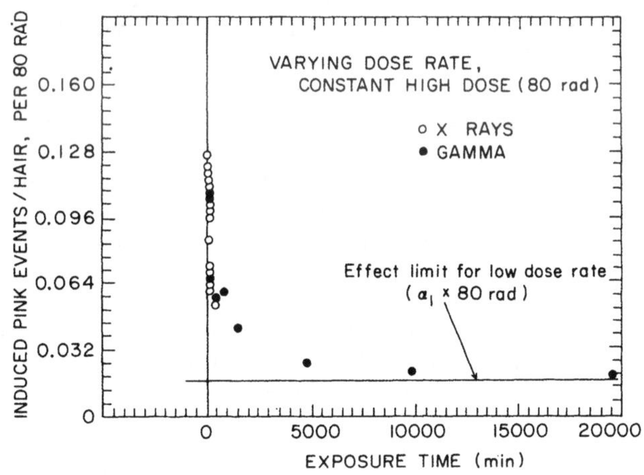

Fig. 12. Effect of dose rate on the effectiveness of a single large gamma-ray dose of about 80 rad for the induction of pink mutations in Tradescantia. The horizontal line represents the expected limiting low dose-rate value for 80 rads. The effect at 80 rads decreases appreciably as the exposure time is increased, to approach asymptotically the limiting "αD" value for gamma radiation.

DISCUSSION

A key element in the drastically different paradigm for LLE versus HLE we have described is the HSEF, which provides as the analogue for the quantal effects on organs the quantal effects on cells. For example, both cell transformation and loss of organ function can be quantified through the use of HSEF. However, perhaps of equal importance, the HSEF shows that whether a response relationship is threshold or linear (that is, without threshold) depends entirely on the choice of variables on the two coordinates. All else being equal, these observations are not unexpected because neither the type of quantal effect caused by the agent nor the susceptibility or sensitivity of an individual cell or organ to quantal change should depend on whether a given kind and size of physical insult is received by means of stochastic or ordered processes. Sharp distinctions were made above between the collection and handling of response data obtained with the stochastic mode of agent transfer, for which Φ is the independent variable, and the effectively orderly mode of agent transfer seen with HLE and characteristic of the concept of dose, for which dose can be used as the independent variable. In these same sharp distinctions lie several of the principal operational differences between the Public Health subdisciplines, epidemiology, accident statistics, and safety (PHS) and the medical subdisciplines, pharmacology and toxicology (PTM). However, these operational differences do not constitute the essential differentiating characteristics of the two disciplines with respect to applicable functions. Rather, the principal distinguishing feature of PHS is that it normally deals with the health impact on a large exposed population of rare events and agent transfer to a few individuals, so that the variables needed are 1) either the fluence of agent sources or the small number or incidence of individuals hit and "dosed" and thus injured, and 2) a PTM-type HSEF to separate out the smaller group of hit and injured with a given severity of injury (for example, those with a given quantal effect). On the other hand, PTM is distinguished by its focus on the individual, so that the variables needed are the amount of agent received by every involved individual and the resulting amount of injury to the hit or dosed cell or organ. The severity of effect can then be characterized in terms of the probability of a quantal response for a given above-threshold hit size or dose.

Clearly LLE fits only in the PHS category, and is unusual in being among the few public health situations in which Φ is accurately quantifiable. Proportional surrogates must be used in most other exposure situations--an exception is lightning, as shown by Orville et al. (1983). However, radiation exposure is unique in that it shows clearly a smooth and extended transition from the clearly single-hit to the multihit range (Fig. 5), and thus shows that the discipline of public health and safety (the region of LLE) and that of pharmacology-toxicology (the region of HLE) are closely related. We would also point out that A in Fig. 5 is a single curve with two quite separate meanings, 1) an exposure-quantal response function for single-cell effects extended into the multihit region, and 2) a function for dose versus the amount of organ impairment, if the curve A is a measure of the single-cell quantal effect of cell lethality. A multihit component is almost nonexistent in essentially all analogous macrocollision situations, because the individual hit is soon removed from the scene so that a second hit cannot occur until complete recovery has taken place. Although the proportional functions of PHS are well known, this also explains why actual linear, no-threshold curves are not: namely, linear functions are too trivial to warrant plotting.

This latter fact has permitted the disciplines of public health and medicine to remain separated, with relatively little interaction or effort to explore the conceptual relationship between the two. This complete separation of the macro public health and medicine-pharmacology situations has contributed to the difficulty in articulating clearly that radiation exposure does involve the two quite separate disciplines, medicine and pharmacology for HLE and effects on the single organ, and public health, epidemiology and accident statistics for LLE and single-cell late effects. Lack of definition has contributed substantially to the confusion and misunderstandings that have surrounded LLE especially because of the resulting tendency to regard LLE as no more than a quantitative extension of HLE dose-response relationships into the LLE region. Consequently one is left with the mistaken impression that radiation is an agent so unique that any dose, however small, can have serious or even lethal consequences. Such an interpretation can be disquieting and even frightening, because it is the general and correct view that for any agent a small dose will not result in an increase in the risk of serious harm.

The approach also makes it clear that the public health situation is the general case with respect to agent delivery and health impact, while the medical-pharmacological-toxicological situation is a special case. That is, each physical variable as well as the individuals who are hit or dosed are all fixed by orderly processes in the pharmacological special case, while they are all stochastic in the Public Health situation. The biological variable, individual variation in sensitivity, is left as the only stochastic variable in the pharmacological case.

Our new approach differs from all other models in that no assumptions have been made with respect to molecular or subcellular mechanisms that may play a role in tumor induction or cell transformation. That is to say, only the complete expression of quantal effects has been utilized in deriving the functional relationship, the HSEF.

By replacing the complicated apparatus of "absorbed dose-response" curves and RBE with the effects of radiation quality in terms of the single HSEF, a number of radiobiological observations are readily explicable. Thus the linear, no-threshold relationship is seen to be merely an artifact of data presentation. Also, it presents the impact on a population of a given exposure in terms of the number or incidence of individual organisms with a single-cell quantal effect, rather than the amount of injury to an individual cell as the result of a given total dose to the cell target, as is implied by the current use of "absorbed dose" with LLE. Clearly the no threshold is for the risk of a quantal effect to the individual(s) having the exposed cell population, and not for an actual effect on any individual. Increments of risk are linearly additive with LLE and cumulative (that is, increasing at a rate greater than linear) only with HLE.

Our approach also makes it clear that neither absorbed dose-response curves nor RBE can in any way reflect differences in sensitivity of the individual cells to different amounts of the agent energy. This is particularly so since absorbed dose does not reflect even the effectiveness of the same number of agent sources from radiations of different quality; it measures different numbers of agent sources per unit absorbed dose. The latter fact also explains why the function of RBE versus mean LET, in general, goes through a maximum for eukaryotic cell systems. The initial increase is dominated by an increase in the mean hit-size and the corresponding rapid increase in effectiveness of producing a quantal response. When the maximum response has been achieved, and 100 percent of cells has responded quantally, the function must decrease because the only remaining variable is the decreasing number of agent sources per unit absorbed dose, with further increase in the mean LET.

In using a proportional counter to obtain the number of cells hit for a given Φ and the spectrum of hit sizes, emphasis is focused on the "agent transfer capability" of charged particles in the radiation field. It is only this quality, with LLE, that is relevant to the amount or distribution of amounts of agent received, and therefore the individual or population impact of the exposure.

The new approach may be of substantial value in defining the radiobiological basis for radiation protection standards, using expected excess incidence in the exposed population as the criterion for setting limits. With the dosimetry reevaluations for the exposed Japanese (Bond and Thiessen, 1981), the dose-response curve for external neutrons, thought to be derivable from the Hiroshima data, seems no longer to exist in a usable form. Also, the excess incidence curves for low-LET radiation from these Japanese and other sources yield only upper limits for the true risk coefficient. Thus no RBE values as a basis for Q can be obtained for external high-LET radiation of humans, and data on RBE from animals are so dependent on species, strain and tumor type as to render quite tenuous the use of these values for humans. The present method offers a new approach to this problem. By determining the HSEF for the same or different quantal effects in the same cell types, across species and within one species, patterns may be discernible that will permit relatively confident extrapolation to human tumorigenesis from radiations of any quality or mixture of qualities.

REFERENCES

Bond, V. P., 1979, Quantitative risk in radiation protection standards, Radiat. Environ. Biophys., 17:1.

Bond, V. P., 1982, The conceptual basis for evaluating risk from low-level radiation exposure, in: "Critical Issues in Setting Radiation Dose Limits," No. 3, pp. 25-65, National Council on Radiation Protection and Measurements, Bethesda, MD.

Bond, V. P., 1984, A stochastic basis for curve shape, RBE and temporal dependence, in: "Radiation Carcinogenesis, Epidemiology and Biological Significance," Vol. 26, pp. 378-402, J. D. Boice, Jr. and J. F. Fraumeni, Jr., eds. Raven Press, New York.

Bond, V. P., and Thiessen, J. W., eds., 1981, "Reevaluations of Dosimetric Factors, Hiroshima and Nagasaki," Technical report CONF-810928 (DE 81026279). National Technical Information Service, Springfield, VA.

Bond, V. P., and Varma, M. N., 1983, A stochastic, weighted hit size theory of cellular radiobiological action, in: "Radiation Protection," pp. 423-438, J. Booz and H. G. Ebert, eds., Comm. of European Communities, 8395, Luxemborg.

Finney, D. J., 1964, "Probit Analysis," 2nd ed., Cambridge University Press, London.

ICRU Report 33, 1980, "Radiation Quantities and Units," International Commission on Radiation Units and Measurements, Bethesda, MD.

Kellerer, A. M., 1976, Microdosimetry and its implication for the primary processes in radiation carcinogenesis, in: "Biology of Radiation Carcinogenesis," pp. 1-12, J. M. Yuhas, R. W. Tennant and J. D. Regan, eds., Raven Press, New York.

NCRP National Council on Radiation Protection and Measurements, 1980, "Influence of Dose and its Distribution in Time on Dose-Response Relationships for Low-LET Radiation," NCRP Report No. 64, National Council on Radiation Protection and Measurements, Washington, DC.

Nauman, C. H., Underbrink, A. G., and Sparrow, A. H., 1975, Influence of radiation dose rate on somatic mutation induction in Tradescantia stamen hairs, Radiat. Res. 62:79.

Orville, R. E., Henderson, R. W., and Bosart, L. F., 1983, An east coast lightning detection network, Bull. Am. Meteorol. Soc., 64:1029.

Richmond, D. R., Bowen, I. G., and White, G. S., 1961, Effects of impact on mice, rats, guinea pigs and rabbits, Aerosp. Med., 32:789.

Rossi, H. H., 1968, Microscopic energy distribution in irradiated matter, in: "Radiation Dosimetry," pp. 43-92, F. H. Attix and W. C. Roesch, eds., Academic Press, New York.

Rossi, H. H., 1979, The role of microdosimetry in radiobiology, Radiat. Environ. Biophys., 17:29.

Sparrow, A. H., Underbrink, A. G., and Rossi, H. H., 1972, Mutations induced in Tradescantia by small doses of x rays and neutrons: analysis of dose-response curves, Science, 176:916.

Turner, J., and Hollister, H., 1965, Relationship of the velocity of a charged particle to its relative biological effectiveness, Nature (London), 207:36.

Underbrink, A. G., and Sparrow, A. H., 1974, The influence of experimental end points, dose, dose rate, neutron energy, nitrogen ions, hypoxia, chromosome volume and ploidy level on RBE in Tradescantia stamen hairs and pollen, in: "Biological Effects of Neutron Irradiation," pp. 185-214, IAEA-SM-179/31, International Atomic Energy Agency, Vienna.

Underbrink, A. G., Kellerer, A. M., Mills, R. E., and Sparrow, A. H., 1976, Comparison of x-ray and gamma ray dose-response curves for pink somatic mutations in Tradescantia clone 02, Radiat. Environ. Biophys., 13:295.

Varma, M. N., and Bond, V. P., 1983, Empirical evaluation of cell critical volume dose vs. cell response function for pink mutations in Tradescantia, in: "Radiation Protection," pp. 439-450, J. Booz and H. G. Ebert, eds., Comm. of European Communities, 8395, Luxemborg.

PARTICIPANTS

ABRAHAMSON, Seymour A.
Zoology Research Building
U. of Wisconsin
Madison, WI 53706

BAILAR, John S., III
Office of Disease Prevention and Health Promotion
Room 2132
330 C Street, S.W.
Washington, DC 20201

BAN, Sadayuki
Biology Dept.
Brookhaven National Lab
Upton, NY 11973

BECKNER, William M.
National Council on Radiation Protection
Woodmont Avenue
Bethesda, MD 20205

BENDER, Michael A.
Medical Dept.
Brookhaven National Lab
Upton, NY 11973

BERTRAM, John S.
Grace Cancer Drug Center
Roswell Park Memorial Inst.
666 Elm Street
Buffalo, NY 14263

BOGDAN, Kenneth G.
N.Y. State Dept. of Health
Div. of Health Risk Control
Bur. of Tox. Subs. Assess.
Corning Tower Building, ESP
Albany, NY 12237

BOND, Victor P.
Directors Office
Brookhaven National Lab
Upton, NY 11973

BORG, Donald C.
Medical Dept.
Brookhaven National Lab
Upton, NY 11973

BRIDGES, Bryn A.
Med. Research Council Cell Mutation Unit
U. of Sussex, Falmer
Brighton BN1 9RR, UK

BROWN, Charles C.
Biostatistics Branch
Room 5C03
Landow Building
National Cancer Institute
National Institutes of Health
Bethesda, MD

CHEN, David J.-C.
Genet. Group, LS-3, MS-M886
Los Alamos National Lab
Los Alamos, NM 87545

CLIFTON, Kelly H.
Dept. of Human Oncology
Radiobiology Section
Wisc. Clinical Cancer Center
U. of Wisconsin
600 Highland Avenue
Madison, WI 53792

DE MARCO, Charles J.
Xerox Corporation
Bldg. 317, 800 Phillips Rd.
Webster, NY 14580

DIAMOND, Alan M.
Lab of Molec. Carcingenesis
Dana Farber Cancer Instit.
Harvard Medical School
44 Binney Street
Boston, MA 02115

DROZDOFF, Vladimir
Section 5501
Memorial Sloan Kettering
Cancer Center
1275 York Avenue
New York, NY 10021

EDINGTON, Charles W.
Office of Health and Environmental Research (ER-70)
U.S. Dept. of Energy
Washington, DC 20545

EHLING, Udo H.
Institute for Genetics
Soc. for Rad. and Env. Res.
Post Oberschleissheim
D-8042 Neuherberg, FRG

EICHHORN, Henry C.
Water Quality Engr. Div.
U.S. Army Envir. Hyg. Agency
Aberdeen Proving Ground
MD 21010

EVANS, H. John
Medical Research Council
Clinical and Population
Cytogenetics Unit
Western General Hospital
Crewe Road
Edinburgh EH4 2XU, UK

FRY, R.J. Michael
Nuclear Division
Oak Ridge National Lab
Box Y
Oak Ridge, TN 37830

GOUGH, Michael
Off. of Tech. Assessment
United States Congress
Washington, DC 20510

HALL, Eric J.
Radiation Research Lab
Columbia University
630 West 168th Street
New York, NY 10032

HAYNES, Robert H.
Dept. of Biology, York U.
4700 Keele Street, Downsview
Ontario M3J 1P3, Canada

HEI, Tom K.
Radiology Research Lab
Columbia University
630 West 168th St.
New York, NY 10032

HENRY, Sara H.
Dept. of Health and
Human Services
Bureau of Foods
Div. of Toxicol., HFF-159
U.S. Food and Drug Admin.
Washington, DC 20201

HOWARD, Andrew
Lovelace Inhalation Toxicol.
Research Inst.
P.O. Box 5890
Albuquerque, NM 87185

HUBERMAN, Eliezer
Div. of Biol. and Med. Res.
Argonne National Lab
9700 South Cass Avenue
Argonne, IL 60439

HUGHES, Donald H.
The Proctor and Gamble Co.
Ivorydale Tech Center
5299 Spring Grove Ave.
Cincinnati, OH 45217

ISHIKAWA, Takatoshi
Dept. of Exper. Pathology
Cancer Institute, 1-37-1
Kami-Ikebukuro Toshima-ku
Tokyo 170, Japan

JACOBSEN, J. Sten
U. of Med. and Dent. of N.J.
100 Bergen Street
Newark, NJ 07103

JABLON, Seymour
Comm. of Life Sciences
Natl. Research Council
2101 Constitution Ave.
Washington, DC 20418

KERBEL, Robert S.
Dept. of Pathology
Queen's University and
Natl. Cancer Inst. of Can.
Kingston, Ontario
K7L 3N6, Canada

KILPPER, Robert W.
Xerox Corporation
Joseph C. Wilson Center for Technology
Rochester, NY 14644

KUTZMAN, Raymond S.
Medical Department
Brookhaven National Lab
Upton, NY 11973

LAND, Charles E.
Envir. Epidemiology Branch
Room 3C16 Landow Building
Natl. Cancer Instit.
Natl. Institutes of Health
Bethesda, MD 20205

LEE, William R.
Dept. of Zoology
Louisiana State U.
Baton Rouge, LA 70803

LIBER, Howard L.
Dept. of Nut. and Food Sci.
Room E-18-666
Mass. Inst. of Technol.
Cambridge, MA 02139

LONGFELLOW, David G.
Chem. and Physical Carcinogenesis Branch
Div. of Cancer Etiology
Room 8C29 Landow Building
National Cancer Institute
Natl. Institutes of Health
Bethesda, MD 20205

LUNDGREN, David L.
Lovelace Inhalation Toxicol. Research Inst.
P.O. Box 5890
Albuquerque, NM 87185

MARSH, Gary M.
Ctr. for Env. Epid./Biostat.
Grad. School of Public Health
U. of Pittsburgh
Pittsburgh, PA 15261

MARTNER, John E.
Dept. of Exper. Therapeutics
Grace Cancer Drug Center
Roswell Park Mem. Inst.
666 Elm Street
Buffalo, NY 14263

MAZUMDAR, Sati
Dept. of Biostatistics
U. of Pittsburgh
Pittsburgh, PA 15261

MENDELSOHN, Mortimer L.
Biomed. and Environmental Science Division, L-425
Lawr. Livermore Natl. Lab
U. of California
P.O. Box 5507
Livermore, CA 94550

MERMELSTEIN, Robert
Xerox Corporation
Building 317
800 Phillips Road
Webster, NY 14580

MILLER, Valerie M.
Medical Department
Brookhaven National Lab
Upton, NY 11973

MORRISON, Paul F.
Biomedical Engineering
Natl. Institutes of Health
Bethesda, MD 20205

MURRAY, James L.
Room 8C09 Landow Building
Natl. Cancer Institute
Natl. Institutes of Health
Bethesda, MD 20205

NAUMANN, Charles H.
U.S. Envir. Prot. Agency, RD-689, 401 M Street, S.W.
Washington, DC 20460

NEEL, James V.
Dept. of Human Genetics
U. of Mich. Med. School
4708 Medical Sciences II
Box 015
Ann Arbor, MI 48109

PAINTER, Robert B.
Lab of Radiobiology and Environmental Health
U. of California
San Francisco, CA 94143

PANKETH, Joe
Texas Air Control Board
6330 Highway 290 East
Austin, TX 78723

PATEL, Dhun B.
N.J. Dept. of Envir. Prot.
Off. of Sci. and Res., CN 402
190 West State Street
Trenton, NJ 08625

PETO, Richard
Clinical Trial Serv. Unit
Radcliffe Inf., CTSU Freepost
Oxford OX2 6BR, UK

PORTIER, Christopher J.
Natl. Institute of Environmental Health Sciences
MD-B302, P.O. Box 12233
Res. Triangle Park, NC 27709

REDDY, Prem
Room 1D15, Building 37
Natl. Cancer Institute
Natl. Institutes of Health
Bethesda, MD 20205

RICCI, Paolo F.
Elec. Power Research Inst.
3412 Hillview Avenue
P.O. Box 10412
Palo Alto, CA 94303

SCHNEIDERMAN, Marvin A.
Dept. of Prev. Medicine
Uniform Services
U. of the Health Sciences
Bethesda, MD 20814

SETLOW, Richard B.
Biology Department
Brookhaven National Lab
Upton, NY 11973

SHELLABARGER, Claire J.
Medical Department
Brookhaven National Lab
Upton, NY 11973

SHIBOSKI, Stephen
N.J. Dept. of Envir. Prot.
Office of Sci. and Research
CN 402, 190 West State St.
Trenton, NJ 08625

SONNENBLICK, B.P.
Dept. of Zool. and Physiol.
Newark Coll. of Arts and Sci.
Rutgers St. U. of N.J.
195 University Avenue
Newark, NJ 07102

STARR, Thomas B.
Chem. Ind. Inst. for Toxicol.
P.O. Box 12137
Research Triangle Park
NC 27709

STRNISTE, Gary F.
Genetics Group, LS-3, MS-M886
Los Alamos National Lab
Los Alamos, NM 87545

SUTHERLAND, Betsy M.
Biology Department
Brookhaven National Lab
Upton, NY 11973

THILLY, William G.
Dept. of Nutrition and Food Science
Room E-18-666
Mass. Inst. of Technol.
Cambridge, MA 02139

TICE, Raymond R.
Medical Department
Brookhaven National Lab
Upton, NY 11973

TROSKO, James E.
Dept. of Pediatrics/Human Development
B240 Life Sciences Building
Michigan State University
East Lansing, MI 48824

VAN RYZIN, John R.
Dept. of Biostatistics
School of Public Health
Columbia University
600 West 168th Street
New York, NY 10032 and

VAN RYZIN, JOHN R.
Applied Math Department
Brookhaven National Lab
Upton, NY 11973

VARMA, Matesh N.
Safety and Environmental Protection Division
Brookhaven National Lab
Upton, NY 11973 and

VARMA, Matesh N.
U.S. Department of Energy
Washington, DC 20545

ACHHOLZ, Bruce W.
Room 8C09 Landow Building
Natl. Cancer Institute
Natl. Institutes of Health
Bethesda, MD 20205

ANG, Ju-Jun
Biology Department
Brookhaven National Lab
Upton, NY 11973

IENER, Myra
Toxicology Department
FMC Corporation
Chem. Res. and Devel. Center
Princeton, NJ 08540

WOODHEAD, Avril D.
Biology Department
Brookhaven National Lab
Upton, NY 11973

ZELIKOFF, Judith
A.J. Lanza Laboratories
NYU Medical Center
550 First Avenue
New York, NY 10016

INDEX

MIX
Papier aus verantwortungsvollen Quellen
Paper from responsible sources
FSC® C105338

If you have any concerns about our products,
you can contact us on
ProductSafety@springernature.com

In case Publisher is established outside the EU,
the EU authorized representative is:
Springer Nature Customer Service Center GmbH
Europaplatz 3, 69115 Heidelberg, Germany

Printed by Libri Plureos GmbH
in Hamburg, Germany